Artificial Intelligence and Sustainable Energy Systems

Artificial Intelligence and Sustainable Energy Systems

Volume I

Editors

Luis Hernández-Callejo
Sergio Nesmachnow
Sara Gallardo Saavedra

MDPI • Basel • Beijing • Wuhan • Barcelona • Belgrade • Manchester • Tokyo • Cluj • Tianjin

Editors

Luis Hernández-Callejo
University of Valladolid
Spain

Sergio Nesmachnow
Universidad de la República
Uruguay

Sara Gallardo Saavedra
Universidad de Valladolid
Spain

Editorial Office
MDPI
St. Alban-Anlage 66
4052 Basel, Switzerland

This is a reprint of Topic published online in the open access journal *Applied Sciences* (ISSN 2076-3417), *Entropy* (ISSN 1099-4300), *Sustainability* (ISSN 2071-1050), *Electronics* (ISSN 2079-9292), and *Energies* (ISSN 1996-1073) (available at: https://www.mdpi.com/topics/Artificial Intelligence Energy_Systems).

For citation purposes, cite each article independently as indicated on the article page online and as indicated below:

LastName, A.A.; LastName, B.B.; LastName, C.C. Article Title. *Journal Name* **Year**, *Volume Number*, Page Range.

Volume I
ISBN 978-3-0365-7644-2 (Hbk)
ISBN 978-3-0365-7645-9 (PDF)

Volume I-III
ISBN 978-3-0365-7642-8 (Hbk)
ISBN 978-3-0365-7643-5 (PDF)

Contents

About the Editors

Luis Hernández-Callejo

Luis Hernández-Callejo is an electrical engineer at the Universidad Nacional de Educación a Distancia (UNED, Spain), a computer engineer at UNED, and a PhD candidate at the Universidad de Valladolid (Spain). Professor and researcher at the Universidad de Valladolid. His areas of interest are renewable energy, microgrids, photovoltaic energy, wind energy, smart cities, and artificial intelligence. He has participated in numerous research projects, directed many doctoral theses, and is the author of hundreds of scientific articles.

Sergio Nesmachnow

Sergio Nesmachnow is a full professor at the Faculty of Engineering, Universidad de la República, Uruguay. He is a level III researcher (the maximum level) of the National System of Researchers in Uruguay and a visiting professor at renowned universities and research centers in America and Europe. He has more than 400 publications in scientific journals and international conferences and is responsible for more than 50 research projects.

Sara Gallardo Saavedra

Sara Gallardo Saavedra is a professor and researcher at the Campus Duques de Soria of the University of Valladolid, Spain. Her research focuses on the detection, characterization, and classification of defects in photovoltaic (PV) modules through the use of thermography, electroluminescence, I-V curves, and visual analysis. She has participated in numerous national and international R+D+I projects, carrying out active dissemination of the results with high regularity in the scientific production, including scientific publications in high impact factor journals, book chapters, and contributing to congresses on advanced maintenance in PV. She has made a predoctoral and a postdoctoral stay in the Unit of Solar PV Energy in the Energy Department of the Energy Research Center, Environment, and Technology (CIEMAT) in Madrid, and the researcher has collaborated with different institutions such as the University of Gavle in Sweden, the Universidad del Valle in Colombia, the National Polytechnic Institute of Mexico, and the University of Cuenca in Ecuador.

Preface to "Artificial Intelligence and Sustainable Energy Systems"

The problems that affect humanity are numerous and occur in different areas. Energy sustainability, climate change, and the effects derived from pollutants and viruses are some of the most relevant problems. The main objective of researchers is to provide solutions to these and other problems.

In recent years, the use of artificial intelligence has increased considerably. Artificial intelligence is used in different areas: energy, sustainability, medicine, health, mobility, industry, etc. Therefore, it is necessary to continue advancing in the application of artificial intelligence to the aforementioned problems. Energy is a precious commodity, and it is increasingly difficult to dispose of it in a sustainable way. In this sense, renewable energy sources are essential, although the use of conventional energy cannot be forgotten. Therefore, sustainable energy systems, integrating renewable and non-renewable energy sources, smart systems, and new business models, are crucial.

Therefore, in this book, the best accepted and published articles on the topic "Artificial Intelligence and Sustainable Energy Systems" are presented. All articles refer to the themes indicated above.

Luis Hernández-Callejo, Sergio Nesmachnow , and Sara Gallardo Saavedra
Editors

sustainability

MDPI

Article

Synthetic Dataset of Electroluminescence Images of Photovoltaic Cells by Deep Convolutional Generative Adversarial Networks

Héctor Felipe Mateo Romero [1,*], Luis Hernández-Callejo [2,*], Miguel Ángel González Rebollo [1], Valentín Cardeñoso-Payo [3], Victor Alonso Gómez [4], Hugo Jose Bello [5], Ranganai Tawanda Moyo [6] and Jose Ignacio Morales Aragonés [4]

1 Departamento Fisica de la Materia Condensada, Universidad de Valladolid, 47011 Valladolid, Spain
2 Departamento Ingeniería Agrícola y Forestal, Universidad de Valladolid, 47002 Valladolid, Spain
3 Departamento Informatica, Universidad de Valladolid, 47011 Valladolid, Spain
4 Departamento de Física, Universidad de Valladolid, 47002 Valladolid, Spain
5 Departamento de Matemática Aplicada, Universidad de Valladolid, 47011 Valladolid, Spain
6 Department of Mechanical Engineering, Durban University of Technology, P.O. Box 1334, Durban 4000, South Africa
* Correspondence: hectorfelipe.mateo@uva.es (H.F.M.R.); luis.hernandez.callejo@uva.es (L.H.-C.); Tel.: +34-975129418 (L.H.-C.)

Abstract: Affordable and clean energy is one of the Sustainable Development Goals (SDG). SDG compliance and economic crises have boosted investment in solar energy as an important source of renewable generation. Nevertheless, the complex maintenance of solar plants is behind the increasing trend to use advanced artificial intelligence techniques, which critically depend on big amounts of data. In this work, a model based on Deep Convolutional Generative Adversarial Neural Networks (DCGANs) was trained in order to generate a synthetic dataset made of 10,000 electroluminescence images of photovoltaic cells, which extends a smaller dataset of experimentally acquired images. The energy output of the virtual cells associated with the synthetic dataset is predicted using a Random Forest regression model trained from real IV curves measured on real cells during the image acquisition process. The assessment of the resulting synthetic dataset gives an Inception Score of 2.3 and a Fréchet Inception Distance of 15.8 to the real original images, which ensures the excellent quality of the generated images. The final dataset can thus be later used to improve machine learning algorithms or to analyze patterns of solar cell defects.

Keywords: generative adversarial neural networks; photovoltaics; artificial intelligence; synthetic data; electroluminescence

Citation: Romero, H.F.M.; Hernández-Callejo, L.; Rebollo, M.Á.G.; Cardeñoso-Payo, V.; Gómez V.A.; Bello, H.J.; Moyo, R.T.; Aragonés, J.I.M. Synthetic Dataset of Electroluminescence Images of Photovoltaic Cells by Deep Convolutional Generative Adversarial Networks. *Sustainability* **2023**, *15*, 7175. https://doi.org/10.3390/su15097175

Academic Editors: Jack Barkenbus and Firoz Alam

Received: 24 February 2023
Revised: 23 April 2023
Accepted: 24 April 2023
Published: 25 April 2023

1. Introduction

A number of factors (energy crisis, wars, climate change, etc.) are causing a rise in renewable energies use. Solar energy can be easily and affordably converted either into thermal energy by means of thermal panels or into electrical energy, using photovoltaic panels (PV) [1]. Industrial plants generating electricity from solar energy, commonly known as solar farms, are generally composed of a high number of photovoltaic (PV) panels made of PV cells. As the number of installations increases, the maintenance of solar farms becomes a nontrivial problem [2]. The energy produced depends on different conditions, such as the state of the panels, the climate, or the time of year. Solar panels are also vulnerable to phenomena that can reduce or nullify their performance. These issues make necessary a system to control and optimize production, since manual human labor is not enough as the number of panels gets higher.

Artificial intelligence is usually applied to solve difficult control and optimization problems. When applied to PV systems, different AI methods such as fuzzy logic, metaheuristics, or neural networks have been used to solve problems [3]. The most important problems include [4] max power point tracking, output power forecasting, parameter estimation, and defect detection.

Defect detection is one of the most interesting problems and researched topics in PV systems [5,6]. It is usually tackled with different kinds of neural networks, since they have great performance, although, as an important drawback, they need high amounts of data to perform better than other machine learning methods [7]. This could be a major impossibility for problems where it is difficult to harvest new data. Most of the approaches to detect the state of PV panels use electroluminescence images of the cells as an input, which is an invasive method that makes it difficult to carry out measurements and gather data.

Data augmentation is the most common method to deal with image data scarcity by means of the introduction of slight modifications (rotations, flips, and minor deformations) to the original images in order to create new images [8–10]. More recent papers promote the use of more complex AI techniques, such as Generative Adversarial Networks (GANs), to generate synthetic images [11,12]. GANs are state-of-the-art algorithms for data generation [13]. They have also been applied to PV systems for solving different problems [14].

The generation of synthetic EL images of PV cells using GANs has also been proposed in other works [11,12]. These works present synthetic datasets created with different GAN architectures trained with EL images of cells with different kinds of defects. Although the datasets in these works have not been made public, from their work, it can be deduced that the synthetic images are just labeled from visual inspection in order to train standard defect/normal classifiers, ignoring the output energy output performance of synthetic cells, since they cannot be measured.

In this paper, we present a new approach to deal with the image data scarcity problem. Starting from a small set of electroluminescence images of PV cells obtained under experimental conditions to be described, a synthetic dataset of images was created using Generative Adversarial Networks (GANs). For each synthetic image, a scalar value that represents the performance of their energy production is associated using machine learning techniques trained with the energy production of the original PV cells. It is also a continuation of the paper presented in [15], where we augmented and improved the dataset and applied new metrics to ensure the appropriate quality of the generated data.

To ensure reproducibility, the resulting dataset was made publicly available, so it can be used to improve the performance of AI models, to analyze the characteristics, properties, and defects of the cells, or to compare with other methods of generation of synthetic images.

This paper is structured as follows: Section 2 briefly reviews the Generative Adversarial Networks, Section 3 presents the methodology that was followed to generate the dataset, in Section 4, the synthetic dataset is described, and finally, the conclusions of the paper are presented in Section 5.

2. Generative Adversarial Networks

Generative Adversarial Networks (GANs) [16] are one of the most important and popular technologies nowadays and have been applied to different fields [17]. They can be applied to semisupervised and unsupervised learning. A GAN is usually defined as a pair of neural networks that are competing against each other. The network known as the Generator is the one designed with the job of trying realistic new data to try to deceive the other network, known as the Discriminator. This network has to decide if the data that it receives were forged.

The Generator does not have access to the real data, which is an important feature of these algorithms. The Generator has to learn how to create the data based on the feedback from the Discriminator. The Discriminator has access to both kinds of data, but it does not know which kind of data is an image before carrying out the prediction. The networks change their weights depending on the results of the deception; the Generator uses them to improve the forgeries, while the Discriminator tries to improve its recognition of forgeries. Figure 1 shows a diagram of the behavior of the algorithm.

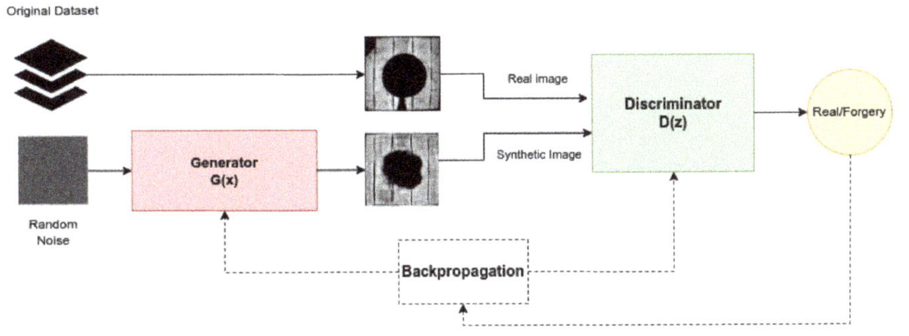

Figure 1. Diagram of a GAN.

The basic principle of operation of a GAN can be expressed as a two-player minimax game played between D and G, with a value function $V_{GAN}(G, D)$ given by the following mathematical expression [16], a binary cross-entropy function, commonly used in binary classification problems:

$$\min_G \max_D V_{GAN}(G, D) = \mathbb{E}_{\mathbf{x} \sim p_{data}(\mathbf{x})}[\log D(x)] + \mathbb{E}_{\mathbf{z} \sim p_z(\mathbf{z})}[\log(1 - D(G(z)))]. \quad (1)$$

The first term in (1) represents the expected value of entropy given by the Discriminator over the real data and the second is the entropy given by the Generator over the fake data obtained.

GANs are usually composed of Deep Feed-forward Networks, but more complex architectures can be used in order to improve the generative capacities of the algorithm and the quality of the forged data. One of the most used architectures are Convolutional Neural Networks.

Deep Convolutional GAN

Research on GANs has led to new interesting architectures which substantially improve the performance of the networks and the quality of the forged data [18]. For this paper, we implemented the architecture known as Deep Convolutional GAN [19]. This architecture is based on convolutional layers, but it also provides a set of constraints in order to provide more stabilized training and better quality in the output. The most important guidelines are the following:

- Use of batch normalization in the Generator and the Discriminator.
- Removal of fully connected hidden layers in both networks.
- Usage of convolutional transposed layers instead of the stridden convolutional layers. This is only applied in the Generator network.

3. Methodology

The correct preparation of the real and synthetic image datasets is a complex process which requires several steps to complete. In this section, we present the methodology that was followed in this research, which implies four relevant stages: manual acquisition of real EL images of PV cells beside their electrical characteristics (IV curves); data preprocessing to prepare for synthetic image generation using GAN; maximum power output assignment to synthetic cells from regression models trained with real images; and model and result validation.

3.1. *Real Images Acquisition*

Data availability is one of the most critical issues when using deep learning techniques. For precise PV cell characterization, we need two different kinds of data: the electroluminescence (EL) images of the photovoltaic cells and their corresponding IV curves. For the first, there are some public datasets available in the bibliography [20], but they do not include information about the IV curve, since it is not easy to measure the individual curve of a single PV cell. To solve this problem, we had to obtain the data using a specific technique developed previously [21] and following a manual process in the laboratory.

The capturing of the EL images was performed using a Hamamatsu InGaAs Camera C12741-03. The cell was isolated from external light in other to avoid the light produced by other kinds of radiations. Figure 2a presents the camera for taking the pictures. EL was chosen as the image capture technique, since it is widely used for detecting defects on PV modules [22] and it is the most used technique in other related works with PV systems. In order to test the different levels of luminous emission of the cell, various values of current were used to power the LED array when obtaining the IV curve and to the PV cell when capturing the EL image.

To measure the IV curve, we used a device specifically designed to measure the values of current and voltage of a single cell [21]. This device (Figure 2b) provided voltage and current values to build the IV curve of the cells. The IV curves can be used to calculate the max power point and the performance of the cells. Figure 2c shows the setting used for taking the IV curves. The original paper for this device validates the accuracy of the measurements.

(a) (b) (c)

Figure 2. Devices used to obtain the data: (**a**) InGaAs Camera; (**b**) IV curve tracer; and (**c**) setting to measure the IV curves.

The number of damaged cells was extremely limited in quantity and variety. The solution for this issue was to create artificial shadows in order to improve the amount and variety of images. The different shadows were created taking into account the most important defects found in solar farms. Figure 3 presents a representative image of each kind of artificial shadow that was used.

The final acquired dataset was composed of 602 different EL images with their corresponding IV curves. In order to allow the repeatability of the experiments, the dataset was made publicly available (The dataset can be downloaded from https://github.com/hectorfelipe98/Synthetic-PV-cell-dataset. The repository includes the original images, the synthetic dataset, and a CSV file that maps each image with its relative power and its class).

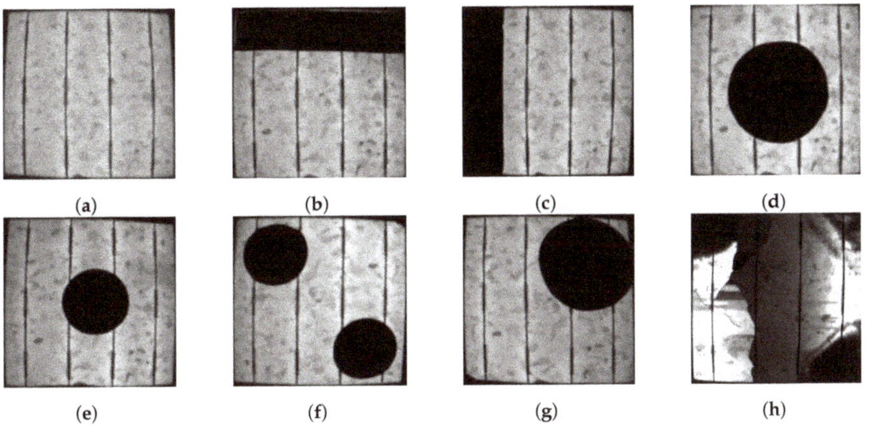

Figure 3. Shadows and defects used to create the dataset: (**a**) shadow/defect 0: original image; (**b**) shadow/defect 1: long horizontal line; (**c**) shadow/defect 2: long vertical line; (**d**) shadow/defect 3: big central circle; (**e**) shadow/defect 4: small central circle; (**f**) shadow/defect 5: two small circles in the corners; (**g**) shadow/defect 6: big circle in one corner; and (**h**) shadow/defect 7: other defects.

3.2. *Image Preprocessing*

The captured EL images were not suitable to be directly used with machine learning algorithms, since there were factors affecting image quality to be addressed first. For this reason, some algorithms were used to improve both the quality and normalization of the images. This section explains how the different problems found in image quality were solved.

- Dead pixels and luminous noise: The insulation of the system when capturing the images was not perfect. Some leaks could be found in some cases. This produced luminous noise when the image was taken. Another problem could be found in the camera. Due to its extended use, the camera had some dead pixels. Even though these phenomena are not easily visible to the human eye, they degenerate the quality of the obtained images. In order to solve these two problems, an image was taken before the real measure and without turning on the lights from the LED array. This provided information about the luminous noise and the dead pixel of the camera. Having this information, it is possible to perform a subtraction of the noise from the original image. This process was performed for each of the EL images.
- Images with poor lighting: The scale of the histograms of the images can produce problems in some libraries or programs which cannot recognize which values correspond with the white color. Figure 4a,b shows how the values are distributed in only a small fraction of the possible values, which makes the image look extremely dark. The solution to this problem corresponds to performing a min–max normalization, subtracting the minimum values of the image, and dividing it by the maximum value. The resulting image can be found in Figure 4c,d.
- Black surrounding contours: Figure 4c shows how images are surrounded by a black area. The reason for this phenomenon is caused by a limited focusing of the camera, which is capturing part of the walls of the insulated area. The solution to this issue consisted of performing a change in perspective. The first step consisted of applying different filters in order to delete the details, obtaining the maximum contour polygon of the cell. The next step consisted of performing a Hough transform to find the corners of the cells, using the polygon calculated previously as the base. In some cells, it was necessary to make some tweaks manually, as they presented strange defects that notably changed the aspect of the cell. The results of the transformation can be found in Figure 5.

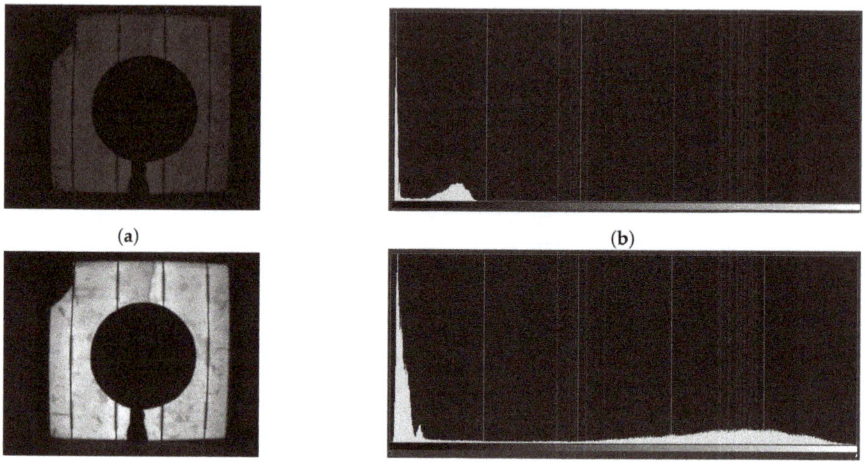

(**a**) (**b**) (**c**) (**d**)

Figure 4. Image before and after applying color normalization: (**a**) original image; (**b**) original histogram; (**c**) image after the min–max normalization; and (**d**) histogram after the normalization.

Figure 5. Cell after removing the surrounding image contour corresponding to wall portions.

3.3. *Generation of the Synthetic Images*

As announced, a Deep Convolutional Generative Adversarial Network was chosen to generate extra images from the original recorded set. The model is composed of two different networks: the Generator and the Discriminator. This section explains the architecture and parameters of both networks and the process of training.

3.3.1. Generative Network

The generative network was created following the principles of the DCGN. It has 3 different Convolutional Transpose Layers (Deconvolutional) in order to generate patterns. The use of these layers in combination with batch normalization [23] improves the generative capacities of the network and improves the stability of the training, The network uses Leaky Relu [24] as an activation function, since it usually outperforms the standard Relu. The architecture of this networks is represented in Figure 6. The input to the network

(Figure 7a) is a random noise array from a normal distribution. The output is a 200 × 200 image (Figure 7b. This size was chosen in order to reduce the computation cost of the algorithm while still obtaining images with a large amount of information. Other important hyperparameters can be found in Table 1.

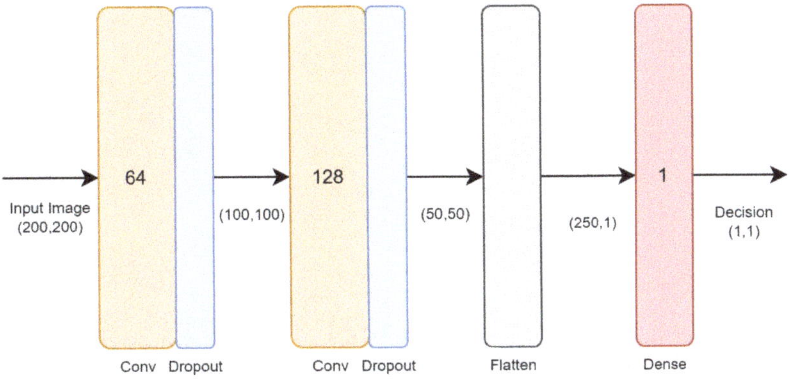

Figure 6. Architecture of the generative network.

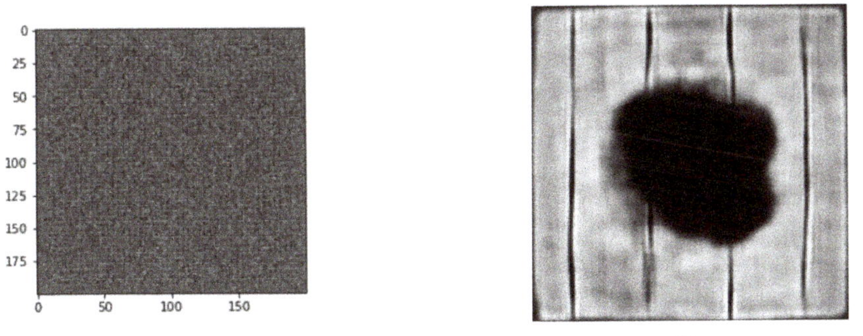

(**a**) (**b**)

Figure 7. GAN's output images before and after training. (**a**) Output image before training. (**b**) Output image after training.

Table 1. Hyperparameters for both networks

Activation Function	Loss Function	Learning Rate	Epochs	Batch Size	Output Size
Leaky Relu	Cross Entropy	5×10^{-5}	800	4	200 × 200

3.3.2. Discriminator Network

The Discriminator network also followed the principles of DCGAN. The network uses different Convolutional Neural Networks to find patterns from the images. The typical feed-forward part of the network was also removed, with only the output layer remaining. The use of dropout layers and batch normalization improves the generalization capacity of the networks and stabilizes the training. The architecture is represented in Figure 8. The input corresponds to a 200 × 200 image and the output to a binary value that determines if the image is a real cell or a forgery. The other important hyperparameters were the same as the Generator, so they can be found in Table 1.

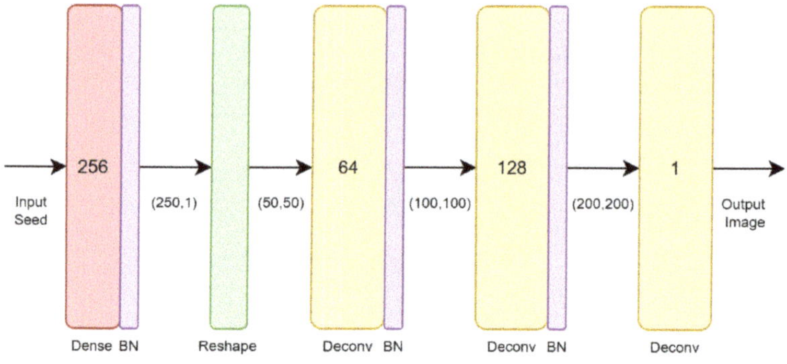

Figure 8. Architecture of the Discriminator network.

3.3.3. Training the GAN

The training was performed simultaneously in both networks with all the available samples (602). The training loop starts with the creation of synthetic images by the Generator using random seeds. After that, the Discriminator is provided with a mix of real and synthetic images. That input data are used to train the network, the loss is computed for each of the networks, based on the results of the Discriminator. Figure 9 shows the evolution of the loss for both networks. In the first epochs, the loss of the Discriminator network is quite high. It has not learned the patterns of the original images, so it cannot identify which images are real or forgered, even if the forged images are extremely similar to noise. After a while, the Discriminator reduces its loss since it can differentiate which images are real; this provokes an increase in the loss of the Generator. After that, the loss of the Generator steadily decreases its value since it starts to learn the patterns to create images similar to the original ones. Until epoch 400, it continues to learn when it obtains its lowest value. At the same time, the Discriminator is unable to identify which images are real, making its loss its maximum value. After epoch 400, the values have almost no change, so it can be concluded that the training should end at this point.

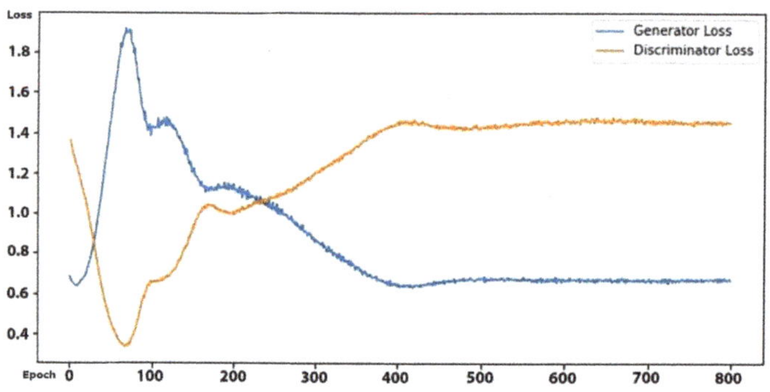

Figure 9. Evolution of the Generator and Discriminator loss.

The training was performed with a CPU AMD Ryzen 7 5800H, 16 GB of RAM, and a GPU Nvidia Geforce GTX 1650. It took 2 h and 41 min to complete the training.

After completing the training, the Generator network was used to generate the synthetic dataset. A total of 10,000 different images were created using 10,000 different random seeds. Figure 10 presents a selection of some generated images.

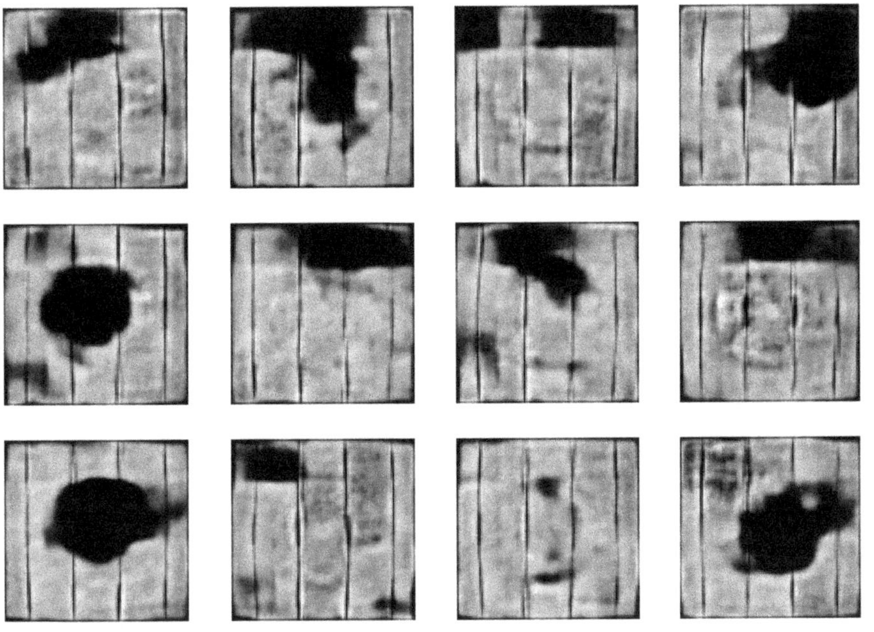

Figure 10. Samples of the synthetic images generated by our GAN-based method.

3.4. Assigning Maximum Power to Cells

The labeling of the images was not a trivial process, since it required associating each image (real or synthetic) with an output maximum power value, which represents the performance of the cell. This section explains how this association was made both for original and GAN-generated images.

3.4.1. Original Images

As explained before, each EL image has an associated IV curve. The IV curve provides information about the performance of the cell, which can only be obtained after taking the following steps:

- Compute the power–voltage curve of each cell from its IV curve.
- The maximum value of the curve is saved as the max power point (MPP).
- We gather the cells into 6 groups, in terms of the irradiance used to excite the cell before emission, since the MPP depends on the amount of light projected onto the cell.
- For each group, the 5 highest values of MPP are considered, and their mean is calculated.
- This average of the maximum values is understood as the expected MPP of a cell without any defects or shadows in that group.
- The MPP of each cell is divided by the averge MPP of the cell group it belongs to.

The resulting value will measure the relative performance of the cell, normalizing its MPP with the maximum value of the cells of the same group. This maximum is the mean of the 5 highest values, since it reduces the effects of potential incorrect measurement. Values of relative power near 1 will correspond with cells in good condition, with only a few or no defects. Low values correspond with underperforming cells, mostly due to their defects.

3.4.2. Synthetic Images

The labeling of the synthetic dataset was a completely different problem. As explained, the values of the original images were calculated based on their IV curve. This process cannot be replicated in the synthetic images, since they are not real cells; so, they cannot be measured.

To solve this problem, we formulated it as a regression problem that can be solved using a machine learning model. The model is trained with the full dataset of original images, together with their normalized power (MPP) calculated from IV curves, as explained in the previous subsection (602 samples). The chosen model is Random Forest [25], since it provided a low error in the original dataset and showed excellent generalization power to associate the MPP to the synthetic images. The implementation of the algorithm in the Sklearn library [26] was used for this work. The tuning of the hyperparameters of the RF model was carried out using the Grid Search method in the Sklearn library (GridSearchCV), obtaining the optimal values shown in Table 2.

Table 2. Estimation of Random Forest hyperparameters using GridSearchCV.

Parameter	Range	Optimum Value
n_estimators	$[20, 500]$	200
max_depth	$[0, 10]$	10
min_samples_split	$[1, 10]$	1
min_samples_leaf	$[1, 10]$	1
min_weight_fraction_leaf	$[0, 0.8]$	0

Since Random Forest is not suitable to work directly on raw images, some features were extracted from the images. The features are based on typical statistics (mean, standard deviation, etc.) and other characteristics directly extracted from the histogram (amounts peaks, peaks width, peaks height, amount of colors, etc.) A complete list can be found in Table 3. Feature selection (FS) is an important step in the preparation of machine learning models. We used correlation-based FS. As depicted in Figure 11, the cross-correlation between all the original sets of features shows that almost no feature is highly correlated with the others, except for the standard deviation and the variance, which are completely dependent on each other, meaning one of them can thus be safely removed from the final set of features.

The dataset was split into two sets: training (67%) and validation (33%). We decided to only use two due to our data limitation. The target variable was the relative power of each cell, standardized between 0 and 1.

Table 3. Features for Random Forest Regressor.

Mean	Median	Mode	Variance	Std
roughness	blacks	burned whites	others	peaks_number
peaks distance	peak 0 height	peak 0 width	peak 1 height	peak 1 width

The model obtained a Mean Absolute Error (MAE) of 0.041 and a Mean Squared Error (MSE) of 0.0038 in the validation dataset. The distribution of the predictions of the model can be found in Figure 12. The low error and the similarity in the distribution confirm the validity of the model. The distribution of the prediction for the synthetic dataset can also be observed. Finally, the images were divided into two groups according to their predicted power (class 0 $>$ 0.8 and class 1 $<=$ 0.8). In total, 6963 images were classified as class 0 and 3037 as class 1.

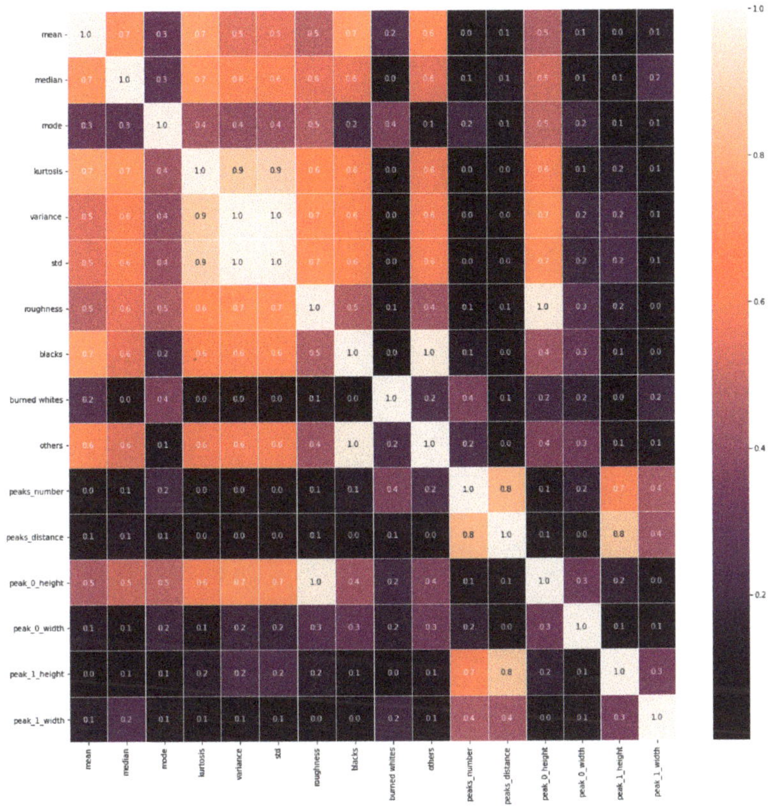

Figure 11. Correlation heatmap of the initial set of features.

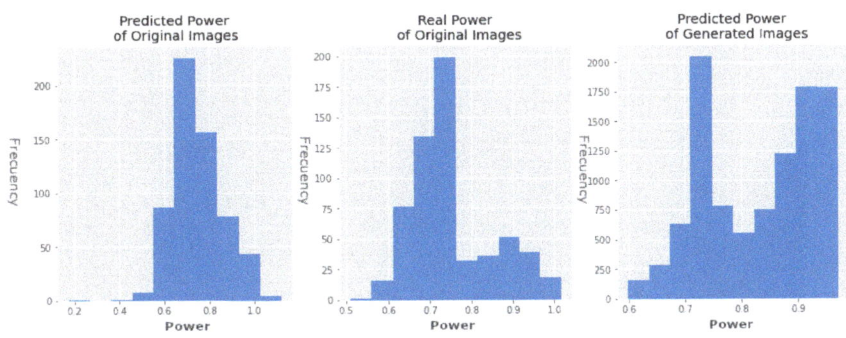

Figure 12. Histograms of real and predicted normalized power of the original and generated dataset.

4. Results

The resulting dataset was divided into two different folders, one for each class: Class 0 (6963 samples, Figure 13a) represented the images whose relative power is at least 0.8, and the images in that class can be considered as functional PV cells. Class 1 (3037 samples, Figure 13b) represented the images with a power of less than 0.8, and the images in that class can be considered as underperforming PV cells.

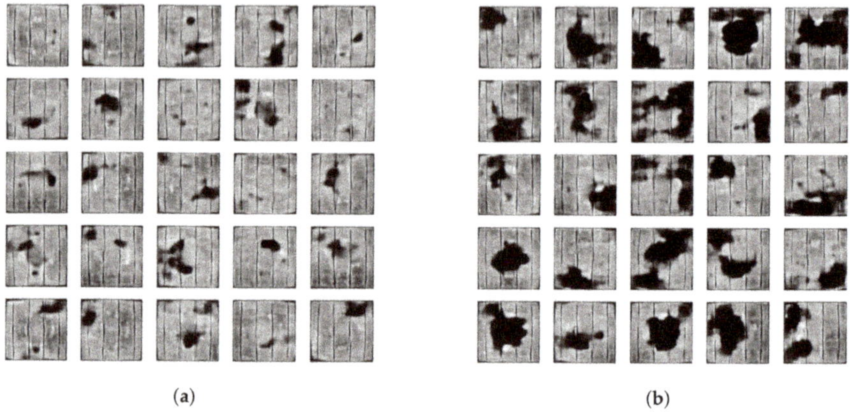

Figure 13. Sample of images of both classes. (**a**) Sample of images of class 0. (**b**) Sample of images of class 1.

4.1. Visual Analysis

For ensuring the quality and similarity of the images, we propose two different methods: in this section, an analysis based on visual characteristics and histogram, and in the next section, an analysis based on different metrics.

As can be observed in Figure 13, the generated images present a similar structure while presenting new patterns of shadows different from the original ones (Figure 3). This is an interesting feature produced by the generative capacity of the GAN, since it can combine the different kinds of shadows presented in the original images in order to create new kinds of patterns. This improves the variety of shadows presented in the dataset.

Figure 14 presents the distribution of the features selected previously for labeling (Table 3). For each feature, the relationship between the values of the feature and the relative power of the cell is presented. Synthetic images are represented with orange dots, and the original images are represented with blue dots. In most features, the original dataset images appear as a subset of the synthetic dataset, with some exceptions caused mostly by underrepresented cases. This means that synthetic images not only present the characteristics of the original images, but they also present new cases of defects or shadows while maintaining the most important characteristics. This is mostly produced by the generative properties of the GAN, which can create new patterns combining the patterns of the input data; these new patterns improve the diversity of the dataset. This can lead to an improvement in performance in the machine learning methods that use this dataset. Another interesting finding is that the most underrepresented cases in the original data do not appear in the synthetic data; this is also provoked by the properties of the GAN, since it needs a considerable amount of samples to find patterns.

4.2. Histogram Analysis

The histogram of the images gives a lot of information about them. Figure 15a presents the mean histogram of all images of class 0 of the original dataset and the mean for all of the pictures of class 0 of the synthetic one. It can be seen that the images in this class present a large amount of light gray–white pixels (Values near 200), but they also can present some minor defects or shadows, as can be seen by the number of black pixels (values near 0). A difference between both datasets can be seen: the synthetic dataset images have higher but narrower peaks and are sometimes a bit moved to the left.

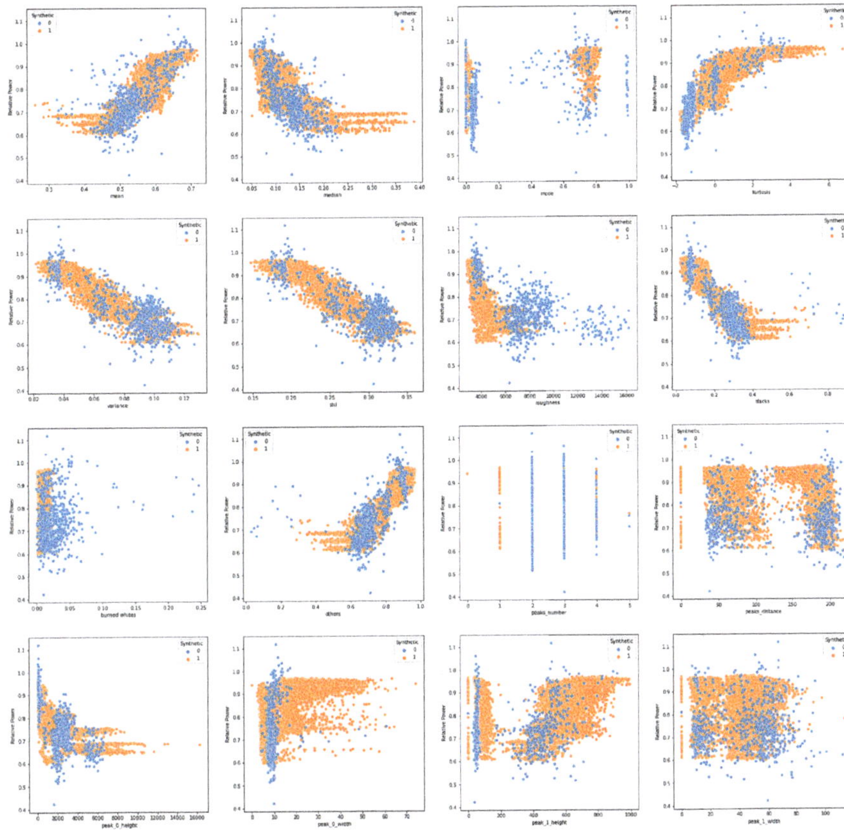

Figure 14. Distribution of the relative power generated by a cell as a function of the value of each of the sixteen features used to characterize the images: orange dots—synthetic images; blue dots—original images.

Figure 15a presents the same information for the images in class 1. The images in this class present a large number of dark pixels due to their defects and shadows. The amount of lighter pixels is considerably lower. In the synthetic images, the peak of black pixels is higher, but its width is narrower. The light pixels are extremely similar to the original ones.

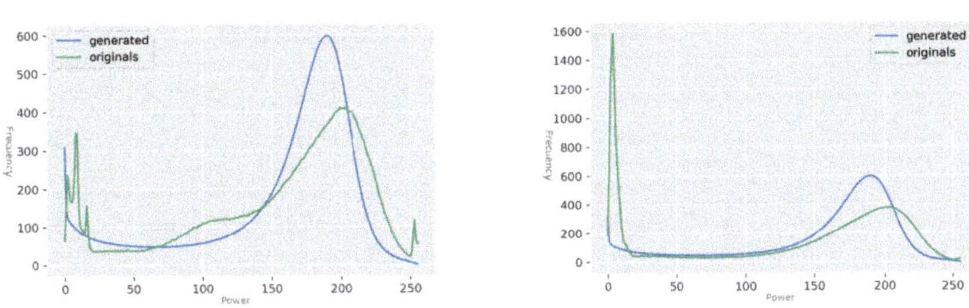

(**a**) Histogram of the images of class 0

(**b**) Histogram of the images of class 1

Figure 15. Different generated images of both classes.

As shown, the aspect of the histograms of both datasets is quite similar. The minor differences are mostly produced due to the augmented variety of patterns of defects and shadows.

Figure 16 presents two different cells of class 1: one original and one synthetic with similar aspects. Visual inspection of both histograms shows that they have a similar structure, presenting the same amount of peaks and even placed in a similar position. Nevertheless, synthetic images tend to show a more symmetrical histogram and a shift from maximum to lower intensities, since the extreme intensity values are less frequent than in real images.

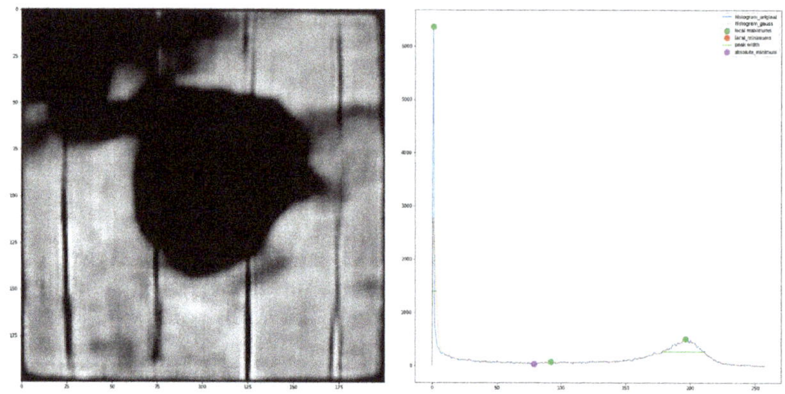

(**a**) A defective synthetic cell with its histogram

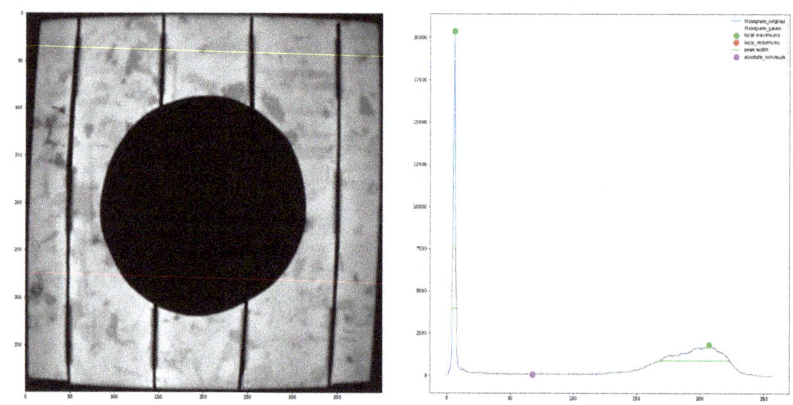

(**b**) A defective original cell with its histogram

Figure 16. Comparison of a defective synthetic cell and a defective original cell.

Figure 17 presents the same for two images of class 0. Both images have almost no apparent defects. Their histograms have similar shapes, as can be seen by the number of peaks and their placement.

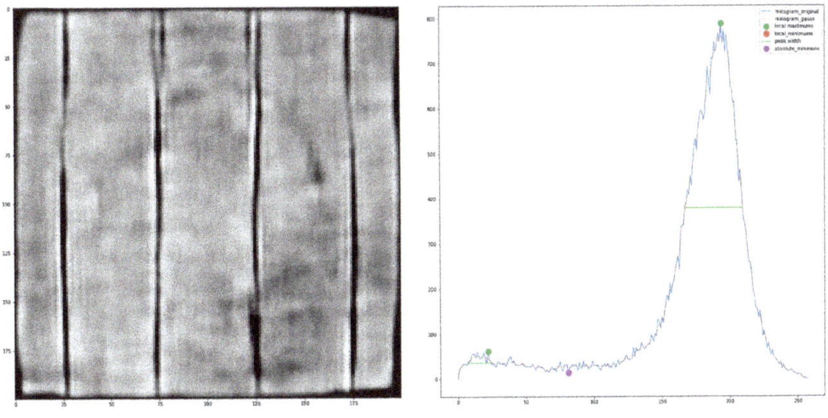

(**a**) A defective synthetic cell with its histogram

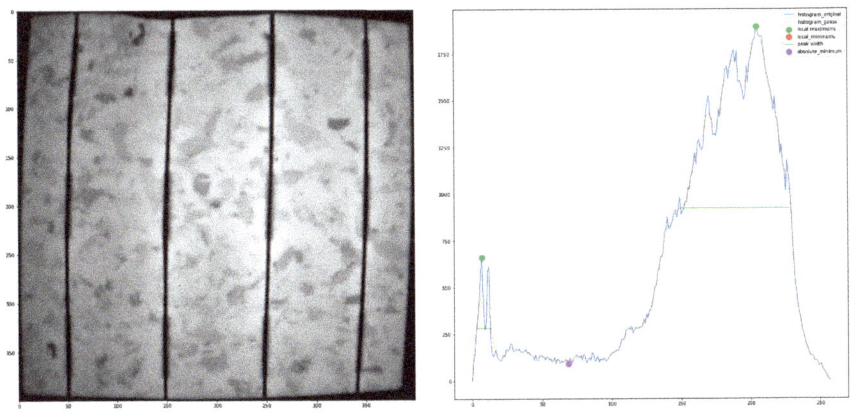

(**b**) A defective original cell with its histogram

Figure 17. Comparison of a defective synthetic cell and a defective original cell.

Figure 18a presents a comparison of the histograms of the position of the maximum, found in the right half of the histograms (gray/white colors); it can be seen that both histograms have a similar shape. Figure 18b presents the histograms of the number of pixels that have low values (up to 10% of the maximum values); it can be seen how the synthetic images do not completely imitate the original ones. A similar case can be seen in Figure 18c, which represents the histograms of the number of pixels in the last decile; it shows how even if the shapes are pretty similar, there is a shift to the left. It seems this GAN method has some flaws when finding the patterns around the most extreme values. This issue is not critical but shows that this method still has some room to improve.

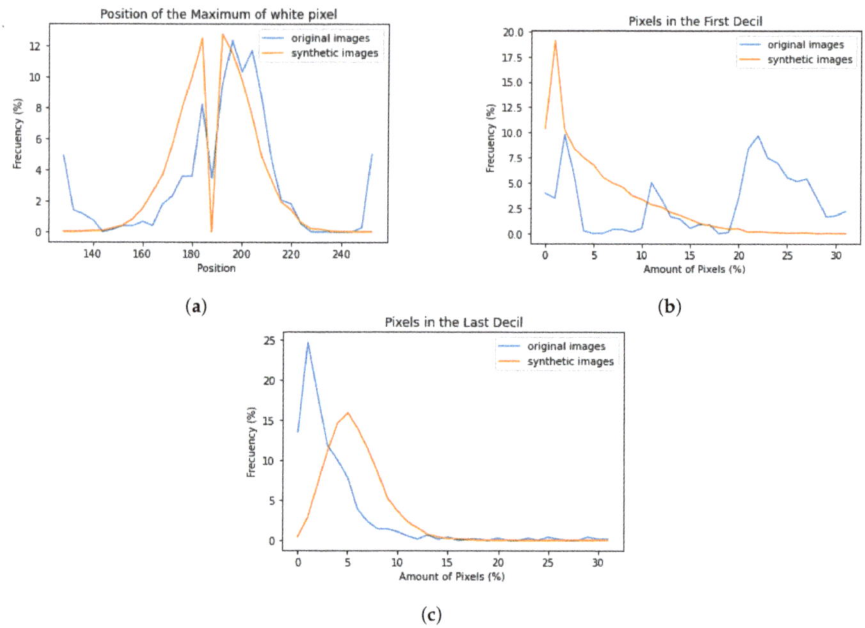

Figure 18. Comparison of the different aspects of the histograms of both original and synthetic images: (**a**) histogram of the position of the peak of gray/white colors for both original and synthetic images; (**b**) histogram of the number of dark pixels (first decile) for both original and synthetic images; and (**c**) histogram of the number of white pixels (last decile) for both original and synthetic images.

4.3. Image Quality Metrics

Previous works related to the synthetic generation of EL images of PV cells have not addressed the issue of ensuring the quality of their data by providing objective metrics. The Inception Score (IS) and the Fréchet Inception Distance (FID) are the most important metrics to ensure the quality of synthetic images. In the next paragraphs, both metrics are explained and used. A summary of the results of these metrics can be found in Table 4.

Table 4. Metrics for ensuring the quality of the synthetic dataset (Original: O, Synthetic: S, and Noise: N).

Metric	Mean Score	Std.
IS (O)	2.144	0.055
IS (S)	2.341	0.407
IS (N)	1.050	0.002
FID (O-O)	0.431	0.095
FID (S-S)	0.150	0.025
FID (N-N)	2.511	0.016
FID (O-S)	15.80	0.075
FID (O-N)	293.8	0.366
FID (S-N)	296.7	0.735

4.3.1. Inception Score

This metric was first proposed in 2016 [27] for evaluating the quality of generated artificial images. The score is computed based on the results of a pretrained InceptionV3 model [28] applied to the generated images. This score is maximized when two conditions are met: The value of the labels is the same for each image, in other words, the entropy of the distribution is minimized. The other condition is that the images are diverse, meaning that the labels are evenly distributed across all possible labels.

For our problem, we used a custom implementation of the metric based on Python and Tensorflow. We compared the results of three different datasets: the original dataset, the synthetic dataset, and a dataset composed only of noise. For each experiment, we split the evaluated dataset into 10 subsets and computed the IS of each one; after that, we computed the mean and the standard deviation. This process reduces the memory cost of the algorithm and the effects of randomness.

The original dataset obtained a mean score of 2.1440 with a standard deviation of 0.0559, the synthetic dataset obtained a mean score of 2.3418 with a standard deviation of 0.4079, and the noise dataset obtained a mean score of 1.0506 with a standard deviation of 0.0026. The results show that both datasets do not obtain a great result with this metric, but their results are better than the results of the noise. This is provoked mostly by the fact that Inceptionv3 was not trained to deal with EL images. It can also be observed that the results of both datasets are similar, with an error of 9.2%. The fact that [26] produced similar IS values for both shows that they have a high similarity. This proves the quality of the synthetic dataset but also shows that it has some room to improve.

4.3.2. Fréchet Inception Distance

This metric was first proposed in 2017 [29] to evaluate the quality of synthetic images. In contrast to the IS, this metric compares the distribution of the synthetic data with the distribution of the original data, measuring the similarity between the two datasets.

As in the other case, we used a custom implementation based on Python and Tensorflow. We compared the same three datasets as in the other metric. To compare each dataset with itself, we divided the datasets into two halves after shuffling them. We repeated the process five times in order to reduce the effects of the randomization. The results show that the values of the datasets with themselves were near 0 (0.43118 originals, 0.15095 synthetics, and 2.5117 noise).

We also compared the distance between the three datasets: 15.808 between originals and synthetics, 293.82 between originals and noise, and 296.77 between synthetics and noise. It can be seen that both originals and synthetics are considerably far from noise. It is true that the distance between them is not a value near 0, but it is still a low value, which shows that the difference between both datasets is not high. This is reinforced when this difference is compared with the difference between the noise and both datasets. The difference from 0 is mostly provoked due to the new patterns of shadows and defects that are generated due to the combinations of the different shadows of the original images, thanks to the generative capacity of the DCGAN models.

5. Conclusions and Future Work

The creation of synthetic electroluminescence images of photovoltaic cells is not a trivial problem. Different factors need to be taken into account in order to create high-quality images. The gathering of data is by its nature a manual process, requiring the EL image and its IV curve. The IV curve was measured individually for each cell, which is an innovation from other papers in the bibliography. The obtained images need proper preprocessing in order to perform well in the models. The labeling of the generated data is also a complex problem, since it is not possible to measure the output power or the IV curve of a not-real cell; so, we designed a model to assign a value of performance. This model was trained with the values of the real images, and its low error values show its good performance.

The algorithm used for creating the new images, DCGAN, has shown great performance, not only creating images that are very similar to the originals but also creating new patterns of defects and shadows. This similarity between the original images and the synthetics images was proved using different methods: A visual inspection of the cells shows that they present the characteristics of a real PV cell, and the histogram also proves their similarity, since they share the most important aspects of its shape. Image quality metrics such as Inspection Score (2.3) and French Inception Distance (15.8) also prove their similarity.

The results of this paper prove the quality of the synthetic images, but the dataset can be improved. The most direct way of improving the dataset would be increasing the amount of data. The inclusion of new kinds of defects and shadows would improve the generative capacities of the GAN. Another interesting option would be trying to use different kinds of PV cells, such as monocrystalline PV cells, since our dataset only consists of polycrystalline PV cells. This would improve the usefulness of the dataset in machine learning problems that use other kinds of PV cells.

Another way of improving the quality of the dataset would be increasing the size of the networks. The size of the current networks are based on the limitations in hardware and budget. Bigger networks could reduce the amount of epochs and improve the quality of the generated images. The use of the most innovative architectures could also improve the dataset.

Author Contributions: Conceptualization, H.F.M.R., L.H.-C. and V.C.-P.; methodology, H.F.M.R., V.C.-P. and M.Á.G.R.; validation, H.F.M.R. and V.A.G.; writing—original draft preparation, H.F.M.R. and H.J.B.; writing—review and editing, H.F.M.R., L.H.-C., V.C.-P., J.I.M.A. and R.T.M.; project administration, L.H.-C. All authors have read and agreed to the published version of the manuscript.

Funding: This study was supported by the University of Valladolid with the predoctoral contracts of 2020, co-funded by Santander Bank.

Institutional Review Board Statement: Not applicable.

Informed Consent Statement: Not applicable.

Data Availability Statement: The synthetic dataset can be found in https://github.com/hectorfelipe98/Synthetic-PV-cell-dataset.

Acknowledgments: This study was supported by the Universidad de Valladolid with ERASMUS+ KA-107. We also appreciate the help of other members of our departments.

Conflicts of Interest: The authors declare no conflict of interest. The funders had no role in the design of the study; in the collection, analyses, or interpretation of data; in the writing of the manuscript; or in the decision to publish the results.

Abbreviations

The following abbreviations are used in this manuscript:

EL	Electroluminescence
PV	Photovoltaic
GAN	Generative Adversarial Network
DCGAN	Deep Convolutional Generative Adversarial Network
IV	Intensity–Voltage
LED	Light-Emitting Diode
MPP	Max Power Point
MAE	Mean Absolute Error
MSE	Mean Squared Error
IS	Inception Score
FID	Fréchet Inception Distance

References

1. Rathore, N.; Panwar, N.; Yettou, F.; Gama, A. A Comprehensive review on different types of solar photovoltaic cells and their applications. *Int. J. Ambient Energy* **2019**, *42*, 1–48. [CrossRef]
2. Hernández-Callejo, L.; Gallardo-Saavedra, S.; Alonso-Gómez, V. A review of photovoltaic systems: Design, operation and maintenance. *Sol. Energy* **2019**, *188*, 426–440. [CrossRef]

3. Mellit, A.; Kalogirou, S.A. Artificial intelligence techniques for photovoltaic applications: A review. *Prog. Energy Combust. Sci.* **2008**, *34*, 574–632. [CrossRef]
4. Mateo Romero, H.F.; Gonzalez Rebollo, M.A.; Cardenoso-Payo, V.; Alonso Gomez, V.; Redondo Plaza, A.; Moyo, R.T.; Hernandez-Callejo, L. Applications of Artificial Intelligence to Photovoltaic Systems: A Review. *Appl. Sci.* **2022**, *12*, 10056. [CrossRef]
5. Pillai, D.S.; Blaabjerg, F.; Rajasekar, N. A Comparative Evaluation of Advanced Fault Detection Approaches for PV Systems. *IEEE J. Photovoltaics* **2019**, *9*, 513–527. [CrossRef]
6. Hong, Y.Y.; Pula, R.A. Methods of photovoltaic fault detection and classification: A review. *Energy Rep.* **2022**, *8*, 5898–5929. [CrossRef]
7. Sarker, I.H. Deep Learning: A Comprehensive Overview on Techniques, Taxonomy, Applications and Research Directions. *SN Comput. Sci.* **2021**, *2*, 420. [CrossRef]
8. Demirci, M.Y.; Beşli, N.; Gümüşçü, A. Efficient deep feature extraction and classification for identifying defective photovoltaic module cells in Electroluminescence images. *Expert Syst. Appl.* **2021**, *175*, 114810. [CrossRef]
9. Su, B.; Chen, H.; Chen, P.; Bian, G.; Liu, K.; Liu, W. Deep Learning-Based Solar-Cell Manufacturing Defect Detection with Complementary Attention Network. *IEEE Trans. Ind. Inform.* **2021**, *17*, 4084–4095. [CrossRef]
10. Balzategui, J.; Eciolaza, L.; Arana-Arexolaleiba, N. Defect detection on Polycrystalline solar cells using Electroluminescence and Fully Convolutional Neural Networks. In Proceedings of the 2020 IEEE/SICE International Symposium on System Integration, SII 2020, Honolulu, HI, USA, 12–15 January 2020; pp. 949–953. [CrossRef]
11. Shou, C.; Hong, L.; Ding, W.; Shen, Q.; Zhou, W.; Jiang, Y.; Zhao, C. Defect Detection with Generative Adversarial Networks for Electroluminescence Images of Solar Cells. In *Proceedings of the 2020 35th Youth Academic Annual Conference of Chinese Association of Automation (YAC)*; IEEE: Piscataway, NJ, USA, 2020; pp. 312–317. [CrossRef]
12. Luo, Z.; Cheng, S.Y.; Zheng, Q.Y. GAN-Based Augmentation for Improving CNN Performance of Classification of Defective Photovoltaic Module Cells in Electroluminescence Images. *IOP Conf. Ser. Earth Environ. Sci.* **2019**, *354*, 012106. [CrossRef]
13. Creswell, A.; White, T.; Dumoulin, V.; Arulkumaran, K.; Sengupta, B.; Bharath, A.A. Generative Adversarial Networks: An Overview. *IEEE Signal Process. Mag.* **2017**, *35*, 53–65. [CrossRef]
14. Schreiber, J.; Jessulat, M.; Sick, B. Generative Adversarial Networks for Operational Scenario Planning of Renewable Energy Farms: A Study on Wind and Photovoltaic. In Proceedings of the Artificial Neural Networks and Machine Learning—ICANN 2019: Image Processing: 28th International Conference on Artificial Neural Networks, Munich, Germany, 17–19 September 2019; pp. 550–564. [CrossRef]
15. Mateo Romero, H.; González Rebollo, M.; Cardeñoso-Payo, V.; Gomez, V.A.; Moyo, R.; Hernandez Callejo, L.; Bello, H. Synthetic Dataset of Electroluminescence images of Photovoltaic cells by Deep Convolutional Generative Adversarial Networks. In Proceedings of the V Ibero-American Congress of Smart Cities, Cuernavaca, Mexico, 13–17 November 2022; pp. 855–869.
16. Goodfellow, I.J.; Pouget-Abadie, J.; Mirza, M.; Xu, B.; Warde-Farley, D.; Ozair, S.; Courville, A.; Bengio, Y. Generative Adversarial Networks. *arXiv* **2014**, arXiv.1406.2661.
17. Xue, J.H.; Titterington, D.M. Do unbalanced data have a negative effect on LDA? *Pattern Recognit.* **2008**, *41*, 1558–1571. [CrossRef]
18. Pan, Z.; Yu, W.; Yi, X.; Khan, A.; Yuan, F.; Zheng, Y. Recent Progress on Generative Adversarial Networks (GANs): A Survey. *IEEE Access* **2019**, *7*, 36322–36333. [CrossRef]
19. Radford, A.; Metz, L.; Chintala, S. Unsupervised Representation Learning with Deep Convolutional Generative Adversarial Networks. *arxiv* **2015**, arXiv:1511.06434.
20. Deitsch, S.; Christlein, V.; Berger, S.; Buerhop-Lutz, C.; Maier, A.; Gallwitz, F.; Riess, C. Automatic classification of defective photovoltaic module cells in electroluminescence images. *Sol. Energy* **2019**, *185*, 455–468. [CrossRef]
21. Morales-Aragonés, J.I.; Gómez, V.A.; Gallardo-Saavedra, S.; Redondo-Plaza, A.; Fernández-Martínez, D.; Hernández-Callejo, L. Low-Cost Three-Quadrant Single Solar Cell I-V Tracer. *Appl. Sci.* **2022**, *12*, 6623. [CrossRef]
22. Gallardo-Saavedra, S.; Hernández-Callejo, L.; Alonso-García, M.d.C.; Santos, J.D.; Morales-Aragonés, J.I.; Alonso-Gómez, V.; Moretón-Fernández, Á.; González-Rebollo, M.Á.; Martínez-Sacristán, O. Nondestructive characterization of solar PV cells defects by means of electroluminescence, infrared thermography, I–V curves and visual tests: Experimental study and comparison. *Energy* **2020**, *205*, 117930. [CrossRef]
23. Ioffe, S.; Szegedy, C. Batch Normalization: Accelerating Deep Network Training by Reducing Internal Covariate Shift. *arXiv* **2015**, arXiv:1502.03167.
24. Xu, B.; Wang, N.; Chen, T.; Li, M. Empirical Evaluation of Rectified Activations in Convolutional Network. *arxiv* **2015**, arXiv:1505.00853.
25. Breiman, L. Random Forests. *Mach. Learn.* **2001**, *45*, 5–32. [CrossRef]
26. Pedregosa, F.; Varoquaux, G.; Gramfort, A.; Michel, V.; Thirion, B.; Grisel, O.; Blondel, M.; Prettenhofer, P.; Weiss, R.; Dubourg, V.; et al. Scikit-learn: Machine Learning in Python. *J. Mach. Learn. Res.* **2011**, *12*, 2825–2830.
27. Salimans, T.; Goodfellow, I.; Zaremba, W.; Cheung, V.; Radford, A.; Chen, X.; Chen, X. Improved Techniques for Training GANs. In *Proceedings of the Advances in Neural Information Processing Systems*; Lee, D., Sugiyama, M., Luxburg, U., Guyon, I., Garnett, R., Eds.; Curran Associates, Inc.: Red Hook, NY, USA, 2016; Volume 29.

28. Szegedy, C.; Vanhoucke, V.; Ioffe, S.; Shlens, J.; Wojna, Z. Rethinking the Inception Architecture for Computer Vision. In Proceedings of the 2016 IEEE Conference on Computer Vision and Pattern Recognition (CVPR), Las Vegas, NV, USA, 27–June 30 2016; pp. 2818–2826. [CrossRef]
29. Heusel, M.; Ramsauer, H.; Unterthiner, T.; Nessler, B.; Hochreiter, S. GANs Trained by a Two Time-Scale Update Rule Converge to a Local Nash Equilibrium. *Adv. Neural Inf. Process. Syst.* **2017**, *30*. [CrossRef]

entropy

MDPI

Article

An Adaptive Hybrid Model for Wind Power Prediction Based on the IVMD-FE-Ad-Informer

Yuqian Tian [1], Dazhi Wang [1,*], Guolin Zhou [1], Jiaxing Wang [1], Shuming Zhao [1] and Yongliang Ni [2]

1 College of Information Science and Engineering, Northeastern University, Shenyang 110819, China
2 China North Vehicle Research Institute, Beijing 100072, China
* Correspondence: prodzw@126.com

Abstract: Accurate wind power prediction can increase the utilization rate of wind power generation and maintain the stability of the power system. At present, a large number of wind power prediction studies are based on the mean square error (MSE) loss function, which generates many errors when predicting original data with random fluctuation and non-stationarity. Therefore, a hybrid model for wind power prediction named IVMD-FE-Ad-Informer, which is based on Informer with an adaptive loss function and combines improved variational mode decomposition (IVMD) and fuzzy entropy (FE), is proposed. Firstly, the original data are decomposed into K subsequences by IVMD, which possess distinct frequency domain characteristics. Secondly, the sub-series are reconstructed into new elements using FE. Then, the adaptive and robust Ad-Informer model predicts new elements and the predicted values of each element are superimposed to obtain the final results of wind power. Finally, the model is analyzed and evaluated on two real datasets collected from wind farms in China and Spain. The results demonstrate that the proposed model is superior to other models in the performance and accuracy on different datasets, and this model can effectively meet the demand for actual wind power prediction.

Keywords: wind power prediction; improved variational mode decomposition; fuzzy entropy; adaptive loss function; Informer

Citation: Tian, Y.; Wang, D.; Zhou, G.; Wang, J.; Zhao, S.; Ni, Y. An Adaptive Hybrid Model for Wind Power Prediction Based on the IVMD-FE-Ad-Informer. *Entropy* **2023**, *25*, 647. https://doi.org/10.3390/e25040647

Academic Editors: Luis Hernández-Callejo, Sergio Nesmachnow and Sara Gallardo Saavedra

Received: 6 March 2023
Revised: 5 April 2023
Accepted: 11 April 2023
Published: 12 April 2023

1. Introduction

The global energy-shortage problem is becoming more and more serious, and it is essential to accelerate the pace of energy structure transformation based on the increasing proportion of renewable energy. Wind power, as an economical and environmentally friendly emerging renewable energy source, has been vigorously developed by various countries, and its application prospects are promising [1,2]. However, with the random fluctuation of wind power, it has strong uncontrollability, resulting in a decrease in the dispatching efficiency of the power grid and an imbalance between energy supply and demand [3]. Therefore, achieving high-accuracy and high-reliability prediction of wind power in practical applications can minimize energy loss and make the power grid operate more stably and safely.

In recent years, a large number of scholars have studied wind power prediction models, which can be mainly divided into physical models [4], statistical models [5], artificial intelligence (AI) models [6], and hybrid models [7]. The physical models are based on the method of fluid mechanics, which uses numerical weather prediction data to calculate the wind turbine output curve and then calculate wind power from it [8]. However, the fluid mechanics method has the disadvantages of the high complexity of building the model and massive computational cost. The statistical models are based on the mapping relationship between historical data and future data [9,10]. Rajagopalan et al. [11] proposed an autoregressive moving average (ARMA) model for ultra-short-term wind power forecasting and achieved superior results. The autoregressive integrated moving average (ARIMA) model is a widely used statistical model that is based on ARMA with the addition of difference operation [12,13]. However, this method requires a large-scale dataset, making it difficult to mine the nonlinear relationship of complex data.

AI models are current technical trends and are widely used in the field of large-scale and multi-dimensional data prediction [14]. AI models are mainly divided into machine-learning models and deep-learning models [15]. For example, an echo state network (ESN) [16] was applied to wind speed forecasting and improved the prediction performance. Khan et al. [17] used the Naive Bayes Tree (NB) to extract the probabilities of each feature of wind power, successfully predicting wind power values from hours to years. Machine-learning methods are based on rigorous mathematical theories that enable rapid computation in high-dimensional spaces. Owing to weak generalization ability, machine-learning methods are prone to overfitting, and it is difficult to achieve good prediction effects. In contrast, deep-learning models unify feature-learning tasks and prediction tasks into one model, making them more suitable than shallow machine-learning models to solve wind power prediction problems in complicated uncertainty scenarios [18]. Tian et al. [19] used a model based on the attention mechanism and demonstrated its efficacy in wind power prediction. Liu et al. [20] presented a novel deep convolutional neural network (CNN) capable of automatically extracting hidden information from multi-dimensional data and efficiently implementing multi-step prediction. Hu et al. [21] applied a model integrated with a deep-learning framework and basic ESN network for energy prediction, which enhanced the model's memory capacity with a stacked hierarchy of reservoirs. Although these methods have achieved some success in wind power prediction, the fact that a single model cannot fully exploit the time series information leads to limited prediction performance [22].

Hybrid models are created by multiple intelligent algorithms or prediction models that combine the advantages of different models to achieve an improvement in prediction accuracy. Hybrid prediction models consist of combined multiple models and stacking models based on data processing [23,24]. Chen et al. [25] designed a weighted combination prediction composed of six long short-term memory networks (LSTM), and its prediction effect is better than that of a single prediction model. Xiong et al. [26] proposed a multi-scale hybrid prediction model that combines attention mechanism, CNN, and LSTM to adequately capture the high-dimensional features in wind farm data. Zheng et al. [27] established a hybrid model combining bidirectional long-short-term memory (Bi-LSTM) and CNN, which adopted a unique feature extraction method of space and then time. Although the combined prediction method of multiple models exhibits high prediction accuracy, it suffers low computational efficiency and narrow application scenarios [28]. Considering the nonlinear implicit relationship in the time series of wind power data, the stacking model based on data processing is proposed to improve the prediction accuracy by mining deep features through data decomposition. For example, Wu et al. [29] proposed a multi-step prediction method using variational modal decomposition (VMD) and chain ESN, which achieved multi-steps prediction at multiple time scales. Yang et al. [30] employed the VMD method to decompose the wind speed data, which was then utilized as input for an optimized LSTM network to perform predictions. Ren et al. [31] proposed a hybrid model of empirical-mode decomposition (EMD) and support-vector regression (SVR) for wind power prediction. Lv et al. [32] decomposed wind speed data into 3-dimensional input features using singular spectrum analysis (SSA) and fed them into a convolutional long-short-term memory (ConvLSTM) network, which effectively enhanced the local correlation between multivariate data. Khazaei and Ehsan [33] used prediction methods combining wavelet transform (WT) decomposition with the AI model, and the results showed that the model has high accuracy. Hybrid models based on data processing have a simple structure and strong feasibility, but the accuracy of the prediction greatly depends on the effectiveness of data decomposition. Over-decomposition of data can result in redundant components and reduce the efficiency of calculation, while the insufficient decomposition of data can lead to mode mixing, which fails to meet the needs of high-precision prediction [34].

Most prediction models use mean square error (MSE) as a loss function, which needs to meet the condition that prediction errors obey a Gaussian distribution. However, the use of MSE as a loss function in models that are insensitive to outliers in wind power data with high randomness may result in large errors [35]. In response to this issue, some researchers have improved the loss function of the model to minimize the impact of errors on the prediction results. Hu et al. [36] proposed a loss function without fixed distribution, which effectively solves the problem of prediction-gradient descent at wind power intervals. Duan et al. [35] designed a loss function with non-Gaussian distributed errors and combined it with an LSTM model to predict wind power. The loss function is significantly important in the wind power prediction process as it determines the training direction and accuracy of the model [37]. Although these improved loss functions have positive effects on wind power prediction, the models still exhibit limitations in terms of their adaptability and robustness.

Based on the above analysis, a robust and adaptive IVMD-FE-Ad-Informer hybrid model for wind power prediction is proposed in this paper, which aims to improve the precision of data decomposition and the predictive performance of non-stationary wind power data. The main contributions of this paper are outlined as follows: (1) Considering the difficulty in selecting the number of VMD, the IVMD algorithm improved by the maximum information coefficient (MIC) decomposes the original wind power data into K optimal sub-series, which effectively reduces the difficulty of wind power prediction. (2) fuzzy entropy (FE) is used to reconstruct sub-series into new elements of similar complexity together, alleviating the burden of model operation. (3) An adaptive loss function is innovatively introduced into the Informer network to solve the problem of traditional MSE's insensitivity to randomly fluctuating wind power data. This novel model can reduce the impact of outliers in non-smooth wind power data. (4) Ablation experiments and comparative experiments are performed on datasets collected from both different wind farms to verify the effectiveness and stability of the model. The prediction results show that the proposed model framework is reasonable, and it exhibits significantly better prediction performance and accuracy compared to other models.

The specific contents of this paper are as follows: Section 2 specifically describes the basic methodologies of hybrid models; Section 3 presents the construction framework and evaluation indicators of the IVMD-FE-Ad-Informer model; Section 4 constructs four experiments to verify the accuracy and validity of the proposed model; and finally, this paper is summarized in Section 5.

2. Methodologies

2.1. Variational Mode Decomposition

VMD [38] is a commonly used data decomposition method that converts wind power sequences from the time domain to the frequency domain and subsequently decomposes them into K intrinsic mode functions (IMFs). Firstly, build the variational constraint equation:

$$\begin{cases} \min\left\{ \sum\limits_{K=1}^{K} \left\| \partial_t \left[\left(\delta(t) + \frac{i}{\pi t} \right) * u_K(t) \right] e^{-iw_K t} \right\|_2^2 \right\} \\ \text{s.t. } \sum\limits_{K=1}^{K} u_K = x(t) \end{cases} \tag{1}$$

where $*$ is the convolution calculation symbol, $x(t)$ is the wind power sequence, w_K and u_K are the central frequency and band components of the kth IMF value, $\delta(t)$ is the impulse function, and ∂_t is used to denote the derivative of the function.

To simplify the variational constraint equation to a simple unconstrained problem, the Lagrange function $\lambda(t)$ and the penalty factor α are introduced:

$$L(\{u_K\}, \{w_K\}, \lambda) = \alpha \sum_{K=1}^{K} \left\| \partial_t \left[\left(\delta(t) + \frac{i}{\pi t} \right) * u_K(t) \right] e^{-iw_K t} \right\|_2^2 + \left\| x(t) - \sum_{K=1}^{K} u_K(t) \right\|_2^2 + \left\langle \lambda(t), x(t) - \sum_{K=1}^{K} u_K(t) \right\rangle \tag{2}$$

Then the optimal solution of the unconstrained problem is solved using the alternating direction method of multiplication with the following iterative procedure:

$$\hat{u}_K^{n+1}(w) = \frac{\hat{x}(w) - \sum_{i \neq K} \hat{u}_j(w) + (\hat{\lambda}(w)/2)}{1 + 2\alpha(w - w_K)^2} \tag{3}$$

$$w_K^{n+1} = \frac{\int_0^\infty w \left| \hat{u}_K^{n+1}(w) \right|^2 dw}{\int_0^\infty \left| \hat{u}_K^{n+1}(w) \right|^2 dw} \tag{4}$$

Finally, after applying the above process, the original wind power series is decomposed into the K sub-series.

2.2. Fuzzy Entropy

Fuzzy entropy (FE) [39] is a dynamical method for analyzing the complexity of time series. The FE value changes smoothly with changes in the set parameters, which makes it more robust to noise and more resistant to interference. Firstly, for time series with the length of n, the FE algorithm is introduced into the fuzzy membership function, and the specific formula is as follows:

$$D(x) = \exp\left[- \ln(2)\left(\frac{x}{r}\right)^2 \right] \tag{5}$$

where r is the similarity tolerance, $x = d_{ij}^m$, d_{ij}^m is the distance between vectors that reconstruct the time series into m-dimensional phase space, and $i, j = 1,2..., n - m + 1, i \neq j$.

Averaging over each i in D_{ij}^m yields, the average similarity function is as follows:

$$\phi^m(r) = \frac{1}{N - m + 1} \sum_{i=1}^{N-m+1} \left(\frac{1}{N - m} \sum_{j=1, j \neq i}^{N-m+1} D_{ij}^m \right) \tag{6}$$

Therefore, the FE expression is as follows:

$$\text{FuzzyEn}(m, r, n) = \ln \phi^m(r) - \ln \phi^{m+1}(r) \tag{7}$$

2.3. Informer

The Informer network is a variant of the Transformer that effectively addresses the long-sequence prediction problem [40]. Improvements of the Informer include: using a probsparse self-attention mechanism to reduce the complexity of matrix computation; introducing a self-attention distillation mechanism to extract the main features of time series, which effectively reduces memory usage; using a decoder to directly output the predicted values generatively to achieve the purpose of long-series prediction. The structure of the Informer model is shown in Figure 1.

The traditional self-attention mechanism consists of query, key, and value, and the expression is as follows:

$$f_A(Q, K, V) = \text{softmax}\left(\frac{QK^\top}{\sqrt{d}} \right) V \tag{8}$$

where $Q \epsilon \mathbb{R}^{L_Q \times d}$, $K \epsilon \mathbb{R}^{L_K \times d}$, $V \epsilon \mathbb{R}^{L_V \times d}$, d is the input dimension.

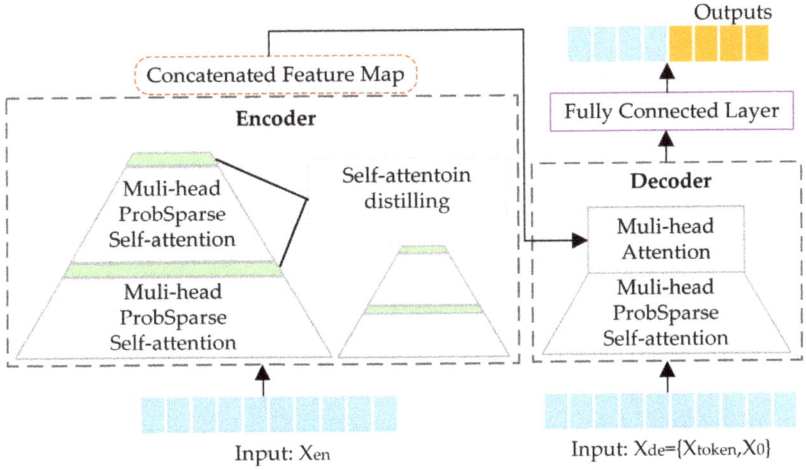

Figure 1. The structure of the Informer.

As the matrix multiplication involved in Equation (8) is computationally huge, the probsparse self-attention mechanism is introduced to select the important elements in Q to calculate the attention values.

$$f_A(Q,K,V) = \text{softmax}\left(\frac{\overline{Q}K^{\top}}{\sqrt{d}}\right)V \tag{9}$$

where $\overline{Q}$ is obtained through probabilistic sparsity of Q and controlled by a constant sampling factor c and the number of $\overline{Q}$ is $c * \ln L_K$.

Therefore, the similarity and importance between query and key are measured by Kullback–Leibler divergence, as follows:

$$k(q_i,k_j) = \ln\sum_{l=1}^{L_K} e^{\frac{q_i k_j^{\top}}{\sqrt{d}}} - \frac{1}{L_K}\sum_{j=1}^{L_K}\frac{q_i k_j^{\top}}{\sqrt{d}} - \ln L_K \tag{10}$$

where the relevance of q_i to k_j is proportional to the magnitude of $k(q_i,k_j)$. If $p(k_j|q_i)$ is close to a uniform distribution, i.e., $p(k_j|q_i) = 1/L_K$, indicating that q_i has the same similarity to all k_j, then q_i is deemed as a redundant vector and can be dropped.

Based on this, the sparsity evaluation formula that defines the i-th query is:

$$M(q_i,K) = \ln\sum_{l=1}^{L_K} e^{\frac{q_i k_j^{\top}}{\sqrt{d}}} - \frac{1}{L_K}\sum_{j=1}^{L_K}\frac{q_i k_j^{\top}}{\sqrt{d}} \tag{11}$$

The self-attentive distillation mechanism is introduced in the encoder. The width of the feature map is reduced to half its length after the distillation layer, which can reduce the overall memory usage and effectively solve the problem of long input. The concrete representation is as follows:

$$X_{j+1}^t = f_{\text{MP}}\left(\text{ELU}\left(f_{\text{Conv}}\left(\left[X_j^t\right]_{AB}\right)\right)\right) \tag{12}$$

where f_{MP} represents the maximum pooling layer function, f_{Conv} denotes the convolutional layer function, and $[\cdot]_{AB}$ is the attention unit.

The input of the decoder uses the time shield technique, and its input vector is as follows:

$$X_{\text{dec}}^{\text{in}} = f_{\text{Concat}}\,(X_{\text{token}}\,, X_0) \in \mathbb{R}^{(L_{\text{token}} + L_{\text{p}}) \times d_{\text{model}}} \tag{13}$$

where $X_{\text{token}} \in \mathbb{R}^{L_{\text{token}} \times d_{\text{model}}}$ is the input start token, L_{token} is the length of the start token, $X_0 \in \mathbb{R}^{L_p \times d_{\text{model}}}$ is the 0-value matrix, and L_p is the length of the part to be predicted.

2.4. Adaptive Loss Function

The adaptive loss function [41] obtains a generalized loss function by introducing robustness as a continuous parameter. During the training process, the adaptive loss function automatically adjusts the robustness parameters around the minimization loss algorithm, thereby enhancing the prediction accuracy. The generalized loss function formula is as follows:

$$f(z,\beta,c) = \frac{|\beta-2|}{\beta}\left(\left(\frac{(z/c)^2}{|\beta-2|}+1\right)^{\beta/2}-1\right) \tag{14}$$

where z is the difference between the true value and the predicted value, $c > 0$ serves as a scale factor that controls the curvature of the quadratic function at x = 0, and β is a variable parameter that controls the robustness.

By analyzing Equation (14), the adaptive loss function changes with the change of β. For different β, the adaptive loss function formula is as follows:

$$L(z,\beta,c) = \begin{cases} \frac{1}{2}(z/c)^2 & \text{if } \beta = 2 \\ \log\left(\frac{1}{2}(z/c)^2+1\right) & \text{if } \beta = 0 \\ \sqrt{(z/c)^2+1}-1 & \text{if } \beta = 1 \\ 1-\exp\left(-\frac{1}{2}(z/c)^2\right) & \text{if } \beta = -\infty \\ \frac{|\beta-2|}{\beta}\left(\left(\frac{(z/c)^2}{|\beta-2|}+1\right)^{\alpha/2}-1\right) & \text{otherwise} \end{cases} \tag{15}$$

It can be seen that the adaptive loss function can be a variety of loss functions, such as the MSE, Cauchy, Charbonnier, and Welsch loss functions, by adjusting the value of the variable parameter β.

3. Proposed Model

3.1. Improved VMD

With a solid mathematical theoretical foundation, VMD can effectively separate the components of complex signals and greatly suppress mode mixing. However, the decomposition parameter of VMD is given in advance, which limits the performance of data decomposition. In order to overcome the shortcomings of VMD in a parameter setting, this paper proposes the incorporation of the decomposition method and MIC [42] to determine the most suitable number of decompositions K. The degree of decomposition is determined by calculating the MIC value between the original sequence y and the reconstructed sequence y′, and the MICyy′ value is positively correlated with the number of decomposition numbers K. The closer MICyy′ value is to 1, the less information is lost during VMD decomposition, indicating a more adequate decomposition.

3.2. IVMD-FE-Ad-Informer Model Framework

In consideration of the high volatility of wind power data, this paper introduces the improved VMD and FE methods to the Informer network with adaptive loss function, and the framework is shown in Figure 2. In the data processing stage, the original data are decomposed into K IMFs by IVMD. Next, FE is used to calculate the complexity of each IMFs, and the IMFs with similar values are reconstructed into new elements. In the model-building stage, the input variables for the model are obtained through feature selection using the MIC algorithm and then input into a robust Ad-Informer prediction model. In the results analysis stage, the wind power forecasting results are obtained by linearly superposing the predicted values of each element, followed by visualizing the forecasting curve.

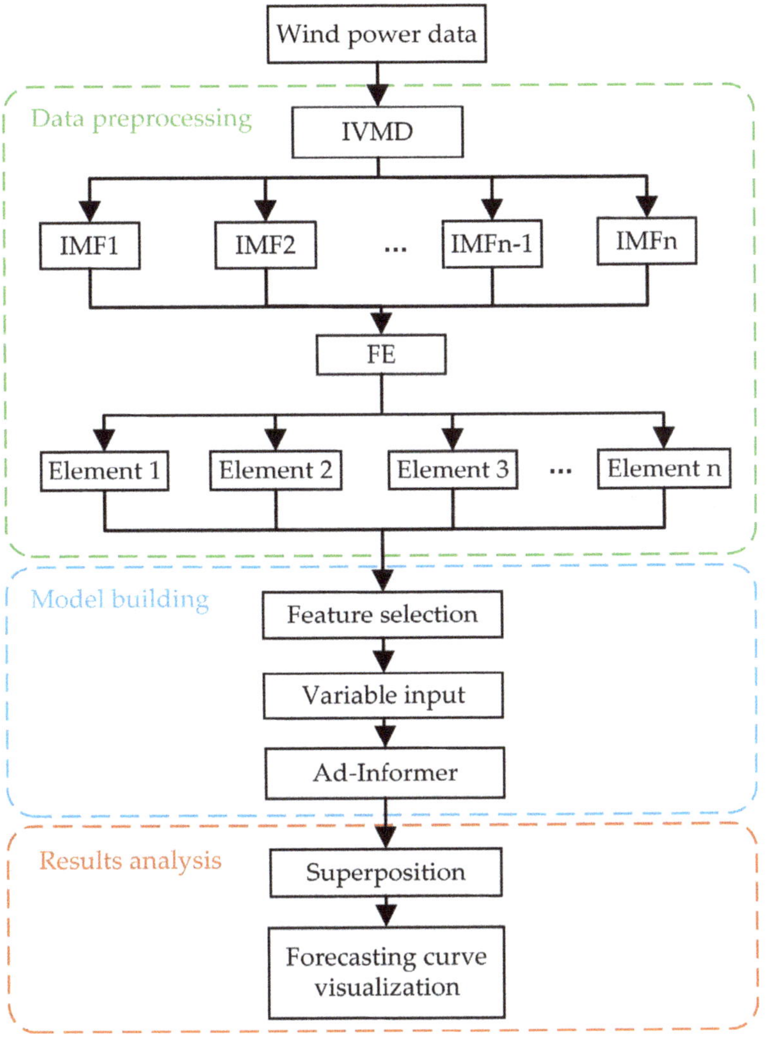

Figure 2. The framework of IVMD-FE-Ad-Informer model.

3.3. Evaluation Indexes

The mean absolute error (MAE), root mean square error (RMSE), and coefficient of determination (R^2) are used as evaluation indicators for the prediction performance of IVMD-FE-Informer and other benchmark models. The mathematical formula is as follows:

$$MAE = \frac{1}{N}\sum^{N}\left|q_{\text{true}}(t) - q_{\text{pred}}(t)\right| \tag{16}$$

$$RMSE = \sqrt{\frac{1}{N}\sum_{t=1}^{N}\left(q_{\text{true}}(t) - q_{\text{pred}}(t)\right)^2} \tag{17}$$

$$R^2 = 1 - \frac{\sum_{t=1}^{N}\left(q_{\text{true}}(t) - q_{\text{pred}}(t)\right)^2}{\sum_{t=1}^{N}\left(q_{\text{true}}(t) - \overline{q}\right)^2} \tag{18}$$

where $q_{\text{true}}(t)$ and $q_{\text{pred}}(t)$ denote the true and predicted values of wind power at time t, respectively, $\overline{q}$ is the mean value of q_{true}, and N is the number of samples in the dataset.

4. Experiment and Analysis

In this section, four sets of experiments are conducted on datasets with different sampling intervals, capacities, and regions. Experiment 1 aims to describe the specific details of the data processing. Experiment 2 is designed as an ablation experiment to verify the prediction performance of the hybrid model. Experiment 3 mainly aims to design a comparative experiment to verify the viability and superiority of each module. Experiment 4 aims to verify the applicability and stability of the proposed model on different datasets. All experiments are run in Python 3.7 and Pytorch environment with Intel(R) Core (TM) i5-12500H CPU @ 4.50 GHz, 12 Cores, NVIDIA GeForce RTX 3050 GPU, a memory capacity of 16 Gb, and Windows 11 operating system.

4.1. Data Description

The experiments are mainly conducted on two complete datasets without missing values in this paper. Dataset A is based on a wind farm in Gansu, China, which was selected from 1 July to 30 September 2019, with a sampling interval of 15 min. Dataset A contains wind power, wind speeds at different heights (10 m, 30 m, 50 m, 70 m, and hub height), air temperature, air pressure, and humidity features. Dataset B was collected from the Sotavento Galicia wind farm in Spain from 18 January to 12 March 2020, with a sampling interval of 10 min. Dataset B contains only wind power, wind speed, and wind direction features. The wind power curves from different datasets are show in Figure 3.

The prediction process is the same for different datasets; in fact, dataset A is used for Experiments 1 to 3, and dataset B is used for Experiment 4. The datasets are divided into the training set, validation set, and test set in a ratio of 7:2:1, and the results of each experiment are obtained by taking the average of 10 iterations. The characteristics of the datasets, including number, maximum value (Max), minimum value (Min), mean, standard deviation (Std), and coefficient of variation (COV), are shown in Table 1.

Table 1. The characteristics of datasets.

Dataset	Number	Max (MW)	Min (MW)	Mean (MW)	Std (MW)	COV
Dataset A	8832	120.43	0	23.51	24.74	1.0523
Dataset B	7920	2.803	0	0.8976	0.7885	0.8784

4.2. Experiment 1: The Specific Details of Data Processing

The data processing part mainly includes data decomposition, new elements reconstruction, and feature selection, and in this part, the operation process and the selection of parameters for data processing will be specifically discussed.

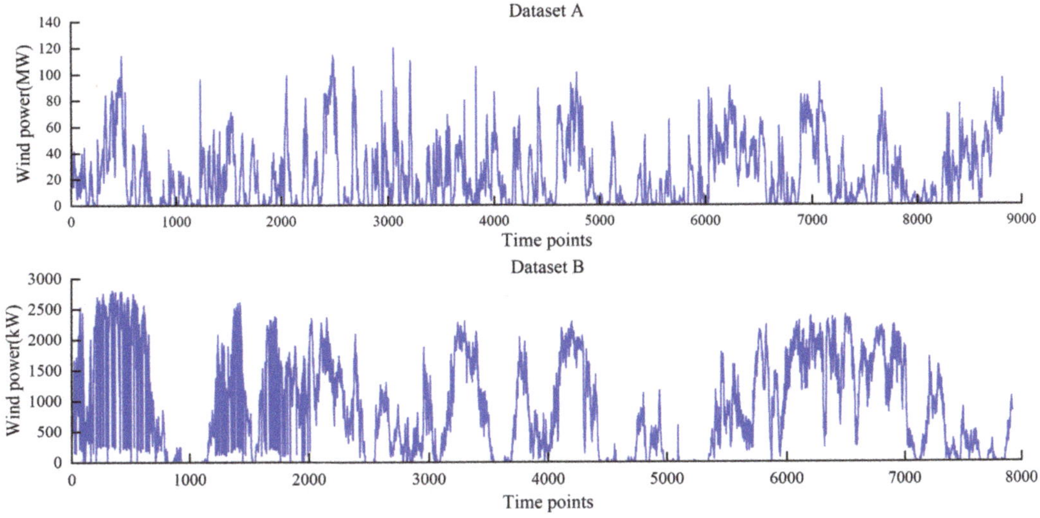

Figure 3. The curve of datasets.

4.2.1. Data Decomposition

The IVMD algorithm solves the traditional VMD problem of K selection by calculating the MICyy′ value. The original wind power data are fed into the IVMD model, which is decomposed into K IMFs. Based on the results of MICyy′ corresponding to different values of K as indicated in Figure 4, it can be observed that the value of MICyy′ remains stable and constant for $K = 16$. The IMFs curve after IVMD decomposition and its corresponding spectrum diagram are shown in Figure 5. By observing the principal frequencies of different IMFs from Figure 5, it can be concluded that the IVMD algorithm proposed in this paper can effectively separate each IMF accurately.

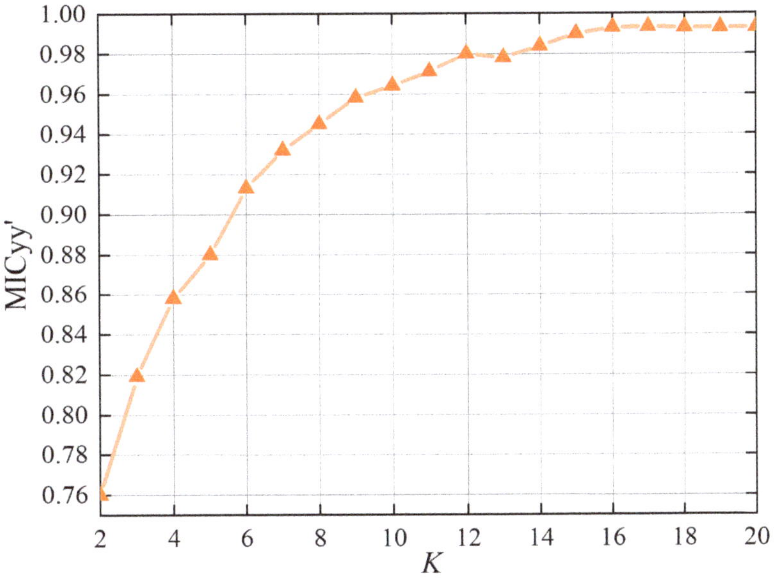

Figure 4. The curve of MICyy′ vs. K.

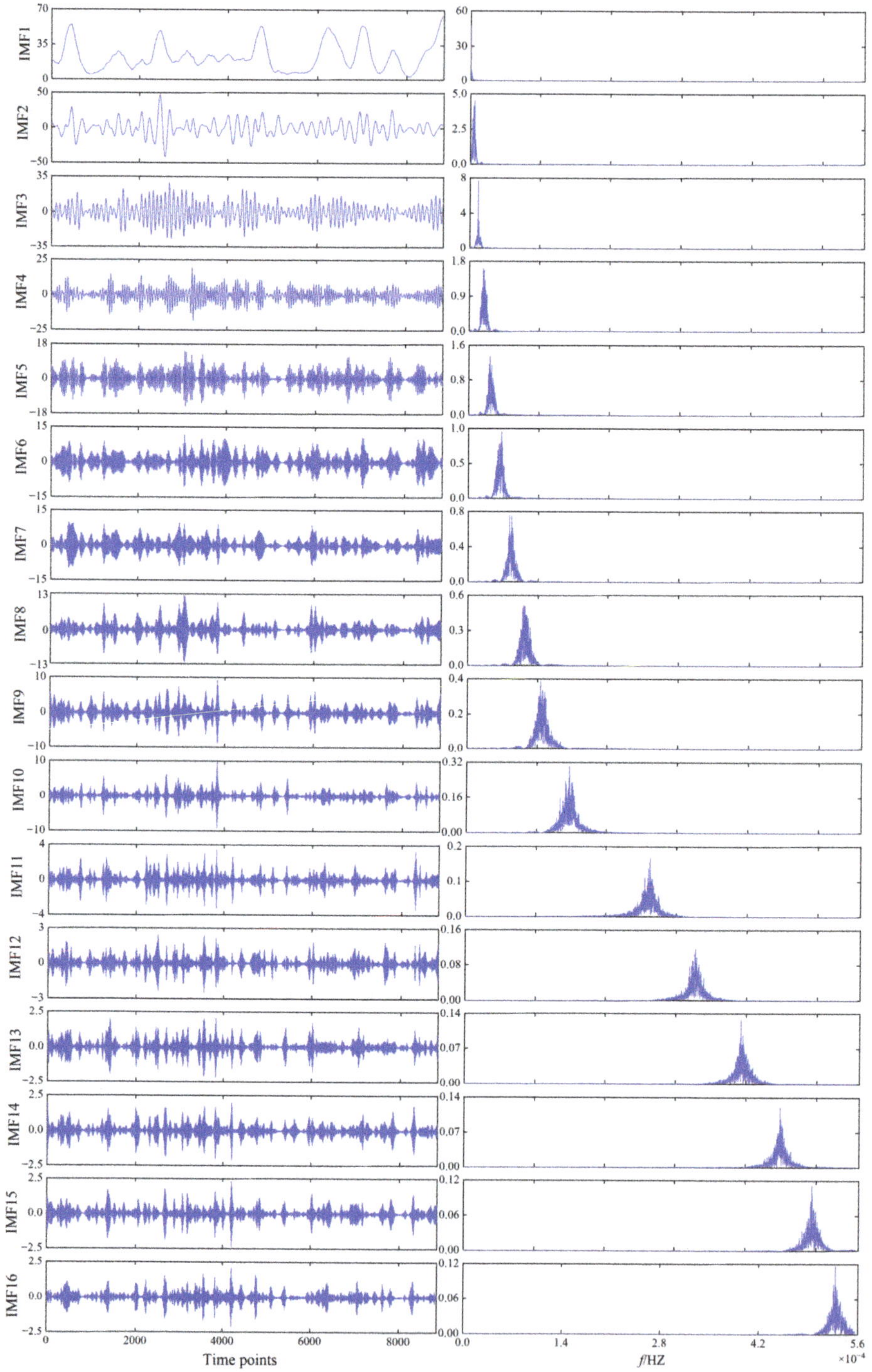

Figure 5. The results of the IVMD algorithm. The IMFs curves are shown (**left**), and the spectral densities corresponding to the IMFs are shown (**right**).

4.2.2. New Elements Reconstruction

The original wind power data are decomposed into 16 IMFs, and if all the sub-series are directly fed into the prediction model, it will increase the operational burden of the prediction model. Therefore, the complexity of these IMFs will be evaluated by FE, and then the IMFs with similar complexity will be reconstructed into new elements. After conducting extensive experiments, the values of m = 2 and r = 0.25std are found to be the optimal settings for achieving the best accuracy and running time of the model. The FE values of each IMF are shown in Figure 6, and the reconstructed new elements based on these FE values are shown in Table 2.

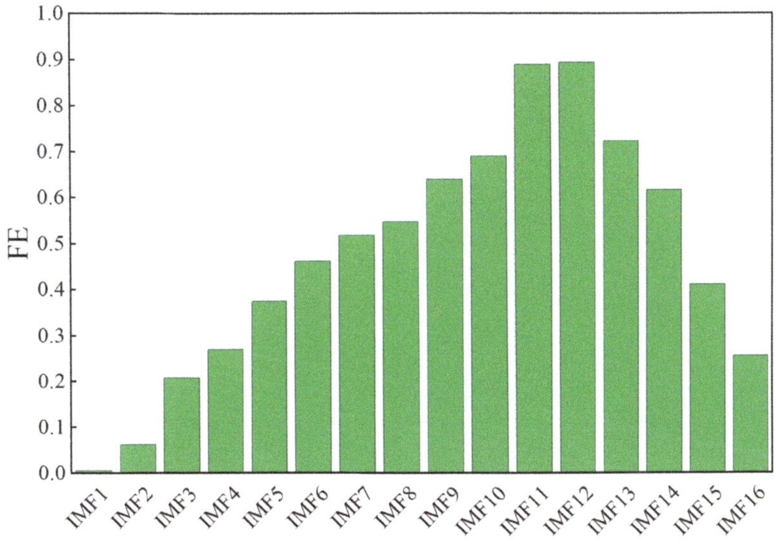

Figure 6. The FE value of each IMF.

Table 2. New elements reconstruction.

Reconstruction Elements	IMFs
Element 1	IMF1, IMF2
Element 2	IMF3, IMF4, IMF16
Element 3	IMF5, IMF6, IMF7, IMF8, IMF15
Element 4	IMF9, IMF10, IMF13, IMF14
Element 5	IMF11, IMF12

4.2.3. Feature Selection

The computational efficiency and generalization ability of the model can be improved by removing some irrelevant or redundant features from the original dataset. Therefore, MIC is used to analyze the correlation between meteorological features and each element and extract typical features reflecting each element through MIC value. The confusion matrix of MIC is given in Figure 7. It can be found from Figure 7 that the influence characteristics of each element are different, reflecting the overall correlation and local characteristics, respectively. In order to select the features with the highest relevance to build the input variables, the MIC thresholds of each element are set to 0.5. The input feature selection results are shown in Table 3.

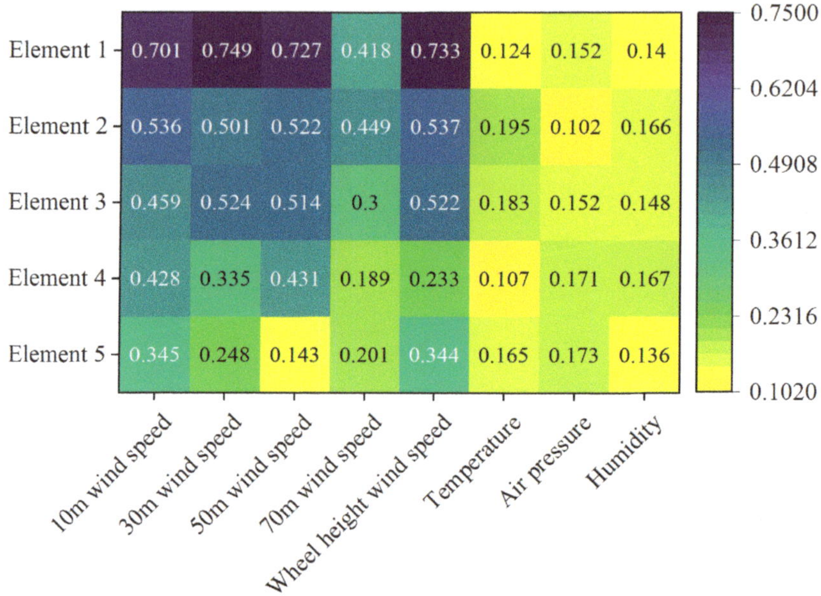

Figure 7. The confusion matrix of MIC.

Table 3. The feature selection results.

Element	Input Variables
Element 1	Element 1, 10 m, 30 m, 50 m, wheel height wind speed
Element 2	Element 2, 10 m, 50 m, wheel height wind speed
Element 3	Element 3, 30 m, 50 m, wheel height wind speed
Element 4	Element 4
Element 5	Element 5

4.3. Experiment 2: Ablation Experiment

The purpose of conducting ablation experiments is to verify whether the complex hybrid model has improved the prediction accuracy as compared to simple combinatorial models and single models. The selected benchmark models are IVMD-FE-Ad-Informer, Ad-Informer, and Informer, of which the Informer model uses the MSE loss function. The parameters of AD-Informer are obtained using the grid search method, where the robustness parameter β is adaptively adjusted using the Adam optimizer. The specific parameters are shown in Table 4. The input size of the encoder and decoder is equal to the number of input variables of the model. The prediction curve results of each sub-mode after the training of the IVMD-FE-Ad-Informer model are shown in Figure 8, and the wind power prediction results can be obtained by superimposing them. The final forecasting curves of the ablation experiment are shown in Figure 9, and forecasting errors are shown in Table 5. Figure 9 not only portrays the overall trend of the test set but also amplifies the values from position 300 to 480 in order to offer an in-depth analysis of the predicted results. The main reason is that the wind power data within the test set from the 300th to the 480th position displays more sudden changes and a wider range of variation, thus providing a more comprehensive evaluation of the predictive performance of the proposed model.

Table 4. Parameter setting of the Ad-Informer.

Parameters	Values
Input sequence length	96
Start token length	24–96
Prediction sequence length	24–96
Num of encoder layers	3
Num of decoder layers	2
Input size of encoder	5-1
Input size of decoder	5-1
Decoder output	1
Num of heads	8
Dimension of model	512
Probsparse attention factor	5
Early stopping patience	5
Learning rate	0.0001
Dropout	0.05
Epochs	100
Scale factor	1.2
Optimizer	Adam
Gpu	Cuda0

From Figure 9, it can be seen that the Ad-Informer model is significantly closer to the true value than the Informer model at the inflection point, indicating that the proposed adaptive function can effectively mitigate the impact of errors at the abrupt points. Compared with the Ad-Informer model, the IVMD-FE-Ad-Informer model is closer to the real value, indicating that the data processing method can reduce the time delay in the process of prediction. According to Table 5, the Ad-Informer model requires less time than the Informer model, mainly attributed to the automatic adjustment of the adaptive loss function, which enables the model to obtain the optimal loss during the training process and enhance its robustness. The hybrid model proposed in this paper shows significant improvements over the AD-Informer and Informer models, with a decrease of 45.09% and 59.67% in MAE, 44.4% and 55.44% in RMSE, and an increase of 11.42% and 22.72% in R2, respectively. By comparing the considered models, it can be seen that IVMD-FE-Ad-Informer decomposes the original wind power data into finer granularity, which can better explore the internal features of wind power, resulting in a significant improvement in both prediction accuracy and performance.

Table 5. Forecasting errors of ablation experiment.

Model	MAE (MW)	RMSE (MW)	R^2	Time (s)
IVMD-FE-Ad-Informer	3.19	4.67	0.956	1633.21
Ad-Informer	5.81	8.40	0.858	177.13
Informer	7.91	10.48	0.779	165.559

4.4. Experiment 3: Comparative Experiment

To verify the superiority of each module, EMD-FE-Ad-Informer, IVMD-FE-Informer, IVMD-FE-LSTM, LSTM, and ANN are used as benchmark models in the comparison experiments, and the parameter settings of ANN and LSTM are the same as [19,43]. EMD decomposes wind power data into 11 IMFs by trail-and-error method, and then these IMFs are reconstructed into three new components (IMF1~IMF3, IMF4~IMF6, and IMF7~IMF11) using FE. The forecasting curves of different models are shown in Figure 10. The forecasting errors are shown in Table 6. The boxplots of the forecasting errors for each model are given in Figure 11.

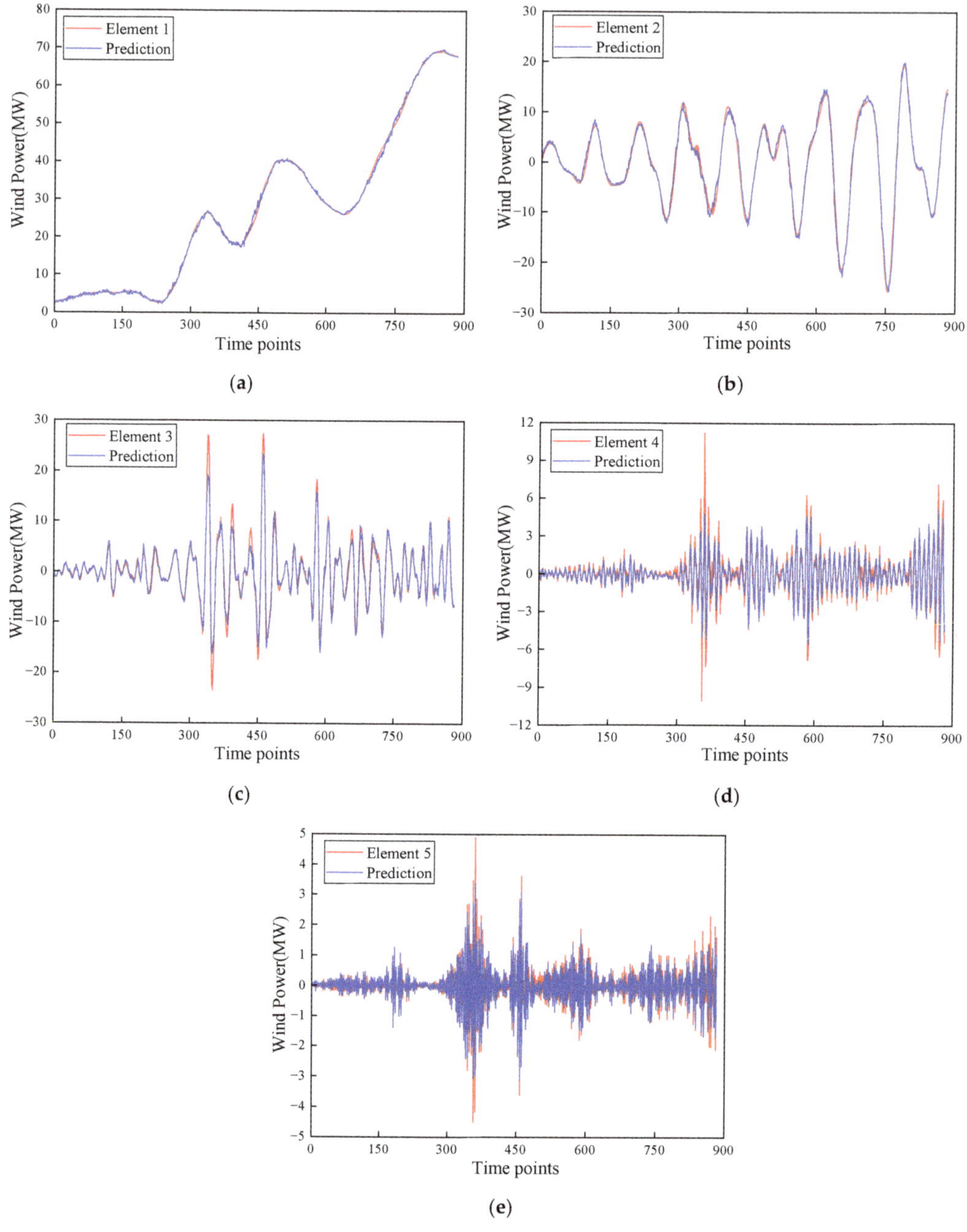

Figure 8. The prediction curve results of each sub-mode. (**a**) Element 1 prediction curve; (**b**) Element 2 prediction curve; (**c**) Element 3 prediction curve; (**d**) Element 4 prediction curve; (**e**) Element 5 prediction curve.

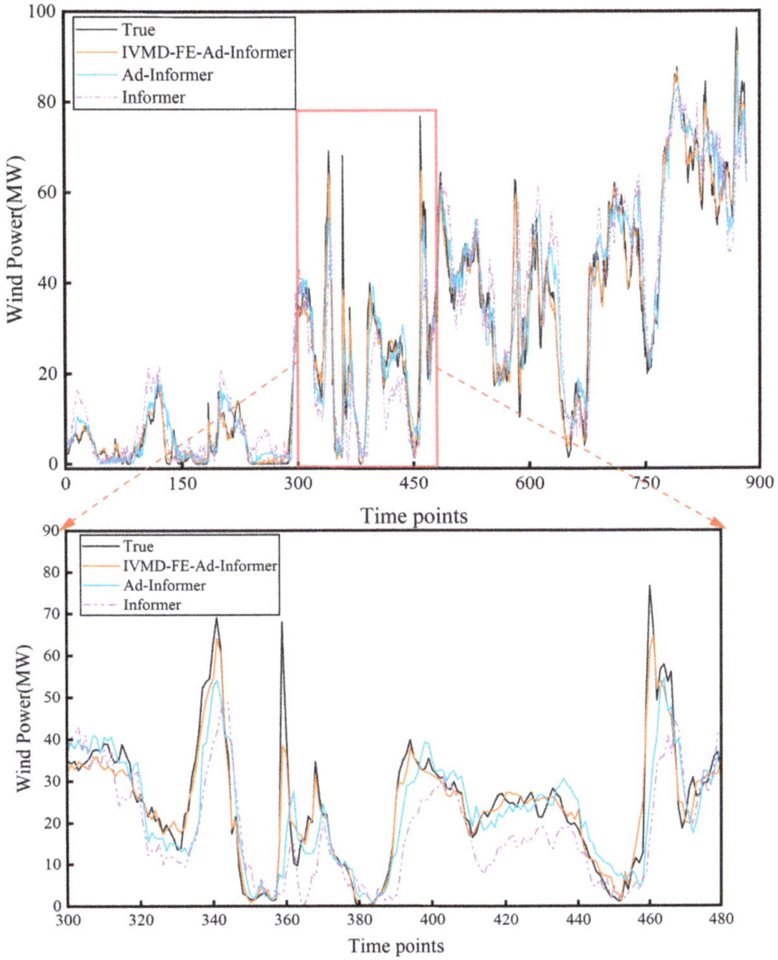

Figure 9. The forecasting curves of the ablation experiment. The overall forecasting trends are shown at the (**top**), and the local enlargement is shown at the (**bottom**).

Table 6. Forecasting errors of different models.

Model	MAE (MW)	RMSE (MW)	R^2	Time (s)
IVMD-FE-Ad-Informer	3.19	4.67	0.956	1633.21
EMD-FE-Ad-Informer	4.96	7.31	0.905	1362.47
IVMD-FE-Informer	5.81	8.40	0.889	1878.63
IVMD-FE-LSTM	6.63	9.79	0.808	1732.57
LSTM	7.86	10.95	0.759	305.11
ANN	8.04	11.58	0.731	81.02

Based on the results in Table 6, the IVMD-FE-Ad-Informer model outperforms the single prediction model and the other hybrid models across all evaluation metrics. It can be concluded that MAE decreased by about 35.68–60.32%, RMSE decreased by about 36.11–59.67%, and R^2 increased by about 5.64–30.78%. According to Table 6 and Figure 10, it can be inferred that the IVMD algorithm has superior data decomposition ability compared to the traditional EMD algorithm under similar data processing. This improved ability enables the IVMD algorithm to more effectively reduce non-smooth features in the original data, resulting in smoother data and improved wind power prediction accuracy. Furthermore, the prediction accuracy of Ad-Informer is much higher than that of Informer and LSTM for the same data processing method, with R^2 of 0.925, 0.889, and 0.808, respectively. While IVMD-FE-Ad-Informer is relatively time-consuming due to the implementation of the Ad-Informer prediction module five times after the IVMD-FE data preprocessing, it demonstrates a closer resemblance to the actual curve and produces the smallest forecasting errors. It can be indicated that the model proposed in this paper is an optimal combined model with high prediction performance.

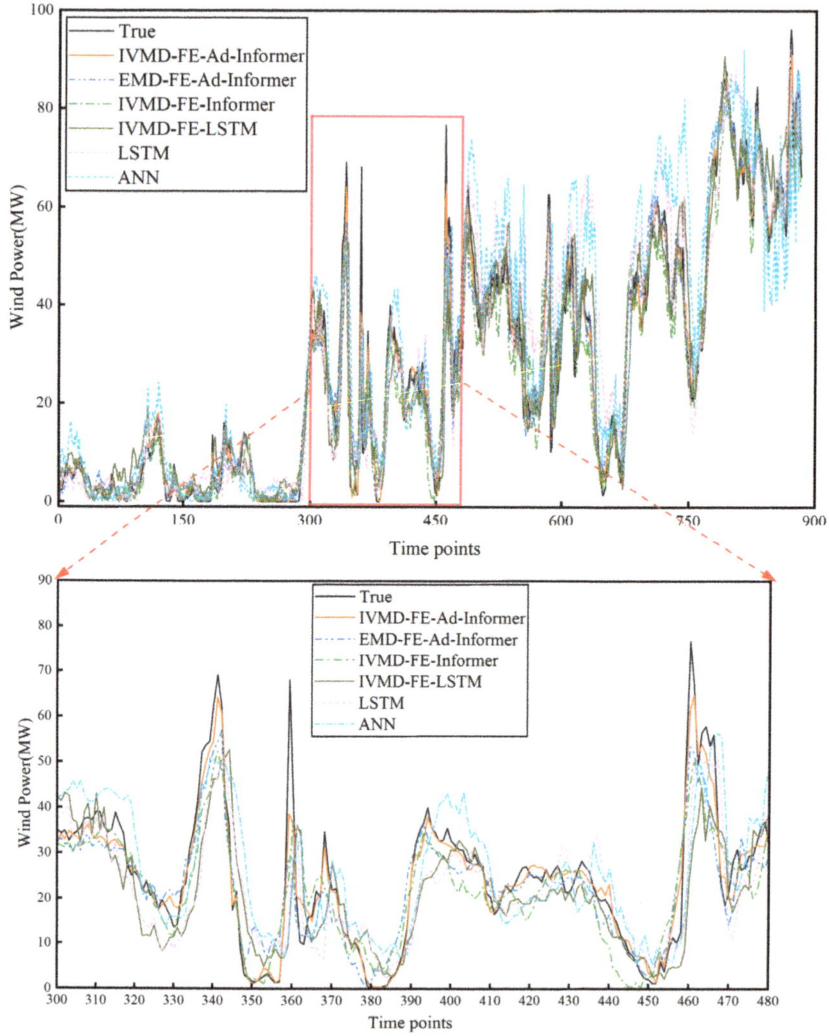

Figure 10. The forecasting curves of different models. The overall forecasting trends are shown at the (**top**), and the local enlargement is shown at the (**bottom**).

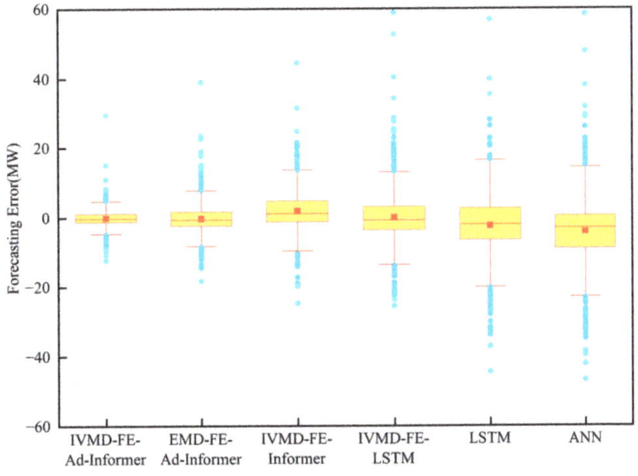

Figure 11. The boxplots of different models.

4.5. Experiment 4: The Stability of IVMD-FE-Ad-Informer Forecasting

The experimental results demonstrate that IVMD-FE-Ad-Informer outperforms other benchmark models on dataset A and exhibits considerable wind power prediction ability. However, the statistical distributions of wind power data vary across different time intervals, regions, and capacities, which may lead to the phenomenon of unstable forecasting. Therefore, the stability and applicability of the model still need further discussion. In this section, EMD-FE-Ad-Informer, Ad-Informer, LSTM, and ANN are used as benchmark models on dataset B, which are collected from the Sotavento Galicia wind farm in Spain at 10 min sampling intervals. The parameter-setting method of this experiment is the same as Experiment 3, and the specific parameter settings of each algorithm are shown in Table A1, Appendix A. The forecasting curves of dataset B are shown in Figure 12, and the forecasting errors are shown in Table 7.

Table 7. Forecasting errors of different datasets.

Model	MAE (kW)	RMSE (kW)	R^2	Time (s)
IVMD-FE-Ad-Informer	83.01	60.43	0.962	1076.34
EMD-FE-Ad-Informer	115.89	70.75	0.914	671.47
Ad-Informer	144.51	105.46	0.866	156.91
LSTM	186.63	131.04	0.762	228.88
ANN	197.16	140.62	0.746	62.15

According to Figure 12 and Table 7, the results obtained from dataset B are comparable to those from dataset A, indicating that the model proposed in this paper has high stability and generalization ability on different datasets. From Figure 12, IVMD-FE-Ad-Informer exhibited the closest fit to the true values among all the considered models, with the EMD-FE-Ad-Informer following closely behind. It can also be seen from Table 7 that the IVMD-FE-Ad-Informer has the best prediction performance with regard to MAE, RMSE, and R^2, which are 83.01 kW, 60.43 kW, and 0.962, respectively. The results further confirm that the IVMD algorithm is a superior and effective method for wind power data decomposition.

Based on the experimental results of the two different datasets, it is apparent that the IVMD-FE-Ad-Informer outperforms other benchmark models in terms of all evaluation metrics and has the closest fit of prediction curves to the true values. Meanwhile, the COV value is introduced for further analysis of the influence of prediction accuracy on different datasets. This value is a typical indicator of the degree of data fluctuation, with more volatile data having a higher COV value [44]. It also can be concluded that the accuracy of the proposed model prediction is inversely related to the degree of fluctuation in the original data. For example, when using Ad-Informer to forecast wind power on dataset A, the R^2 is 0.866, whereas, on dataset B with a higher COV, R^2 is slightly lower at 0.858. Furthermore, the superiority of the proposed model in terms of prediction performance becomes more prominent as the original wind power sequence contains more nonlinear features. The outstanding contribution is the development of an adaptive loss function, which can accurately identify and predict violent changes in wind power, thereby effectively mitigating the impact of outliers.

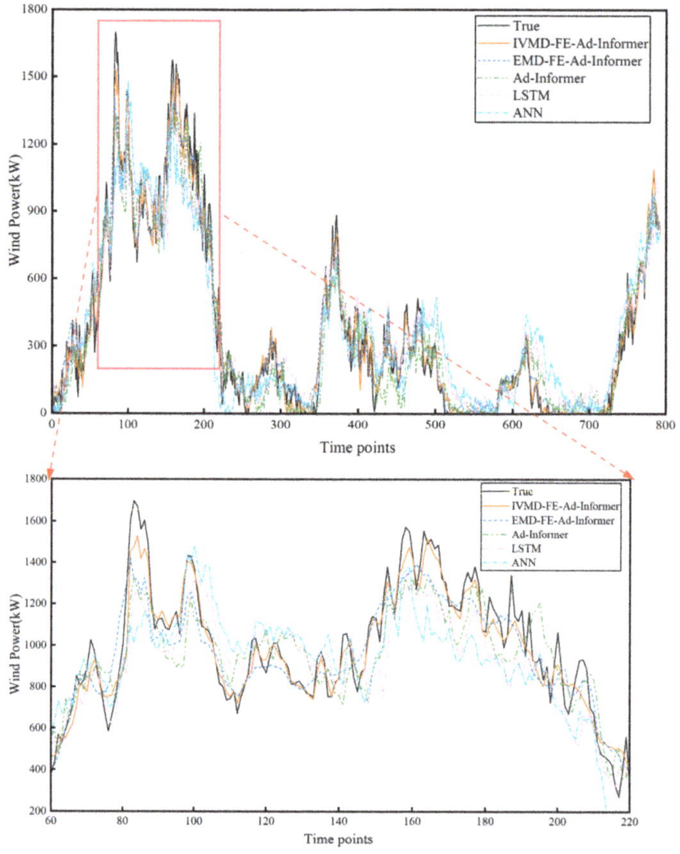

Figure 12. The forecasting curves of different datasets. The overall forecasting trends are shown at the (**top**), and the local enlargement is shown at the (**bottom**).

5. Conclusions

The actual operation of wind farms is influenced by various factors such as weather conditions, season variation, and atmospheric circulation, which can lead to numerous outliers and non-smooth features in the wind power data. The presence of such factors brings many obstacles to achieving the further improvement of accuracy and performance of wind power prediction. Thus, an adaptive hybrid model for wind power prediction based on improved VMD, FE, and Informer in conjunction with adaptive loss function is proposed in this paper. The IVMD-FE-Ad-Informer model is a promising hybrid model that enables adaptive forecasting of stochastically fluctuating wind power data, and its main advantages are summarized as follows:

1. The IVMD-FE-Ad-Informer is a hybrid model that demonstrates high accuracy and better robustness by integrating the advantages of multiple technologies, outperforming the basic EMD- FE-Ad-Informer, Ad-Informer, LSTM, and ANN. The results of the proposed model obtained from the Spanish and Chinese datasets demonstrate a significant improvement compared to benchmark models, with a maximum reduction of 57.89% in MAE, 57.03% in RMSE, and a maximum increase of 30.78% in R^2;
2. Compared with traditional data decomposition methods, VMD improved by MIC can better mine the nonlinear features of the original data, which effectively improves the data quality and reduces the difficulty of prediction;
3. Based on a comprehensive analysis of experimental results, the adaptive loss function has a rapid response to non-Gaussian distributed wind power data, which can react quickly to outliers and predict variation trends;
4. By prediction experiments on wind farm datasets with different sampling intervals, capacities, and regions, the proposed model shows the best prediction results and closest proximity to the true value. It can be demonstrated that IVMD-FE-Ad-Informer has remarkable generalization ability and broad prospects in wind power prediction.

As can be seen from the above, the hybrid wind power prediction model that combines the advantages of several algorithms has higher prediction accuracy and better robustness. However, there are some problems in this study that need to be improved in the future. Firstly, in this paper, only the correlation factor is considered in feature selection, while F-score and sensitivity factors are not taken into account. In future work, the analysis of the relationship between other variables and wind power using F-score and sensitivity will be conducted to reduce the redundancy of massive data. Then, the parameter selection in this paper may not be precise enough, and to address this issue, the optimization algorithm will be introduced to overcome the sensitive defect of deep-learning network-parameter selection.

Author Contributions: Y.T. and D.W. described the proposed framework and wrote the whole manuscript; Y.T. implemented the simulation experiments; G.Z. and J.W. collected data; Y.N. and S.Z. revised the manuscript. All authors have read and agreed to the published version of the manuscript.

Funding: This research is supported in part by the Natural Science Foundation of China under Grant 52077027 Study and in part by the Liaoning Province Science and Technology Major Project No. 2022021000014.

Institutional Review Board Statement: Not applicable.

Data Availability Statement: The data used to support the findings of this study are available from the corresponding author upon request.

Acknowledgments: The authors thank the chief editor and the reviewers for their valuable comments on how to improve the manuscript.

Conflicts of Interest: The authors declare no conflict of interest.

Appendix A

Table A1. The parameter settings of each algorithm on dataset B.

Algorithms	Parameters	Values
IVMD	K	17
FE	m	2
	r	0.25std
IVMD-FE	New mode	4(mode1 = IMF1, IMF2, IMF17; mode2 = IMF3, IMF4, IMF15, IMF16; mode3 = IMF5, IMF12~IMF14; mode4 = IMF6~IMF11)

Table A1. *Cont.*

Algorithms	Parameters	Values
EMD	K	11
EMD-FE	New mode	3(mode1 = IMF1, IMF6, IMF7; mode2 = IMF2~IMF5; mode3 = IMF8~IMF11)

References

1. Lin, Z.; Liu, X. Wind Power Forecasting of an Offshore Wind Turbine Based on High-Frequency Scada Data and Deep Learning Neural Network. *Energy* **2020**, *201*, 117693. [CrossRef]
2. Wang, K.; Zhang, Y.; Lin, F.; Wang, J.; Zhu, M. Nonparametric Probabilistic Forecasting for Wind Power Generation Using Quadratic Spline Quantile Function and Autoregressive Recurrent Neural Network. *IEEE Trans. Sustain. Energy* **2022**, *13*, 1930–1943. [CrossRef]
3. Naik, J.; Dash, P.K.; Dhar, S. A Multi-Objective Wind Speed and Wind Power Prediction Interval Forecasting Using Variational Modes Decomposition Based Multi-Kernel Robust Ridge Regression. *Renew. Energy* **2019**, *136*, 701–731. [CrossRef]
4. Ahmed, A.; Khalid, M. A Review on the Selected Applications of Forecasting Models in Renewable Power Systems. *Renew. Sustain. Energy Rev.* **2019**, *100*, 9–21. [CrossRef]
5. Wang, Y.; Zou, R.; Liu, F.; Zhang, L.; Liu, Q. A Review of Wind Speed and Wind Power Forecasting with Deep Neural Networks. *Appl. Energy* **2021**, *304*, 117766. [CrossRef]
6. Jiang, Z.; Che, J.; He, M.; Yuan, F. A Cgru Multi-Step Wind Speed Forecasting Model Based on Multi-Label Specific Xgboost Feature Selection and Secondary Decomposition. *Renew. Energy* **2023**, *203*, 802–827. [CrossRef]
7. Ye, J.; Xie, L.; Ma, L.; Bian, Y.; Xu, X. A Novel Hybrid Model Based on Laguerre Polynomial and Multi-Objective Runge-Kutta Algorithm for Wind Power Forecasting. *Int. J. Electr. Power Energy Syst.* **2023**, *146*, 108726. [CrossRef]
8. Kisvari, A.; Lin, Z.; Liu, X. Wind Power Forecasting—A Data-Driven Method Along with Gated Recurrent Neural Network. *Renew. Energy* **2021**, *163*, 1895–1909. [CrossRef]
9. Jin, H.; Li, Y.; Wang, B.; Yang, B.; Jin, H.; Cao, Y. Adaptive Forecasting of Wind Power Based on Selective Ensemble of Offline Global and Online Local Learning. *Energy Convers. Manag.* **2022**, *271*, 116296. [CrossRef]
10. Zheng, H.; Hu, Z.; Wang, X.; Ni, J.; Cui, M. Vmd-Cat: A Hybrid Model for Short-Term Wind Power Prediction. *Energy Rep.* **2023**, *9*, 199–211. [CrossRef]
11. Rajagopalan, S.; Santoso, S. Wind Power Forecasting and Error Analysis Using the Autoregressive Moving Average Modeling. In Proceedings of the Paper presented at the General Meeting of the IEEE-Power-and-Energy-Society, Calgary, AB, Canada, 26–30 July 2009.
12. Kaytez, F. A Hybrid Approach Based on Autoregressive Integrated Moving Average and Least-Square Support Vector Machine for Long-Term Forecasting of Net Electricity Consumption. *Energy* **2020**, *197*, 117200. [CrossRef]
13. Yang, Y.; Lu, J. Foreformer: An Enhanced Transformer-Based Framework for Multivariate Time Series Forecasting. *Appl. Intell.* **2022**. [CrossRef]
14. Xiang, L.; Li, J.; Hu, A.; Zhang, Y. Deterministic and Probabilistic Multi-Step Forecasting for Short-Term Wind Speed Based on Secondary Decomposition and a Deep Learning Method. *Energy Convers. Manag.* **2020**, *220*, 113098. [CrossRef]
15. Rangel-Martinez, D.; Nigam, K.D.P.; Ricardez-Sandoval, L.A. Machine Learning on Sustainable Energy: A Review and Outlook on Renewable Energy Systems, Catalysis, Smart Grid and Energy Storage. *Chem. Eng. Res. Des.* **2021**, *174*, 414–441. [CrossRef]
16. Hu, H.; Wang, L.; Tao, R. Wind Speed Forecasting Based on Variational Mode Decomposition and Improved Echo State Network. *Renew. Energy* **2021**, *164*, 729–751. [CrossRef]
17. Khan, M.; He, C.; Liu, T.; Ullah, F. A New Hybrid Approach of Clustering Based Probabilistic Decision Tree to Forecast Wind Power on Large Scales. *J. Electr. Eng. Technol.* **2021**, *16*, 697–710. [CrossRef]
18. Yin, H.; Ou, Z.; Huang, S.; Meng, A. A Cascaded Deep Learning Wind Power Prediction Approach Based on a Two-Layer of Mode Decomposition. *Energy* **2019**, *189*, 116316. [CrossRef]
19. Tian, C.; Niu, T.; Wei, W. Developing a Wind Power Forecasting System Based on Deep Learning with Attention Mechanism. *Energy* **2022**, *257*, 18. [CrossRef]
20. Liu, X.; Yang, L.; Zhang, Z. Short-Term Multi-Step Ahead Wind Power Predictions Based on a Novel Deep Convolutional Recurrent Network Method. *IEEE Trans. Sustain. Energy* **2021**, *12*, 1820–1833. [CrossRef]
21. Hu, H.; Wang, L.; Lv, S.X. Forecasting Energy Consumption and Wind Power Generation Using Deep Echo State Network. *Renew. Energy* **2020**, *154*, 598–613. [CrossRef]
22. Lu, P.; Ye, L.; Zhao, Y.; Dai, B.; Pei, M.; Tang, Y. Review of Meta-Heuristic Algorithms for Wind Power Prediction: Methodologies, Applications and Challenges. *Appl. Energy* **2021**, *301*, 117446. [CrossRef]
23. Del Ser, J.; Casillas-Perez, D.; Cornejo-Bueno, L.; Prieto-Godino, L.; Sanz-Justo, J.; Casanova-Mateo, C.; Salcedo-Sanz, S. Randomization-Based Machine Learning in Renewable Energy Prediction Problems: Critical Literature Review, New Results and Perspectives. *Appl. Soft Comput.* **2022**, *118*, 108526. [CrossRef]

24. Meng, A.; Chen, S.; Ou, Z.; Ding, W.; Zhou, H.; Fan, J.; Yin, H. A Hybrid Deep Learning Architecture for Wind Power Prediction Based on Bi-Attention Mechanism and Crisscross Optimization. *Energy* **2022**, *238*, 121795. [CrossRef]
25. Chen, J.; Zeng, G.Q.; Zhou, W.; Du, W.; Lu, K.D. Wind Speed Forecasting Using Nonlinear-Learning Ensemble of Deep Learning Time Series Prediction and Extremal Optimization. *Energy Convers. Manag.* **2018**, *165*, 681–695. [CrossRef]
26. Xiong, B.; Lou, L.; Meng, X.; Wang, X.; Ma, H.; Wang, Z. Short-Term Wind Power Forecasting Based on Attention Mechanism and Deep Learning. *Electr. Power Syst. Res.* **2022**, *206*, 107776. [CrossRef]
27. Zhen, H.; Niu, D.; Yu, M.; Wang, K.; Liang, Y.; Xu, X. A Hybrid Deep Learning Model and Comparison for Wind Power Forecasting Considering Temporal-Spatial Feature Extraction. *Sustainability* **2022**, *12*, 9490. [CrossRef]
28. Tascikaraoglu, A.; Uzunoglu, M. A Review of Combined Approaches for Prediction of Short-Term Wind Speed and Power. *Renew. Sustain. Energy Rev.* **2014**, *34*, 243–254. [CrossRef]
29. Wu, Z.; Zeng, S.; Jiang, R.; Zhang, H.; Yang, Z. Explainable Temporal Dependence in Multi-Step Wind Power Forecast Via Decomposition Based Chain Echo State Networks. *Energy* **2023**, *270*, 126906. [CrossRef]
30. Yang, S.; Yang, H.; Li, N.; Ding, Z. Short-Term Prediction of 80-88 Km Wind Speed in near Space Based on Vmd-Pso-Lstm. *Atmosphere* **2023**, *14*, 315. [CrossRef]
31. Ren, Y.; Suganthan, P.N.; Srikanth, N. A Novel Empirical Mode Decomposition with Support Vector Regression for Wind Speed Forecasting. *IEEE Trans. Neural Netw. Learn. Syst.* **2016**, *27*, 1793–1798. [CrossRef]
32. Lv, S.X.; Wang, L. Multivariate Wind Speed Forecasting Based on Multi-Objective Feature Selection Approach and Hybrid Deep Learning Model. *Energy* **2023**, *263*, 126100. [CrossRef]
33. Khazaei, S.; Ehsan, M.; Soleymani, S.; Mohammadnezhad-Shourkaei, H. A High-Accuracy Hybrid Method for Short-Term Wind Power Forecasting. *Energy* **2022**, *238*, 122020. [CrossRef]
34. Duan, J.; Wang, P.; Ma, W.; Tian, X.; Fang, S.; Cheng, Y.; Chang, Y.; Liu, H. Short-Term Wind Power Forecasting Using the Hybrid Model of Improved Variational Mode Decomposition and Correntropy Long Short-Term Memory Neural Network. *Energy* **2021**, *214*, 118980. [CrossRef]
35. Luo, X.; Sun, J.; Wang, L.; Wang, W.; Zhao, W.; Wu, J.; Wang, J.H.; Zhang, Z. Short-Term Wind Speed Forecasting Via Stacked Extreme Learning Machine with Generalized Correntropy. *IEEE Trans. Ind. Inform.* **2018**, *14*, 4963–4971. [CrossRef]
36. Hu, J.; Lin, Y.; Tang, J.; Zhao, J. A New Wind Power Interval Prediction Approach Based on Reservoir Computing and a Quality-Driven Loss Function. *Appl. Soft Comput.* **2020**, *92*, 106327. [CrossRef]
37. Yin, H.; Ou, Z.; Zhu, Z.; Xu, X.; Fan, J.; Meng, A. A Novel Asexual-Reproduction Evolutionary Neural Network for Wind Power Prediction Based on Generative Adversarial Networks. *Energy Convers. Manag.* **2021**, *247*, 114714. [CrossRef]
38. Dragomiretskiy, K.; Zosso, D. Variational Mode Decomposition. *IEEE Trans. Signal Process.* **2014**, *62*, 531–544. [CrossRef]
39. Chen, W.; Zhuang, J.; Yu, W.; Wang, Z. Measuring Complexity Using Fuzzyen, Apen, and Sampen. *Med. Eng. Phys.* **2009**, *31*, 61–68. [CrossRef]
40. Zhou, H.; Zhang, S.; Peng, J.; Zhang, S.; Li, J.; Xiong, H.; Zhang, W. Informer: Beyond Efficient Transformer for Long Sequence Time-Series Forecasting. *AAAI* **2021**, *35*, 11106–11115. [CrossRef]
41. Barron, J.T. A General and Adaptive Robust Loss Function. In Proceedings of the IEEE/CVF Conference on Computer Vision and Pattern Recognition, Long Beach, CA, USA, 15–20 June 2019.
42. Lin, G.; Lin, A.; Gu, D. Using Support Vector Regression and K-Nearest Neighbors for Short-Term Traffic Flow Prediction Based on Maximal Information Coefficient. *Inf. Sci.* **2022**, *608*, 517–531. [CrossRef]
43. Memarzadeh, G.; Keynia, F. A New Short-Term Wind Speed Forecasting Method Based on Fine-Tuned Lstm Neural Network and Optimal Input Sets. *Energy Convers. Manag.* **2020**, *213*, 15. [CrossRef]
44. Zhang, Y.; Tao, P.; Wu, X.; Yang, C.; Han, G.; Zhou, H.; Hu, Y. Hourly Electricity Price Prediction for Electricity Market with High Proportion of Wind and Solar Power. *Energies* **2022**, *15*, 1345. [CrossRef]

sustainability

MDPI

Article

Prediction of Faults Location and Type in Electrical Cables Using Artificial Neural Network

Ana-Maria Moldovan * and Mircea Ion Buzdugan

Faculty of Building Services Engineering, Technical University of Cluj-Napoca, 28 Memorandumului Str., 400114 Cluj-Napoca, Romania
* Correspondence: ana.moldovan@insta.utcluj.ro

Abstract: Detecting and locating faults in electrical cables has been a permanent concern regarding electrical power distribution systems. Over time, several techniques have been developed aiming to manage these faulty situations in an efficient way. These techniques must be fast, accurate, but, above all, efficient. This paper develops a new approach for detecting, locating, classifying, and predicting faults, particularly in different types of short-circuits in electrical cables, based on a robust artificial neural network technique. The novelty of this approach lies in the ability of the method to predict fault's location and type. The proposed method uses the Matlab and Simulink platform and comprises four consecutive stages. The first one is devoted to the development of the Simulink model. The second one implies a large number of simulations in order to generate the necessary dataset for training and testing the artificial neural network model (ANN). The following stage uses the ANN to classify the location and the type of potential faults. Finally, the fourth stage consists of predicting the location and the type of future faults. In order to reduce the time and the resources of the simulation process, a virtual machine is used. The study reveals the efficiency of the method, and its ability to successfully predict faults in real-world electrical power systems.

Keywords: electrical cables; detecting, locating and predicting faults; artificial neural network; Classification Learner app

Citation: Moldovan, A.-M.; Buzdugan, M.I. Prediction of Faults Location and Type in Electrical Cables Using Artificial Neural Network. *Sustainability* **2023**, *15*, 6162. https://doi.org/10.3390/su15076162

Academic Editors: Luis Hernández-Callejo, Sergio Nesmachnow and Sara Gallardo Saavedra

Received: 7 March 2023
Revised: 29 March 2023
Accepted: 30 March 2023
Published: 3 April 2023

1. Introduction

Today, people's lives are entirely dependent on the sustainability of electrical power systems. This supposes that the continuity of the supply of the electrical power distribution systems is mandatory. In this respect, electrical cables have the important role of linking all components of a power system.

The presented method depicted in the following sections aims to contribute to a higher degree of sustainability of the distribution power systems, accelerating the maintenance process in fault cases, due to its accuracy in predicting the location and the type of faults in the energy cables.

A fault in a cable directly affects the sustainability of the system, and the duration of a power outage, being crucial to ensure the cables' integrity during their entire operation time [1–3]. However, if any defect occurs in a cable, the reaction must be as fast as possible to reduce to a minimum the duration of its clearing time [4,5].

Methodologies of detecting faults in electrical cables have evolved with the advancements in technology, with several methods being implemented: time domain reflectometry technique, impedance-based method, knowledge-based method, traveling wave methods or hybrid methods [6]. Each of them has its benefits and its limitations [7,8]. For instance, time domain reflectometry can be successfully used in the case of a single cable, being useless for systems that have more than two branches [7,9,10].

Algorithms based on artificial intelligence (AI) propose solutions that are able to manage these more complex systems [11–13]. The artificial neural network technique

(ANN) provides efficient pattern recognition algorithms that can be applied in predicting, locating and classifying faults [14,15]. The ANN technique is able to solve nonlinear problems, based on learned experiences, which implies different possible configurations of the electrical distribution systems [16–20]. At the same time, the main ANN algorithms features are robustness, generalizability, and noise immunity [21–23].

A complex method of detecting and locating faults in electrical cables should return their exact type and location [24–26]. For a three-phase electrical power system, the fault types are interruptions or short-circuits, with the last ones being: single line to ground fault, line to line fault, double line to ground fault, three-line fault and three line to ground fault [27–29].

This paper presents an efficient method of detecting, locating, and predicting the different types of short-circuits in electrical cables. In order to develop this method, a model of distribution electrical system has been modeled in Simulink, followed by the use of the ANN technique of the Classification Learner app available in Matlab.

The model selected for analysis contains a three-phase source block of 20 kV and several distributed parameters line blocks with a total of 22 km of cables. The ANN technique is based on data generated by the Simulink model of the electrical distribution system. The method aims to reach a high rate of validation accuracy of the trained model delivered by the Classification Learner app. After running a large number of simulations, more precisely 6150 simulations, a 98% rate of validation accuracy was obtained.

The generated data represent the training dataset for the ANN algorithm which has an important impact over the accuracy of the method, since the performance of the method increases with the complexity of the dataset. The case presented below highlights the complexity and advantages of using ANNs methods in predicting, detecting, locating, and classifying short-circuits in complex distribution electrical power systems.

The article is structured in five sections. The present section is the introductory one, highlighting the importance of detecting faults in electrical power distributions systems. In the second one, entitled Materials and Methods, the working principle of the method is presented. Then, the Results section comprises four consecutive stages. The first stage is devoted to the development of the Simulink model. The second stage presents a large number of simulations in order to generate the necessary dataset for training and testing the ANN model. The third one uses the ANN to classify the location and the type of potential faults. Finally, the fourth stage consists of predicting the location and the type of future faults. The fourth and fifth sections are devoted to the discussion, conclusions and further work.

2. Materials and Methods

The need to detect and locate faults in electrical systems has generated different methods in the attempt to solve these problems. In the present section, a new approach involving a simulated model of a distribution electrical power system, combined with the benefits of ANN applications, will be presented.

As mentioned above, the method comprises four consecutive stages. The first one is devoted to the development of the model, and the second one presents a large number of simulations in order to generate a dataset necessary for training and testing the ANN model, while the third stage uses ANN to classify the location and the type of faults. Finally, the fourth stage consists of predicting the location and the type of potential future faults.

After modeling the distribution electrical system in Simulink (R2022a), several simulations were performed for different types of short-circuits in different locations of the cables. The results of all simulations have been saved in a database which became the training dataset for the Classification Learner app from Matlab (R2022a). The input data for the Classification Learner app are the measured values of the voltage and current, the responses being either the location of the faults or both their location and type. Based on the trained neural network, the location and the type of a further fault can be predicted.

The model was created using the blocks contained in the Simscape (R2022a) electrical library, a library dedicated to electrical power systems. The developed simulation model contains a three-phase source block, distributed parameters line blocks, three-phase voltage-intensity (VI) measurement blocks, three-phase load blocks and a three-phase fault block able to induce faults in different locations of the cables. The model and these blocks will be detailed in the next section.

Since the process of simulating the model which provides the training set for the neural network is a time-consuming one, it had to be automated. The automation consisted in writing a Matlab code which ran these simulations, and at each simulation, modifying the parameters and saving the data delivered by the measurement blocks.

Once the database is accessible, the whole set of simulation results is introduced in the Classification Learner app from Matlab. At this point, the training process may start. In this last Matlab application, different types of algorithms based on artificial intelligence are available, such as decision trees, discriminant analysis, naive bayes classifiers, support vector machines, nearest neighbor classifiers, ensemble and neural network classifiers [30]. The Classification Learner app allows the training of all these algorithms based on the accuracy of validation, enabling the possibility of choosing the most efficient one. For the present case, the most accurate algorithm turned out to be the medium neural network model. This model can be exported into Matlab workspace, being later used in the prediction of the trained model response for another set of measurements, corresponding to a further fault due to the versatility and complexity of the Matlab and Simulink platform.

The presented method is synthesized in the process diagram in Figure 1.

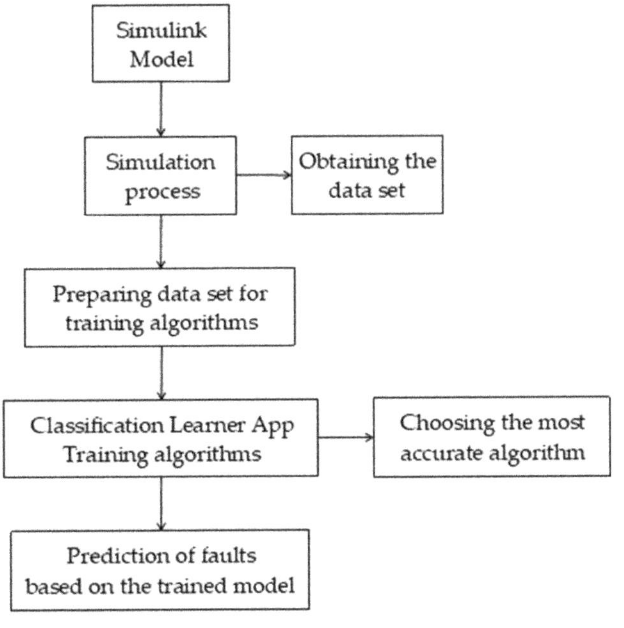

Figure 1. The process diagram of the method presented.

3. Results

As stated above, the case under analysis proposes a method of detecting and locating faults in electrical systems based on the medium neural network algorithm, which can be successfully used in solving faults detection, location, and prediction.

3.1. Simulink Model

This first stage of the presented method is devoted to the development of the Simulink model of a distribution electrical system. The development of the model concept consists of inserting different types of short-circuits in different locations of the system and observing their influence on the measured voltage–current pairs. Based on it, the training dataset necessary in the third stage will be used.

As mentioned in Section 2, in order to develop the simulation model, the Simscape electrical library was used. The selected model of the distribution electrical system presented in Figure 2 contains: a three-phase source block, depicted in green; six subsystems for the six lines L1 to L6, depicted in dark green, which contains the distributed parameters line blocks; eight three-phase voltage-intensity VI measurement blocks B1 to B8, depicted in blue; three-phase load blocks, noncolored; a three-phase fault block, bordered in red; and powergui, the environment block for Simscape electrical specialized power system models, set to discrete simulation type with the sample time of 2×10^{-6} s.

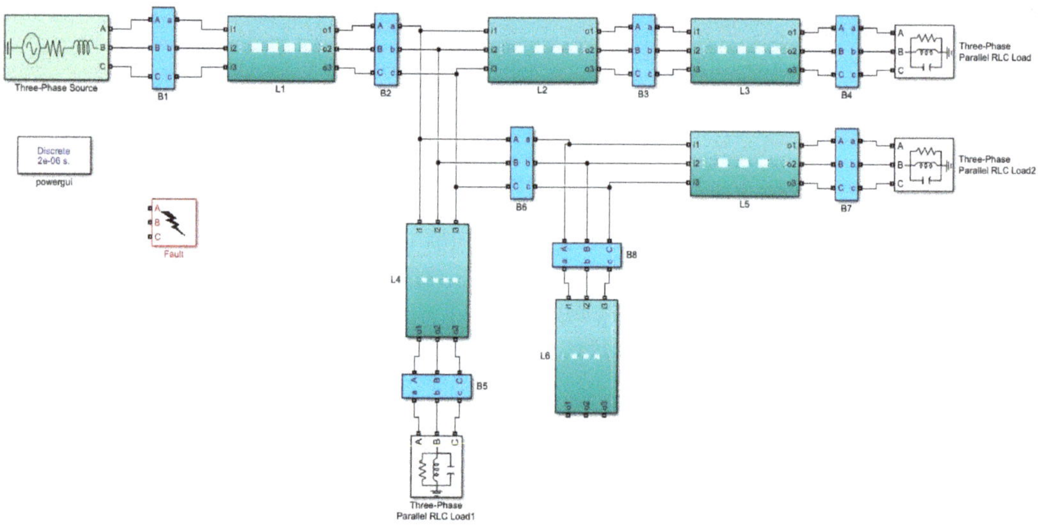

Figure 2. Distribution electrical system—Simulink model.

To introduce the fault block in different locations, each line is divided into three or four sectors of 1 km length, totaling 22 sectors (see Figure 3 and Table 1). From Figure 4, one can see that the sectors are modeled using three phase distributed parameter line blocks.

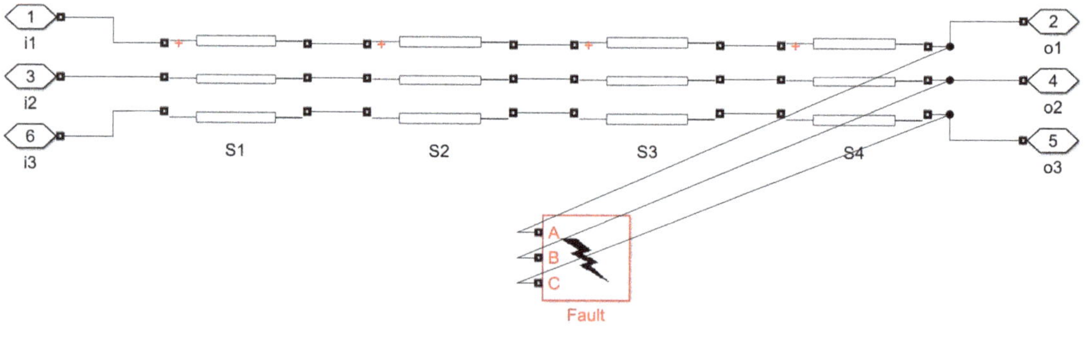

Figure 3. Fault block connected to L1—Sector 4.

Table 1. Subsystems of Lines.

Line	1	2	3	4	5	6
Number of Sectors	4	4	4	4	3	3

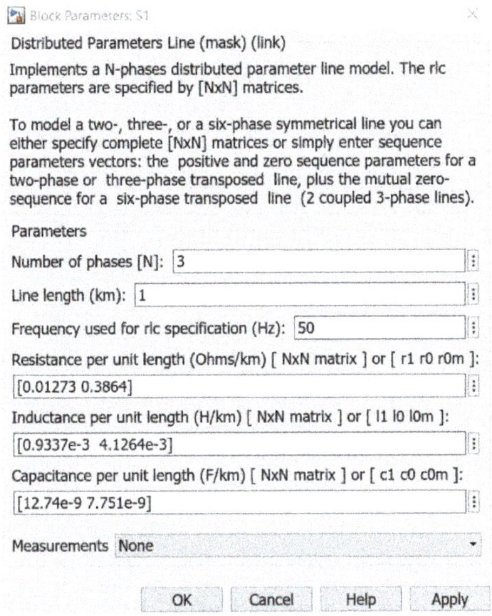

Figure 4. Distributed parameter line block.

The three-phase fault block can be set for twelve types of faults, the fault resistance, the ground resistances, and the no-fault situation included (see Table 2). The letters a, b, and c indicate the three power lines, while g indicates the ground plane.

Table 2. Types of faults.

Fault	no	ag	bg	cg	ab	bc	ac	abg	bcg	acg	abc	abcg

The three-phase voltage-intensity VI measurement blocks are mandatory to collect the values of the voltage–current pairs. As an example, Figure 5 presents the three-phase VI measurement block B1.

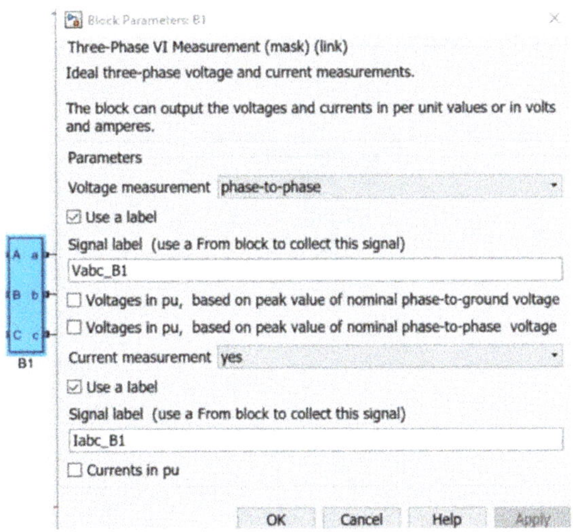

Figure 5. Three-phase VI measurement block—B1.

All measurements collected from the Simulink model are exported into the Matlab data acquisition workspace. Its workflow is described in the structure depicted in Figure 6a,b. For the sake of clarity, a cropped detail is presented in Figure 7.

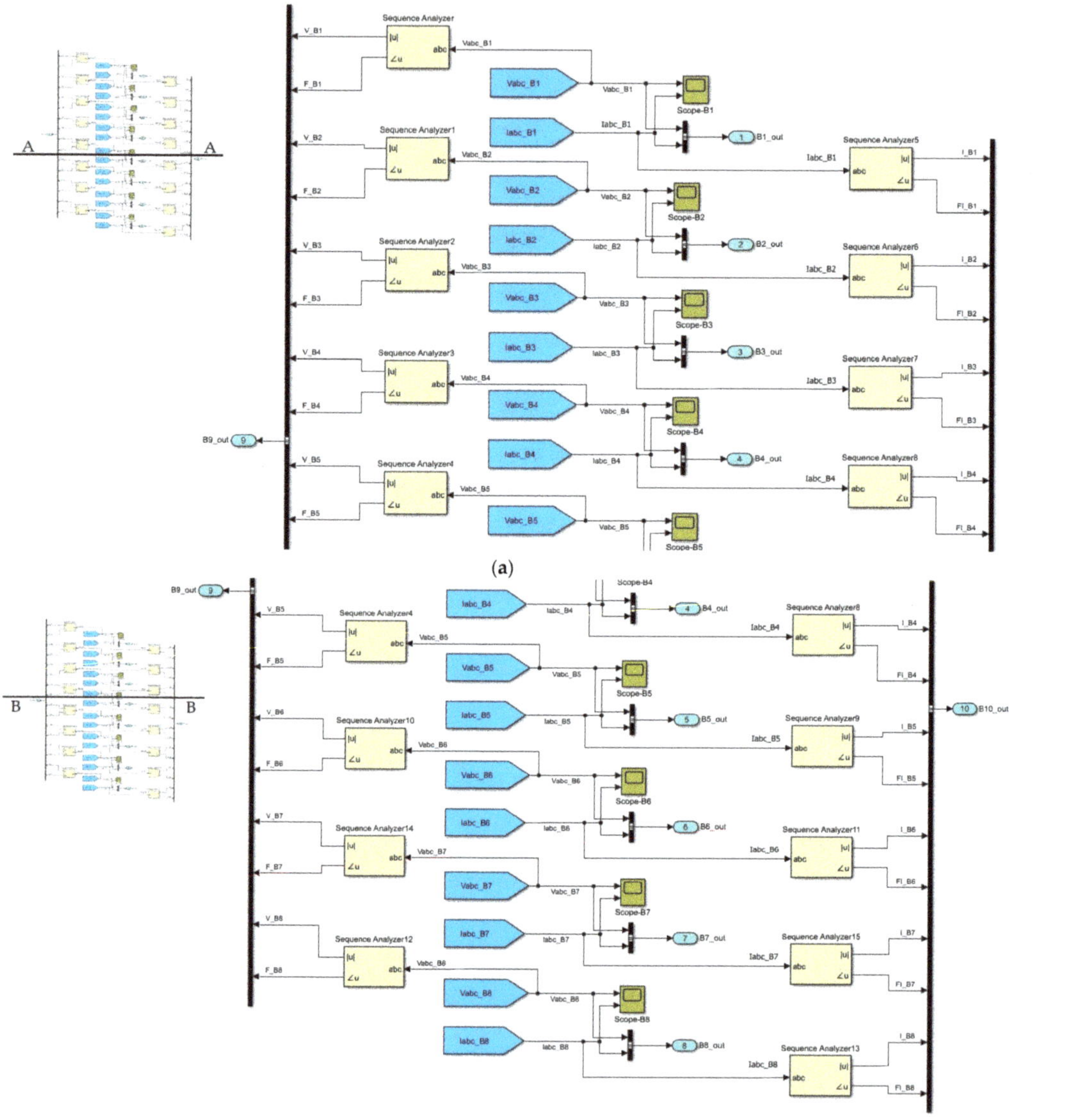

Figure 6. Data acquisition from Simulink model. (**a**) first half of the structure (A-A); (**b**) second half of the structure (B-B).

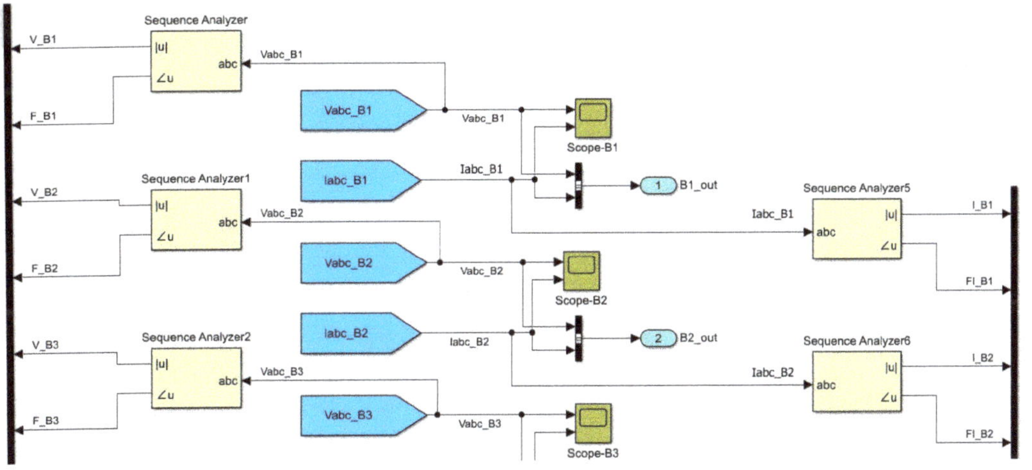

Figure 7. Data acquisition from Simulink model (cropped detail).

The voltage and current values are passed from the three-phase VI measurement blocks to the sequence analyzer, and then are exported to the Matlab variables. At each simulation, data are saved in a table which in the end will become the training dataset for the artificial intelligence algorithm.

3.2. Simulation Process

After implementing the model of the distribution electrical system, the simulation process of different types of short-circuits may start for different values of fault and ground resistances in different locations of the system.

The fault block is moved along the system and is positioned at the end of each sector of all the six lines, which totals 22 positions. For each of these positions and for 25 situations of different chosen values for the fault and ground resistances, all the twelve types of faults are simulated. Performing all these simulations requires that for each simulation the location of the fault block, the type of fault, the values of fault or ground resistances be changed. The time needed for running a simulation is approximately two minutes, not considering the process of changing the parameters. This means that over the course of an hour, less than 30 simulations can be performed. Due to the large number of necessary simulations, and the time required to perform them, automation was almost mandatory.

The automation process has been achieved by implementing a Matlab program, which runs the Simulink model. Consequently, at each run, either the faults block position or the faults block parameters are automatically changed. Thus, the automation reduces the time of each simulation to 1.5 min./simulation, having benefits over the duration of the entire simulation process. Data from the measurement blocks are saved, and used later for the training algorithm of ANN.

The code of the program is presented in Appendix A.

Once the running process is completed, the 6150 simulations performed led to a dataset, which represents the input of the Classification Learner app. Examples of these data can be seen in Tables 3 and 4, which contain examples of the voltage–current values, provided by the eight measurement blocks along with the faults' location and type.

Table 3. Data from simulations—voltage measurements.

V_B1	F_B1	V_B2	F_B2	V_B3	F_B3	V_B4	F_B4	V_B5	F_B5	V_B6	F_B6	V_B7	F_B7	V_B8	F_B8	Fault
27,055.78	21.49	15,580.69	−10.52	15,569.99	−11.19	15,561.13	−11.86	15,571.83	−11.19	15,580.69	−10.52	15,574.22	−11.02	15,580.69	−10.52	normal
18,171.13	16.65	10,185.01	−16.34	10,175.79	−17.06	10,168.00	−17.78	10,177.02	−17.06	10,185.01	−16.34	10,179.12	−16.88	10,185.01	−16.34	L1/S1/ab
21,772.26	17.81	11,901.15	−16.32	11,884.51	−17.03	11,869.45	−17.74	11,885.95	−17.03	11,901.15	−16.32	11,889.87	−16.85	11,901.15	−16.32	L1/S4/abg
18,535.94	17.04	10,107.14	−16.74	10,071.28	−17.43	10,036.67	−18.12	10,072.50	−17.43	10,107.14	−16.74	10,081.27	−17.26	10,107.14	−16.74	L1/S2/abc
18,049.38	17.43	9543.07	−17.40	9507.63	−18.09	9473.39	−18.78	9508.79	−18.09	9543.07	−17.40	9517.47	−17.92	9543.07	−17.40	L1/S3/abcg
26,086.87	19.90	14,907.98	−12.67	14,817.87	−13.79	14,809.03	−14.45	14,899.10	−13.34	14,907.98	−12.67	14,901.49	−13.17	14,907.98	−12.67	L2/S3/ag
22,056.05	22.85	12,037.82	−9.73	11,419.66	−11.16	11,385.26	−11.86	12,007.42	−10.43	12,037.82	−9.73	12,015.14	−10.25	12,037.82	−9.73	L2/S4/bc
24,523.68	20.02	13,801.96	−12.93	13,711.52	−13.84	13,690.30	−14.46	13,781.46	−13.55	13,801.96	−12.93	13,786.77	−13.39	13,801.96	−12.93	L2/S1/ac
23,704.60	19.76	13,229.93	−13.43	13,025.51	−14.78	13,000.68	−15.41	13,206.50	−14.06	13,229.93	−13.43	13,212.52	−13.91	13,229.93	−13.43	L2/S2/acg
24,365.01	21.04	13,688.99	−11.52	13,357.55	−12.78	13,186.54	−13.79	13,674.71	−12.22	13,688.99	−11.52	13,678.42	−12.04	13,688.99	−11.52	L3/S2/bg
25,762.93	21.45	14,670.19	−10.81	14,515.01	−11.80	14,363.95	−12.80	14,653.02	−11.47	14,670.19	−10.81	14,657.48	−11.31	14,670.19	−10.81	L3/S4/cg
21,336.88	22.09	11,538.24	−10.91	10,839.79	−12.85	10,642.02	−13.90	11,509.08	−11.60	11,538.24	−10.91	11,516.48	−11.43	11,538.24	−10.91	L3/S1/bcg
19,859.98	19.21	10,529.43	−15.11	10,520.14	−15.83	10,512.29	−16.54	10,087.15	−17.18	10,529.43	−15.11	10,523.51	−15.65	10,529.43	−15.11	L4/S2/ab
20,159.75	19.86	10,725.32	−14.26	10,693.38	−14.93	10,662.72	−15.61	9922.10	−17.64	10,725.32	−14.26	10,702.41	−14.77	10,725.32	−14.26	L4/S4/abc
23,886.24	20.15	13,352.04	−12.90	13,176.45	−13.44	13,151.85	−14.07	13,328.82	−13.52	13,352.04	−12.90	13,229.38	−14.42	13,352.04	−12.90	L5/S3/ac
19,908.07	18.69	10,574.29	−15.89	10,382.83	−17.06	10,352.96	−17.75	10,545.93	−16.58	10,574.29	−15.89	10,267.29	−17.60	10,574.29	−15.89	L5/S2/abg
18,752.83	18.50	9767.32	−16.67	9722.06	−17.48	9688.60	−18.17	9734.01	−17.36	9767.32	−16.67	9518.96	−18.10	9767.32	−16.67	L5/S1/abcg
25,953.14	19.64	14,816.10	−13.01	14,805.45	−13.68	14,796.51	−14.34	14,807.22	−13.68	14,816.10	−13.01	14,809.61	−13.51	14,816.10	−13.01	L6/S1/ag
26,087.22	18.75	14,955.34	−14.04	14,943.03	−14.74	14,932.66	−15.44	14,944.86	−14.74	14,955.34	−14.04	14,947.64	−14.56	14,955.34	−14.04	L6/S3/ab
20,906.62	21.96	11,237.47	−11.20	11,205.23	−11.89	11,174.43	−12.59	11,206.53	−11.89	11,237.47	−11.20	11,214.37	−11.72	11,237.47	−11.20	L6/S3/bcg

Table 4. Data from simulations—current measurements.

I_B1	FI_B1	I_B2	FI_B2	I_B3	FI_B3	I_B4	FI_B4	I_B5	FI_B5	I_B6	FI_B6	I_B7	FI_B7	I_B8	FI_B8	Fault
467.06	−11.19	467.06	−11.22	155.61	−11.77	155.61	−11.86	155.72	−11.19	155.74	−10.88	155.74	−11.02	0.19	79.48	normal
986.80	−71.99	306.09	−17.09	101.96	−17.68	101.96	−17.77	102.05	−17.05	102.08	−16.72	102.07	−16.87	0.13	72.64	L1/S1/ab
683.43	−63.32	357.35	−17.05	118.99	−17.63	118.98	−17.72	119.15	−17.02	119.20	−16.69	119.19	−16.84	0.15	70.38	L1/S4/abg
1130.02	−77.99	303.45	−17.46	100.87	−18.02	100.83	−18.11	101.19	−17.43	101.33	−17.11	101.27	−17.25	0.13	59.89	L1/S2/abc
1174.33	−78.30	286.58	−18.11	95.25	−18.68	95.21	−18.77	95.56	−18.08	95.71	−17.77	95.65	−17.91	0.12	58.55	L1/S3/abcg
434.18	−31.17	434.26	−31.20	148.20	−14.36	148.20	−14.45	149.09	−13.33	149.11	−13.03	149.11	−13.17	0.18	77.20	L2/S3/ag
755.49	−56.98	755.62	−56.99	114.14	−11.77	114.10	−11.87	120.29	−10.44	120.41	−10.13	120.36	−10.27	0.15	71.03	L2/S4/bc
635.90	−62.07	636.05	−62.08	137.14	−14.38	137.13	−14.46	138.03	−13.55	138.10	−13.27	138.08	−13.40	0.15	72.26	L2/S1/ac
686.53	−64.04	686.68	−64.05	130.30	−15.32	130.28	−15.41	132.32	−14.07	132.41	−13.78	132.38	−13.91	0.15	70.37	L2/S2/acg
552.69	−36.89	552.79	−36.91	325.72	−57.67	131.99	−13.79	136.85	−12.22	136.90	−11.90	136.89	−12.04	0.17	76.22	L3/S2/bg
529.59	−33.87	529.67	−33.89	282.59	−57.47	143.75	−12.80	146.61	−11.47	146.67	−11.17	146.66	−11.31	0.17	76.21	L3/S4/cg
816.81	−61.74	816.95	−61.75	692.95	−76.62	106.72	−13.90	115.34	−11.62	115.46	−11.30	115.41	−11.44	0.14	69.85	L3/S1/bcg
827.00	−69.65	827.15	−69.65	105.38	−16.44	105.38	−16.54	101.14	−17.17	105.49	−15.50	105.49	−15.65	0.13	73.96	L4/S2/ab
920.05	−75.40	920.19	−75.40	107.05	−15.53	107.02	−15.62	99.68	−17.63	107.46	−14.64	107.41	−14.78	0.13	64.71	L4/S4/abc
673.96	−63.11	674.11	−63.13	131.79	−13.99	131.76	−14.08	133.52	−13.53	393.81	−76.21	132.55	−14.41	0.15	71.04	L5/S3/ac
934.87	−73.82	935.01	−73.83	103.97	−17.65	103.94	−17.75	105.85	−16.58	524.97	−82.92	103.09	−17.60	0.13	64.11	L5/S2/abg
1086.03	−77.37	1086.16	−77.38	97.38	−18.07	97.34	−18.17	97.79	−17.36	964.12	−87.53	95.66	−18.08	0.12	59.99	L5/S1/abcg
434.45	−34.11	434.54	−34.13	148.07	−14.25	148.07	−14.34	148.18	−13.68	187.24	−67.36	148.20	−13.51	156.01	−117.51	L6/S1/ag
614.49	−18.20	614.51	−18.22	149.37	−15.32	149.37	−15.42	149.49	−14.72	316.61	−21.34	149.52	−14.54	169.11	−27.41	L6/S3/ab
862.27	−63.42	862.41	−63.43	112.05	−12.51	112.01	−12.60	112.34	−11.91	743.09	−77.07	112.41	−11.73	703.76	−85.42	L6/S3/bcg

Through the sequence analyzer from the data acquisition of the Simulink model, the magnitude and phase angle of the three-phase signals are obtained. For instance, measurement block B1 provides values of the voltage magnitude V_B1, the voltage phase angle F_B1, the current magnitude I_B1 and the current phase angle FI_B1 (see also Figure 6a,b).

3.3. Classification Learner App

Once the simulation process is completed, the large amount of obtained data is used in the Classification Learner App from Matlab to train the artificial intelligence (AI) algorithms. This application can classify data based on the training dataset and return a single response for a further situation [30].

To start the training session, it is necessary to set the parameters observed in Figure 8. The table named "DataTable", containing the results of 6150 simulations, becomes the dataset variable in the Classification Learner App. Data from this table are divided into two types of data, namely predictors and response. Predictors are represented by the values of voltage–current pairs measured at each simulation and the response represents the fault location of the exposed situation. The validation scheme was set as a cross-validation with five folds. After setting these parameters, the session starts by clicking the Start Session button.

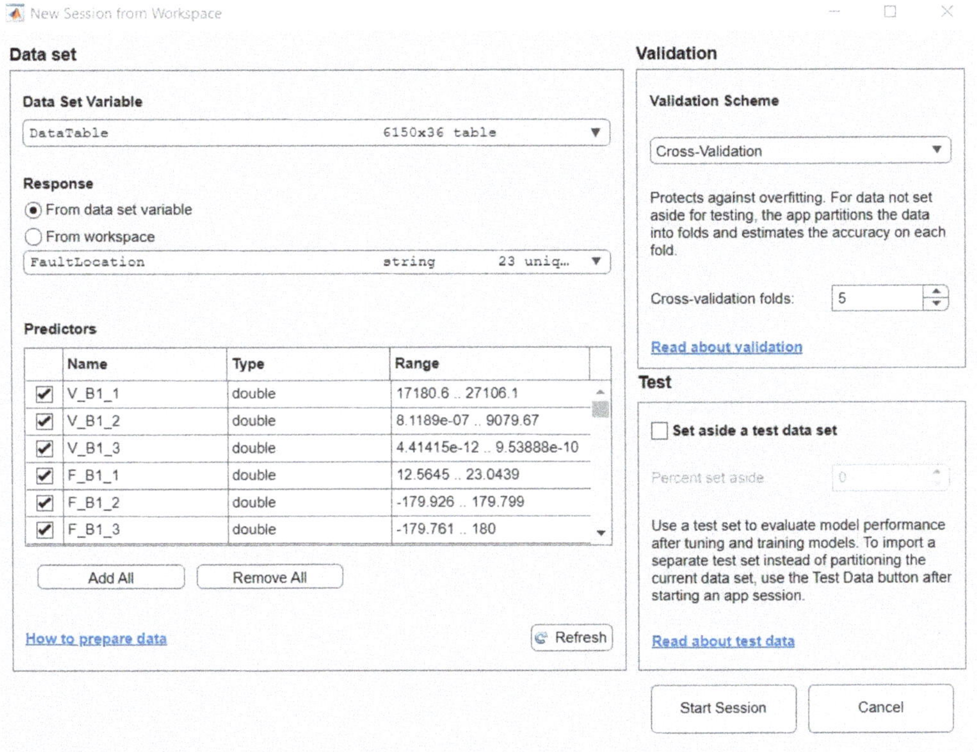

Figure 8. Classification Learner app—training dataset.

Simulations have been also performed for the cases in which the response is both the location and the type of the fault.

In the Classification Learner app, the model of the training algorithm can be set. To obtain the best accuracy validation values, the option of training all models algorithm may also be chosen.

After the analysis of several training models, Figure 9 reveals that the most efficient model sorted in terms of accuracy is the medium neural network algorithm. One can see that the accuracy validation in the case of fault location was 98% and 94.7% in the case of both location and type of the fault.

If the response is only the fault location (example L1/S1 for line 1/sector 1), which implied 23 unique answers, the validation accuracy will be better than in the case where the response is both the location and the type of the fault (example L1/S1/ab for line1/sector 1/type of fault ab) which implied 243 unique answers. These two cases will be presented comparatively in the next stage, where the fault location and the fault location and type based on the trained model are predicted.

In evaluating and observing the performance of a trained model, the analysis of the confusion matrix and of the receiver operating characteristic (ROC) curve are two useful tools.

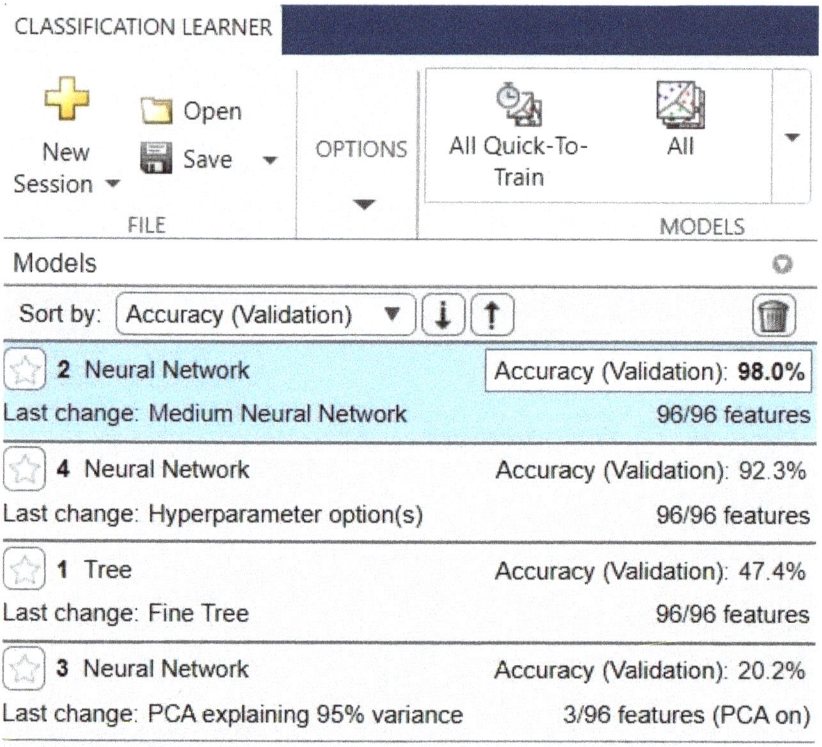

Figure 9. Classification Learner app.

The confusion matrix contains the predicted classes in its columns and the true classes in its rows; therefore, a 100% validation accuracy assumes a perfect principal diagonal confusion matrix. Values situated outside the principal diagonal of the matrix indicate situations that are not well-predicted and need supplementary data training set [30].

In Figures 10 and 11, the validation confusion matrices for the two studied cases are presented. From Figure 10, which presents the case of 23 unique responses, one can observed that the only situation with 100% accuracy is the no-fault (normal) situation. In Figure 11, one can also observe the shape of the principal diagonal in the case of 243 unique responses. Due to the complexity of the simulation, the values situated outside the principal diagonal are not visible in the resolution of Figure 11. The 243 classes containing the location and type of faults are presented in Table 5.

From the confusion matrix depicted in Figure 10, it can be observed that for lines 5 and 6 of the Simulink model, the prediction response is poorer, which is noticeable from the larger number of values that deviate from the principal diagonal. For the other lines, the situation is better, with fewer cases where the predicted class does not coincide with the true class.

Even though some values are far from the principal diagonal, their low values indicate that the probability of a wrong prediction is unlikely. For instance, for the predicted class L4/S4, there is a single situation in which the true class is in fact L1/S2.

Unfortunately, the confusion matrix for the case with 243 unique responses is not useful to indicate the classes that were not well-predicted. This difficulty can be alleviate using other tools provided by the Classification Learner app, one of them being the ROC curve, an efficient way of comparing trained models.

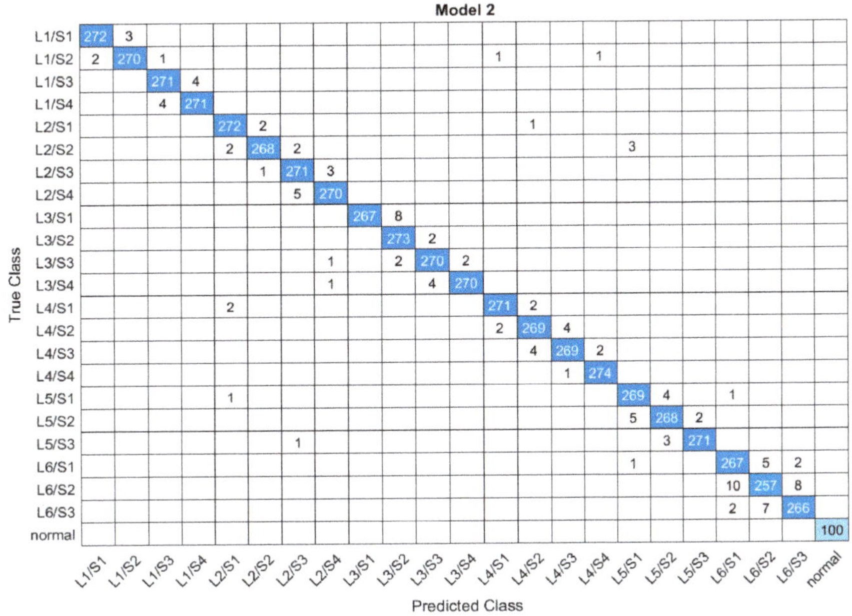

Figure 10. Validation confusion matrix for medium neural network model—accuracy 98.0%—for location of the fault.

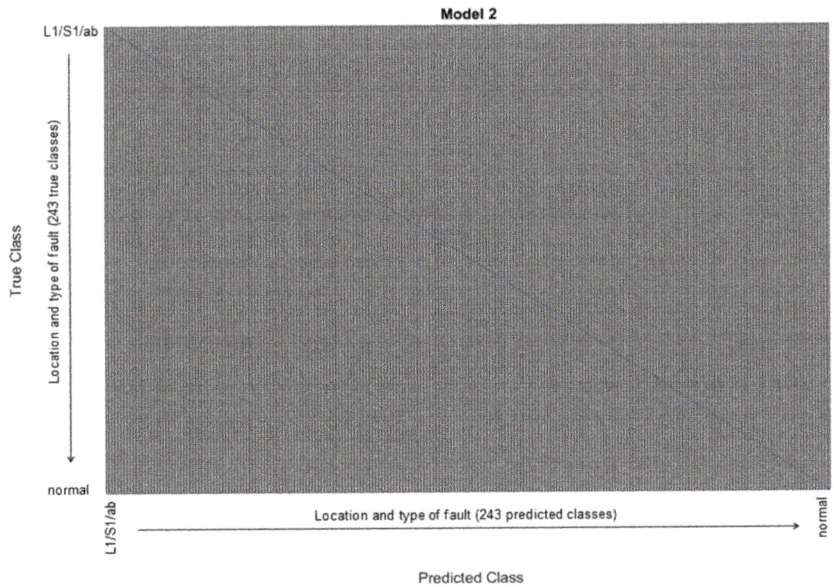

Figure 11. Validation confusion matrix for medium neural network model—accuracy 94.7%—for location and type of fault.

Table 5. Location and type of faults.

Location and Type of Faults										
L1/S1/ab	L1/S3/ab	L2/S1/ab	L2/S3/ab	L3/S1/ab	L3/S3/ab	L4/S1/ab	L4/S3/ab	L5/S1/ab	L5/S3/ab	L6/S2/ab
L1/S1/abc	L1/S3/abc	L2/S1/abc	L2/S3/abc	L3/S1/abc	L3/S3/abc	L4/S1/abc	L4/S3/abc	L5/S1/abc	L5/S3/abc	L6/S2/abc
L1/S1/abcg	L1/S3/abcg	L2/S1/abcg	L2/S3/abcg	L3/S1/abcg	L3/S3/abcg	L4/S1/abcg	L4/S3/abcg	L5/S1/abcg	L5/S3/abcg	L6/S2/abcg
L1/S1/abg	L1/S3/abg	L2/S1/abg	L2/S3/abg	L3/S1/abg	L3/S3/abg	L4/S1/abg	L4/S3/abg	L5/S1/abg	L5/S3/abg	L6/S2/abg
L1/S1/ac	L1/S3/ac	L2/S1/ac	L2/S3/ac	L3/S1/ac	L3/S3/ac	L4/S1/ac	L4/S3/ac	L5/S1/ac	L5/S3/ac	L6/S2/ac
L1/S1/acg	L1/S3/acg	L2/S1/acg	L2/S3/acg	L3/S1/acg	L3/S3/acg	L4/S1/acg	L4/S3/acg	L5/S1/acg	L5/S3/acg	L6/S2/acg
L1/S1/ag	L1/S3/ag	L2/S1/ag	L2/S3/ag	L3/S1/ag	L3/S3/ag	L4/S1/ag	L4/S3/ag	L5/S1/ag	L5/S3/ag	L6/S2/ag
L1/S1/bc	L1/S3/bc	L2/S1/bc	L2/S3/bc	L3/S1/bc	L3/S3/bc	L4/S1/bc	L4/S3/bc	L5/S1/bc	L5/S3/bc	L6/S2/bc
L1/S1/bcg	L1/S3/bcg	L2/S1/bcg	L2/S3/bcg	L3/S1/bcg	L3/S3/bcg	L4/S1/bcg	L4/S3/bcg	L5/S1/bcg	L5/S3/bcg	L6/S2/bcg
L1/S1/bg	L1/S3/bg	L2/S1/bg	L2/S3/bg	L3/S1/bg	L3/S3/bg	L4/S1/bg	L4/S3/bg	L5/S1/bg	L5/S3/bg	L6/S2/bg
L1/S1/cg	L1/S3/cg	L2/S1/cg	L2/S3/cg	L3/S1/cg	L3/S3/cg	L4/S1/cg	L4/S3/cg	L5/S1/cg	L5/S3/cg	L6/S2/cg
L1/S2/ab	L1/S4/ab	L2/S2/ab	L2/S4/ab	L3/S2/ab	L3/S4/ab	L4/S2/ab	L4/S4/ab	L5/S2/ab	L6/S1/ab	L6/S3/ab
L1/S2/abc	L1/S4/abc	L2/S2/abc	L2/S4/abc	L3/S2/abc	L3/S4/abc	L4/S2/abc	L4/S4/abc	L5/S2/abc	L6/S1/abc	L6/S3/abc
L1/S2/abcg	L1/S4/abcg	L2/S2/abcg	L2/S4/abcg	L3/S2/abcg	L3/S4/abcg	L4/S2/abcg	L4/S4/abcg	L5/S2/abcg	L6/S1/abcg	L6/S3/abcg
L1/S2/abg	L1/S4/abg	L2/S2/abg	L2/S4/abg	L3/S2/abg	L3/S4/abg	L4/S2/abg	L4/S4/abg	L5/S2/abg	L6/S1/abg	L6/S3/abg
L1/S2/ac	L1/S4/ac	L2/S2/ac	L2/S4/ac	L3/S2/ac	L3/S4/ac	L4/S2/ac	L4/S4/ac	L5/S2/ac	L6/S1/ac	L6/S3/ac
L1/S2/acg	L1/S4/acg	L2/S2/acg	L2/S4/acg	L3/S2/acg	L3/S4/acg	L4/S2/acg	L4/S4/acg	L5/S2/acg	L6/S1/acg	L6/S3/acg
L1/S2/ag	L1/S4/ag	L2/S2/ag	L2/S4/ag	L3/S2/ag	L3/S4/ag	L4/S2/ag	L4/S4/ag	L5/S2/ag	L6/S1/ag	L6/S3/ag
L1/S2/bc	L1/S4/bc	L2/S2/bc	L2/S4/bc	L3/S2/bc	L3/S4/bc	L4/S2/bc	L4/S4/bc	L5/S2/bc	L6/S1/bc	L6/S3/bc
L1/S2/bcg	L1/S4/bcg	L2/S2/bcg	L2/S4/bcg	L3/S2/bcg	L3/S4/bcg	L4/S2/bcg	L4/S4/bcg	L5/S2/bcg	L6/S1/bcg	L6/S3/bcg
L1/S2/bg	L1/S4/bg	L2/S2/bg	L2/S4/bg	L3/S2/bg	L3/S4/bg	L4/S2/bg	L4/S4/bg	L5/S2/bg	L6/S1/bg	L6/S3/bg
L1/S2/cg	L1/S4/cg	L2/S2/cg	L2/S4/cg	L3/S2/cg	L3/S4/cg	L4/S2/cg	L4/S4/cg	L5/S2/cg	L6/S1/cg	L6/S3/cg
										normal

The ROC curve is a plot tool that provides the false positive rate and the true positive rate of each predicted class. The area under curve (AUC) is an indicator of the quality of the classifier. The AUC values range between 0 and 1, with a higher value indicating a better performance of the classifier [30].

In Figures 12 and 13, the ROC curves for some representative cases studied are presented. Figure 12 presents the ROC curve for the medium neural network model with an accuracy validation of 98.0% and Figure 13 presents the ROC curve for the medium neural network model with an accuracy validation of 94.7%. In both figures, the highest and the lowest values of AUC are shown.

For the model which predicts only the location of the fault, has an accuracy validation of 98.0%, and encompasses 23 situations, the maximum value of AUC (1.00) is reached for several classifiers. Figure 12a presents the ROC curve in one of these situations, while Figure 12b presents the lower value of AUC, which, in this case, is of 0.99 and occurs for Line 6-Sector 2.

For the model which predicts both location and type of the fault, has an accuracy validation of 94.7%, and encompasses 243 situations, the maximum value of AUC (1.00) is also reached for several classifiers. Figure 13a presents the ROC curve in one of these situations, while Figure 13b presents the lower value of AUC, which, in this case, is of 0.94 and also occurs for Line 6-Sector 2.

Both situations reveal a good accuracy validation prediction.

By analyzing the ROC curves and comparing their results with the ones of the confusion matrices, it is obvious that the two plotting tools offer the same results but in different ways. Since the confusion matrix presents an overview of all classes, the ROC curve presents specific results for each class.

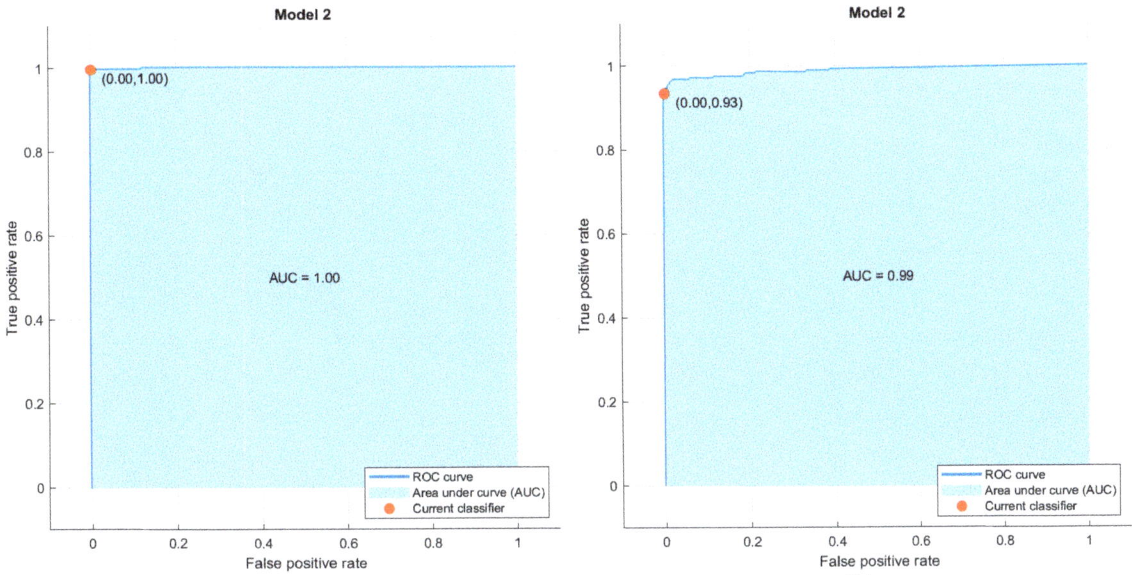

(a) (b)

Figure 12. ROC curve for medium neural network model—accuracy 98.0% (**a**) the AUC 1.00 for Line 4-Sector 4, (**b**) the AUC 0.99 for Line 6-Sector 2.

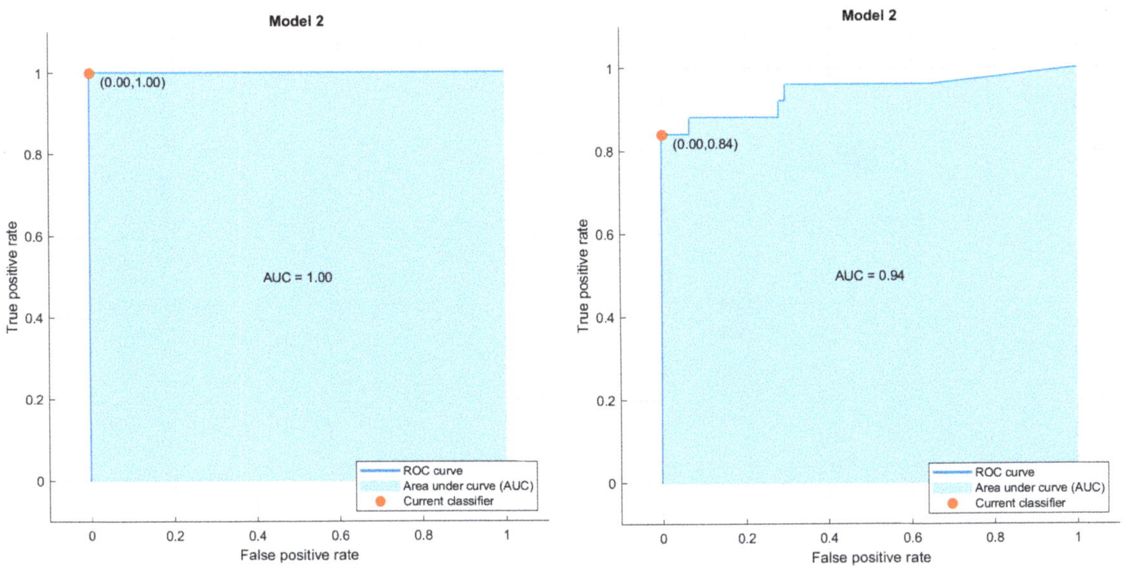

(a) (b)

Figure 13. ROC curve for medium neural network model accuracy—94.7%—(**a**) the AUC 1.00 for Line 4-Sector 4, (**b**) the AUC 0.94 for Line 6-Sector 2.

3.4. Prediction of Faults Based on the Trained Model

After analyzing the different types of trained models, these can be exported from the Classification Learner app to the Matlab workspace as a new variable ("trainedModel"). This variable can be used to predict responses for other faults which may occur in the same distribution system, and which have not been considered in the previous training dataset. As an example, Tables 6 and 7 contain four situations in which different fault and ground resistances have been used.

Table 6. Data for predictions—voltage measurements.

V_B1	F_B1	V_B2	F_B2	V_B3	F_B3	V_B4	F_B4	V_B5	F_B5	V_B6	F_B6	V_B7	F_B7	V_B8	F_B8	Fault
23,387.39	20.70	13,006.09	−12.14	12,881.25	−13.06	12,864.69	−13.77	12,989.73	−12.85	13,006.09	−12.14	12,993.94	−12.67	13,006.09	−12.14	L2/S1/bg
20,579.33	20.40	11,010.63	−13.41	10,238.55	−16.38	9846.89	−18.38	10,981.53	−14.08	11,010.63	−13.41	10,988.92	−13.91	11,010.63	−13.41	L3/S2/abc
25,874.47	19.42	14,763.31	−13.30	14,752.70	−13.97	14,743.79	−14.63	14,754.46	−13.97	14,763.31	−13.30	14,756.85	−13.80	14,763.31	−13.30	L1/S4/ag
19,363.64	19.39	10,176.50	−15.19	10,143.97	−15.87	10,112.66	−16.55	9503.63	−18.32	10,176.50	−15.19	10,153.10	−15.70	10,176.50	−15.19	L4/S3/abcg

Table 7. Data for predictions—current measurements.

I_B1	FI_B1	I_B2	FI_B2	I_B3	FI_B3	I_B4	FI_B4	I_B5	FI_B5	I_B6	FI_B6	I_B7	FI_B7	I_B8	FI_B8	Fault
23,387.39	20.70	13,006.09	−12.14	12,881.25	−13.06	12,864.69	−13.77	12,989.73	−12.85	13,006.09	−12.14	12,993.94	−12.67	13,006.09	−12.14	L2/S1/bg
20,579.33	20.40	11,010.63	−13.41	10,238.55	−16.38	9846.89	−18.38	10,981.53	−14.08	11,010.63	−13.41	10,988.92	−13.91	11,010.63	−13.41	L3/S2/abc
25,874.47	19.42	14,763.31	−13.30	14,752.70	−13.97	14,743.79	−14.63	14,754.46	−13.97	14,763.31	−13.30	14,756.85	−13.80	14,763.31	−13.30	L1/S4/ag
19,363.64	19.39	10,176.50	−15.19	10,143.97	−15.87	10,112.66	−16.55	9503.63	−18.32	10,176.50	−15.19	10,153.10	−15.70	10,176.50	−15.19	L4/S3/abcg

Measurements generated for the predictions must be introduced in a variable ("Ttest") with the same structure as ("DataTable") as the variable used in the training dataset. After processing the variable to be tested, using the prediction function for the trained model, the response based on the trained model will be obtained.

Figures 14 and 15 present the responses of the two cases studied, e.g., for the medium neural network model with an accuracy of 98.0% and for the medium neural network model with an accuracy of 94.7%.

```
>> yfit = trainedModel.predictFcn(Ttest)

yfit =

  4×1 cell array

    {'L2/S1'}
    {'L3/S2'}
    {'L1/S4'}
    {'L4/S3'}
```

Figure 14. Predicted results for test examples (for medium neural network model accuracy 98.0%—for the fault location).

```
>> yfit = trainedModel1.predictFcn(Ttest)

yfit =

  4×1 cell array

    {'L2/S1/bg'  }
    {'L3/S2/abc' }
    {'L1/S4/ag'  }
    {'L4/S3/abcg'}
```

Figure 15. Predicted results for test examples (for medium neural network model accuracy 94.7%—for location and type of the fault).

By applying the prediction function to Tables 6 and 7, which contain the voltage and current measurement data, two cell arrays are created, namely two column vectors (see Figures 14 and 15).

It can be observed that in the first case, the response contains only the location of the fault, and in the second case, the response includes in addition the type of the fault. By comparing the result of the prediction function with the last column ("Fault") of the ("Ttest") table, one can observe that the trained model operates properly.

4. Discussion

The paper presents a solution for detecting and locating faults in electrical distribution systems using a Simulink model combined with the ANN algorithms of the Classification Learner app provided by Matlab.

The high performance of the proposed technique emphasizes the potential of using this principle for real-world distribution electrical systems.

After the first stage devoted to the development of the simulation model and the simulations presented in the second stage, in the third stage, it can be observed that the performance of the trained model is correlated to the training dataset. The larger the database is, the higher the performance of the trained model is. In order to have a good accuracy validation value, the number of simulations must be adapted to the complexity of the analyzed system.

Simulating all the cases which can offer a solid test set for the presented case study implies the use of high-level hardware and software resources. These needs can be covered by using local hardware resources; however, much better results can be obtained by using a virtual machine which can significantly shorten the required simulations time. If several virtual machines are simultaneously used, the simulation process can be accelerated, and in the end, the data collected from all of them can be processed and integrated into a single database.

When applying the method presented above for a real-world distribution electrical system, it is mandatory that the Simulink model must contain its very real components. This method can also be used in other situations, for instance, in industrial estates, where the consumers, components of the system and their structure are already well-known.

5. Conclusions

The developed method has a good accuracy and its use in real-world situations is therefore recommended. The trained model which predicted the location of faults (with 23 possible responses) had an accuracy validation of 98%, while the trained model which predicted both location and type of faults (with 243 possible responses) had a slightly lower accuracy validation (94.7%). Both trained models were based on the same data training set (measurements from 6150 of simulations), which revealed that for obtaining a good accuracy validation value, a larger number of responses, and of the data training set were needed.

The major advantage of the presented method lies in its good precision in detecting, locating, and predicting faults. At the same time, running a large number of simulations could be considered a tedious operation. Clearly, the simulations' time was closely related to the complexity of the Simulink model and the number of parameters that needed to be modified at each simulation. Additionally, the developed Simulink model of the electrical power system could also decisively contribute by performing different updates of the system if necessary. It could be used, for example, to simulate the impact of a new structure of the system or to analyze possible future improvements performed on the electrical system.

Methods of faults detecting, locating, and predicting in electrical power systems, based on ANN algorithms, could also solve complex problems encountered at electric lines or branched systems of cables, situations difficult to be managed using classic methods of detecting faults.

In further research, this method could be improved in the direction of generating dynamic changes of the loads simulations, and in developing a friendly graphical user interface to the application.

The presented topic, regarding faults detection and location in electrical systems, was and remains a main and permanent concern of utility companies, which continuously seek to improve and adapt proprietary methods in order to have an optimal operation of their electrical power systems.

Author Contributions: Conceptualization, A.-M.M. and M.I.B.; methodology, A.-M.M.; software, A.-M.M.; validation, A.-M.M. and M.I.B.; formal analysis, M.I.B.; investigation, A.-M.M.; resources, A.-M.M.; data curation, M.I.B.; writing—original draft preparation, A.-M.M.; writing—review and editing, A.-M.M. and M.I.B.; visualization, M.I.B.; supervision, M.I.B.; project administration, A.-M.M. All authors have read and agreed to the published version of the manuscript.

Funding: This research received no external funding.

Institutional Review Board Statement: Not applicable.

Informed Consent Statement: Not applicable.

Data Availability Statement: Not applicable.

Conflicts of Interest: The authors declare no conflict of interest. The funders had no role in the design of the study; in the collection, analyses, or interpretation of data; in the writing of the manuscript; or in the decision to publish the results.

Appendix A

```
T = [];
model = 'name_model';
tstop = 0.1;
%         1   2   3   4   5   6
Line.Name = {'L1', 'L2', 'L3', 'L4', 'L5','L6'};
Line.Sectors = [ 4    4    4    4    3    3 ];
Faults = {'0','ag','bg','cg','ab','bc','ac','abg','bcg','acg','abc','abcg'};
nLine = 1:6;
nFault = 1:12;
Faulty.Lines.Name = {Line.Name{nLine}};
Faulty.Sectors = Line.Sectors(nLine);
Type_Faults = {Faults{nFault}};
open_system(model);
for iFaultyLines = 1:length(Faulty.Lines.Name)
   for iFaultType = 1:length(Type_Faults)
      Fault_Type = Faults(iFaultType);
   Faulty.Lines = [model,'/',Faulty.Lines.Name{iFaultyLines}];
   open_system(Faulty.Lines);
   Nsectors = Faulty.Sectors(iFaultyLines);
   FaultBlock = [Faulty.Lines,'/Fault'];
   try
      addbd = add_block([model,'/Fault'],FaultBlock,'Commented','off','Position', [545
-545    595    -495]);
   catch
      Position = get_param(FaultBlock,'Position');
      delete_block(FaultBlock);
      add_block([model,'/Fault'],FaultBlock,'Commented','off','Position',Position + 10);
   end
if contains(Fault_Type,'g')
   set_param(FaultBlock,'GroundFault','on')
else
   set_param(FaultBlock,'GroundFault','off')
end
if contains(Fault_Type,'a')
   set_param(FaultBlock,'FaultA','on')
else
```

```
    set_param(FaultBlock,'FaultA','off')
end
if contains(Fault_Type,'b')
    set_param(FaultBlock,'FaultB','on')
else
    set_param(FaultBlock,'FaultB','off')
end
if contains(Fault_Type,'c')
    set_param(FaultBlock,'FaultC','on')
else
    set_param(FaultBlock,'FaultC','off')
end
for iSectors = 1:Nsectors
    NameBlock = [Faulty.Lines,'/S',num2str(iSectors)];
    hFault = get_param(FaultBlock,'PortHandles');
    hBlock = get_param(NameBlock,'PortHandles');
    hLines = add_line(Faulty.Lines,hFault.LConn,hBlock.RConn);
    Ron = logspace(log10(1e-4),log10(20),5); %FaultResistance
    Rg = logspace(log10(1e-4),log10(20),5); %GroundResistance
    for iR1 = 1:length(Ron)
      set_param(FaultBlock,'FaultResistance',num2str(Ron(iR1)));
        for iR2 = 1:length(Rg)
            set_param(FaultBlock,'GroundResistance',num2str(Rg(iR2)));
            out = sim(model);
            cellrow{1,1} = out.yout{9}.Values.V_B1.Data(end,:);
            cellrow{1,2} = out.yout{9}.Values.F_B1.Data(end,:);
            cellrow{1,3} = out.yout{9}.Values.V_B2.Data(end,:);
            cellrow{1,4} = out.yout{9}.Values.F_B2.Data(end,:);
            cellrow{1,5} = out.yout{9}.Values.V_B3.Data(end,:);
            cellrow{1,6} = out.yout{9}.Values.F_B3.Data(end,:);
            cellrow{1,7} = out.yout{9}.Values.V_B4.Data(end,:);
            cellrow{1,8} = out.yout{9}.Values.F_B4.Data(end,:);
            cellrow{1,9} = out.yout{9}.Values.V_B5.Data(end,:);
            cellrow{1,10} = out.yout{9}.Values.F_B5.Data(end,:);
            cellrow{1,11} = out.yout{9}.Values.V_B6.Data(end,:);
            cellrow{1,12} = out.yout{9}.Values.F_B6.Data(end,:);
            cellrow{1,13} = out.yout{9}.Values.V_B7.Data(end,:);
            cellrow{1,14} = out.yout{9}.Values.F_B7.Data(end,:);
            cellrow{1,15} = out.yout{9}.Values.V_B8.Data(end,:);
            cellrow{1,16} = out.yout{9}.Values.F_B8.Data(end,:);
            cellrow{1,17} = out.yout{10}.Values.I_B1.Data(end,:);
            cellrow{1,18} = out.yout{10}.Values.FI_B1.Data(end,:);
            cellrow{1,19} = out.yout{10}.Values.I_B2.Data(end,:);
            cellrow{1,20} = out.yout{10}.Values.FI_B2.Data(end,:);
            cellrow{1,21} = out.yout{10}.Values.I_B3.Data(end,:);
            cellrow{1,22} = out.yout{10}.Values.FI_B3.Data(end,:);
            cellrow{1,23} = out.yout{10}.Values.I_B4.Data(end,:);
            cellrow{1,24} = out.yout{10}.Values.FI_B4.Data(end,:);
            cellrow{1,25} = out.yout{10}.Values.I_B5.Data(end,:);
            cellrow{1,26} = out.yout{10}.Values.FI_B5.Data(end,:);
            cellrow{1,27} = out.yout{10}.Values.I_B6.Data(end,:);
            cellrow{1,28} = out.yout{10}.Values.FI_B6.Data(end,:);
            cellrow{1,29} = out.yout{10}.Values.I_B7.Data(end,:);
            cellrow{1,30} = out.yout{10}.Values.FI_B7.Data(end,:);
```

```
                cellrow{1,31} = out.yout{10}.Values.I_B8.Data(end,:);
                cellrow{1,32} = out.yout{10}.Values.FI_B8.Data(end,:);
                FaultLocation = [Faulty.Lines.Name{iFaultyLines},'/S',num2str(iSectors)];
                StringFaultLocation = convertCharsToStrings(FaultLocation);
                if contains (Fault_Type,'0')
                    cellrow{1,33} = ("normal");
                else
                    cellrow{1,33} = strcat(StringFaultLocation,"/",Fault_Type);
                end
                cellrow{1,34} = Ron(iR1);
                cellrow{1,35} = Rg(iR2);
                T1 = cellrow;
                T = [T;T1];
                save(date);
                end
        end
        delete_line(hLines);
    end
        end
    DataTabel = cell2table(T);
    save('data_generation');
    end
```

References

1. Sumit, S.V. Iterative and Non-Iterative Methods for Transmission Line Fault-Location Without using Line Parameters. *Int. J. Eng. Innov. Technol.* **2013**, *3*, 310–314.
2. Wang, L.; Liu, H.; Dai, L.V.; Liu, Y. Novel Method for Identifying Fault Location of Mixed Lines. *Energies* **2018**, *2018*, 11. [CrossRef]
3. Michau, G.; Hsu, C.-C.; Fink, O. Interpretable Detection of Partial Discharge in Power Lines with Deep Learning. *Sensors* **2021**, *21*, 2154. [CrossRef] [PubMed]
4. Muzzammel, R.; Arshad, R.; Raza, A.; Sobahi, N.; Alqasemi, U. Two Terminal Instantaneous Power-Based Fault Classification and Location Techniques for Transmission Lines. *Sustainability* **2023**, *15*, 809. [CrossRef]
5. Ali, S.; Bhargava, A.; Saxena, A.; Kumar, P. A Hybrid Marine Predator Sine Cosine Algorithm for Parameter Selection of Hybrid Active Power Filter. *Mathematics* **2023**, *11*, 598. [CrossRef]
6. Moldovan, A.M.; Oltean, S.; Buzdugan, M.I. Methods of Faults Detection and Location in Electrical Systems. In Proceedings of the 2021 9th International Conference on Modern Power Systems (MPS), Cluj-Napoca, Romania, 16–17 June 2021. [CrossRef]
7. Moldovan, A.M.; Buzdugan, M.I.; Oltean, S. Modeling a Time Domain Reflectometer using Matlab/Simulink for detection of faults in electrical cables. In Proceedings of the 2022 IEEE 20th International Power Electronics and Motion Control Conference (PEMC), Brasov, Romania, 25–28 September 2022; pp. 281–284. [CrossRef]
8. Tariq, R.; Alhamrouni, I.; Rehman, A.U.; Eldin, E.T.; Shafiq, M.; Ghamry, N.A.; Hamam, H. An Optimized Solution for Fault Detection and Location in Underground Cables Based on Traveling Waves. *Energies* **2022**, *15*, 6468. [CrossRef]
9. Lee, C.-K.; Chang, S.J. Fault Detection in Multi-Core C&I Cable via Machine Learning Based Time-Frequency Domain Reflectometry. *Appl. Sci.* **2020**, *10*, 158. [CrossRef]
10. Scarpetta, M.; Spadavecchia, M.; Adamo, F.; Ragolia, M.A.; Giaquinto, N. Detection and Characterization of Multiple Discontinuities in Cables with Time-Domain Reflectometry and Convolutional Neural Networks. *Sensors* **2021**, *21*, 8032. [CrossRef]
11. Tekli, L.; Filipovi, B.; Pavicic, I. Artificial Neural Network Approach for Locating Faults in Power Transmission System. *Eurocon* **2013**, *2013*, 1425–1430.
12. Mahafzah, K.A.; Obeidat, M.A.; Mansour, A.M.; Al-Shetwi, A.Q.; Ustun, T.S. Artificial-Intelligence-Based Open-Circuit Fault Diagnosis in VSI-Fed PMSMs and a Novel Fault Recovery Method. *Sustainability* **2022**, *14*, 16504. [CrossRef]
13. Forootan, M.M.; Larki, I.; Zahedi, R.; Ahmadi, A. Machine Learning and Deep Learning in Energy Systems: A Review. *Sustainability* **2022**, *14*, 4832. [CrossRef]
14. Kebir, S.T.; Cheggaga, N.; Ilinca, A.; Boulouma, S. An Efficient Neural Network-Based Method for Diagnosing Faults of PV Array. *Sustainability* **2021**, *13*, 6194. [CrossRef]
15. Aziz, R.M.; Mahto, R.; Goel, K.; Das, A.; Kumar, P.; Saxena, A. Modified Genetic Algorithm with Deep Learning for Fraud Transactions of Ethereum Smart Contract. *Appl. Sci.* **2023**, *13*, 697. [CrossRef]
16. Jamil, M.; Sharma, S.K.; Singh, R. Fault detection and classification in electrical power transmission system using artificial neural network. *SpringerPlus* **2015**, *4*, 334. [CrossRef]

17. Khorashadi-Zadeh, H.; Aghaebrahimi, M.R. A Novel Approach to Fault Classification and Fault Location for Medium Voltage Cables Based on Artificial Neural Network. *World Acad. Sci. Eng. Technol. Int. J. Energy Power Eng.* **2008**, *2*, 1273–1276.
18. Sharma, A.; Kumari, S.; Bhardwaj, I. Analysis of Underground Cable Fault Location Identification Using Artificial Neural Network. *Int. J. Adv. Sci. Technol.* **2020**, *29*, 5190–5201.
19. Obuma, O.; Madueme, T. Application of artificial neural network (ANN) to enhance power systems protection: A case of the Nigerian 330 kV transmission line. *Electr. Eng.* **2018**, *100*, 1467–1479. [CrossRef]
20. Hasija, K.; Vadhera, S.; Kumar, A.; Kishore, A. Detection and location of faults in underground cable using Matlab/Simulink/ANN and OrCad. In Proceedings of the 2014 6th IEEE Power India International Conference (PIICON), Delhi, India, 5–7 December 2014; pp. 1–5. [CrossRef]
21. Hernandez-Mejia, J.C. Time Domain Reflectometry (TDR). In *Cable Diagnostic Focused Initiative (CDFI) Phase II*; Georgia Tech Research Corporation: Atlanta, GA, USA, 2016.
22. Warlyani, P.; Yadav, A.; Thoke, A.S.; Patel, R. Fault Classification and Faulty Section Identification in Teed Transmission Circuits Using ANN. *Int. J. Comput. Electr. Eng.* **2011**, *3*, 807–811. [CrossRef]
23. Yadav, A.; Thoke, A.S. Transmission line fault distance and direction estimation using artificial neural network. *Int. J. Eng. Sci. Technol.* **2011**, *3*, 110–121. [CrossRef]
24. Mantach, S.; Lutfi, A.; Tavasani, H.M.; Ashraf, A.; El-Hag, A.; Kordi, B. Deep Learning in High Voltage Engineering: A Literature Review. *Energies* **2022**, *15*, 5005. [CrossRef]
25. Wang, M.-H.; Sian, H.-W.; Lu, S.-D. Hybrid Methodology Based on Symmetrized Dot Pattern and Convolutional Neural Networks for Fault Diagnosis of Power Cables. *Processes* **2022**, *10*, 2009. [CrossRef]
26. Al-Katheri, A.A.; Al-Ammar, E.A.; Alotaibi, M.A.; Ko, W.; Park, S.; Choi, H.-J. Application of Artificial Intelligence in PV Fault Detection. *Sustainability* **2022**, *14*, 13815. [CrossRef]
27. Nsaif, Y.M.; Lipu, M.S.H.; Hussain, A.; Ayob, A.; Yusof, Y.; Zainuri, M.A.A.M. A Novel Fault Detection and Classification Strategy for Photovoltaic Distribution Network Using Improved Hilbert–Huang Transform and Ensemble Learning Technique. *Sustainability* **2022**, *14*, 11749. [CrossRef]
28. Hichri, A.; Hajji, M.; Mansouri, M.; Abodayeh, K.; Bouzrara, K.; Nounou, H.; Nounou, M. Genetic-Algorithm-Based Neural Network for Fault Detection and Diagnosis: Application to Grid-Connected Photovoltaic Systems. *Sustainability* **2022**, *14*, 10518. [CrossRef]
29. Omar, A.M.S.; Osman, M.K.; Ibrahim, M.N.; Hussain, Z.; Abidin, A.F. Fault classification on transmission line using LSTM network. *Indones. J. Electr. Eng. Comput. Sci.* **2020**, *20*, 231–238. [CrossRef]
30. Available online: https://www.mathworks.com/help/ (accessed on 6 March 2023).

electronics

MDPI

Article

A Power System Timing Data Recovery Method Based on Improved VMD and Attention Mechanism Bi-Directional CNN-GRU

Kangmin Xie, Jichun Liu * and Youbo Liu

School of Electrical Engineering, Sichuan University, Chengdu 610065, China
* Correspondence: jichunliu@scu.edu.cn

Abstract: The temporal data of the power system are expanding with the growth of the power system and the proliferation of automated equipment. However, data loss may arise during the acquisition, measurement, transmission, and storage of temporal data. To address the insufficiency of temporal data in the power system, this study proposes a sequence-to-sequence (Seq2Seq) architecture to restore power system temporal data. This architecture comprises a radial convolutional neural unit (CNN) network and a gated recurrent unit (GRU) network. Specifically, to account for the periodicity and volatility of temporal data, VMD is employed to decompose the time series data output into components of different frequencies. CNN is utilized to extract the spatial characteristics of temporal data. At the same time, Seq2Seq is employed to reconstruct each component based on introducing a feature timing and multi-model combination triple attention mechanism. The feature attention mechanism calculates the contribution rate of each feature quantity and independently mines the correlation between the time series data output and each feature value. The temporal attention mechanism autonomously extracts historical–critical moment information. A multi-model combination attention mechanism is introduced, and the missing data repair value is obtained after modeling the combination of data on both sides of the missing data. Recovery experiments are conducted based on actual data, and the method's effectiveness is verified by comparison with other methods.

Keywords: neural networks; VMD; data reconfiguration; attention mechanisms

Citation: Xie, K.; Liu, J.; Liu, Y. A Power System Timing Data Recovery Method Based on Improved VMD and Attention Mechanism Bi-Directional CNN-GRU. *Electronics* **2023**, *12*, 1590. https://doi.org/10.3390/electronics12071590

Academic Editors: Luis Hernández-Callejo, Sergio Nesmachnow and Sara Gallardo Saavedra

Received: 18 February 2023
Revised: 23 March 2023
Accepted: 27 March 2023
Published: 28 March 2023

1. Introduction

Power grid spatial and temporal character is becoming more complicated with the development of the power system, and the automation equipment rapidly expands with large-scale power systems [1,2]. At the same time, measuring data are increasing. They are starting to resemble big data due to the rapid advancement of power system measurement technologies and the ongoing reduction in measurement costs [3]. The transmission, storage, and analysis of massive amounts of big data for power grids have emerged as a significant area of research in recent years thanks to the rapid advancement of big data technology [4,5]. It is possible to estimate the status of the power system and equipment to a significant extent as well as to optimize operation and accident analysis through the analysis of massive and multiple types of time series data [6,7].

It goes without saying that obtaining authentic and accurate data is crucial for data processing. Still, since signal attenuation, interference, and occasionally failing electronic acquisition equipment cause data to be lost during data acquisition, measurement, transmission, and storage, it is impossible to obtain accurate time series data. In addition to complicating the analysis of prediction outcomes or trend development based on extensive data analysis, missing data can also have an impact on system state estimate, stability, and other critical features based on network data analysis [8,9]. The power system measurement configuration itself has a certain redundancy at the beginning of the design, and for some of the missing time series data, under the premise of satisfying the observable state estimation, it can be replaced by pseudo-measurements or by data that are similar in time and space without causing an unacceptable impact on the overall system state accuracy. In addition, the state estimation through the system can be used as a basis for filling in the missing data [10]. However, when there is a large amount of missing time series data, state estimation by redundancy is not possible. Then, the missing data need to be repaired by mathematical or engineering means.

Numerous solutions to the missing data issue have been put forth by domestic and international researchers. The pre-processing category and the post-assessment category are the two main categories. On the basis of system timing data and system topology, the latter primarily builds the system state equation to recover the data. In the literature [11], a real-time dynamic parameter estimation method based on vector measurement unit (PMU) data with extended Kalman filtering (EKF) is proposed. In order to reduce the computational difficulty of this method, the literature [11] uses a model decoupling technique. However, this method is only applicable to the real-time estimation of the states and parameters related to electromechanical dynamics. The literature [12] proposes a robust detection method using temporal correlation and statistical consistency of time series data, offering three innovative matrices that capture measurement correlation and statistical consistency by processing predicted states and reliable information inserted from phasor measurement units. Pre-processing techniques are primarily employed to recover missing data from known data. The two main categories of preprocessing ideas are: (1) analyzing the characteristics of the data in the missing data domain to complete the data as described in the literature [13–16] and (2) analyzing the overall trend and overall structure of the data and completing the data [17–19]. Ref. [13] used a Lagrangian interpolation polynomial method for adaptive estimation of incomplete and missing data, but this method is limited to the case where there are few missing data.There are also some scholars who convert the measured value into a Page matrix and then use low-rank matrix estimation based on the optimal singular value threshold to reconstruct the original signal [14]. The literature [15] first proved that the power quality data have the property of approximately low rank. Based on this, a multi-parametric joint rank optimization model is designed, and the alternating direction multiplier method is applied to decompose it into several subproblems for solving separately. At the same time, the optimal selection strategy of adaptive iteration steps is proposed to speed up the model solution for the problem that the traditional alternating direction multiplier method solves slowly. Ref. [16] uses the singular value threshold algorithm to complete the missing data twice and analyzes its error on the basis of the completion. However, the above two methods are not effective for complex missing data. According to Ref. [17], a shallow coder is used to learn the data features, and after processing, the data are supplemented by weighting the data structure. According to Ref. [18], forward and reverse GRU networks are used separately to learn the existing data, and their combined results are then weighted to achieve the goal of completing the data. The above two methods have a large gap in the reconstruction effect of different types of data. Ref. [19] has constructed an improved generative countermeasure network learning time series data with complex time and space relations. According to the data's redundancy and inherent physical and mathematical relations, the data can be restored to a considerable extent. However, this method consumes many resources and could be more conducive to practical use. Table A1 in the Appendix A shows a summary of information for similar work.

Based on the above background and the time series and multidimensional correlation characteristics of power system timing data, a method for recovering missing data from power system measurements based on dual radial gated cyclic units is proposed. The method learns the spatiotemporal characteristics of the historical data, obtains sufficient generalization capability for the time series data, constructs a mapping of the existing data to the missing data, and makes this mapping select the relatively valuable information in the existing data to repair the missing data in real time through a triple attention mechanism.

In order to make full use of the existing information, this paper proposes a joint neural network approach, i.e., to build neural networks on both sides of the missing data separately and finally to obtain the weighted repair results by combining two neural networks. Finally, a comparison between simulated and actual data shows that this data-driven method of repairing missing data in power system measurements, which does not rely on the power system topology, can maintain a high accuracy rate under different numbers of absent conditions.

2. Power System Time Series Missing Data Recovery Model Structure

The method of time series data recovery of power systems based on VMD and triple attention mechanism bi-directional CNN-GRU is shown in Figure 1, which is divided into four main steps:

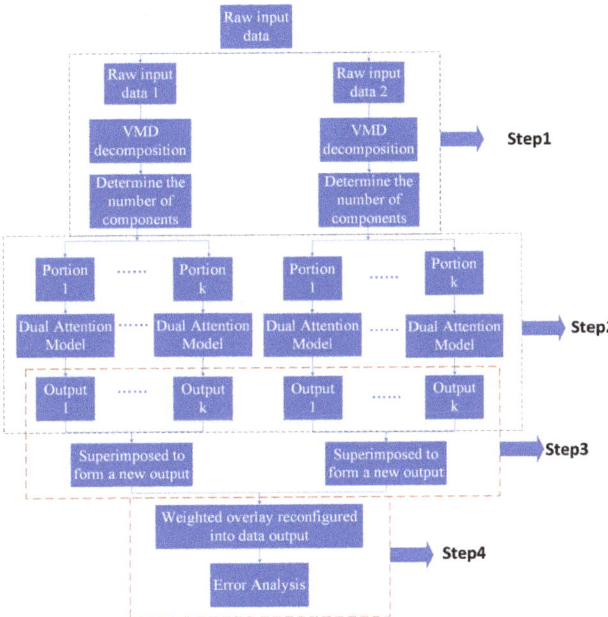

Figure 1. Power system timing data recovery model based on improved VMD and double radial GRU with a triple attention mechanism.

- The mode number k in the VMD technique is ascertained via the dual-threshold filtration method. Subsequently, the acquired k value and VMD methodology are employed to disintegrate the time-series data bilaterally, yielding k modal numbers.
- A CNN-GRU forecasting model is erected for each component, incorporating a dual attention mechanism. The input coding assay incorporates a feature attention mechanism to excavate the correlation between the time series data and the corresponding feature quantity. On the other hand, the output coding assay integrates a temporal attention mechanism to unearth the pertinent relationship between temporal data and missing data on time scales.
- The data of each component are superimposed and reconstructed to complete the data recovery results on one side.
- The data on either side of the absent data within the power system possess excellent completeness; hence, both sides of the missing data are individually modeled. Subsequently, the ultimate data restoration outcome is attained by incorporating the data restoration results of one side through adaptive weight allocation. The results are then scrutinized.

3. Decomposition of Quantitative Time Series Data Based on Improved VMD

VMD is an adaptive, quasi-orthogonal decomposition method as well as a more cutting-edge signal processing method that was proposed in 2014. The essence of the algorithm is Wiener filtering for noise reduction [20]. The goal is to decompose a time series data X adaptively into several Intrinsic Mode Functions (IMFs) with finite bandwidth x_k. In order to calculate the spectral bandwidth of each component, this goal can be achieved by the following three steps: (1) Obtain the one-sided spectrum by Hilbert transform. (2) Transfer the spectrum of each quantitative data component to the baseband region by mixing an index tuned to the respective estimated center frequency. (3) Estimate the bandwidth of the decomposed signal by H1 Gaussian smoothing. For the input signal, the constrained variational model signal is represented as follows.

$$\min_{\{x_k\},\{\omega_k\}} \left\{ \sum_k \left\| \partial_t \left[\left(\delta(t) + \frac{j}{\pi t} \right) * x_k(t) \right] e^{-j\omega_k t} \right\|_2^2 \right\} \qquad \text{s.t.} \sum_k x_k = X \tag{1}$$

The center frequency of the corresponding component is represented by the letter $\{\omega_k\}$ in the equation above. The original input signal X should be represented by the superposition of all components. Equation (1) can be changed into the following form by adding the quadratic penalty term and the Lagrange multiplier λ to reconstruct the constraint:

$$\mathcal{L}(\{x_k\},\{\omega_k\},\{\lambda\}) = \alpha \sum_k \left\| \partial_t \left[\left(\delta(t) + \frac{j}{\pi t} \right) * x_k(t) \right] e^{-j\omega_k t} \right\|_2^2 + \left\| X(t) - \sum_k x_k(t) \right\|_2^2 + \left\langle \lambda(t), X(t) - \sum_k x_k(t) \right\rangle \tag{2}$$

where α is the penalty factor, which ensures the signal reconstruction's accuracy even when noise interference is present. The Lagrangian multiplier, λ, firmly guarantees that the constraints are upheld. The components and their center frequencies can be obtained from the saddle points of the above extended Lagrangian equation using the alternate direction method of multipliers (ADMM).

$$\hat{u}_k^{n+1}(\omega) = \frac{\hat{f}(\omega) - \sum_{i \neq k} \hat{u}_i(\omega) + \frac{\hat{\lambda}(\omega)}{2}}{1 + 2\alpha(\omega - \omega_k)^2} \tag{3}$$

$$\omega_k^{n+1} = \frac{\int_0^\infty \omega |\hat{u}_k(\omega)|^2 d\omega}{\int_0^\infty |\hat{u}_k(\omega)|^2 d\omega} \tag{4}$$

where $\hat{u}_k^{n+1}$ represents the residual of the current Wiener filter, and the frequency domain mode $\hat{u}_k(\omega)$ can be Fourier inverted to produce the time domain mode $x_k(t)$ before its real part is taken [21].

The final number of modes the VMD algorithm produces is determined by the modal number k, which has a non-negligible position in the algorithm. The center frequencies of the decomposed components are typically compared to determine whether there is under- or over-resolution, which has a relatively high degree of subjectivity. To a certain extent, the double-threshold screening method can prevent this issue [22].

The VMD analysis's components have narrow bandwidth properties, meaning that most of the modes are centered around the center frequency and have high corresponding amplitudes. Two thresholds—amplitude thresholds T_2 and and frequency interval thresholds T_1 —are established based on the aforementioned characteristics. By examining the spectral properties of the input timing signal and the threshold T_1, it is possible to divide the entire spectrum into several frequency bands, with each band being used as a potential component. The frequency bands above are measured for their corresponding amplitude using the threshold T_2, and those whose amplitude satisfies the criteria are kept, while those with insufficient amplitude are ignored. The following four steps can be used to divide the application of the double threshold screening method to ascertain the modal number k: (1) Based on the spectral properties of the input data being examined, choose the appropriate frequency interval threshold T_1 and amplitude threshold T_2. (2) Look for local maxima in the spectrum, and using the frequency interval threshold T_1, divide the local maxima into corresponding frequency bands. (3) Examine the valid frequencies divided into each band; the valid frequency bands are those whose amplitudes meet the T_2 amplitude threshold. (4) The number of modes equals the number of legal frequency bands.

4. Dual Attention Model

Time series data of power systems are generated in chronological order. Considering the need for data analysis and storage, the power system timing data are often composed of discrete time series data. There is usually some connection between each power system timing data and other power time series data [23]. For the missing power system timing data, the reconstruction of power system data can be realized by analyzing the more complete time series data on both sides of it and mining the inner correspondence law between the complete power time series data and the missing power system timing data on the basis of considering the temporal sequence characteristics [24].

Figure 2 presents a comprehensive schematic of the dual attention model implementation, which comprises three primary components: the CNN layer, the GRU layer, and the attention layer. The components derived from the enhanced VMD algorithm are normalized using MinMaxScaler to confine the overall value range between 0 and 1 and subsequently fed into the CNN layer. During data input, the individual features of the input are weighted using the feature attention mechanism, thereby reinforcing the features that exert a significant impact on the outcome. On the output side, the temporal attention mechanism is employed to enhance the model accuracy by capitalizing on the correlation of the data on the time scale.

4.1. CNN-GRU Neural Network

Through a convolutional kernel, convolutional neural networks extract features from the input data locally, and convolutional neural network units learn the patterns in the input data window [25]. Convolutional neural networks gain two significant characteristics: First, the patterns studied by a convolutional neural network are translation invariant. A specific pattern learned by a convolutional neural network in a local data segment can identify this pattern in an arbitrary place.

Due to the stability of electricity consumption habits, the power system data possesses apparent repeatability. For example, the curve of regional load data in the same period of two days has substantial similarity, and the photovoltaic generation curve also shows strong similarity in two days with similar climate environments. This similarity provides a reasonable basis for convolutional neural networks. Second, the convolutional neural network can learn patterns' spatial hierarchy; the first convolutional layer can learn smaller local patterns, and the second convolutional layer further reconstructs the first convolutional layer's patterns to form larger patterns. This feature can make the convolutional neural network can learn more and more complex and abstract data [26]. Regarding temporal data processing, one-dimension (1D) convolution is commonly utilized. The operational principle of one-dimension convolution is depicted in Figure 3. The 1D convolutional layer is adept at detecting local patterns in a sequence. As the same input transformation is applied to each sequence segment, patterns discovered at one position in the temporal data can be subsequently recognized at other positions, rendering the 1D convolutional neural network translation invariant (concerning time translations).

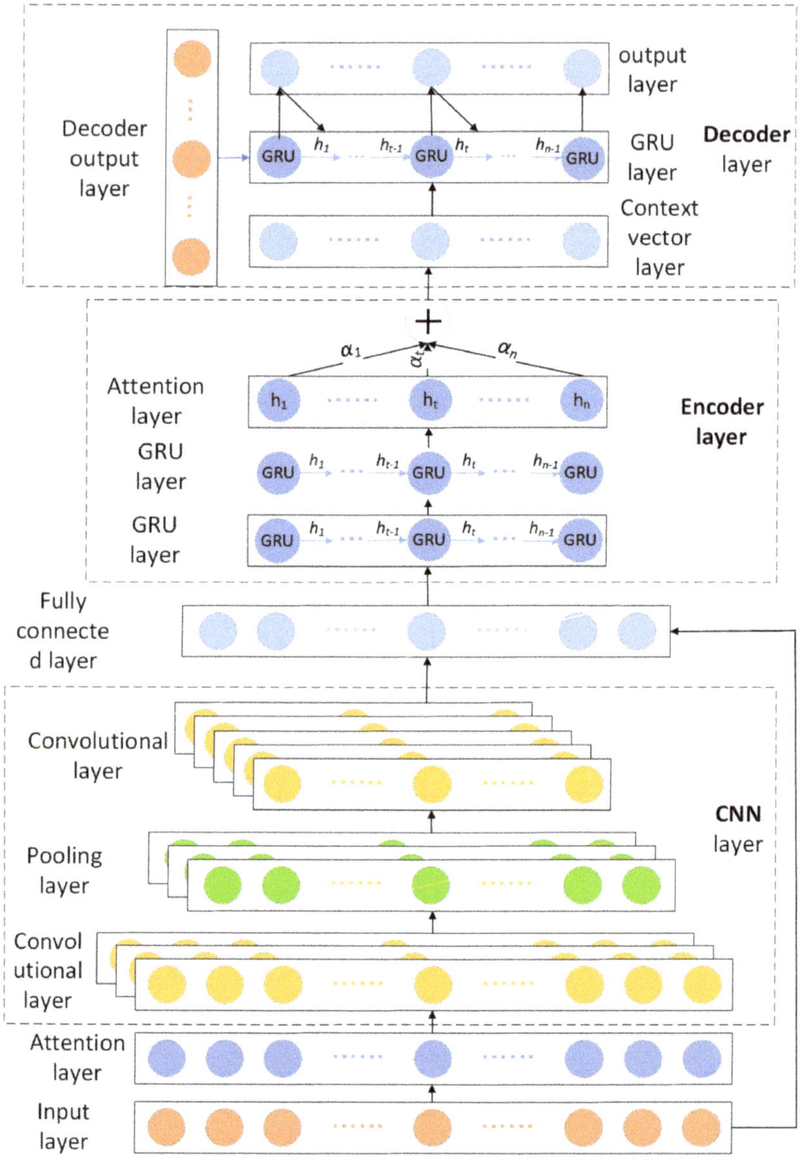

Figure 2. Dual attention CNN-GRU model.

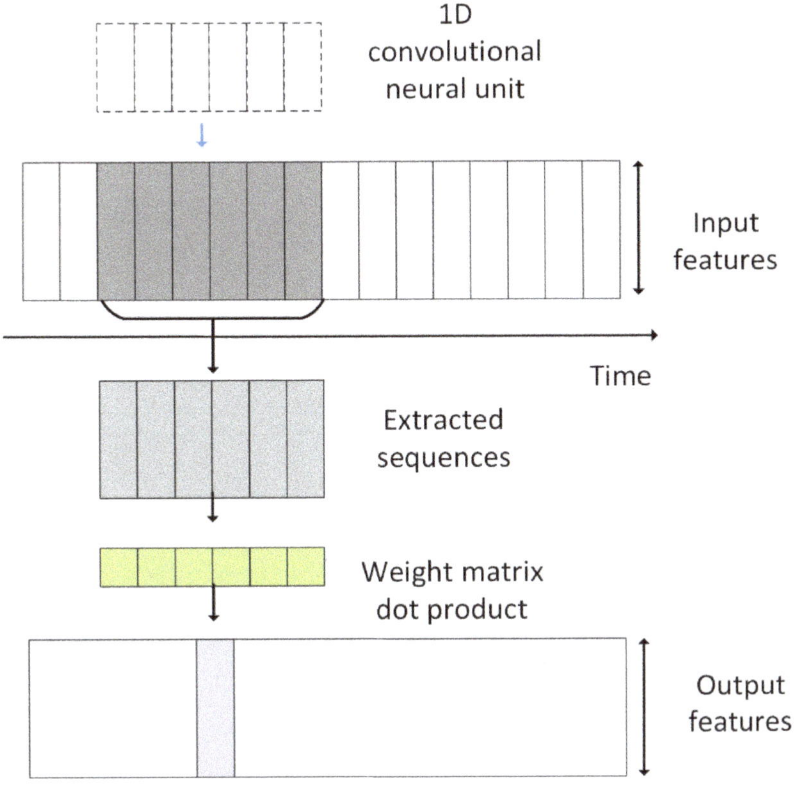

Figure 3. Working principle of one-dimensional convolution neural network.

By sampling the feature map through the maximum pooling layer, it can reduce the number of input features in the upper convolutional layer by sampling the input signal, which makes the model structure more streamlined, and the number of parameters to be computed decreases significantly. The maximum pooling layer extracts the essential information from the upper layer and transfers it to the lower convolutional layer, which allows the convolutional layer to have a larger and larger observation window (the window covers the proportion of the original input size), thus making the convolutional layer neural network have a spatial hierarchy.

A variant of the long short-term memory (LSTM) neural network, the GRU is a relatively new neural network structure primarily proposed by Junyoung Chung et al. at the International Conference on Machine Learning in 2015 [27]. In the GRU, the cell state and hidden state are combined with forgetting and input gates to create a single update gate. When applied to large-scale data, the original three gate structures are combined into just two gates, the parameters are decreased while maintaining the characteristics of LSTM, and the computational speed is thus noticeably increased.

The structure of the GRU shows the two gates that the GRU has: the reset gate rt and the update gate Z_t [28]. The update gate is a linear transformation of the input signal x_t at time step t and the state h_{t-1} at the previous time step, respectively, which are added together, and the information obtained is activated by a sigmoid function. An update gate determines how much of the signal from the past is going to be passed to the future. The reset gate is similar to the update gate in that the input signal x_t at time step t and the state h_{t-1} at the previous time step are linearly transformed and added together. The resulting information is activated by a sigmoid function. However, the essence is to decide how much information needs to be forgotten. After obtaining the reset gate, the linearly transformed reset gate and the linearly transformed input are added. The result is put into the hyperbolic tangent activation function to obtain the current required memory content $\hat{h}_t$. A unit of 1 is subtracted from the output of the update gate to obtain a difference. Then, the product of this difference and the current memory content is multiplied by the product of the state of the previous time step and the result of the update gate to obtain the state of the present time step. The specific formula for GRU is shown below.

$$\begin{aligned} z_t &= \sigma_g(W_z x_t + U_z h_{t-1} + b_z) \\ r_t &= \sigma_g(W_r x_t + U_r h_{t-1} + b_r) \\ \hat{h}_t &= \phi_h(W_h x_t + U_h(r_t \odot h_{t-1}) + b_h) \\ h_t &= (1 - z_t) \odot h_{t-1} + z_t \odot \hat{h}_t \end{aligned} \tag{5}$$

where x_t denotes the input vector; h_t denotes the output vector; $\hat{h}_t$ denotes the current desired memory content; z_t denotes the update gate; r_t denotes the reset gate; W, U and b represent the parameter matrices and vectors; σ_g denotes the sigmoid function; ϕ_h denotes the hyperbolic tangent function; and $\odot$ denotes the Hadamard product.

4.2. Attentional Mechanisms

The processing of visual signals by attentional mechanisms is specific to human vision. Human vision requires daily access to enormous amounts of image data as one of the most crucial information acquisition channels. The brain typically concentrates on the image's critical components while ignoring the image's comparatively minor components when processing this information. This mechanism has the potential to speed up and improve the processing of visual data significantly [29]. The attention mechanism in this paper is similar to this in that the important parts of the signal are selected and given a relatively large weight, resulting in a greater increase in the output accuracy of the whole system.

Let the input timing and the corresponding characteristics be:

$$x = [x_1, x_2, \ldots, x_T] = [x^{(1)}, x^{(2)}, \ldots, x^{(n)}]^T \tag{6}$$

The expansion can be represented by the following matrix.

$$x = \begin{bmatrix} x_1^{(1)} & x_1^{(2)} & \cdots & x_1^{(n)} \\ x_2^{(1)} & x_2^{(2)} & \cdots & x_2^{(n)} \\ \vdots & \vdots & & \vdots \\ x_T^{(1)} & x_T^{(2)} & \cdots & x_T^{(n)} \end{bmatrix} \in \mathbf{R}^{T \times n} \tag{7}$$

$x_t = \left[x_t^{(1)}, x_t^{(2)}, \cdots, x_t^{(n)}\right](1 \le t \le T)$ is the feature set of the above n features at moment t. $x^{(m)} = \left[x_1^{(m)}, x_2^{(m)}, \cdots, x_T^{(m)}\right](1 \le m \le n)$ is the value of the mth relevant eigenvalue at moment $t(1 \le t \le T)$.

In order to obtain the association of each feature variable with the current time series, i.e., for the present time series, the importance of its corresponding feature quantity, the feature attention method is used for calculation. The attention weights corresponding to the feature quantities at the current moment are calculated by associating the feature variables at the moment t with the corresponding output variables at moment $t-1$, h_{t-1} and the state variables, s_{t-1}.

$$e_t^{(m)} = V_e^{\mathrm{T}} \tanh\left(W_e[h_{t-1}; s_{t-1}] + U_e x^{(m)} + b_e\right) \tag{8}$$

V_e, W_e, U_e are the weight matrices, respectively. b_e is the corresponding bias term.

After obtaining the weight values, they need to be normalized so that the sum of the weight values corresponding to moment t is 1.

$$\alpha_t^{(m)} = \frac{\exp\left(e_t^{(m)}\right)}{\sum_{i=1}^{n} \exp\left(e_t^{(i)}\right)} \tag{9}$$

After obtaining the weighting coefficients, the input eigenvalues are multiplied by them to obtain the weighted input:

$$\widetilde{x}_t = \left[\alpha_t^{(1)} x_t^{(1)}, \alpha_t^{(2)} x_t^{(2)}, \cdots, \alpha_t^{(n)} x_t^{(n)}\right] \tag{10}$$

The obtained adaptive weighted input $\widetilde{x}_t$ afterward is fed into the subsequent model instead of the original input x_t. This method can dynamically extract the correlation between the feature values and the corresponding time series. The state h_t of the hidden layer at each moment needs to be updated at the next moment.

$$h_t = f_1(h_{t-1}, \tilde{x}_t) \tag{11}$$

f_1 is the GRU network unit.

After the adaptive feature attention results are computed, the output results obtained by the feature attention mechanism are used as the input of the temporal attention mechanism in the next stage. The time series attention mechanism focuses attention on the input series and obtains the adaptive time principal series output by weighted average for the input time series [30]. Figure 4 illustrates the principle of the attention mechanism implementation. The attentional mechanism is calculated as follows:

$$c_i = \sum_{s=1}^{\mathrm{T}} \alpha_{ts} \overline{\boldsymbol{h}}_s \tag{12}$$

$$\alpha_{ts} = \frac{\exp\left(score\left(h_t, \overline{h}_s\right)\right)}{\sum_{s'=1}^{S} \exp\left(score\left(h_t, \overline{h}_{s'}\right)\right)} \tag{13}$$

where h_t is the output of the decoder corresponding to time t and $\overset{h}{h}\ {}_{-s}$ is the source hidden state of the encoder. Here, score expression is specifically calculated as follows:

$$score\left(h_t, \overline{h}_s\right) = h_t^{\top} \overline{h}_s \tag{14}$$

The core idea is to make the context vector c_t, which is otherwise invariant in the seq2seq structure, dynamic by reorganizing it several times. The traditional context vector c_t selects the output of the last time step as the final output because the data can be regarded as the process of gradually extracting features after GRU processing. The output of the last time step often contains important information about the state of the past time steps. However, for data with long time steps and strong information correlation, this model often leads to the vital information of a past time step being ignored or not highlighting the information strongly associated with the prediction results [31]. Through the temporal attention mechanism, the vector c_t is dynamized, and c_t is no longer just the output of the last time step but also a dynamic combination of individual time steps. Different weight coefficients are given for different predicted contents, thus giving different c_t, which finally achieves the role of extracting necessary information in the time series.

Fuse the real-time updated context vector c_t with the output h_t of time step t as the input of the decoder:

$$\tilde{h}_t = \tanh(W_c[c_t; h_t] + b_c) \tag{15}$$

where W_c and b_c are the weight and bias of the fused input, respectively; tanh is the hyperbolic tangent function; $\tilde{h}_t$ is the encoder output.

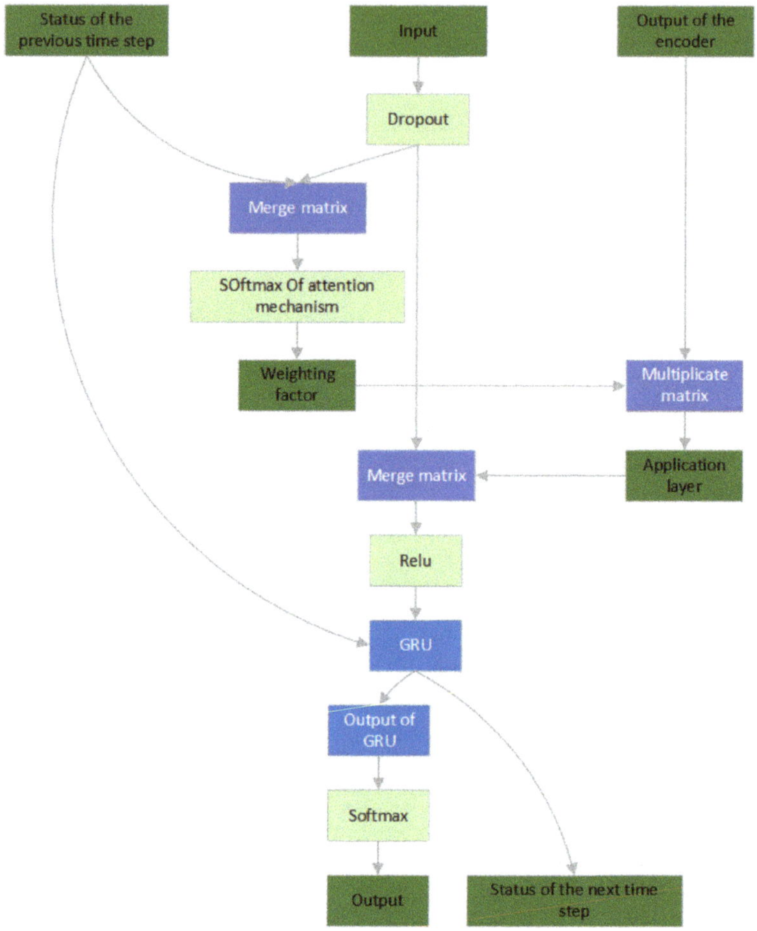

Figure 4. Schematic diagram of realization of attention mechanism.

5. Example Analysis

5.1. Data Pre-Processing and Error Indicators

In this paper, three datasets are used to validate the model proposed in this paper: Singapore power load data, a public wind power dataset in the United States, and a public photovoltaic (PV) output dataset in Australia. The first 80% of the data is used as the training set, the first 80% to 95% is used as the validation set, and the last 5% is used as the test set. The wind power generation data set has a total of 50,500 points, with a sampling period of 10 min. Again, the first 80% of the data is used as the training set, the first 80% to 95% of the data is used as the validation set, and the last 5% of the data set is used as the test set. The photovoltaic output at night is always zero, so there is no meaningful training for this period, and the photovoltaic data in the night part of the data have been removed during the data processing. The photovoltaic output data set has 98,000 points with a sampling period of 5 min. Here, 80% of the data is used as the training set, the first 80% to 95% is used as the validation set, and the last 5% is used as the test set. In order to make the neural network training more efficient beforehand, the MinMaxScaler in sklearn was used to normalize the data by min–max between (0, 1), which are calculated as:

$$MAE = \frac{1}{m}\sum_{n}^{i=1}|(y_i - \hat{y}_i)| \tag{16}$$

$$MSE = \frac{1}{m}\sum_{m}^{i=1}(y_i - \hat{y}_i)^2 \tag{17}$$

where m denotes the total number of model output results; y_i and $\hat{y}_i$ mean the actual value of the ith point in the output results and the model output value, respectively. Smaller MAE and MSE indicate better model fit.

Adam is chosen as the optimizer (adaptive moment estimation), and the Adam optimization algorithm can achieve adaptive gradient selection, which can jump out some local minima.

5.2. Model Configuration

The convolutional layers are stacked in two layers. The number of neurons in each layer is 128, the convolutional window is 64, and the activation function is selected as Relu. The pooling layer is selected as the maximum pooling. The encoder chooses two layers of GRU stacking: 128 neurons for the first layer and 64 neurons for the second layer. Tanh is selected for both activation functions, the discard rate is set to 0.1 for the first layer and 0.24 for the second layer, and the decoder chooses single-step GRU decoding with 64 neurons. Each GRU step is output through a unit fully connected layer. The training step size is 0.0075.

5.3. Data Set Comparison

In this paper, a total of five models are introduced for comparison, and the five models are LSTM, multi-layer perceptron (MLP), Seq2Seq, CNN and the model of this paper using one-sided reconstruction. All models use the same normalization procedure for data output, and the input of the models is the corresponding historical sequence data to ensure the scientific accuracy and validity of the comparison method. The data in the test set were taken from 32 sampling points and 128 sampling points for data reconstruction, and the evaluation results of each quantitative data index are shown in Tables 1 and 2. In order to make the comparison results more intuitive and comparable, the comparison results are directly normalized using the model output results, which can avoid the problem of the inability to compare because of different units and large differences in the size of the original data.

Table 1. MAE and MSE of each model with reconstructed length of 32 sampling points.

		Load Data Set	Wind Power Dataset	Photovoltaic Power Dataset
Model of this paper	MAE/(e-2)	1.96	5.38	4.48
	MSE/(e-4)	5.97	37.29	31
One-sided modeling	MAE/(e-2)	2.9	9.58	8.28
	MSE/(e-4)	11.6	74.06	61.7
LSTM	MAE/(e-2)	2.27	8.25	6.87
	MSE/(e-4)	7.49	61.68	51.4
CNN	MAE/(e-2)	2.48	8.54	7.12
	MSE/(e-4)	8.57	65.04	54.2
Seq2seq	MAE/(e-2)	2.14	7.12	5.84
	MSE/(e-4)	6.57	44.09	36.7
MLP	MAE/(e-2)	2.85	8.54	7.21
	MSE/(e-4)	7.98	59.46	49.8

Table 2. MAE and MSE of each model with reconstructed length of 128 sampling points.

Model of This Paper		Load Data Set	Wind Power Dataset	Photovoltaic Power Dataset
	MAE/(e-2)	2.74	9.95	6.97
One-sided modeling	MSE/(e-4)	11.14	83.53	64.48
	MAE/(e-2)	4.09	16.77	13.33
LSTM	MSE/(e-4)	21.46	165.89	125.25
	MAE/(e-2)	3.28	20.54	10.85
CNN CNN	MSE/(e-4)	15.22	108.56	114.11
	MAE/(e-2)	3.59	21.26	10.96
Seq2seq	MSE/(e-4)	17.22	101.46	123.58
	MAE/(e-2)	3.08	12.67	9.46
MLP	MSE/(e-4)	13.25	106.7	81.11
	MAE/(e-2)	4.29	15.79	11.82

5.3.1. Reconstructed Data Length of 32 Sampling Time Points Reconstruction Effect Comparison

Table 1 presents a summary of the Mean Squared Error (MSE) and Mean Absolute Error (MAE) for various models with a reconstructed data length of 32. The experimental results from three data sets, namely, the load data set, wind power generation data set, and photovoltaic power generation data set, are arranged from left to right. As the summary table employs normalized data results, it does not include any units. Figures 5–7 depict the schematic diagram of the load data reconstruction results, wind power generation data reconstruction results, and photovoltaic power generation data reconstruction results, respectively, with a reconstructed data length of 32 sampling time points. To facilitate comparison and highlight the characteristics of model reconstruction, a period with more salient features was selected from each of the three datasets.

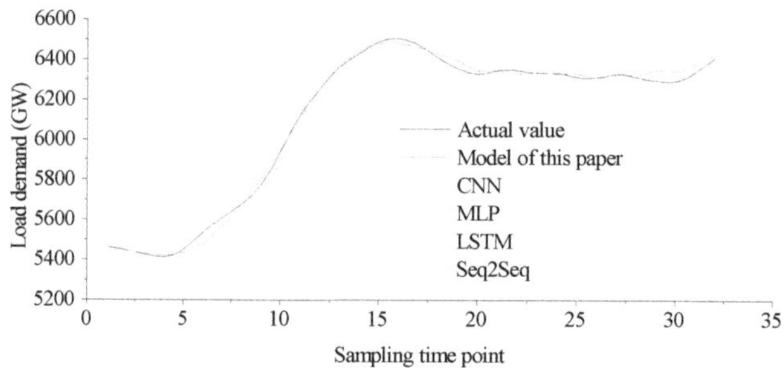

Figure 5. Results of reconstructing 32 load time sampling points with different models.

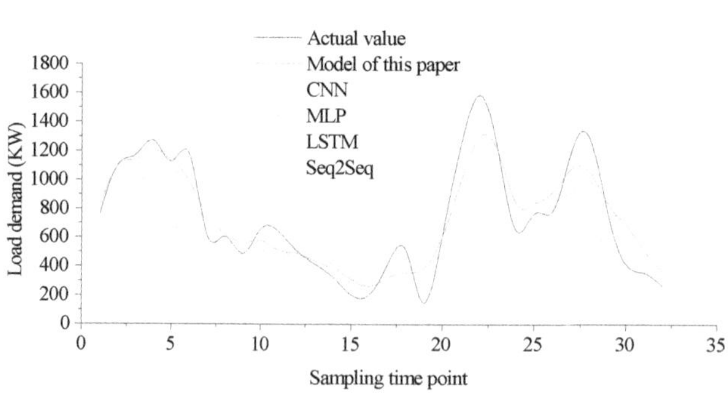

Figure 6. Results of reconstructing 32 wind power time sampling points with different models.

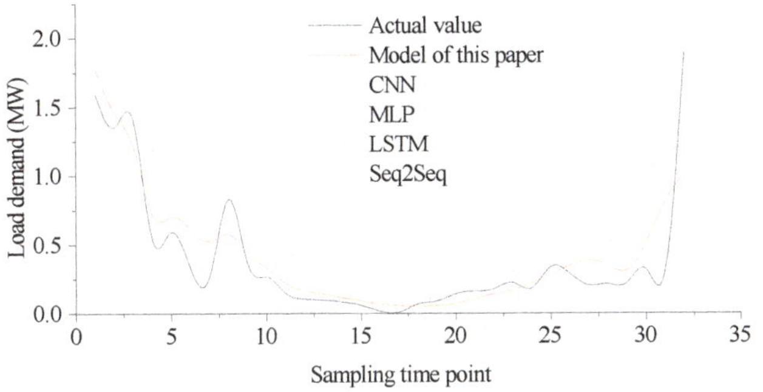

Figure 7. Results of reconstructing 32 photovoltaic power time sampling points with different models.

For the loaded data set, we can see that when the reconstructed data length is 32 sampling time points, the average MAE of the proposed model in this paper decreases by 32.41%, 13.66%, 20.97%, 8.41%, and 31.23% relative to the unilateral reconstruction model, LSTM, CNN, Seq2Seq, and MLP, respectively. The mean MSE of the proposed model decreases by 48.53%, 20.29%, 30.34%, 9.13%, and 25.19% relative to the one-sided reconstruction model, LSTM, CNN, Seq2Seq, and MLP, respectively.

For the wind power dataset, we can see that when the reconstructed data length is 32 samples, the mean MAE and mean MSE of the model proposed in this paper decreased by 43.83%, 34.79%, 37.01%, 24.44%, 15.42%, 37.00%, and 49.65%, 39.54%, 42.67%, 15.42%, and 37.29%, respectively.

For the photovoltaic generation dataset, we can see that when the reconstructed data length is 32 samples, the MAE mean and MSE mean of the proposed model in this paper decreased by 45.89%, 34.79%, 37.08%, 23.29%, 37.86%, and 49.76%, respectively, when compared to the unilateral reconstruction model, LSTM, CNN, Seq2Seq, and MLP. 39.69%, 42.80%, 15.53%, and 37.75%.

The reconstructed data results graph demonstrates that the model proposed in this paper is more closely aligned with the actual trend for data reconstruction. Specifically, the proposed model better reconstructs the data mutation part. An extreme value point near the fifteenth sampling point is observed for the load reconstruction results, and the proposed model fits this extreme value point better. Furthermore, the rise and fall of the load reconstruction results are also in close proximity to the actual data. The effect is more pronounced for the photovoltaic power generation data and wind power generation data with apparent changes. The wind power data reconstruction results show that the data are more fluctuating, and the data are relatively very variable, with changes occurring repeatedly within 32 sampling points. The proposed model can react quickly at the abrupt change points, such as the 6th sampling point, 19th sampling point, 22nd sampling point, and 28th sampling point, and keep up with the data change trend in time. The relative change of photovoltaic data is located between the load and wind power generation. However, photovoltaic data have an obvious characteristic that the data will occasionally suddenly drop to very low or rise from very low to a higher position. This characteristic is mainly due to the photovoltaic power generation's dependence on weather, which has a significant impact on light intensity. For all three data sets in this paper, the proposed model's reconstructive ability on relatively short missing data is significantly stronger than the comparison model.

5.3.2. Reconstructed Data Length of 128 Sampling Time Points Reconstruction Effect Comparison

Table 2 presents a summary of the MSE and MAE for each model with a reconstructed data length of 128. It includes experimental results from three datasets. Appendix A Figure A1 displays a schematic diagram of the reconstructed load data with a reconstructed data length of 128 sampling time points. Appendix A Figure A2 exhibits a schematic diagram of the reconstructed wind power generation data with a reconstructed data length of 128 sampling time points. Appendix A Figure A3 illustrates a schematic diagram of the reconstructed photovoltaic power generation data with a reconstructed data length of 128 sampling time points. The three datasets were selected for their distinct performance characteristics over a period to facilitate comparison and emphasize the features of model reconstruction.

For the loaded dataset, when the reconstructed data length is 128 samples, the mean MAE of the model proposed in this paper decreases by 33.01%, 16.56%, 23.68%, 11.04%, and 36.13% relative to the one-sided reconstruction model, LSTM, CNN, Seq2Seq, and MLP, respectively. The mean MSE of the model proposed in this paper decreased by 48.09%, 26.81%, 35.31%, 15.92%, and 37.49% relative to the one-sided reconstruction model, LSTM, CNN, Seq2Seq, and MLP, respectively.

For the wind power dataset, when the reconstructed data length is 128 samples, the mean MAE and mean MSE of the model proposed in this paper decreased by 40.67%, 51.56%, 53.20%, 21.47%, 36.99%, and 49.65%, respectively, when compared to the unilateral reconstruction model, LSTM, CNN, Seq2Seq, and MLP. 23.06%, 17.67%, 21.72%, and 44.69%.

For the photovoltaic generation data set, when the reconstructed data length is 128 samples, the mean MAE and mean MSE of the model proposed in this paper decreased by 47.71%, 35.76%, 36.41%, 26.31%, 41.03%, and 48.52%, respectively, when compared to the unilateral reconstruction model, LSTM, CNN, Seq2Seq, and MLP, generating values of 43.49%, 47.82%, 20.50%, 47.79%.

We observe that the model proposed in this paper more closely approximates the actual trend in data reconstruction. Specifically, in the case of data mutations, the model presented here performs better. Similar to the analysis with 32 samples, in the load reconstruction graph, although the change structure of 128-sample data is more prominent than that of 32-sample data, our model can track changes more effectively. Likewise, this model outperforms other models in tracking wind and photovoltaic reconstructed graphs. As with the analysis of 32-sample reconstructions, our model excels at handling large data changes over short periods of time in wind reconstruction; for instance, it fits more closely around the 40th and 60th sample points. The same conclusion applies to photovoltaic data reconstruction. It is evident that our model also surpasses other models in reconstructing data at 128 sampling time points. For all three datasets examined in this paper, our proposed model significantly outperforms comparison models in reconstructing relatively long missing data.

5.3.3. Reconstructing Data Error Distribution Analysis

Figures 8–10 display error analysis plots for the reconstruction results of our model and other comparison models at 32 sampling time points. Appendix A Figures A4–A6 present error analysis plots for the reconstruction results of our model and other comparison models at 128 sampling time points. From the perspective of error for different lengths of missing data, the range and median error of the error distribution for 128 sampling time points in all three datasets are significantly higher than those for 32 sampling time points. For instance, the median absolute values of load errors for our model, CNN, and MLP increase from 0.07, 0.126, and 0.25 to 0.31, 0.49, and 0.55, respectively. In terms of absolute error value distribution data, the reconstruction results of our proposed model are more concentrated around the median compared to several other models.

Figure 8 and Appendix A Figure A4 display the error plots of load data reconstruction results for 32 and 128 sampling points, respectively. The analysis of the load data reveals that the reconstruction error of 128 sampling points for all models is significantly larger than the error of 32 sampling points. For the proposed model, most of the errors increase from the 0–50 GW interval to the 0–115 GW interval. However, compared to other models, the error distribution of the proposed model is still relatively concentrated, which is consistent with the conclusion that the MAE and MSE are relatively smaller in the previous analysis. The performance of the proposed model is more outstanding in the error distribution interval 10–90%, error distribution interval 25–75%, and median error distribution, especially the 25–75% error distribution index, which is lower than other models. This indicates that most of the errors of the proposed model are concentrated around 0. Compared to other models, the main reason for the superior performance of the CNN model in the 32 sampling case is related to its working principle. CNN itself does not have the concept of time flow and uses the form of convolutional kernel to identify data features. Even if the recognition window is increased, as long as the features identified by the CNN convolutional kernel can still be applied, the error distribution of the CNN model does not change much.

Figure 9 and Appendix A Figure A5 depict error plots of wind power generation data reconstruction outcomes for 32 and 128 sampling points, respectively. Figure 10 and Appendix A Figure A6 portray error plots of reconstruction outcomes for photovoltaic generation data with 32 and 128 sampling points, respectively. In both cases, wind power data and photovoltaic data, we observe a similar phenomenon to the load data, where the error distribution of each model increases significantly. However, the model proposed in this paper is more dispersed and concentrated in comparison to the other models both for the 32-sample-point reconstruction and the 128-sample-point reconstruction. With regard to the 25–75% error distribution metric, it is evident that the proposed model outperforms other models.

Nonetheless, there are also some individual characteristics. For the wind power data, the rise in error distribution from 32 to 128 sampling points is not as evident as that of the load data set. This is primarily because the wind power data set is more intricate than the load data set, and the changing trend is more challenging to capture. Therefore, even if it is extended to 128 sampling points, the increase in error is limited, given that the error of 32 sampling points is already significant. As for the photovoltaic data, the error variation is between the load data set and the wind power data set, owing to its inherent regularity. However, there are some random variations in the data set.

The absolute value of Seq2Seq error in reconstructing the 128 data sampling points of the model load data set in this paper appears to be more concentrated than the model proposed in this paper. This is probably due to the load data set used being the load data of Singapore, which is relatively stable, and the trend changes are relatively evident. In addition, the data set has fewer dramatic fluctuations at certain points in time, making the data itself very reconfigurable. For the Seq2Seq model, this can be more perfect in grasping the data structure characteristics, and even without adding additional data processing means, the data reconfiguration effect is already excellent. However, for wind power generation data and photovoltaic power generation data with significant fluctuations and non-obvious trend changes, it has been challenging for the Seq2Seq model to capture the complete intrinsic pattern of the data, resulting in a significantly enlarged error distribution.

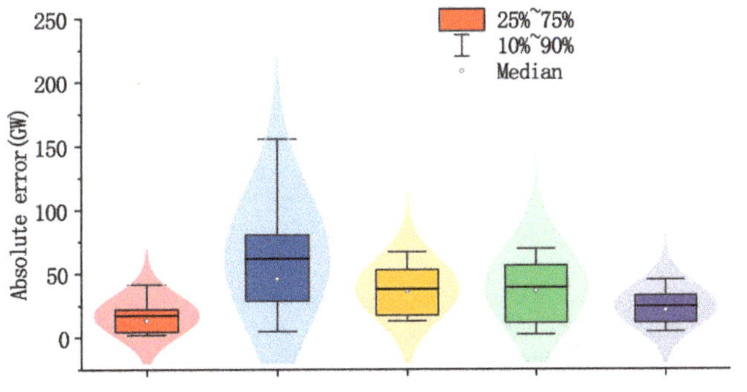

Figure 8. Error results of reconstructing 32 time sampling points of load with different models.

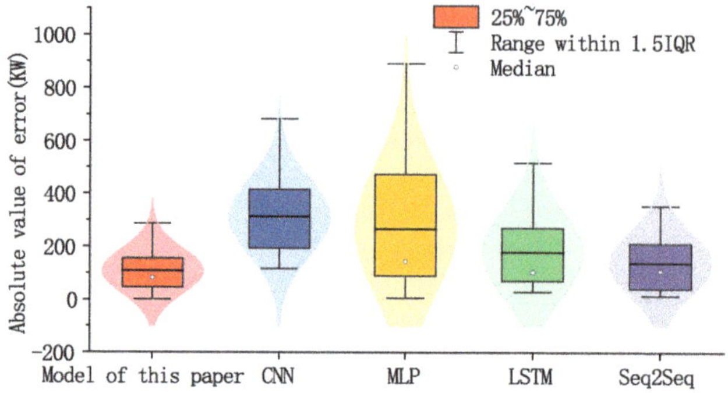

Figure 9. Error results of reconstructing 32 wind power generation time sampling points with different models.

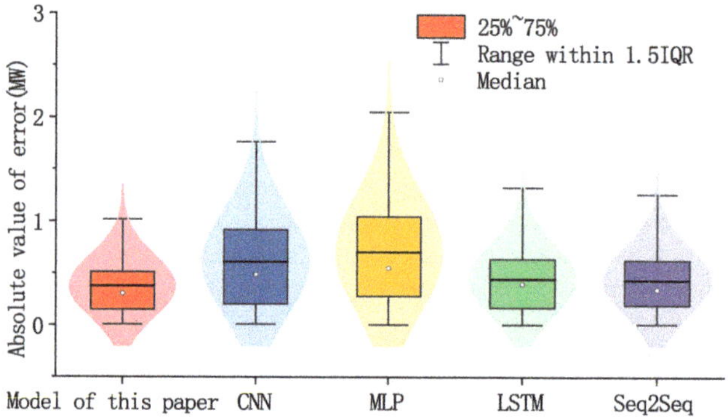

Figure 10. Error results of reconstructing 32 photovoltaic power generation time sampling points with different models.

5.3.4. 128 Sample Points Reconstruction Results Compared to 32 Sample Points Reconstruction Results

It is more challenging to reconstruct 128 sampled time points of data than to reconstruct 32 sampled time points of data. Comparing Tables 1 and 2, we can observe a significant increase in the MAE and MSE metrics for both models. This is because (1) the same data set is evidently less rich for 128 sampling points than for 32 sampling points, resulting in a model that is less comfortable with 128 sampling points. (2). Reconstructing 128 sampling points using the same deep learning model may lead to model limitations, as the model may not be able to handle more data points or capture more complex relationships in the data set, resulting in higher MAE and MSE metrics.

Figure 11 illustrates the percentage of MAE increase for 128 sample points compared to 32 sample points. Figure 12 illustrates the percentage increase of MSE for 128 samples compared to 32 samples. From Figure 11, it is evident that the MAE of each model increased significantly for 128 samples compared with 32 samples, and the most significant increase was 150% for LSTM and CNN. Furthermore, by comparing the three data sets, the MAE of this model is the smallest in most cases, indicating that this model makes the best use of the data. Similarly, it can be observed from Figure 12 that the performance of this model is superior in the other two datasets, except for the wind power dataset. The possible reasons for the low MSE of CNN and LSTM models in wind power data are: (1) The MSE of these two models is already relatively large in the province at 32 sampling points, which means that the model itself does not utilize the data to a high degree. (2) The regularity of the wind power data set itself is more difficult to find, which leads to the model itself not resolving the data set well, resulting in the reconstruction length. When the length of the reconstruction increases, there is a more obvious decrease in the degree of utilization.

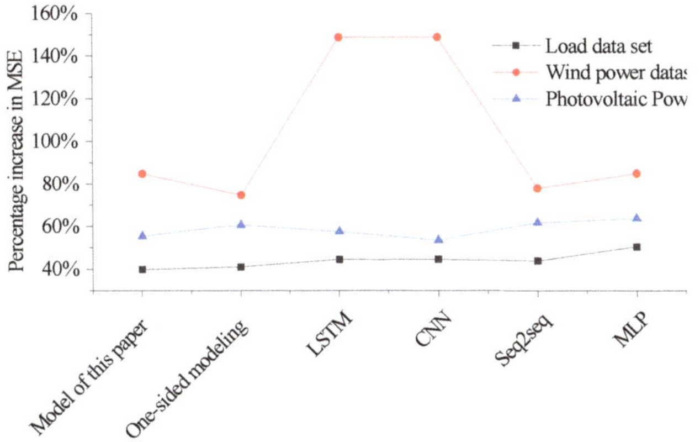

Figure 11. Percentage increase in MAE at 128 sampling points compared to 32 sampling points.

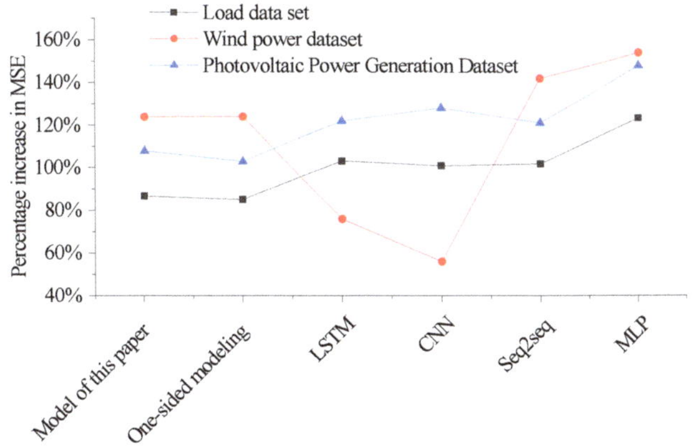

Figure 12. Percentage increase in MSE at 128 sampling points compared to 32 sampling points.

By comparing the load with 32 sampling time points and 128 sampling time points, it is evident that the prediction model in this paper has a better fit to the true value, and the accuracy decreases more slowly as the data length increases. Compared with other models, the model proposed in this paper not only reconstructs the trend of the missing data more accurately but also performs better for the part of the load with large fluctuations and can analyze the law of sudden changes to a greater extent, which improves the overall model accuracy. Comparing the wind power generation data and photovoltaic power generation data, which are more random and fluctuating, it is clear that the model proposed in this paper fits the sudden changes in the data significantly better than the other models, and it not only fits the changes in the trend of the data but also captures the drastic local changes more accurately, making the output of this model closer to the actual situation. For example, in the case of photovoltaic power generation data, although the overall trend can be captured more obviously, the overall data have the characteristic of fluctuating sharply up and down around the trend, and the model in this paper can capture the larger fluctuations more accurately, so that the details can be supplemented to make the overall error smaller.

6. Conclusions

This paper proposes a method to recover power system timing data based on improved VMD and attention mechanism bi-directional CNN-GRU, which initially processes the data by VMD so that the data can be better divided into multiple groups based on frequency centroids, extracts the temporal characteristics of the time series data by using a CNN model and then realizes the Seq2Seq structural model combined with multiple attention mechanisms for the reconstruction of the data. This paper compares the unilateral reconstruction model with LSTM, CNN, Seq2Seq, and MLP. It analyzes the characteristics of the model proposed in this paper compared to other models in terms of three indicators: MAE, MSE, and reconstruction result error. Although the result analysis is sensitive to the data scenarios, the following conclusions can be drawn:

- The model in this paper has strong data reconstruction capability, especially for fitting data mutations.
- This model has better performance in long time series data reconstruction compared to other models.
- This model has a more outstanding effect on data reconstruction with drastic changes compared to other models in this paper and has broader application potential.

In the future, we can study the characteristics of different models with a different focus on data analysis and combine multiple models dynamically to achieve more accurate data processing capability.

Author Contributions: Methodology, K.X.; Formal analysis, Y.L.; Resources, K.X.; Writing – original draft, K.X.; Writing– review and editing, J.L. All authors have read and agreed to the published version of the manuscript.

Funding: This research received no external funding.

Conflicts of Interest: The authors declare no conflict of interest.

Appendix A

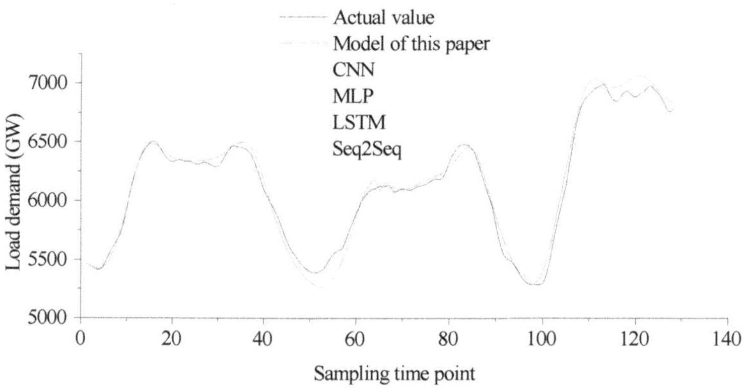

Figure A1. Results of reconstructing 128 load time sampling points with different models.

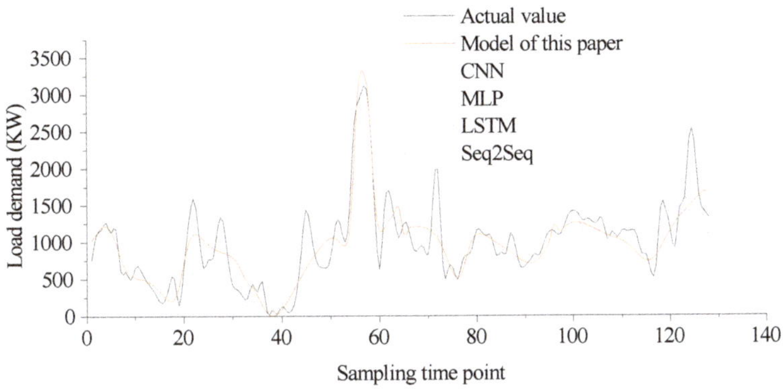

Figure A2. Results of reconstructing 128 wind power time sampling points with different models.

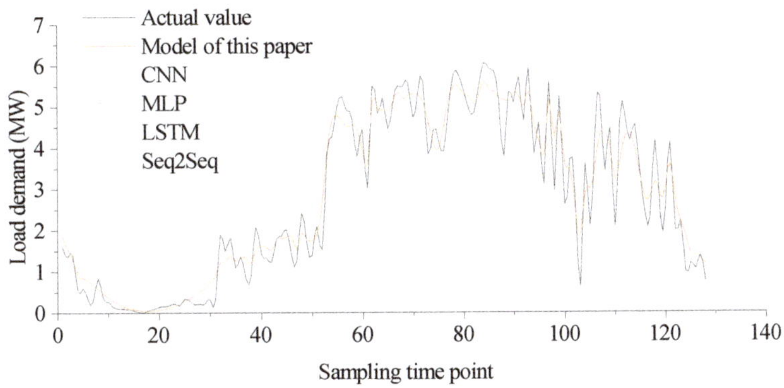

Figure A3. Results of reconstructing 128 temporal sampling points of photovoltaic power generation with different models.

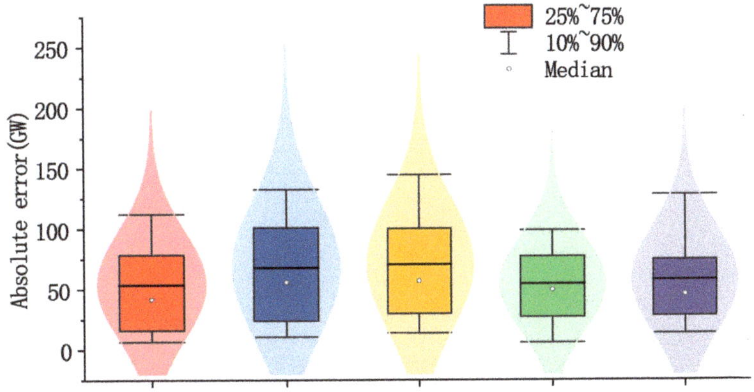

Figure A4. Error results of reconstructing 128 time sampling points of load with different models.

Table A1. Similar work summary table.

Literature	Main Technical Means	Advantage	Disadvantages
[12]	A new method for detecting bad data in power systems using temporal correlation and statistical consistency of measurements is proposed. The method uses three innovative matrices to capture measurement correlation and statistical consistency, and it applies projection statistics to detect bad data.	The computational requirements are not large, and the data categories with high similarity in the power system can be more effectively detected and reconstructed for bad data.	The reconstruction is less effective for data with high dimensionality and complexity, complex change trends, and difficulty in finding patterns.
[13]	In the case of a small amount of synchronous phase volume data missing and using a Lagrangian interpolation polynomial approach to adaptively estimate incomplete and missing data.	Fast computation, practical, good for reconstructing data with small amount of 1D data.	The method has a small range and cannot be used once the missing data becomes long or the data are not one-dimensional.
[14–16]	The core idea is to take advantage of the similarity between the data column where the missing data are located and the complete data column and use this similarity to reconstruct the data through further data processing.	The performance is highly correlated with the degree of data similarity, and the deeper the similarity between the data, the better the method refactoring and vice versa. Supported by explicit mathematical principles.	Large differences between data can seriously affect the effectiveness of data reconstruction, and the larger the amount of data, the more patterns in the data and performance degradation.
[17]	A shallow coder is used to learn the data features and to complement the data by the data structure after the weighting process.	It also has better adaptability for complex data and has a more outstanding reconfiguration effect.	Poor reconfiguration for complex, variable data.
[18,19]	Ref. [18] learns the features of temporal data through a positive and negative GRU network. Ref. [19] then learn the features of the data using generative adversarial networks.	It also has better adaptability for complex data and has a more outstanding reconfiguration effect.	The mathematical mechanism is not clear. The algorithms take longer to compute and are more demanding on computational resources.

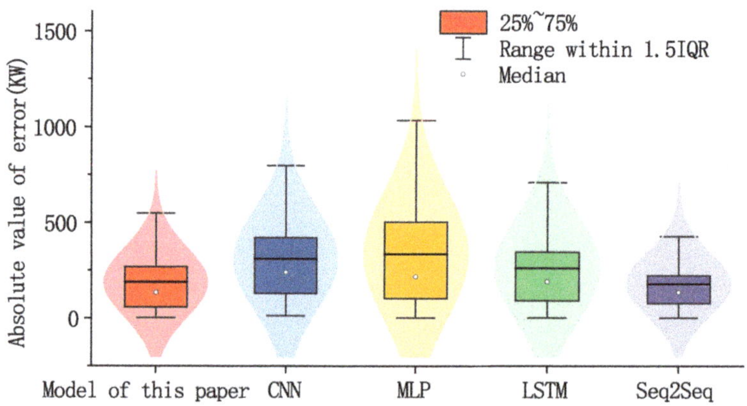

Figure A5. Error results of reconstructing 128 wind power generation time sampling points with different models.

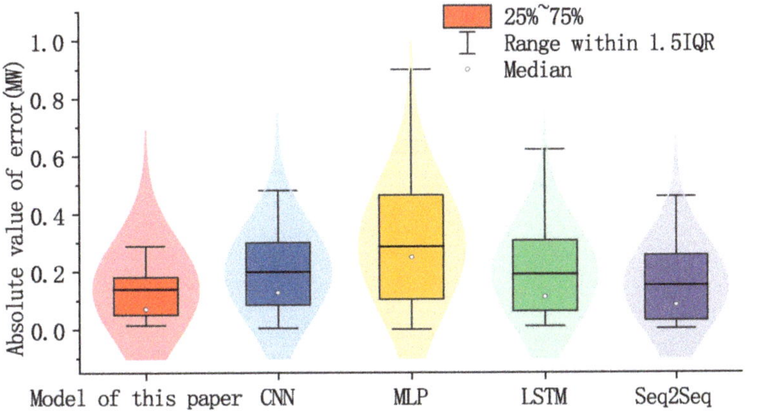

Figure A6. Error results of reconstructing 128 photovoltaic power generation time sampling points with different models.

References

1. Cheng, M.; Zhang, D.; Yan, W.; He, L.; Zhang, R.; Xu, M. Power system abnormal pattern detection for new energy big data. *Int. J. Emerg. Electr. Power Syst.* **2023**, *24*, 91–102. [CrossRef]
2. Majeed Butt, O.; Zulqarnain, M.; Majeed Butt, T. Recent advancement in smart grid technology: Future prospects in the electrical power network. *Ain Shams Eng. J.* **2021**, *12*, 687–695. [CrossRef]
3. Luo, L.; Li, X.; Yang, K.; Wei, M.; Chen, J.; Yang, J.; Yao, L. Intelligent Identification over Power Big Data: Opportunities, Solutions, and Challenges. *CMES - Comput. Model. Eng. Sci.* **2023**, *134*, 1565–1595. [CrossRef]
4. Li, C.; Li, G.; Wang, K.; Han, B. A multi-energy load forecasting method based on parallel architecture CNN-GRU and transfer learning for data deficient integrated energy systems. *Energy* **2022**, *259*, 124967. [CrossRef]
5. Alonso, A.M.; Nogales, F.J.; Ruiz, C. A single scalable LSTM model for short-term forecasting of massive electricity time series. *Energies* **2020**, *13*, 5328. [CrossRef]
6. Zhang, H.; Han, K. A Hybrid Observability Analysis Method for Power System State Estimation. *IEEE Access* **2020**, *8*, 73388–73397. [CrossRef]
7. Zhang, Y.; Shi, X.; Zhang, H.; Cao, Y.; Terzija, V. Review on deep learning applications in frequency analysis and control of modern power system. *Int. J. Electr. Power Energy Syst.* **2022**, *136*, 107744. [CrossRef]
8. Khan, M.F.; Azam, M.; Khan, M.A.; Algarni, F.; Ashfaq, M.; Ahmad, I.; Ullah, I. A Review of Big Data Resource Management: Using Smart Grid Systems as a Case Study. *Wirel. Commun. Mob. Comput.* **2021**, *2021*, 3740476. [CrossRef]
9. Adnan, F.A.; Jamaludin, K.R.; Wan Muhamad, W.Z.A.; Miskon, S. A review of the current publication trends on missing data imputation over three decades: Direction and future research. *Neural Comput. Appl.* **2022**, *34*, 18325–18340. [CrossRef]
10. Nguyen, T.; Nguyen-Duy, K.M.; Nguyen, D.H.M.; Nguyen, B.T.; Wade, B.A. DPER: Direct Parameter Estimation for Randomly missing data. *Knowl.-Based Syst.* **2022**, *240*, 108082. [CrossRef]
11. Fan, L.; Wehbe, Y. Extended Kalman filtering based real-time dynamic state and parameter estimation using PMU data. *Electr. Power Syst. Res.* **2013**, *103*, 168–177. [CrossRef]
12. Zhao, J.; Zhang, G.; La Scala, M.; Wang, Z. Enhanced robustness of state estimator to bad data processing through multi-innovation analysis. *IEEE Trans. Ind. Inform.* **2017**, *13*, 1610–1619. [CrossRef]
13. Huang, C.; Li, F.; Zhan, L.; Xu, Y.; Hu, Q.; Zhou, D.; Liu, Y. Data quality issues for synchrophasor applications Part II: Problem formulation and potential solutions. *J. Mod. Power Syst. Clean Energy* **2016**, *4*, 353–361. [CrossRef]
14. Biswas, S.; Centeno, V.A. A Model-Agnostic Method for PMU Data Recovery Using Optimal Singular Value Thresholding. *IEEE Trans. Power Deliv.* **2022**, *37*, 3302–3312. [CrossRef]
15. Yang, T.; Li, Y.; He, Z.; Han, X.; Pen, H.; Lu, Y. Matrix Completion Theory Based Recovery Algorithm for Power Quality Data in Ubiquitous Power Internet of Things. *Autom. Electr. Power Syst.* **2020**, *44*, 13–21.
16. Hong, W.; Li, Q.; Zhang, Y.; Yang, Y. Quadratic matrix completion based phase sequence identification algorithm for low-voltage distribution network. *Electr. Power Autom. Equip.* **2022**, *42*, 133–138.
17. Osipov, D.; Chow, J. PMU Missing Data Recovery Using Tensor Decomposition. In Proceedings of the 2021 IEEE Power & Energy Society General Meeting (PESGM), Orlando, FL, USA, 16–20 July 2021; p. 1. [CrossRef]
18. Guo, P.; Qi, B.; Liu, S.; Feng, X. Missing air quality data supplementation algorithm incorporating bi-directional GRU. *Small Microcomput. Syst.* **2022**, *43*, 1345–1349. [CrossRef]
19. Wang, C.; Cao, Y.; Zhang, S.; Ling, T. A Reconstruction Method for Missing Data in Power System Measurement Based on LSGAN. *Front. Energy Res.* **2021**, *9*, 651807. [CrossRef]

20. Wang, Y.; Bai, D. Application of wavelet threshold method based on optimized VMD to ECG denoising. In Proceedings of the 2021 IEEE 3rd International Conference on Frontiers Technology of Information and Computer (ICFTIC), Greenville, SC, USA, 12–14 November 2021; pp. 741–744. [CrossRef]
21. Li, C.; Wu, Y.; Lin, H.; Li, J.; Zhang, F.; Yang, Y. ECG Denoising Method Based on an Improved VMD Algorithm. *IEEE Sens. J.* **2022**, *22*, 22725–22733. [CrossRef]
22. Zhang, Y.; Zhao, Y.; Gao, S. A Novel Hybrid Model for Wind Speed Prediction Based on VMD and Neural Network Considering Atmospheric Uncertainties. *IEEE Access* **2019**, *7*, 60322–60332. [CrossRef]
23. Wang, J.N.; Cui, J.F.; Chen, C.L. A Prediction Method of Consumer Buying Behavior Based on Attention Mechanism and CNN-BiLSTM. *J. Netw. Intell.* **2022**, *7*, 375–385.
24. Pan, H.; Ye, Z.; He, Q.; Yan, C.; Yuan, J.; Lai, X.; Su, J.; Li, R. Discrete Missing Data Imputation Using Multilayer Perceptron and Momentum Gradient Descent. *Sensors* **2022**, *22*, 5645. [CrossRef]
25. Alsharekh, M.F.; Habib, S.; Dewi, D.A.; Albattah, W.; Islam, M.; Albahli, S. Improving the Efficiency of Multistep Short-Term Electricity Load Forecasting via R-CNN with ML-LSTM. *Sensors* **2022**, *22*, 6913. [CrossRef]
26. Ghimire, S.; Nguyen-Huy, T.; Deo, R.C.; Casillas-Perez, D.; Salcedo-Sanz, S. Efficient daily solar radiation prediction with deep learning 4-phase convolutional neural network, dual stage stacked regression and support vector machine CNN-REGST hybrid model. *Sustain. Mater. Technol.* **2022**, *32*, e00429. [CrossRef]
27. Chung, J.; Gulcehre, C.; Cho, K.; Bengio, Y. Gated feedback recurrent neural networks. In Proceedings of the International Conference on Machine Learning (PMLR), Lille, France, 7–9 July 2015; Volume 3, pp. 2067–2075.
28. Cao, Y.; Hao, Y.; Li, B.; Xue, J. Depression prediction based on BiAttention-GRU. *J. Ambient Intell. Humaniz. Comput.* **2022**, *13*, 5269–5277. [CrossRef]
29. Xiao, Y.; Zou, C.; Chi, H.; Fang, R. Boosted GRU model for short-term forecasting of wind power with feature-weighted principal component analysis. *Energy* **2023**, *267*, 126503. [CrossRef]
30. Hakami, N.A.; Mahmoud, H.A.H. A Dual Attention Encoder-Decoder Text Summarization Model. *Comput. Mater. Contin.* **2023**, *74*, 3697–3710. [CrossRef]
31. Ying, W. Gated Recurrent Unit Based On Feature Attention Mechanism For Physical Behavior Recognition Analysis. *J. Appl. Sci. Eng.* **2023**, *26*, 357–365.

entropy

MDPI

Article

An Improved Deep Reinforcement Learning Method for Dispatch Optimization Strategy of Modern Power Systems

Suwei Zhai [1], Wenyun Li [2], Zhenyu Qiu [3], Xinyi Zhang [3] and Shixi Hou [3,*]

[1] Electric Power Research Institute of China Southern Power Grid Yunnan Power Grid Co., Ltd., Kunming 650217, China
[2] Yunnan Power Dispatching Control Center of China Southern Power Grid, Kunming 650011, China
[3] College of IOT Engineering, Hohai University, Nanjing 210098, China
* Correspondence: houshixi@hhu.edu.cn

Abstract: As a promising information theory, reinforcement learning has gained much attention. This paper researches a wind-storage cooperative decision-making strategy based on dueling double deep Q-network (D3QN). Firstly, a new wind-storage cooperative model is proposed. Besides wind farms, energy storage systems, and external power grids, demand response loads are also considered, including residential price response loads and thermostatically controlled loads (TCLs). Then, a novel wind-storage cooperative decision-making mechanism is proposed, which combines the direct control of TCLs with the indirect control of residential price response loads. In addition, a kind of deep reinforcement learning algorithm called D3QN is utilized to solve the wind-storage cooperative decision-making problem. Finally, the numerical results verify the effectiveness of D3QN for optimizing the decision-making strategy of a wind-storage cooperation system.

Keywords: wind farm; energy storage system; reinforcement learning; deep neural networks

Citation: Zhai, S.; Li, W.; Qiu, Z.; Zhang, X.; Hou, S. An Improved Deep Reinforcement Learning Method for Dispatch Optimization Strategy of Modern Power Systems. *Entropy* **2023**, *25*, 546. https://doi.org/10.3390/e25030546

Academic Editor: Luis Hernández-Callejo

Received: 26 January 2023
Revised: 15 March 2023
Accepted: 15 March 2023
Published: 22 March 2023

1. Introduction

Since the beginning of the 21st century, higher requirements for energy conservation, emission reduction, and sustainable development have been put forward as a result of the increasing pressure from the use of global resources. Thus, clean energy has gained much attention, which further accelerates the global energy transformation [1–3]. At present, the commonly used clean energy sources include wind energy, solar energy, and tidal energy. Among these clean energy sources, wind energy outperforms with its rich resources, low cost, and relatively mature technology [4,5].

However, because of the great correlation between wind energy and environmental information, its power generation is characterized by randomness, uncontrollability, and volatility, which seriously affects the power balance and threatens the stable and safe operation [6]. Equipping the wind farm with an energy storage system can alleviate the above problems to a certain extent [7–10]. Therefore, how to realize a high-efficient wind-storage cooperative decision-making is a key issue for promoting the full absorption of wind energy [11,12].

Reinforcement learning, also known as a promising information theory, is a machine learning method based on environmental feedback information [13,14]. Its decision theory is very suitable for issues containing complex environments and multiple variables. At present, some studies have proven the feasibility and effectiveness of the energy allocation strategy using reinforcement learning in the field of power system, such as load frequency control on the generation side and market competition strategy [15–18].

Despite several works that have proposed reinforcement learning methods for wind-storage cooperative decision-making, some issues still exist, as follows:

(1) The flexible loads embedded in the wind-storage cooperative framework have not been developed sufficiently in the existing literature. In [11,19–21], the authors did not

focus on the favorable effect of the flexible loads in the proposed wind-storage model. As an example, flexible loads were considered in [22], where the benefits from the suitable management of demand-side flexible loads were validated. However, the detailed formula for when the load in the price response load model should be shifted was not given.

(2) The exploration of reinforcement learning methods for wind-storage cooperative decision-making needs to be enhanced. In [19,20,23,24], a deep Q-learning strategy was considered in wind-storage systems. However, the main mechanism of the deep Q-learning strategy is to select the actions that can obtain the maximum benefits according to the Q values, which are constructed by the state and action. It has been reported that using the same networks to generate the Q values and its maximum estimated value will result in the maximizing deviation issue, which tends to deteriorate the network accuracy.

Motivated by the above analysis, a novel wind-storage cooperative decision-making model including demand-side flexible loads is developed in this paper, which comprehensively considers the direct or indirect control of various power components, improves the reasonable allocation ability of the energy controller, and enhances the economy and stability of the power grid. Moreover, in order to tackle the defects of the traditional deep Q-learning method, the dueling double deep Q-network (D3QN), which is constructed by two networks (the evaluation network and target network), is developed for the wind-storage cooperative decision-making control mechanism in this study.

The remainder of this study is organized as follows: wind-storage cooperative model and D3QN are presented in Section 2. In Section 3, the wind-storage cooperative decision-making algorithm using D3QN is presented. The algorithm evaluation details and the numerical results are presented in Sections 4 and 5. Section 6 presents the conclusions.

2. Wind-Storage Cooperative Model and D3QN

2.1. Wind-Storage Cooperative Decision-Making Model

This study mainly focusses on a wind-storage cooperative model, including wind turbines and energy storage systems, which also is connected to the external power grid.

The architecture of the wind-storage cooperative model is shown in Figure 1. Three layers exist: the electricity layer, information layer, and signal layer. The electricity layer includes a distributed energy resources (DER) based on wind power, an energy storage system (ESS) for the storage and release of wind power energy, a group of thermostatically controlled loads (TCLs), and a group of price responsive loads. The information layer is composed of a two-way communication system between the external power grid, each power module, and the energy controller (EC). Information such as electricity price, as well as the battery charge and discharge status are transmitted in the information layer. The signal layer transmits the control signals sent by the energy controller to each controllable module. The whole system model has three direct control points, namely, the switch control of TCLs, the charging and discharging control of ESS, and the trading control of energy on the external power grid.

At the same time, the whole wind-storage cooperative model can also be regarded as a multi-agent system. Each module in the system is regarded as an autonomous agent, which can interact with the environment and other agents. Moreover, the simple or complex behavior of each agent is controlled by its internal model. The models used in each module of the whole wind-storage cooperation model will be introduced in detail below.

2.1.1. External Power Grid

Because of the intermittent and uncontrollable characteristics of DER, the use of DER alone may not be able to balance the relationship between supply and demand in the power grid. Therefore, the external power grid is considered as the regulatory reserve in this system model. The external power grid can provide electric energy immediately when the wind-storage energy is insufficient, and the external power grid can also accept the excess electricity when the wind energy is in surplus. The transaction price is defined by the real-time price in the power market. The market prices are expressed as (P_t^u, P_t^d), where P_t^u and P_t^d represent the increased and decreased price, respectively.

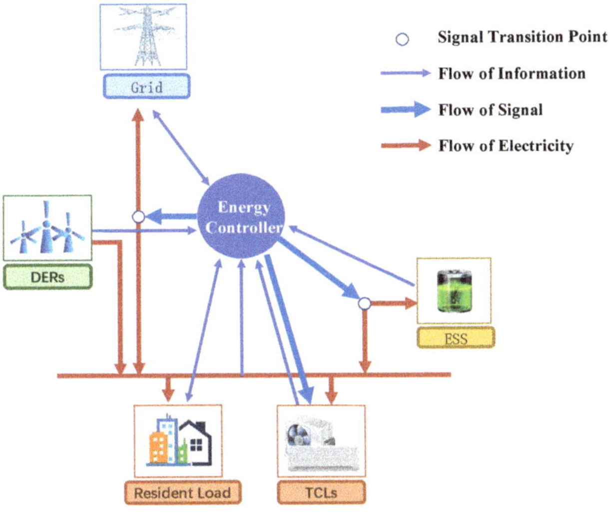

Figure 1. Wind-storage cooperative model.

2.1.2. Distributed Energy Module

Wind turbines are considered as the distributed energy equipment in this study. Specifically, actual wind data from a wind farm in Finland [25] are directly used to construct the model of DER. DER shares the currently generated electric energy information G_t with the energy controller.

2.1.3. Energy Storage System Module

In order to reasonably optimize the allocation of energy and reduce the cost of energy consumption, this study uses the community energy storage system, rather than a separate household storage battery. As a centralized independent energy storage power station invested by a third party, the community energy storage system can integrate and optimize the allocation of the dispersed energy storage resources from the power grid side, power supply side, and user side.

For each time step t, the dynamic model of ESS is defined as follows [26]:

$$B_t = B_{t-1} + \eta_c C_t - \frac{D_t}{\eta_d} \tag{1}$$

where $B_t \in [0, B_{\max}]$ is the electric energy stored by ESS at time t, and $B_{\max}$ is the maximum storage capacity of ESS. η_c and η_d are the charging and discharging efficiency coefficients of energy storage equipment, respectively, and $(\eta_c, \eta_d) \in (0,1]^2$. The variables $C_t \in [0, C_{\max}]$ and $D_t \in [0, D_{\max}]$ represent charge and discharge power, respectively, which are limited by the maximum charge and discharge rate $C_{\max}$ and $D_{\max}$ of ESS, respectively.

The state-of-charge variable of ESS is defined as BEC:

$$BEC_t = \frac{B_t}{B_{\max}} \times 100\% \tag{2}$$

When the energy controller releases the charging signal, ESS obtains the current electricity stored in the battery and verifies the feasibility of the charging operation by referring to the maximum storage capacity $B_{\max}$ and the maximum charging rate $C_{\max}$.

Then, ESS stores the corresponding electricity according to the actual situation and the remaining excessive electricity will be sold to the external power grid. When ESS receives the discharge signal, it verifies the relevant conditions again to judge the operational feasibility and provides the electricity accordingly. If ESS cannot fully provide the requested electricity, the insufficient part will be automatically provided by the external power grid, and the agent will need to pay the relevant costs.

2.1.4. Thermostatically Controllable Load

Thermostatically controllable loads (TCLs) are characterized by their large size, flexible control, and energy conservation. In this study, it is assumed that the vast majority of households are equipped with TCLs, such as air conditioners, water heaters, and refrigerators. These TCLs can be directly controlled in each time unit t and the control signal comes from the TCL aggregator. As EC directly controls TCL equipment, this study defines that TCL will only be charged for power generation costs C_{gen} in order to compensate TCL users. To maintain the comfort of users, each TCL is equipped with a backup controller, which can keep the temperature within an acceptable range. The backup controller receives the on/off operation u_t^i from the TCL aggregator and modifies its action by verifying the temperature constraints. The specific definitions are as follows:

$$u_{b,t}^i = \begin{cases} 0 & if \quad T_t^i > T_{\max}^i \\ u_t^i & if \quad T_{\min}^i < T_t^i < T_{\max}^i \\ 1 & if \quad T_t^i < T_{\min}^i \end{cases} \tag{3}$$

where $u_{b,t}^i$ is the on/off action of the ith TCL backup controller at t, T_t^i is the operating temperature of the ith TCL at t, and $T_{\max}^i$ and $T_{\min}^i$ are the upper and lower temperature boundaries set by the client, respectively. The differential equation of the temperature change in the building is designed as follows [27]:

$$\dot{T}_t^i = \frac{1}{C_a^i}(T_t^0 - T_t^i) + \frac{1}{C_m^i}(T_{m,t}^i - T_t^i) + L_{TCL}^i u_{b,t}^i + q^i \tag{4}$$

$$\dot{T}_{m,t}^i = \frac{1}{C_m^i}(T_t^i - T_{m,t}^i) \tag{5}$$

where T_t^i, $T_{m,t}^i$, and T_t^0 are the indoor air temperature, indoor solid temperature, and outdoor air temperature at t, respectively, C_a^i and C_m^i are expressed as the equivalent heat capacity of indoor air and solid, respectively, q^i is the thermal power provided by indoor temperature control equipment, and L_{TCL}^i is the rated power of TCL.

Finally, the state of charge (SoC) is used to represent the relative position of the current temperature T_t^i within the expected temperature range. The SoC of each TCL at t is defined as follows:

$$SoC_t^i = \frac{(T_t^i - T_{\min}^i)}{(T_{\max}^i - T_{\min}^i)} \tag{6}$$

2.1.5. Resident Price Response Load

Some power demands exist from household that the energy controller cannot directly control in the residential load [28]. This study assumes that the daily electricity consumption of residents is composed of the daily basic electricity consumption and the flexible load affected by the electricity price. The flexible load can operate in advance or later within the acceptable time range and can be transferred according to the power generation situation of DER, such that the resource utilization rate can be improved and the household electricity expenditure can also be reduced. In this module, each household i has a sensitivity factor $\beta_i \in (0,1)$ and a patience parameter λ_i, in which the sensitivity factor β represents the percentage of load that can be operated in advance or later when the price decreases or increases, and the patience parameter λ represents the hours to repay the transferred load.

For example, when the electricity price is high, this part of the load can be cut now and operated after λ_i.

At t, the load L_t^i of household i is modeled by the following formula:

$$L_t^i = L_{b,t} - SL_t^i + PB_t^i \quad (7)$$

$$SL_t^i = L_{b,t} * \beta_i * \delta_t \quad (8)$$

where $L_{b,t}$ represents the daily basic load of residents, $L_{b,t} > 0$, and $L_{b,t}$ follows the daily consumption pattern, which can be inferred from the average daily consumption curve of residential areas. SL_t^i is the shift load (SL) defined by (8), where δ_t represents the electricity price level at t. Therefore, SL_t^i is positive when the price is high, i.e., $\delta_t > 0$, then $SL_t^i > 0$, and when the price is low, i.e., $\delta_t < 0$, then $SL_t^i < 0$. The positive transfer load will be repaid after a certain period of time λ. The negative transfer load is the electricity provided in advance, so it will exist in the future. The loads to be compensated can be formulated as follows:

$$PB_t^i = \sum_{j=0}^{t-1} \omega_{i,j} * SL_j^i \quad (9)$$

where $\omega_{i,j} \in \{0,1\}$ represents the compensation degree for the transferred load at j. Generally, the closer t minus j is to λ_i, the higher $\omega_{i,j}$ is. In addition, the compensation action also should be related to the electricity price, i.e., $\omega_{i,j}$ becomes smaller when $\delta_t > 0$. Therefore, $\omega_{i,j}$ can be designed as follows:

$$\omega_{i,j} = clip\left(\frac{-\delta_t * sign(SL_j^i)}{2} + \frac{t-j}{\lambda_i}, 0, 1\right) \quad (10)$$

$$clip(\mathrm{X,a,b}) = \begin{cases} a & if \quad X < a \\ X & if \quad a \le X \le b \\ b & if \quad X > b \end{cases} \quad (11)$$

Given (10), when $\delta_t > 0$, one can obtain $SL_t^i > 0$ and $sign(SL_j^i) > 0$, then $\frac{-\delta_t * sign(SL_j^i)}{2} < 0$, which means that $\omega_{i,j}$ becomes smaller and the positive transfer load almost cannot be compensated in the case of a high price [29,30].

2.1.6. Energy Controller

In this study, EC can extract the information provided by different modules and the observable environment to determine the best supply and demand balance strategy. EC mainly manages the power grid through four control mechanisms, as shown in Figure 2, including TCL direct control, price level control, energy deficiency action, and energy excess action.

(1) TCL direct control

At each time step t, EC will allocate a certain amount of electric energy for TCLs. Then, they will be distributed to each TCL through a TCL aggregator. The TCL aggregator judges the priority of energy distribution according to the power delivered by EC and the SoC of each TCL, and then determines the on/off action of each TCL: TCL with a lower SoC has a higher priority in energy allocation than TCL with a higher SoC. The TCL aggregator also operates as an information aggregator transmitting the real-time average SoC information of the TCL cluster to EC [31]. The specific transmission process is shown in Figure 3.

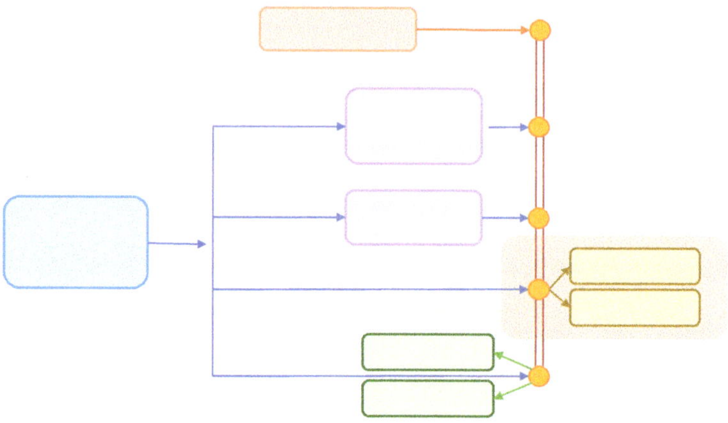

Figure 2. The control mechanism of Energy controller.

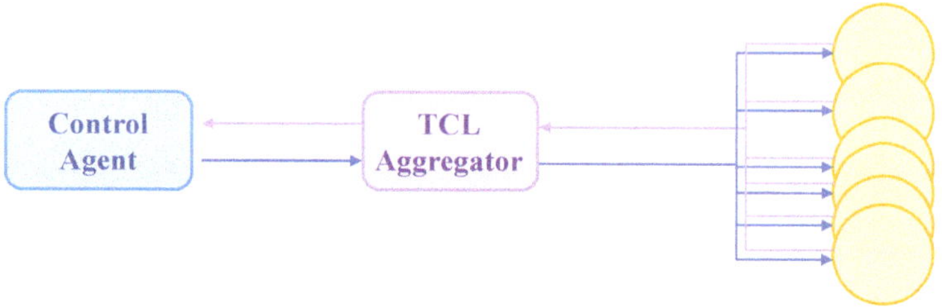

Figure 3. The intermediary role of the TCL aggregator.

(2) Price level control

In order to effectively utilize the elastic benefits of the residential price response load, EC must determine the electricity price level δ_t at each time step t. In order to ensure the competitiveness of the system model proposed in this paper, a pricing mechanism is designed: The price can fluctuate around the median value, but the average price of the daily electricity price P_{avg} cannot exceed 2.9% of the market electricity price provided by power retailers [32]. From a practical point of view, the electricity price at the DR side is discrete, and its fluctuation is affected by the electricity price level δ_t. So, the real-time electricity price is selected from five values:

$$P_t \in (P_{market} + \delta_t * cst) \tag{12}$$

where $\delta_t \in \{-2, -1, 0, 1, 2\}$, cst is the constant to determine the specific increment or reduction in electricity price.

In addition, the model also pays attention to the electricity price level δ_t at each moment. When the sum of the previous electricity price levels is higher than the set threshold, the market electricity price is adjusted to P_{market} instead of the price given by the agent. The effective electricity price level $\delta_{t,eff}$ is defined as follows:

$$\delta_{t,eff} = \begin{cases} \delta_t & if \quad \sum_{j=0}^{t} \delta_t \leq threshold \\ 0 & if \quad \sum_{j=0}^{t} \delta_t > threshold \end{cases} \tag{13}$$

(3) Energy deficiency action

When the power generated from DER cannot meet the power demand, EC can dispatch the energy stored in ESS or purchase energy from an external power grid. EC will determine the energy priority between ESS and an external power grid. In addition, if the high priority energy is ESS but the electricity stored in ESS cannot meet the power demand, the remaining power will be automatically supplied by an external power grid.

(4) Energy excess action

When the electricity generated by local DER exceeds the electricity demand, the excess electricity must be stored in ESS or be sold to an external power grid. In this case, EC also will determine the priority between ESS and the external power grid. If ESS is the preferred option and it has reached the max capacity, the remaining electricity will be automatically transmitted to an external power grid.

2.2. D3QN

In this section, the basic principle of DQN (deep Q-network) and SARSA (state−action−reward−state−action) is presented first.

The train mechanism of DQN can be formulated as follows:

$$Q_{k+1}(s,a) = Q_k(s,a) + \alpha E_k \tag{14}$$

$$E_k = R + \gamma argmax_{a'} Q(s',a') - Q(s,a) \tag{15}$$

Using (14), one can find that the update iteration needs to achieve the approximation of the action-value function value (i.e., $Q_{k+1}(s,a) = Q_k(s,a)$), which means $R + \gamma argmax_{a'} Q(s',a') - Q(s,a) \to 0$. Thus, the DQN network parameters can be updated by minimizing the mean square error loss function in the DQN algorithm.

The difference in the SARSA algorithm lies in how the Q value is updated. Specifically, when the agent with the SARSA algorithm is in the state s, it selects the action a according to the $\varepsilon - greedy$, and then observes the next state s' from the environment, and selects the action a' again. The sequence $\{s,a,r,s',a'\}$ is stored in the empirical replay set, and the calculation of the target Q value also depends on it. The core idea of the SARSA algorithm can be simplified as follows:

$$Q(s,a) \leftarrow Q(s,a) + \alpha[R + \gamma \mathrm{argmax}_{a'} Q(s',a') - Q(s,a)] \tag{16}$$

In the existing study, DQN and SARSA have been developed for the wind-storage cooperative decision-making algorithm. However, both DQN and SARSA use $Q(s,a)$ and $maxQ(s',a')$ produced by the same network to update the Q network parameter ω, which leads to the variation in the timing difference goal and a reduction in the convergence performance. Therefore, in view of the above possible problems, this paper uses the D3QN algorithm to optimize the model decision. The specific improvements are collected as follows:

(1) Referring to the double DQN (DDQN) algorithm, two neural networks with the same structure are constructed as the estimation network $Q(s,a,\omega)$ and the target network $Q'(s,a,\omega')$, respectively. The estimation network is used to select the action corresponding to the maximum Q value, and its network parameters are constantly updated. The target network is used to calculate the target value y, and its network parameters are fixed, but they are updated by using the current estimated network parameters value at regular intervals. The parameters in the target network are fixed for a period of time, which makes the convergence target of the estimated network relatively fixed, which is beneficial to the convergence of the algorithm model, and also avoids the agent selecting the overestimated suboptimal action. The overestimation problem of the DQN algorithm can also be effectively solved.

(2) In this paper, the structure of the deep neural network is adjusted. Referring to dueling DQN based on competitive architecture, the main output is divided into two parts: one part is the state-value function $V(S,\omega,\alpha)$, which represents the current state; the other

part is the advantage function $A(S, A, \omega, \beta)$, which judges the additional value level of each action for the current state. The neural network structure of DQN is shown in Figure 4, and the neural network structure of D3QN is shown in Figure 5.

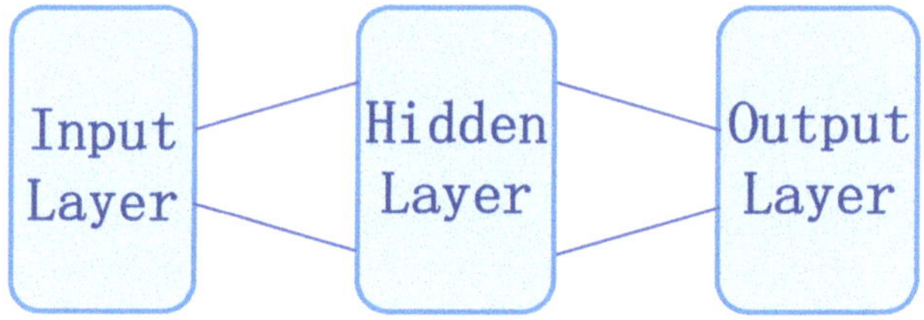

Figure 4. The network structure of DQN.

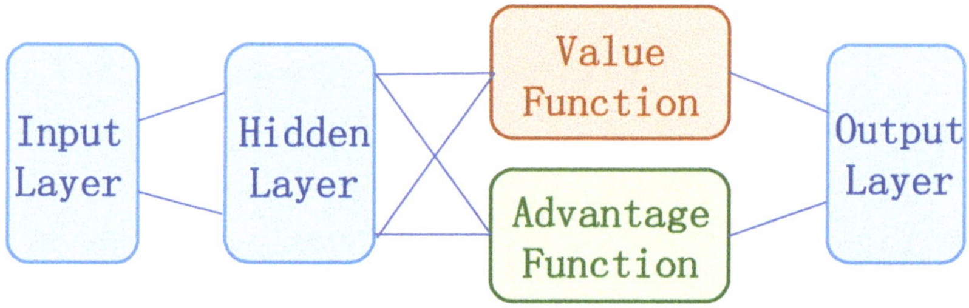

Figure 5. The network structure of D3QN.

Finally, the output of the Q network is obtained by the linear combination of the output of the state-value function network and the advantage function network:

$$Q(S, A, \omega, \alpha, \beta) = V(S, \omega, \alpha) + A(S, A, \omega, \beta) \tag{17}$$

However, (17) cannot identify the respective functions of $V(S, \omega, \alpha)$ and $A(S, A, \omega, \beta)$ in the final output. In order to reflect this identifiability, the advantage function is generally set as the single action advantage function minus the average value of all of the action advantage functions in a certain state, so it can be modified as follows:

$$\begin{aligned} Q(S, A, \omega, \alpha, \beta) = V(S, \omega, \alpha) + \\ A(S, A, \omega, \beta) - \tfrac{1}{A} \sum_{a' \in A} A(S, a', \omega, \beta) \end{aligned} \tag{18}$$

The flow chart of D3QN is shown in Figure 6:

In Figure 6, the D3QN algorithm stores the experience gained from the interaction in the experience pool one by one. After a certain amount is accumulated, the model randomly extracts a certain batch of data from the experience pool in each step to train the neural network. These randomly extracted experiences break the correlation between data, improve the generalization performance, and benefit from the stability of network training. Meanwhile, in Figure 6, the D3QN algorithm constructs two neural networks with the same structure, namely, the estimated network $Q_E(S, A, \omega, \alpha, \beta)$ and the target network $Q_T(S, A, \omega', \alpha', \beta')$. The estimated network is used to select the action and parameter ω is updated constantly. The target network is used to calculate the temporal difference of the target value. Parameter ω' is fixed and replaced with the latest estimated network parameter ω at regular intervals. ω' remains unchanged for a period of time, resulting in a relatively fixed convergence goal of the estimated network Q_E, which is beneficial for convergence. The actions of the maximum function generated by the estimated network and the target network are not necessarily the same. Using Q_E to generate actions and Q_T to calculate the target value can prevent the model from selecting the overestimated sub-optimal actions and can effectively solve the overestimation problem of the DQN algorithm.

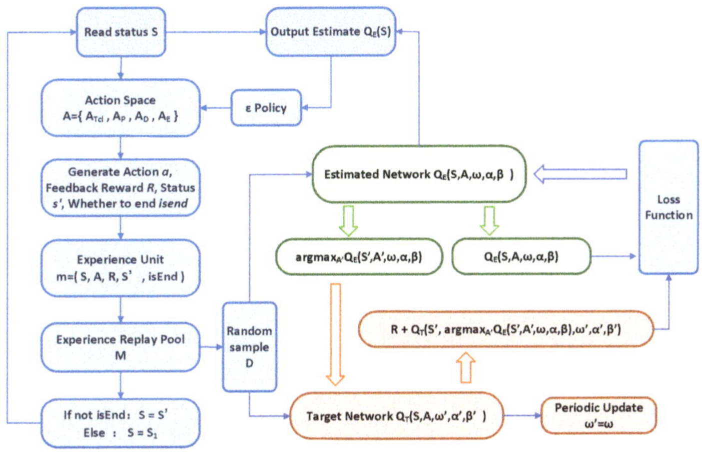

Figure 6. The flow chart of D3QN.

3. Wind-Storage Cooperative Decision-Making Based on D3QN

In this section, wind-storage cooperative model will be converted into a discrete Markov decision-making process (MDP). According to the reinforcement learning mechanism, the one-day state of the model is discretized into 24 states. In addition, the MDP in this paper takes the online environmental information as the state space, the set of command actions executed by the energy controller as the action space, and the income of electricity sellers as the reward function. The interaction process between the energy controller and the system power environment is shown in Figure 7.

3.1. State Space

The state space is composed of the information that the agent needs to use when making decisions at each time step t, including the controllable state component S^C, the external state component S^X, and the time-dependent component S^T. The controllable state information includes all environmental variables that the agent can directly or indirectly affect. In this study, the controllable state information is composed of TCL's average SoC, ESS's charge and discharge state BSC_t, and the pricing counter C_t^b [33]. The external state information consists of all variables, such as the temperature information T_t, the wind power generation G_t, and the electricity price P_t^u. When the algorithm is implemented, the external state information directly uses the real data set, so it is assumed that the controller can accurately predict the values of three variables in the next moment. The time-dependent component information includes the information strongly related to time in the model, where $L_{b,t}$ represents the current load value based on the daily consumption mode, and t represents the hours of the day.

The state space is expressed as follows:

$$s_t \in S = S^C \times S^X \times S^T \tag{19}$$

$$s_t = [SoC_t, BSC_t, C_t^b, T_t, G_t, P_t^u, L_{b,t}, t] \tag{20}$$

In the implementation process, the electricity price is not given directly. Firstly, the initial electricity price is set. When the price should be increased or decreased, the pricing counter C_t^b will be added or subtracted by 1. Then, the electricity price becomes the initial price plus the product between C_t^b and the unit electricity price.

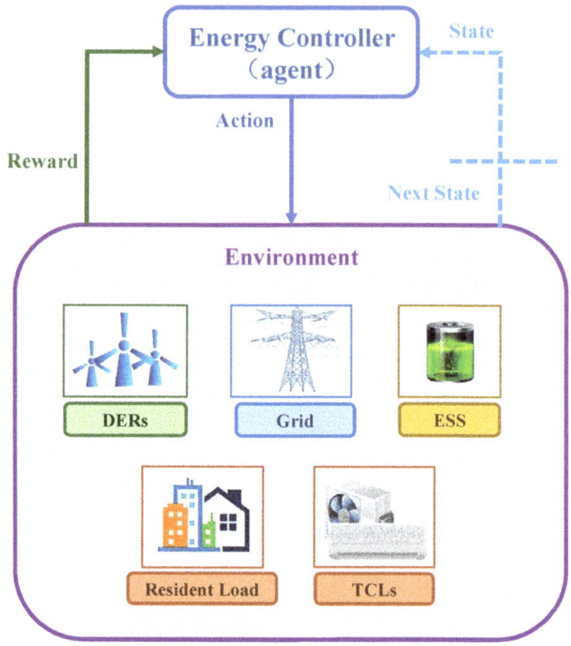

Figure 7. Interaction process between the energy controller and the system environment.

3.2. Action Space

The action space consists of four parts: TCL action space A_{tcl}, price action space A_P, energy shortage action space A_D, and energy excess action space A_E. Among them, the TCL action space consists of four possible actions. The price action space consists of five possible actions. There are two possible actions in the energy shortage and excess action space, that is, the priority between ESS and the external power grid. Therefore, the whole action space contains 80 potential combinations of these actions, which can be expressed as follows:

$$a_t = (a_{tcl}, a_P, a_D, a_E)_t \quad (21)$$

$$a_t \in A = A_{tcl} \times A_P \times A_D \times A_E \quad (22)$$

3.3. Reward Function and Penalty Function

The main form of deep reinforcement learning (DRL) to solve problems is to maximize the reward function. The purpose of using DRL in this paper is to maximize the economic profits of the electricity sellers. Thus, the reward value can be selected as the operating gross profit, i.e., the income from selling electricity to the demand-side and the external power grid minus the cost of wind power generation and purchasing electricity from an external power grid. Therefore, the reward function R_t and penalty function $Costs_t$ are defined as follows:

$$R_t = Rev_t - Costs_t \quad (23)$$

$$Rev_t = P_t \sum_{loads} L_t^i + C_{gen} \sum_{TCLs} L_{TCL}^i u_{b,t}^i + P_t^d E_t^S \quad (24)$$

$$Costs_t = C_{gen} G_t + (P_t^u + C_{tr_{imp}}) E_t^P + C_{tr_{exp}} E_t^S \quad (25)$$

where C_{gen} is the energy price charged to TCL, and it is also the cost of wind power generation. G_t refers to the wind power generation amount. P_t^d and P_t^u are the decreased price and increased price respectively, i.e., the energy price sold to or purchased from an external power grid [25]. E_t^S and E_t^P are the amount of energy sold to or purchased from an external power grid, respectively. $C_{tr_{imp}}$ and $C_{tr_{\exp}}$ are the power transmission costs from the interaction with the external power grid.

4. Implementation Details

Before the algorithm evaluation, implementation details are given in this section.

The computer configuration and environment configuration are collected as Widows11, python3.8, tensorflow1.14; CPU is AMD R7-5800H; GPU is RTX3060; and the memory is 16 GB.

The network structure of the DQN and SARSA algorithms consists of an input layer, two fully connected hidden layers, and an output layer. The activation function of neurons is the ReLU function. In addition, in order to prevent the phenomenon of over fitting after model training, this paper applies the dropout section for neural network training. The number of neurons in the network input layer is the same as the dimension of the system state space, and the number of neurons in the output layer is the same as the dimension of the system action space. The D3QN algorithm adds a competitive network to the structure of the first two algorithms, diverting the abstract value obtained from the full connection layer into two branches. The upper path is the state value function $V(\mathrm{s})$, which represents the value of the state environment itself, and the lower path is the state dependent action advantage function $\mathrm{A}(\mathrm{s}, \mathrm{a})$, which represents the additional value brought by selecting an action in the current state. Finally, these two paths are aggregated to obtain the Q value of each action. This competitive structure can theoretically learn the value of the environmental state without the influence of action, making the practice effect better.

In the training process of the neural network, the discarding rate in dropout is 70%, the sample storage capacity of experience playback set is 500, the scale batch used for each small batch is 200, the reward attenuation coefficient is 0.9, and the target network update interval N is 200. The detailed network structure diagrams of DQN, SARSA, and D3QN are shown in Figures 8 and 9.

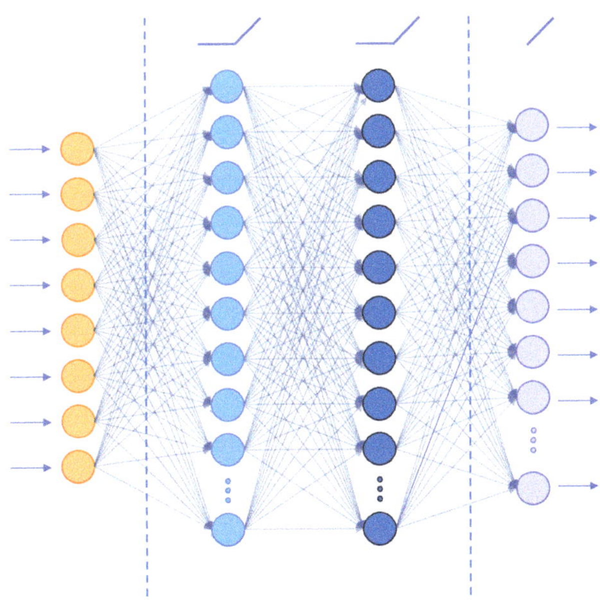

Figure 8. Network structure diagram of the DQN algorithm and SARSA algorithm.

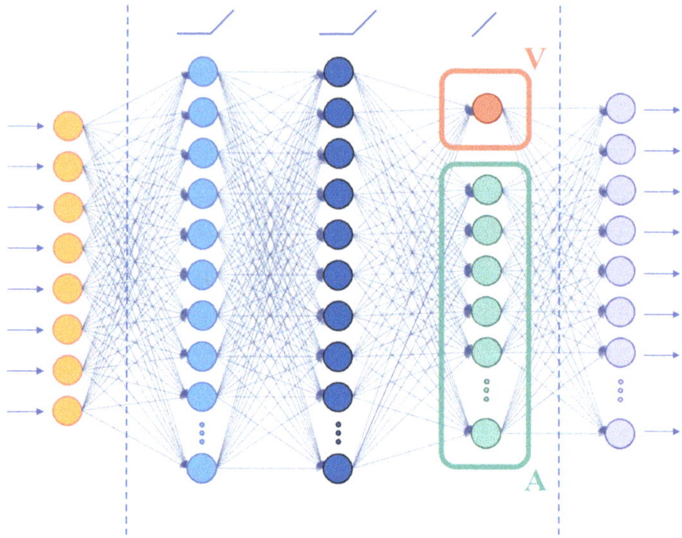

Figure 9. Network structure diagram of the D3QN algorithm.

The proposed decision-making algorithm will be deployed in the cloud server for real-world applications. Generally, the cloud sever possesses enough computational power to execute the DL-based methods.

5. Algorithm Evaluation

In this section, the simulation evaluation is presented to validate the proposed control mechanism. This paper selects the wind power data of a wind farm in Finland. In the wind-storage cooperative model, the control cycle of ESS is 1 day, i.e., 24 intervals. In addition, the parameters involved in the whole system model are summarized in Table 1.

Table 1. Parameters in the system model.

Parameter	Value
ESS	
η_c	0.9
η_d	0.9
$C_{\max}$	250 kW
$D_{\max}$	250 kW
$B_{\max}$	500 kWh
DER	
G_t	1% of the hourly wind power generation (kW)
C_{gen}	32 €/MW
Power grid	
p_t^d	Reduced electricity prices
p_t^u	Increased electricity prices
$C_{tr_{imp}}$	9.7 €/MW
$C_{tr_{\exp}}$	0.9 €/MW

Table 1. *Cont.*

Parameter	Value
TCL	
N_{tcls}	100 (Number of TCL)
T_t^0	Outdoor temperature hourly
C_a^i	$\mathcal{N}(0.004, 0.0008)$
C_m^i	$\mathcal{N}(0.3, 0.004)$
q^i	$\mathcal{N}(0, 0.01)$
L_{TCL}^i	$\mathcal{N}(1.5, 0.01)$ (kW)
$T_{\min}^i$	19
$T_{\max}^i$	35
Load	
N_L	150
N_L	Basic load of residents
λ_i	$\mathcal{N}(10, 6)$ (kW)
β_i	$\mathcal{N}(0.4, 0.3)$
Other parameters	
D	24
δ_t	{−2,−1,0,1,2}
cst	1.5
$threshold$	4
P_{market}	5.48 €/kW
Parameters involved in the algorithm	
N_A	80
A_{tcl}	{0,50,100,150}
A_P	{−2,−1,0,1,2}
A_D	{ESS,Grid}
A_E	{ESS,Grid}
γ	0.9
t	1 h

5.1. Comparisons of Training Results

5.1.1. Penalty Value Curve

The penalty value is composed of the cost of wind power generation, the purchasing power from the external power grid, and power transaction. Figure 10 shows the total cost paid by the wind power producers in each training cycle (episode) during the learning process. The penalty value decreases with the increase in training times and it gradually converges.

It can be seen that the convergence performance of D3QN is superior to its rivals. Although the penalty value using DQN shows a downward and gradual convergence trend, it still vibrates obviously, which is caused by the defects of DQN. D3QN uses two Q networks to calculate the target Q value and the estimated Q value, respectively, which directly reduces the correlation and greatly improves the convergence performance.

5.1.2. Reward Value Curve

Figure 11 shows the reward value curve during the training process, i.e., the income obtained by the wind farm from the external environment in the operation. The specific training time, final reward mean value, and performance improvement rate between the three algorithms are summarized in Table 2. It can be seen that the final reward value of D3QN is higher than that of the other two algorithms, so the overall performance of the system model based on D3QN has been improved.

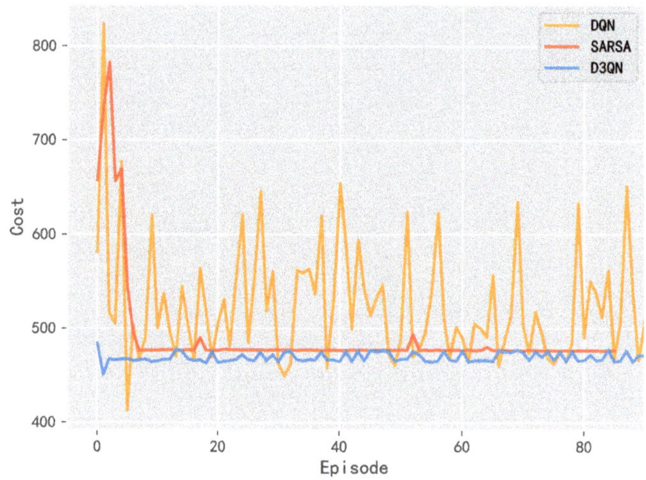

Figure 10. Comparison analysis of the penalty value using DQN, SARSA, and D3QN.

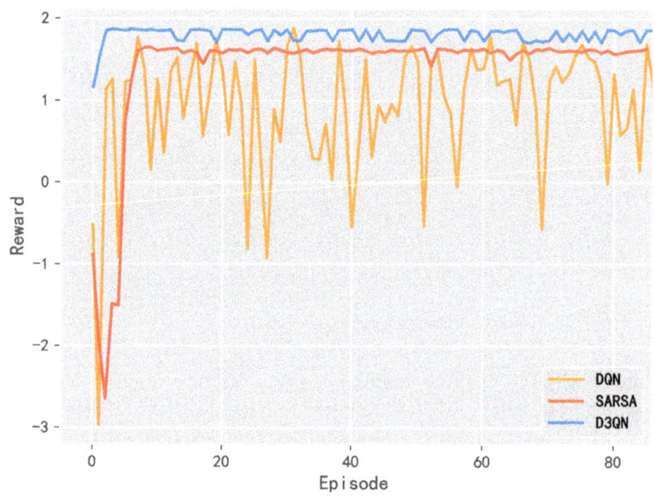

Figure 11. Comparison analysis of reward value curves using DQN, SARSA, and D3QN.

Table 2. Training results between three algorithms.

Algorithm	Training Time (s)	Average Value of Final Reward	Performance Improvement Rate
DQN	196.0111	1.2443	-
SARSA	415.5845	1.6239	30.5%
D3QN	244.1469	1.7909	43.93%

5.2. Comparison of Application Results

5.2.1. 10 Day Revenue Comparison

In order to give a more intuitive understanding of the performance difference for DQN, SARSA, and D3QN, this section selects the data from 10 days in a year, and analyzes the daily total profit obtained by the system model with the three algorithms, as shown in Figure 12.

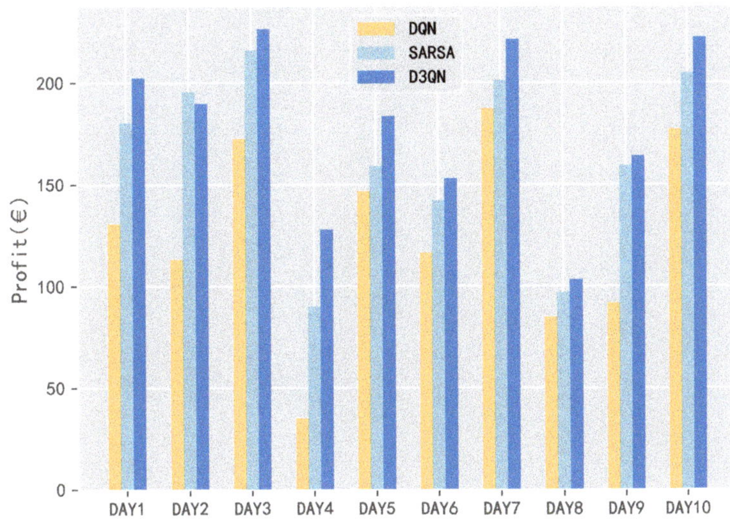

Figure 12. Comparison analysis of the daily income with three DRL algorithms for 10 days in a year.

It can be seen that the daily income using SARSA and D3QN is higher than that of DQN within 10 days. Moreover, the total profit of D3QN is better than that of SARSA in 9 out of 10 days, which also validates the superiority of D3QN.

5.2.2. Daily Electricity Trading Comparison

This section will compare the behavior of the three algorithms in the specific one-day. The one-day data of the environment is shown in Figure 13, including the outdoor temperature, wind power generation, electricity prices, and residential load.

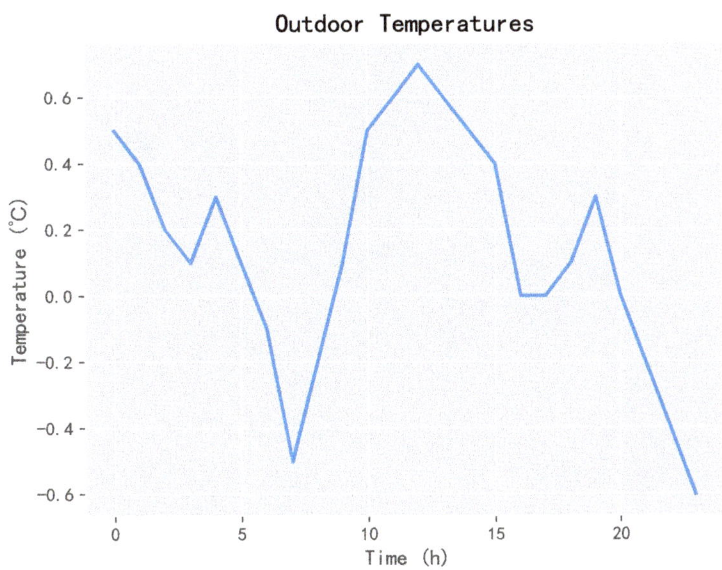

(**a**)

Figure 13. *Cont.*

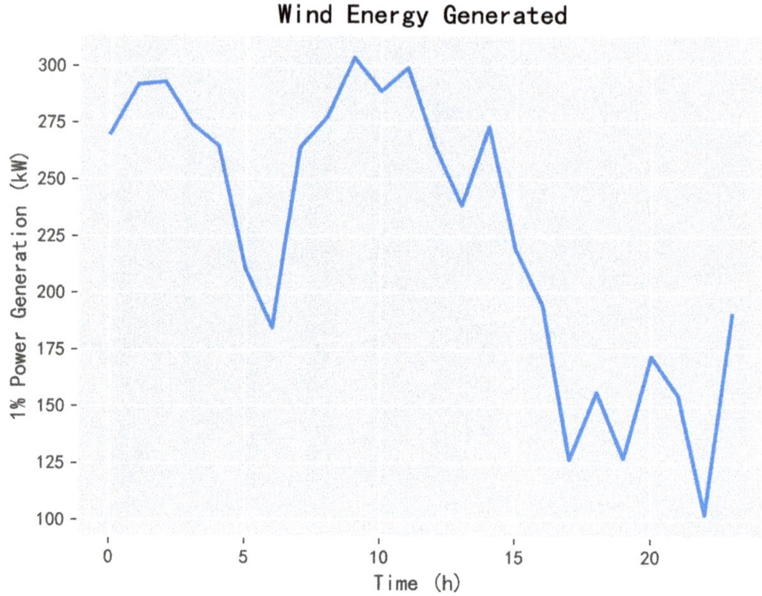

(**b**)

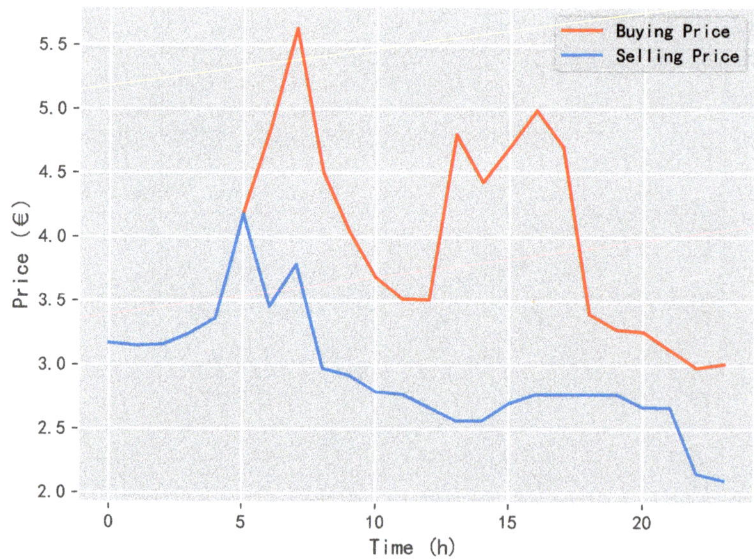

(**c**)

Figure 13. *Cont.*

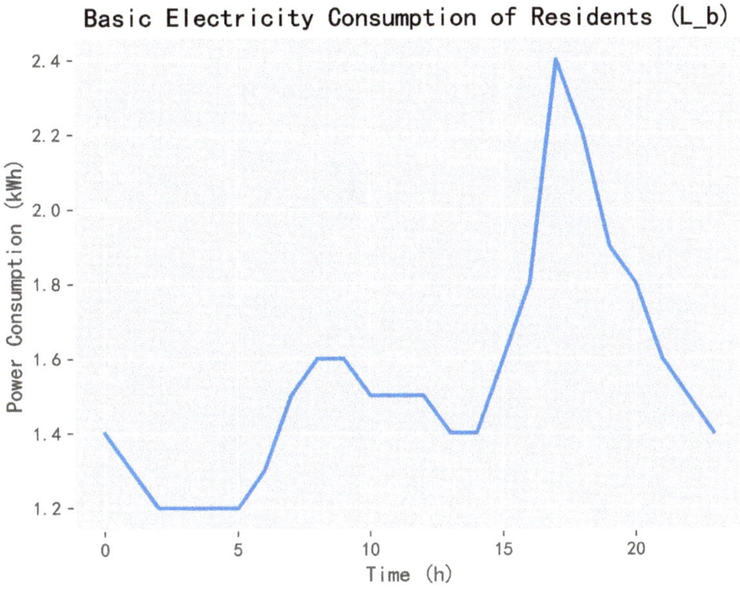

(d)

Figure 13. Environmental data of one-day: (**a**) outdoor temperature, (**b**) energy generated, (**c**) electricity prices, and (**d**) residential loads.

Using DQN, SARSA, and D3QN, one can obtain the energy allocation results of TCLs, the purchased energy, and the sold energy, as shown in Figures 14–16.

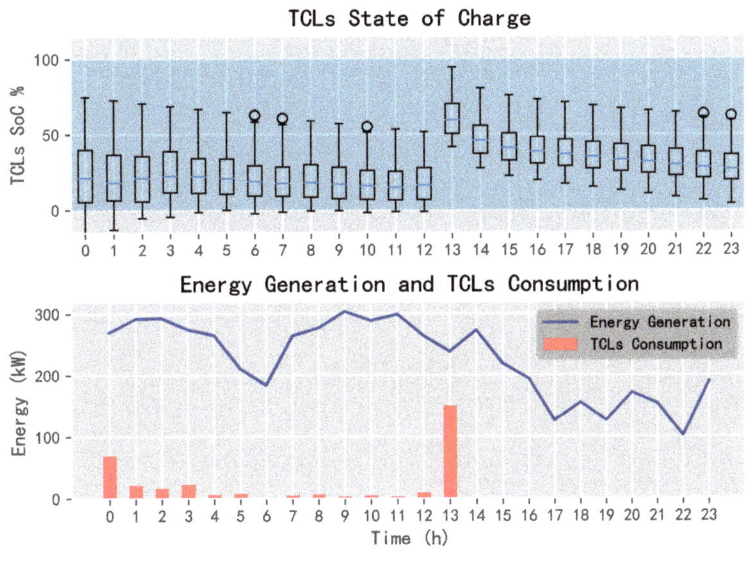

(a)

Figure 14. *Cont.*

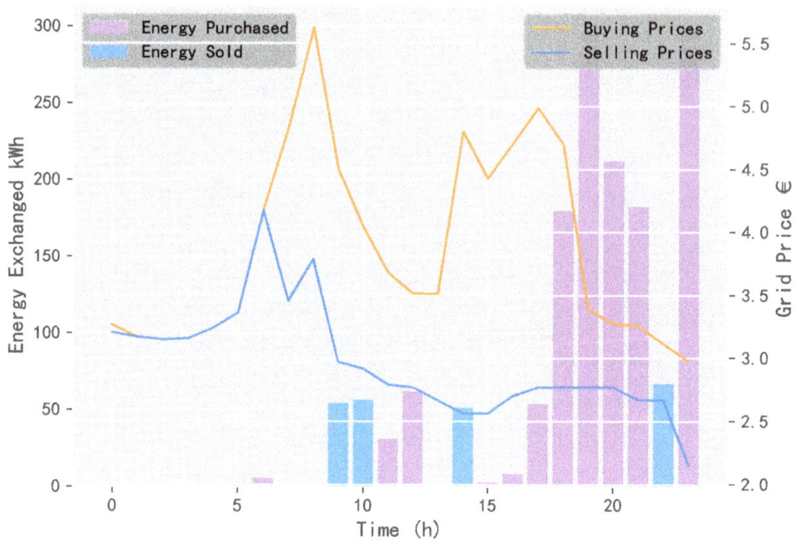

(**b**)

Figure 14. TCLs status and power exchange using DQN: (**a**) TCLs and (**b**) power exchange.

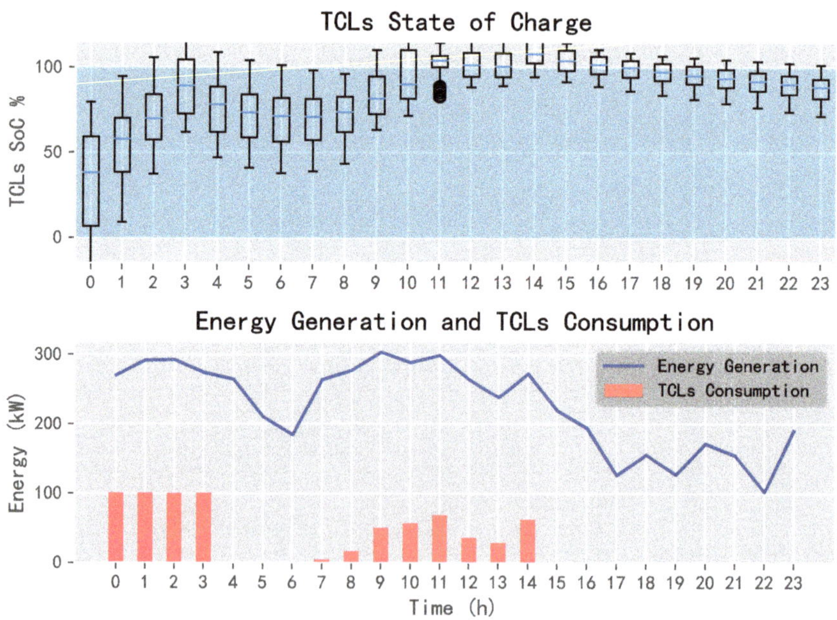

(**a**)

Figure 15. *Cont.*

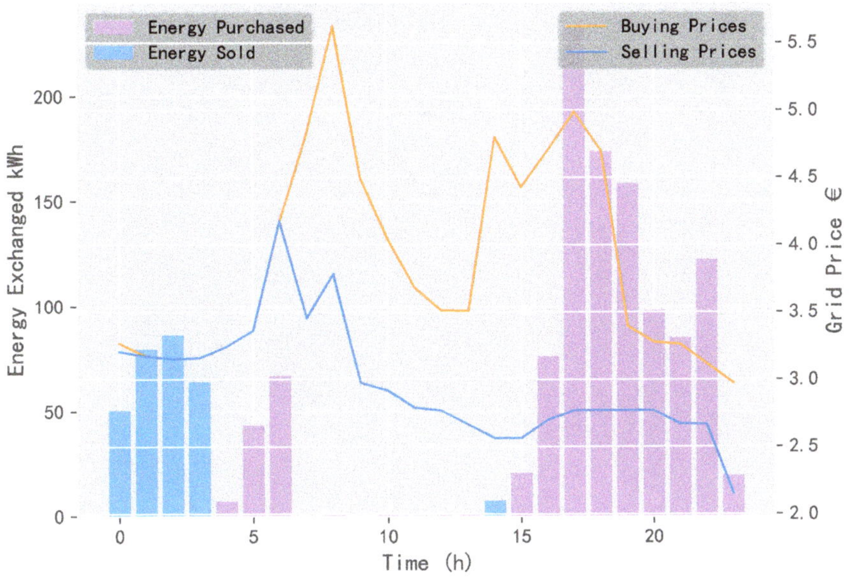

(b)

Figure 15. TCLs status and power exchange using SARSA: (**a**) TCLs and (**b**) power exchange.

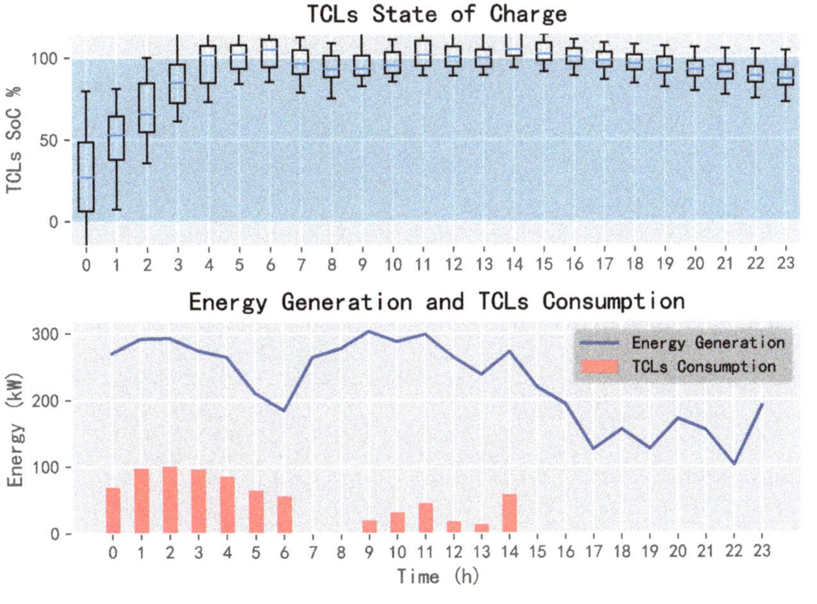

(a)

Figure 16. *Cont.*

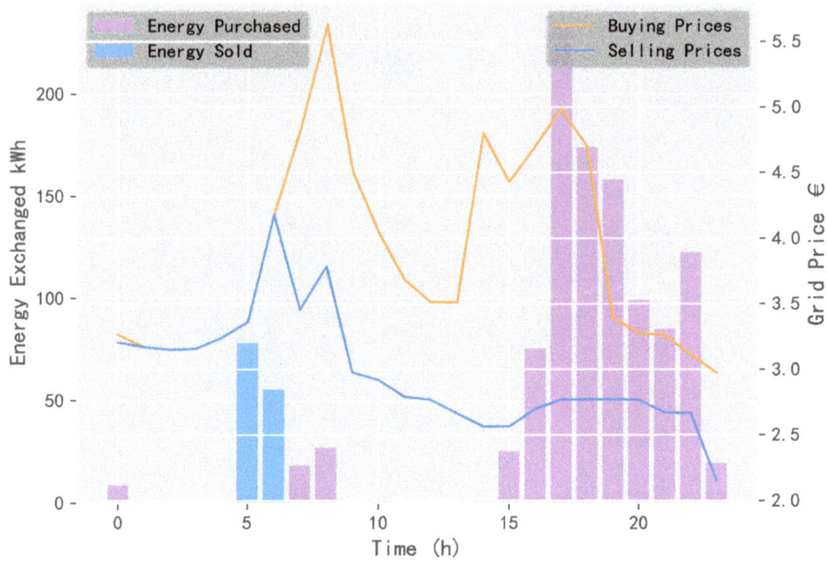

(b)

Figure 16. TCLs status and power exchange using D3QN: (**a**) TCLs and (**b**) power exchange.

In Figures 14–16, the SoC of TCLs reflects the change in indoor temperature for residents. This paper sets the constant temperature range of TCLs as 19~25 °C. When the charging state of TCLs is 0%, it means that the indoor temperature of residents is less than or equal to 19 °C; when the charging state is 100%, it means that the indoor temperature is greater than or equal to 25 °C. It can be seen that SARSA and D3QN can allocate sufficient energy to TCLs when the wind power generation is sufficient, where its state can reach saturation as soon as possible, such that the system can keep the room temperature stable, and gives residents a warm and comfortable experience. In addition, SARSA selects multiple transactions to ensure the income, and D3QN decisively sells a large amount of power to obtain more income when wind energy is sufficient and the electricity price is the highest.

5.2.3. Computational Efficiency Comparison

In order to demonstrate the computational efficiency of the proposed D3QN, the training time, decision-making time, the number of trainable parameters, and performance improvement rate are summarized in Table 3. It takes 196.0111 and 415.5845 s for DQN and SARSA to reach convergence, respectively, while the proposed D3QN takes 244.1469 s. Furthermore, although D3QN possesses the largest number of trainable parameters, the decision-making time of D3QN is close to the other two algorithms, which demonstrates that D3QN can be implemented in real-world applications. From Table 3, one can conclude that the computational cost of D3QN is slightly larger than DQN and SARSA, which is still in an acceptable range. However, it should be noted that it is mainly because of many trainable parameters. Moreover, the performance improvement rate of D3QN is the biggest, which is an important criterion to evaluate different algorithms. Generally, it is worth increasing some computational complexity while the performance can gain enough improvement.

Table 3. Computational efficiency comparison between three algorithms.

Algorithm	Training Time (s)	Decision-Making Time (s)	The Number of Trainable Parameters	Performance Improvement Rate
DQN	196.0111	0.347	8980	-
SARSA	415.5845	0.354	19,080	30.5%
D3QN	244.1469	0.390	27,160	43.93%

6. Conclusions

Considering external conditions such as wind energy resources, demand response load, and market electricity price, this paper puts forward a new research method of wind-storage cooperative decision-making based on the DRL algorithm. The main work of this paper is summarized as follows:

(1) This paper proposes a new wind-storage cooperative model. Based on the conventional model including wind farms, energy storage systems, and external power grids, this paper also takes into account a variety of flexible loads based on demand response, including residential price response loads and thermostatically controllable loads (TCLs). Meanwhile, this model also can be applied to other renewable energy sources, such as photovoltaic power generation, hydroelectric power generation, and thermal power generation.

(2) This paper proposes a new wind-storage cooperative decision-making mechanism using D3QN, which takes the energy controller as the central allocation controller of the system energy, realizing the direct control of TCLs and the indirect control of the residential price response load, and the management of priority between ESS and the external power grid in the case of sufficient or insufficient energy.

(3) It is worth mentioning that the application of the D3QN algorithm is a new attempt in the research field of wind-storage cooperative decision-making. Based on the historical data of wind farm and market electricity prices, the effectiveness of D3QN in dealing with the wind-storage cooperative decision-making problem is verified, and the superior performance of D3QN is also analyzed.

Author Contributions: Conceptualization, S.Z. and W.L.; methodology, S.Z. and W.L.; validation, S.Z., W.L. and X.Z.; writing—original draft preparation, S.Z., W.L. and Z.Q.; writing—review and editing, S.Z. and W.L.; supervision, S.H.; project administration, S.H.; funding acquisition, S.H. All authors have read and agreed to the published version of the manuscript.

Funding: This research was funded by Science and Technology Project of China Southern Power Grid Yunnan Power Grid Co., Ltd., grant number. YNKJXM20220048, China Postdoctoral Science Foundation, grant number. 2021MD703895.

Institutional Review Board Statement: Not applicable.

Data Availability Statement: Data sharing are not applicable.

Conflicts of Interest: The authors declare no conflict of interest.

References

1. Li, J.; Zhao, H. Multi-Objective Optimization and Performance Assessments of an Integrated Energy System Based on Fuel, Wind and Solar Energies. *Entropy* **2021**, *23*, 431. [CrossRef] [PubMed]
2. Bin, L.; Shahzad, M.; Javed, H.; Muqeet, H.A.; Akhter, M.N.; Liaqat, R.; Hussain, M.M. Scheduling and Sizing of Campus Microgrid Considering Demand Response and Economic Analysis. *Sensors* **2022**, *22*, 6150. [CrossRef]
3. Chu, Y.; Fu, S.; Hou, S.; Fei, J. Intelligent Terminal Sliding Mode Control of Active Power Filters by Self-evolving Emotional Neural Network. *IEEE Trans. Ind. Inform.* **2022**. [CrossRef]
4. Almughram, O.; Ben Slama, S.; Zafar, B. Model for Managing the Integration of a Vehicle-to-Home Unit into an Intelligent Home Energy Management System. *Sensors* **2022**, *22*, 8142. [CrossRef]

5. Shi, J.; Lee, W.-J.; Liu, X. Generation Scheduling Optimization of Wind-Energy Storage System Based on Wind Power Output Fluctuation Features. *IEEE Trans. Ind. Appl.* **2018**, *54*, 10–17. [CrossRef]
6. Sun, K.; Xiao, H.; Pan, J.; Liu, Y. VSC-HVDC Interties for Urban Power Grid Enhancement. *IEEE Trans. Power Syst.* **2021**, *36*, 4745–4753. [CrossRef]
7. Kazda, J.; Cutululis, N.A. Model-Optimized Dispatch for Closed-Loop Power Control of Waked Wind Farms. *IEEE Trans. Control Syst. Technol.* **2020**, *28*, 2029–2036. [CrossRef]
8. Zhang, Z.; Zhou, M.; Wu, Z.; Liu, S.; Guo, Z.; Li, G. A Frequency Security Constrained Scheduling Approach Considering Wind Farm Providing Frequency Support and Reserve. *IEEE Trans. Sustain. Energy* **2022**, *13*, 1086–1100. [CrossRef]
9. Yin, X.; Zhao, X. Deep Neural Learning Based Distributed Predictive Control for Offshore Wind Farm Using High-Fidelity LES Data. *IEEE Trans. Ind. Electron.* **2021**, *68*, 3251–3261. [CrossRef]
10. Zhang, K.; Geng, G.; Jiang, Q. Online Tracking of Reactive Power Reserve for Wind Farms. *IEEE Trans. Sustain. Energy* **2020**, *11*, 1100–1102. [CrossRef]
11. Wei, X.; Xiang, Y.; Li, J.; Zhang, X. Self-Dispatch of Wind-Storage Integrated System: A Deep Reinforcement Learning Approach. *IEEE Trans. Sustain. Energy* **2022**, *13*, 1861–1864. [CrossRef]
12. Ding, T.; Zeng, Z.; Qu, M.; Catalão, J.P.S.; Shahidehpour, M. Two-Stage Chance-Constrained Stochastic Thermal Unit Commitment for Optimal Provision of Virtual Inertia in Wind-Storage Systems. *IEEE Trans. Power Syst.* **2021**, *36*, 3520–3530. [CrossRef]
13. Zhang, Z.; Wang, D.; Gao, J. Learning Automata-Based Multiagent Reinforcement Learning for Optimization of Cooperative Tasks. *IEEE Trans. Neural Netw. Learn. Syst.* **2021**, *32*, 4639–4652. [CrossRef]
14. Fei, H.; Zhang, Y.; Ren, Y.; Ji, D. Optimizing Attention for Sequence Modeling via Reinforcement Learning. *IEEE Trans. Neural Netw. Learn. Syst.* **2022**, *33*, 3612–3621. [CrossRef]
15. Jia, Y.; Dong, Z.Y.; Sun, C.; Meng, K. Cooperation-Based Distributed Economic MPC for Economic Load Dispatch and Load Frequency Control of Interconnected Power Systems. *IEEE Trans. Power Syst.* **2019**, *34*, 3964–3966. [CrossRef]
16. Shangguan, X.-C.; He, Y.; Zhang, C.K.; Jin, L.; Yao, W.; Jiang, L.; Wu, M. Control Performance Standards-Oriented Event-Triggered Load Frequency Control for Power Systems Under Limited Communication Bandwidth. *IEEE Trans. Control Syst. Technol.* **2022**, *30*, 860–868. [CrossRef]
17. Chu, Y.; Hou, S.; Wang, C.; Fei, J. Recurrent-Neural-Network-Based Fractional Order Sliding Mode Control for Harmonic Suppression of Power Grid. *IEEE Trans. Ind. Inform.* **2023**, 305. [CrossRef]
18. Sadeghian, O.; Oshnoei, A.; Tarafdar-Hagh, M.; Kheradmandi, M. A Clustering-Based Approach for Wind Farm Placement in Radial Distribution Systems Considering Wake Effect and a Time-Acceleration Constraint. *IEEE Syst. J.* **2021**, *15*, 985–995. [CrossRef]
19. Huang, S.; Li, P.; Yang, M.; Gao, Y.; Yun, J.; Zhang, C. A Control Strategy Based on Deep Reinforcement Learning Under the Combined Wind-Solar Storage System. *IEEE Trans. Ind. Appl.* **2021**, *57*, 6547–6558. [CrossRef]
20. Liu, F.; Liu, Q.; Tao, Q.; Huang, Y.; Li, D.; Sidorov, D. Deep reinforcement learning based energy storage management strategy considering prediction intervals of wind power. *Int. J. Electr. Power Energy Syst.* **2023**, *145*, 108608. [CrossRef]
21. Yang, J.J.; Yang, M.; Wang, M.X.; Du, P.J.; Yu, Y.X. A deep reinforcement learning method for managing wind farm uncertainties through energy storage system control and external reserve purchasing. *Int. J. Electr. Power Energy Syst.* **2020**, *119*, 105928. [CrossRef]
22. Sang, J.; Sun, H.; Kou, L. Deep Reinforcement Learning Microgrid Optimization Strategy Considering Priority Flexible Demand Side. *Sensors* **2022**, *22*, 2256. [CrossRef] [PubMed]
23. Sanaye, S.; Sarrafi, A. A novel energy management method based on Deep Q Network algorithm for low operating cost of an integrated hybrid system. *Energy Rep.* **2021**, *7*, 2647–2663. [CrossRef]
24. Zhu, J.; Hu, W.; Xu, X.; Liu, H.; Pan, L.; Fan, H.; Zhang, Z.; Chen, Z. Optimal scheduling of a wind energy dominated distribution network via a deep reinforcement learning approach. *Renew. Energy* **2022**, *201*, 792–801. [CrossRef]
25. Fingrid. Fingrid Open Datasets. 2019. Available online: https://data.fingrid.fi/open-data-forms/search/en/index.html (accessed on 12 December 2019).
26. Barbour, E.; Parra, D.; Awwad, Z.; González, M.C. Community energy storage: A smart choice for the smart grid? *Appl. Energy* **2018**, *212*, 489–497. [CrossRef]
27. Claessens, B.J.; Vrancx, P.; Ruelens, F. Convolutional neural networks for automatic state-time feature extraction in reinforcement learning applied to residential load control. *IEEE Trans. Smart Grid* **2018**, *9*, 3259–3269. [CrossRef]
28. Nakabi, T.A.; Toivanen, P. Optimal price-based control of heterogeneous thermostatically controlled loads under uncertainty using LSTM networks and genetic algorithms. *F1000Research* **2019**, *8*, 1619. [CrossRef]
29. Zhang, C.; Xu, Y.; Dong, Z.Y.; Wong, K.P. Robust coordination of distributed generation and price-based demand response in microgrids. IEEE Trans. *Smart Grid* **2018**, *9*, 4236–4247. [CrossRef]
30. De Jonghe, C.; Hobbs, B.F.; Belmans, R. Value of price responsive load for wind integration in unit commitment. *IEEE Trans. Power Syst.* **2014**, *29*, 675–685. [CrossRef]
31. Song, M.; Gao, C.; Shahidehpour, M.; Li, Z.; Yang, J.; Yan, H. Impact of Uncertain Parameters on TCL Power Capacity Calculation via HDMR for Generating Power Pulses. *IEEE Trans. Smart Grid* **2019**, *10*, 3112–3124. [CrossRef]

32. Residential Electric Rates & Line Items. 2019. Available online: https://austinenergy.com/ae/residential/rates/residential-electric-rates-and-line-items (accessed on 16 December 2019).
33. Littman, M.L. Markov decision processes. In *International Encyclopedia of the Social & Behavioral Sciences*; Elsevier: Amsterdam, The Netherlands, 2001; pp. 9240–9242.

electronics

MDPI

Article

Probability Density Forecasting of Wind Power Based on Transformer Network with Expectile Regression and Kernel Density Estimation

Haoyi Xiao [1], Xiaoxia He [1,2,*] and Chunli Li [1]

1 College of Science, Wuhan University of Science and Technology, Wuhan 430065, China
2 Hubei Province Key Laboratory of Systems Science in Metallurgical Process, Wuhan University of Science and Technology, Wuhan 430081, China
* Correspondence: hexiaoxia@wust.edu.cn

Abstract: A comprehensive and accurate wind power forecast assists in reducing the operational risk of wind power generation, improves the safety and stability of the power system, and maintains the balance of wind power generation. Herein, a hybrid wind power probabilistic density forecasting approach based on a transformer network combined with expectile regression and kernel density estimation (Transformer-ER-KDE) is methodically established. The wind power prediction results of various levels are exploited as the input of kernel density estimation, and the optimal bandwidth is achieved by employing leave-one-out cross-validation to arrive at the complete probability density prediction curve. In order to more methodically assess the predicted wind power results, two sets of evaluation criteria are constructed, including evaluation metrics for point estimation and interval prediction. The wind power generation dataset from the official website of the Belgian grid company Elia is employed to validate the proposed approach. The experimental results reveal that the proposed Transformer-ER-KDE method outperforms mainstream recurrent neural network models in terms of point estimation error. Further, the suggested approach is capable of more accurately capturing the uncertainty in the forecasting of wind power through the construction of accurate prediction intervals and probability density curves.

Keywords: wind power forecasting; transformer network; expectile regression; kernel density estimation; probability density forecasting

Citation: Xiao, H.; He, X.; Li, C. Probability Density Forecasting of Wind Power Based on Transformer Network with Expectile Regression and Kernel Density Estimation. *Electronics* **2023**, *12*, 1187. https://doi.org/10.3390/electronics12051187

Academic Editors: Luis Hernández-Callejo, Sergio Nesmachnow and Sara Gallardo Saavedra

Received: 4 February 2023
Revised: 24 February 2023
Accepted: 27 February 2023
Published: 1 March 2023

1. Introduction

In response to climate problems, environmental pollution, and the energy crisis, the global focus of energy development and utilization has changed from traditional fossil fuels to clean and renewable energy sources such as wind and solar power [1]. Among these, wind energy is a non-polluting and sustainable energy source with huge storage capacity, stable production, and widespread use, making it one of the most popular sustainable renewable energy sources in the world [2]. According to forecasts, wind energy is estimated to account for a significant share of global electricity generation by 2030 [1], with China, in particular, proposing the development of a new power system based on renewable sources such as wind and solar [3]. Wind power is anticipated to play a pivotal role in the future energy mix with plans to integrate it into power systems around the world. This highlights the enormous potential for future growth in the wind power industry.

However, wind power generation is chiefly influenced by natural wind fluctuations and other meteorological conditions, and its intermittent, stochastic, and unstable nature inevitably produces technical challenges for power system planning and scheduling, as well as safe and stable operations [3]. Comprehensive and precise power network forecasting is necessary for the incorporation of wind farm technology into existing power grids. Successful forecasting is necessary to manage risks and successfully maintain a balanced network with significant wind components as part of the overall electrical grid. The challenges associated with accomplishing this require careful mathematical analysis combined with data verification to merge wind networks into existing power grids. The stochastic issues with wind power differ significantly from more traditional power sources, so data analysis, statistical estimators, stochastic analysis, and predictive methodologies require careful thought.

With the development of wind power generation in recent years, significant research and progress have been made in the field of wind power forecasting (WPF). According to various modeling schemes, WPF can be essentially classified into physical models, statistical models, and artificial intelligence models with machine learning [3–6]. In more detail, physical methods commonly exploit long-term forecasts based on numerical weather predictions (NWPs). Hence, many physical factors are required to achieve the best forecast accuracy [5], and physical models usually exhibit advantages in long-term forecasting [6]. Statistical methods for time-series forecasting include methods such as the Kalman filter (KF), autoregressive integrated moving average (ARIMA), generalized autoregressive conditional heteroskedasticity (GARCH), and its variations [5]. These methodologies are utilized for predicting the future production of wind power based on a large amount of historical data and are more effective than physical methods for short-term wind power forecasting. However, the strict distribution assumptions and smoothness tests on the data result in these statistical models not exhibiting universality and generality. With the rapid development of artificial intelligence in recent years, many machine learning-based prediction approaches such as support vector machine (SVM) [6], random forest (RF) [7], and XGboost [8] have been developed to perform wind speed or wind power prediction. Machine learning approaches usually have large-scale data processing capabilities, more accurate prediction precision, and more remarkable universality and generalization capabilities [3].

Due to the powerful ability of deep learning to learn features and handle complex nonlinear problems, neural network algorithms such as long short-term memory neural networks (LSTMs) [9–12], gated recurrent units (GRUs) [12,13], extreme learning machines (ELMs) [14], and convolutional neural networks (CNNs) [15,16] have been recently extensively employed for short-term wind power prediction. In constructing predictive models for time-series data such as wind power data, recurrent neural network (RNN) frameworks, including LSTMs and GRUs, are particularly effective for modeling sequential data in time-series data prediction tasks such as wind power forecasting. Despite these RNN-based frameworks generally performing well, they exhibit some limitations. The RNNs are often employed to iteratively model sequential data, but these methodologies possess a high training time cost and could result in performance reduction for sequential data with longer time steps. This issue is essentially attributed to the fact that the RNNs can only consider the hidden state of the last moment during processing sequential data [17].

In 2017, Google proposed the transformer network [18], which has already exhibited a momentous impact on the field of natural language processing and the application area of deep learning. The model exclusively relies on the self-attention mechanism to establish global dependencies on sequence data and is capable of mining complex and relevant information from various scales of the sequence [19]. Transformer network-based methodologies have been used by various researchers for wind power prediction [19–21]. The core self-attention mechanism has also been used in combination with recurrent neural networks such as LSTM to construct hybrid models for more accurate wind power prediction [1,3,13,22,23]. The transformer networks are capable of capturing the internal correlation of longer sequences and comprehensively obtaining essential information about wind power data [21].

Most explorations so far have focused on providing deterministic values for point estimates, which are difficult to use in measuring the uncertain characteristics of wind power [24]. On the other hand, interval and probabilistic forecasting of wind power recently attracted considerable attention because it allows the construction of continuous probability density curves and the quantification of uncertainty in wind power output. Thereby, it provides helpful information for power companies, system operators, and related decision-makers and stakeholders [2]. In addition, several investigations have been devoted to interval and probability density forecasting of wind power [25–27]. In [27], the quantile regression neural network (QRNN) approach was implemented for wind power prediction. For this purpose, the prediction results for various conditional quantiles were exploited as input to a nonparametric method of kernel density estimation (KDE) that does not presuppose the data distribution to derive the complete probability density profile of the wind power. The QRNN represents a hybrid model that combines traditional statistics and machine learning. It mainly merges the advantages of quantile regression (QR), such as the ability to estimate the conditional distribution of explanatory variables without considering the distribution type of random variables, with the strong nonlinear fitting capabilities of neural networks.

A nonparametric nonlinear regression model, the so-called expectile regression neural network (ERNN), was proposed in [28]; it builds upon the concept of QRNN by incorporating the expectile regression (ER) framework into the neural network structure. This novel ERNN model is capable of easily predicting the model parameters by standard gradient-based optimization algorithms and direction propagation due to the use of an asymmetric squared loss function, a property that outperforms the QRNN model that uses an asymmetric absolute loss function that is not differentiable at the origin. In addition, the ERNN model can directly output conditional expectation functions that describe the complete distribution of responses based on covariate information and provide more insightful information for decision-making.

The prediction performance of neural networks is commonly influenced by the model structure and hyperparameters [4], and numerous investigators have combined neural network models (NNMs) with modal decomposition techniques [3,20,23,29–31] or optimization algorithms [30–35] to achieve better prediction results. In the current investigation, hence, the transformer (i.e., a model known for its superior performance in sequential data tasks) is utilized as the base model for wind power prediction. Additionally, this effective model is properly combined with the asymmetric loss function of expectile regression and then optimized via the cuckoo search (CS) algorithm [36]. The optimal model structure is then exploited to make wind power predictions at various levels τ. To this end, the KDE model with a Gaussian kernel function in conjunction with the leave-one-out cross-validation (LOOCV) method is employed to obtain the probability density interval estimates for wind power forecasting. The results obtained with the proposed transformer expectile regression and kernel density estimation (Transformer-ER-KDE) model are compared with those of other models and methods for various points and interval estimates by utilizing the wind power data in the time interval of 2022.1–2022.2 provided by the Belgian grid, and its superiority to other models is proved.

The present investigation presents three major contributions in comparison to the preceding ones:

(1) The transformer network, which possesses the best performance in the NLP domain for sequential data tasks, is migrated for wind power prediction. Then, the internal correlations and remote dependencies of more extended sequential data could be captured better than the RNN. The expectile regression in conjunction with a transformer network is utilized for wind power prediction via the ERNN structure. This newly developed model is capable of estimating the NNM-based parameters more easily than the QRNN. Further, it is more sensitive to sample points with larger errors and can output conditional expectation functions that provide more information for decision making. To the best of our knowledge, this is the first expectile regression added to the ERNN structure of the transformer network.

(2) The nonparametric KDE-based approach is implemented to estimate the prediction results of Transformer-ER at a variety of levels, thus allowing the complete wind power probability density estimate to be derived. Since the bandwidth influences

the density function of random variables [27], the leave-one-out cross-validation is employed here for optimal bandwidth selection, fully exploiting the information from the estimation results of various levels τ, while Gaussian kernel functions [3] are commonly utilized to achieve improved probability density estimates.

(3) The probability density estimation results are appropriately derived based on two sets of evaluation criteria for point estimation and interval prediction. The point estimation results, which are attained using the probability density approach, exhibit strong robustness and high accuracy compared with traditional prediction methods [27]. Usually, evaluation metrics, such as prediction interval coverage probability (PICP), prediction interval normalized average width (PINAW), and coverage width-based criterion (CWC), are employed to assess the interval prediction results. The prediction interval estimation error (PIEE) evaluation metrics proposed in [25] are also implemented here for the purpose of evaluating and comparing the probability density interval estimation. Additionally, the PIEE index is incorporated into the CWC composite index to make it more comprehensive and accurate in reflecting the evaluation effect of interval prediction.

2. Related Theories

2.1. Transformer Network

A transformer network is a transduction model that relies entirely on a self-attention mechanism to evaluate its input and output representations without employing RNNs or CNNs [20].

2.1.1. Self-Attention Mechanism

The main advantage of the attention mechanism is its ability to extract relevant information from a large amount of input data in the current task context. Specifically, the self-attention mechanism calculates attention values within a sequence and uses this information to identify structural relationships and connections within the sequence [21].

In self-attention, the input sequence $\boldsymbol{X} \in \mathbb{R}^{l \times d}$ is transformed by matrix operations into $\boldsymbol{Q}$(Query), $\boldsymbol{K}$(Key), and $\boldsymbol{V}$(Value), where l represents the sequence length and d denotes the model dimension:

$$\boldsymbol{Q} = X\boldsymbol{W}_Q, \boldsymbol{K} = \boldsymbol{X}\boldsymbol{W}_K, \boldsymbol{V} = \boldsymbol{X}\boldsymbol{W}_V, \tag{1}$$

where $\boldsymbol{W}_Q \in \mathbb{R}^{d \times d_{qk}}$, $\boldsymbol{W}_K \in \mathbb{R}^{d \times d_{qk}}$, and $\boldsymbol{W}_v \in \mathbb{R}^{d \times d_v}$ are the weight matrix parameters that the neural network is trained to through iterations, $\boldsymbol{Q} \in \mathbb{R}^{l \times d_{qk}}$, $\boldsymbol{K} \in \mathbb{R}^{l \times d_{qk}}$, and $\boldsymbol{V} \in \mathbb{R}^{l \times d_v}$ are evaluated as follows to the output of the self-attentive mechanism:

$$\boldsymbol{A} = Softmax\left(\frac{\boldsymbol{Q}\boldsymbol{K}^T}{\sqrt{d_{qk}}}\right)\boldsymbol{V}. \tag{2}$$

It is evident that $\boldsymbol{Q}\boldsymbol{K}^T$ contains the information of various positions in the whole sequence, and after normalization, it represents the attention weights for each position. Furthermore, the matrix multiplication with $\boldsymbol{V}$ results in the output of attention $\boldsymbol{A} \in \mathbb{R}^{l \times d_v}$. Finally, the output is transformed through the linear transformation as specified in the following form:

$$\boldsymbol{O} = \boldsymbol{A}\boldsymbol{W}_O, \tag{3}$$

where $\boldsymbol{W}_O \in \mathbb{R}^{d_v \times d_{out}}$ represents the linear layer training weight matrix, and the final output would be $\boldsymbol{O} \in \mathbb{R}^{l \times d_{out}}$.

2.1.2. Multi-Head Attention Mechanism

Within the transformer network, the self-attention mechanism is extended to a multi-head attention mechanism, which is calculated in an identical way. The primary difference is that the input sequence $\boldsymbol{X}$ is divided into n subspaces, n heads, and parallel operations of the self-attention mechanism are executed on each subspace. The attention outputs obtained from each head (i.e., $A^1, A^2, \ldots, A^n$) are then concatenated, and the final output O can be obtained through the following linear transformation:

$$O = \mathrm{Concat}\left(A^1, A^2, \ldots\ldots A^n\right) W_O. \tag{4}$$

The operating principle of multi-headed attention is illustrated in Figure 1. Despite the presence of multiple heads, the number of parameters and time complexity are comparable to those of self-attention [20]. The exploitation of multi-head attention allows it to attend to various representation subspaces at various positions, thereby providing enhanced forecasting capabilities. Each subspace makes its own prediction based on its own perspective or a combination of factors, yielding better predictions than a single self-attentive mechanism.

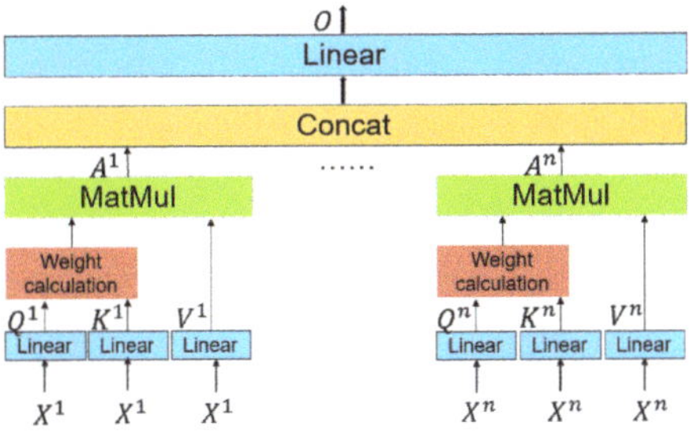

Figure 1. Schematic diagram of the multi-head attention.

2.1.3. Position Encoding

While self-attention considers information from all positions of the sequence data, it may not wholly capture the influence of positional differences. To make full use of the location information of sequence data, this paper incorporates position-encoding information into the sequence data. The position encoding is evaluated in the following form:

$$PE(pos, 2i) = \sin\left(pos/10{,}000^{2i/d}\right), \tag{5}$$

$$PE(pos, 2i+1) = \cos\left(pos/10{,}000^{2i/d}\right), \tag{6}$$

where pos denotes the sequence length index, and i represents the dimensional index from 0 to $d/2$.

2.1.4. Transformer

The structure of the transformer network utilized in the present work is depicted in Figure 2.

The traditional transformer architecture consists of an encoder and a decoder. In the current exploration, only the transformer encoder structure is employed, which is appropriate for regression problems and serves as a general-purpose module for transforming a sequence into a more informative feature representation. The transformer is originally developed for exploitation in the NLP field; hence, minor modifications have been made to its architecture. Instead of a word vector embedding layer, the input data are passed through a linear layer before being encoded based on their position. Similarly, before being output, the prediction results are passed through a linear layer without an activation function rather than a Softmax layer for probabilistic prediction. The remaining elements of the multi-headed attention, two normalization layers, one linear layer, and two residual links, are identical to those in the original transformer.

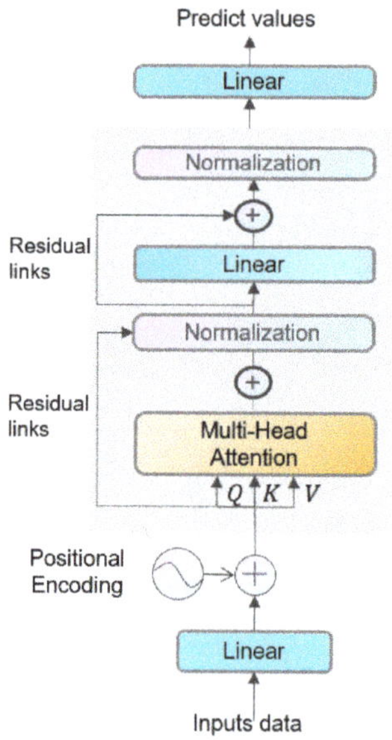

Figure 2. Schematic representation of the proposed transformer network.

2.2. Expectile Regression

Given a response variable $\boldsymbol{Y}$ and a covariate matrix $\boldsymbol{X}$ with observations $(y_i, \boldsymbol{x}_i)$, where i is the sample number such that $i = 1, 2, \ldots, n$ and n denotes the total number of samples, the values y_i of the response variable at the τ level can be derived by the following classical linear expectile regression model:

$$\hat{E}_{y_i}(\tau|\boldsymbol{x}_i) = \boldsymbol{x}_i'\hat{\beta}(\tau), \quad i = 1, 2, \ldots, n \tag{7}$$

$$\hat{\beta}(\tau) = arg \min \sum_{i=1}^{n} \varphi_\tau\left(y_i - \boldsymbol{x}_i'\beta\right). \tag{8}$$

$$\varphi_\tau(u) = \begin{cases} \tau u^2, & u \geq 0 \\ (\tau - 1)u^2, & u < 0 \end{cases} \tag{9}$$

where $\tau \in (0,1)$ is the quantile of a given weight level and denotes the degree of asymmetry of the loss function. $E_{y_i}(\tau|\boldsymbol{x}_i)$ represents the τ-th level of the response variable y_i, and $\hat{\beta}(\tau)$ denotes the regression's coefficient at a given τ for which the estimation can be obtained by solving the optimization problem, as displayed in Equation (8).

$\varphi_\tau(u)$ is an asymmetric loss function that depends on the level τ. When $\tau = 0.5$, the asymmetric squared loss function in Equation (9) above degenerates to the squared loss function $\varphi(u) = u^2$, and the overall expectile regression model degenerates to a simple linear regression model. It has been widely acknowledged that the square loss function, commonly utilized in the training of neural networks through back-propagation, is merely a specific instance of the expectile regression asymmetric loss function.

A neural network can be conceptualized as a nonlinear function denoted by $f(\cdot)$ that serves as a generalized nonlinear model. Given an input $\boldsymbol{x}_i$, the output of this model can be displayed as follows:

$$\hat{E}_{y_i}(\tau|x_i) = f(\boldsymbol{x}_i, w(\tau)), \quad i = 1, 2, \ldots, n \tag{10}$$

where $w(\tau)$ represents the model parameter to be estimated. In the ERNN model, the estimator can be appropriately derived by iterating based on the following loss function:

$$\hat{w}(\tau) = arg\ \min \sum_{i=1}^{n} \varphi_\tau(y_i - f(x_i, w(\tau))), \tag{11}$$

where $\varphi_\tau(u)$ is the same as that given in Equation (9). Unlike the asymmetric absolute value loss function of the QRNN, the empirical loss function of the ERNN model is differentiable and smooth everywhere at various levels of τ. The empirical loss function is also convex, so the standard back-propagation and gradient descent optimization algorithms of neural networks are capable of estimating the ERNN model parameters easily and obtaining the optimal solution $\hat{w}(\tau)$ at different values of τ. Furthermore, it is clear that the ERNN model is derived by replacing the conventional squared loss function employed in general neural networks with an asymmetric quadratic loss function [28].

2.3. Cuckoo Search Algorithm

The cuckoo search algorithm was proposed in 2009 [36] as a bionic intelligent algorithm that would be applicable to optimization problems. Similar to genetic algorithms (GAs), and particle swarm optimization (PSO) algorithms, the CS is also an algorithm for directly searching for the extremum points of the objective function in the feasible domain of the given parameters. The main strategy relies on the Lévy flight to update the position where the nest is located. The Lévy flight step formula is given as follows [37]:

$$s = \frac{u}{|v|^{1/\beta}}, \tag{12}$$

The value of β is usually considered between 1 and 2. In this study, we set $\beta = 1.5$, which is a commonly used value in the literature. Both u and v obey the following normal distribution:

$$u \sim N\left(0, \sigma_u^2\right), \ v \sim N(0,1). \tag{13}$$

$$\sigma_u = \left(\frac{\Gamma(1+\beta) sin\frac{\pi\beta}{2}}{\beta \cdot \Gamma\left(\frac{1+\beta}{2}\right) \cdot 2^{\frac{\beta-1}{2}}} \right)^{\frac{1}{\beta}}. \tag{14}$$

The Lévy flight, which is commonly characterized by a combination of high-frequency small-step movements and low-frequency large-step movements, mimics the random wandering of a cuckoo. This behavior enables the CS algorithm to effectively search for globally optimal solutions while also avoiding being trapped in local optima. Moreover, the incorporation of small steps in the algorithm guarantees a certain level of accuracy in the solution. The position of the nest is updated according to the following relation:

$$x_i^{k+1} = x_i^k + \alpha \times s \otimes x_i^k, \tag{15}$$

where x_i^k denotes the value of the k-th iteration, α represents the scaling factor of the step, s stands for the step of the Lévy flight, and $\otimes$ denotes the dot product. The overall flow of the cuckoo search algorithm is presented in Figure 3.

This exploration takes the hyperparameters of an NNM into account as the search parameters, with the overall ERNN model employed as the adaptation function. The performance of the model in predicting the test set data, as measured by its goodness-of-fit value, is also utilized as the adaptive value. The objective of the current search is to find the optimal set of hyperparameters by maximizing the adaptive value.

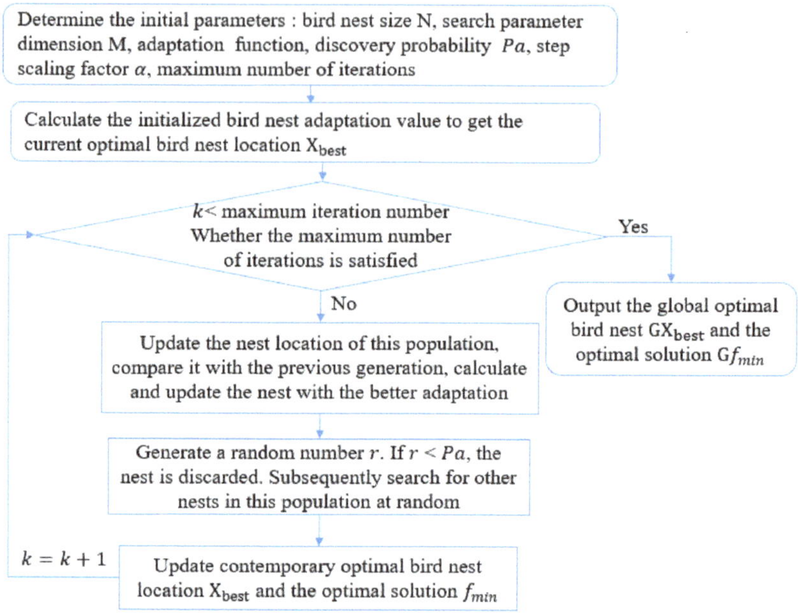

Figure 3. The overall flowchart of the cuckoo search algorithm.

2.4. Kernel Density Estimation

In comparison to the parametric model, kernel density estimation, being a nonparametric method, avoids imposing any prior assumptions on the data distribution, thereby resulting in more accurate estimations. Based on the similarity theory, the obtained conditional quantile is similar to conditional density [27].

2.4.1. KDE-Based Model

The KDE is established based on the sample data to estimate the probability density function. Given the density function of a random variable represented by $f(x)$ and the empirical distribution function denoted by $F(x)$, the basic estimation of $f(x)$ can be provided by the following:

$$f(x) = \frac{F(x+h) - F(x-h)}{2h}, \tag{16}$$

where h represents a non-negative constant. As the value of h approaches zero, an approximate estimation of $f(x)$ can be obtained in the following form:

$$\hat{f}(x) = \frac{1}{Nh}\sum_{i=1}^{N} k\left(\frac{x - x_i}{h}\right), \tag{17}$$

where N denotes the number of samples, h is the bandwidth, and $k(x)$ represents the kernel function. It is worth mentioning that various kernel functions bring different estimation effects. This investigation is aimed to utilize the Gaussian kernel function, which is commonly exploited and known to produce effective results [3]. The function is represented by the following equation:

$$k(x) = \frac{1}{\sqrt{2\pi}} \exp\left(-\frac{x^2}{2}\right). \tag{18}$$

2.4.2. Leave-One-Out Cross-Validation

The bandwidth plays a crucial role in a KDE-based approach. Wide bandwidths are capable of preventing the model from accurately estimating the density of critical features, while a small bandwidth results in an estimation with a higher level of noise. Herein, leave-one-out cross-validation is implemented for the optimal selection of the bandwidth, and mean integrated squared error (MISE) is also utilized to evaluate the error of the kernel density function. The MISE is defined per the following relation:

$$\mathrm{MISE}\left(\hat{f}(x)\right) = E\int\left[\left(\hat{f}(x) - f(x)\right)^2\right]dx. \tag{19}$$

The global error of LOOCV is defined as follows:

$$LV = \frac{1}{N}\sum_{i=1}^{N} MISE_i. \tag{20}$$

The error resulting from the computation of various bandwidths (h) is specified by $LV(h)$. The optimal bandwidth (h_0) is determined by identifying the point at which $LV(h)$ takes its minimum value:

$$h_0 = argmin\ LV(h),\ \ h > 0 \tag{21}$$

LOOCV effectively utilizes all the information of the data, resulting in the calculation of optimal parameters for the sample data. However, the corresponding computational time cost is high, and it is generally utilized in the case of small sample data due to the need for N-training that fits the model and error metric calculations. In the current investigation, the prediction results of the ERNN-based model for different levels of τ are chosen as inputs for kernel density estimation, and then the LOOCV is exploited as the method for bandwidth selection due to the limited number of values for $\tau \in (0,1)$.

3. Methodology Framework and Evaluation Metrics

3.1. Methodology Framework

The framework of the overall WPF is demonstrated in Figure 4. The forecasting process in the present work is divided into the following steps:

(1) Preprocessing of the wind power data, including the division of data into training and test sets, normalization, and the utilization of the sliding window method for the construction of feature and response variables.

(2) Nine distinct models (ER, QRNN, LSTM, GRU, MLP, RNN, Transformer, Transformer-ER, and CS-Transformer-ER) are employed for wind power series prediction, and four commonly used evaluation metrics (MAE, RMSE, MAPE, and R^2) are considered as appropriate measures to compare the performances of the models.

(3) The structure of the optimal Transformer-ER network, as identified by the CS algorithm, is implemented for point prediction at various levels τ, and the error evaluation metrics are calculated for it.

(4) The point prediction results for various levels of τ are utilized as inputs for kernel density estimation, the optimal bandwidth (h) is then determined through LOOCV, and finally, probability density predictions are achieved accordingly.

(5) The results of probability density estimation are appropriately exploited to construct point and interval predictions, and the evaluation metrics of point and interval estimation of various models are separately obtained for comparison.

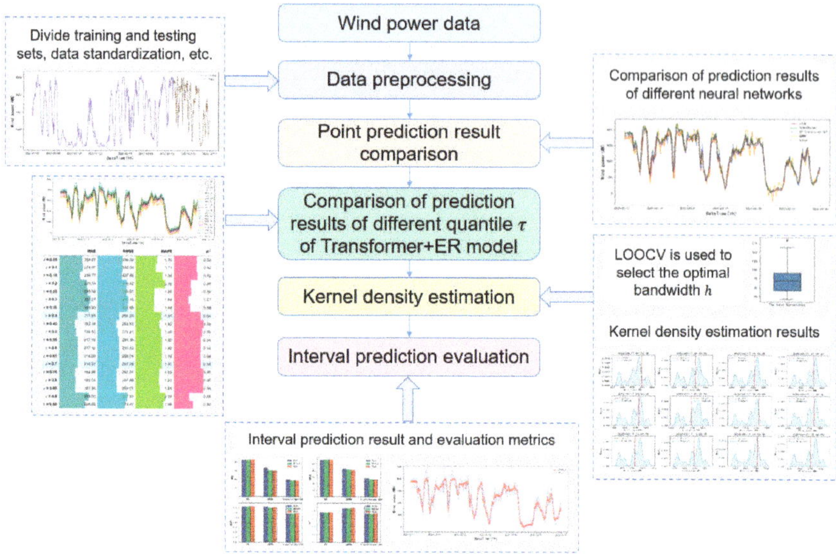

Figure 4. The framework of the overall WPF.

3.2. Point Estimation Evaluation Metrics

In regression problems, four of the most commonly used and reliable evaluation metrics for assessing the point prediction accuracy of different models are mean absolute error (MAE), root mean square error (RSME), mean absolute percentage error (MAPE), and coefficient of determination (R^2). Their calculation formulas are given in the following Equations (22)–(25):

$$\mathrm{MAE} = \frac{1}{n}\sum_{t=1}^{n} \parallel y_i - \hat{y}_i \parallel, \tag{22}$$

$$\mathrm{RMSE} = \sqrt{\frac{1}{n}\sum_{t=1}^{n}(y_i - \hat{y}_i)^2}, \tag{23}$$

$$\mathrm{MAPE} = \frac{1}{n}\sum_{t=1}^{n} \parallel \frac{y_i - \hat{y}_i}{y_i} \parallel, \tag{24}$$

$$R^2 = 1 - \frac{\sum_{t=1}^{n}(y_i - \hat{y}_i)^2}{\sum_{t=1}^{n}(y_i - \overline{y})^2}, \tag{25}$$

where n represents the number of predicted samples, y_i denotes the true value of the response variable, $\hat{y}_i$ is the predicted value, and $\overline{y}$ specifies the mean value of the real data.

3.3. Interval Prediction Evaluation Metrics

The quality of the prediction interval (PI) is a crucial feature in assessing the results of probability density prediction. To evaluate the probability density estimation of the model, herein, the following four metrics are employed for comparison: prediction interval coverage probability (PICP), prediction interval normalized average width (PINAW), prediction interval estimation error (PIEE), and coverage width-based criterion (CWC).

The PICP is a crucial evaluation metric for PI; it represents the probability that future wind power will be within the lower and upper limits of the forecast results, and it is defined by the following equation:

$$\mathrm{PICP} = \frac{1}{n}\sum_{i=1}^{n} C_i, \tag{26}$$

$$C_i = \begin{cases} 1, & y_i \in [L_i, U_i] \\ 0, & y_i \notin [L_i, U_i] \end{cases} \tag{27}$$

where L_i and U_i in order represent the minimum and maximum values of the prediction interval for the i-th sample. The factor C_i denotes a Boolean variable, where $C_i = 1$ if the real value falls within the prediction interval and $C_i = 0$ in other cases. It is evident that a wide PI could result in a high PICP; nevertheless, it has minimal value for power planning and decision making. With this in mind, the PINAW is introduced to evaluate PI; it is defined by the following relation:

$$\text{PINAW} = \sum_{i=1}^{n} \frac{U_i - L_i}{nR}, \tag{28}$$

in which R denotes the difference between the maximum and minimum values of the response variable y to be predicted, and it serves the purpose of standardizing the results to objectively evaluate the width of PI. Lower values of the PINAW imply higher accuracy of the interval prediction results.

The PICP only considers the probability of the real value falling within the prediction interval, without dealing with the error magnitude between the prediction interval and the real value. A relatively novel metric, PIEE [25], provides an understanding of the estimation error of PI. This metric is implemented to more systematically evaluate the risk outside the prediction interval; it is defined as follows:

$$PIEE = \sum_{i=1}^{n} \frac{E_i}{nR}, \tag{29}$$

$$E_i = \begin{cases} y_i - U_i, & y_i > U_i \\ 0, & L_i < y_i < U_i \\ L_i - y_i, & y_i < L_i \end{cases} \tag{30}$$

The PIEE metric enables us to more precisely evaluate the estimation error of the true value outside the model prediction interval. However, as with PICP, a too-wide PI could result in a low PIEE, which is not significant. To ensure a more accurate and comprehensive evaluation, the CWC metric is introduced. A combination of the three metrics PICP, PINAW, and PIEE is employed to construct an improved CWC metric:

$$\text{CWC} = \text{PINAW}\{1 + \gamma_{PICP} \exp[-(1 + \text{PIEE})(\text{PICP} - \mu)]\} \tag{31}$$

$$\gamma_{PICP} = \begin{cases} 0, & PICP \geq \mu \\ 1, & PICP < \mu \end{cases} \tag{32}$$

where the parameter μ represents the basic requirement for interval coverage probability, and a PICP value less than μ leads to an exponential penalty. In the current investigation, we set $\mu = 0.9$. The penalty factor, denoted by $1 + \text{PIEE}$, is exploited in the case of the coverage probability requirement not being satisfied. Additionally, it can be observed that the CWC metric takes into account the coverage probability, average width, and estimation error of the prediction interval and serves as a comprehensive index. A smaller value of the CWC implies a higher quality of the prediction interval.

3.4. Probability Density Prediction Is Constructed as a Point Estimation

In order to compare the estimation of the probability density prediction with that of the point prediction, the mode, median, and mean of the wind power probability density prediction are selected as the point estimation results. The mode corresponds to the peak value of the probability density curve. The median is defined as the middle value of the prediction interval, representing the weighted sum of all probability densities and their predicted values. Hence, this factor takes full advantage of the information from the probability density function [27]. The predicted values of the wind power for the i-th sample, $\hat{y}_{i,1} \leq \hat{y}_{i,2} \leq \ldots \leq \hat{y}_{i,N}$, are denoted by $p_{i,1} \leq p_{i,2} \leq \ldots \leq p_{i,N}$, which are their corresponding probability values. The mode, median, and mean values are calculated by the following Equations (33)–(35):

$$Mode = \hat{y}_{i,\mathrm{argmax}(p_{i,j})},\ j = 1, 2 \ldots, N \tag{33}$$

$$Median = \begin{cases} \hat{y}_{i,\frac{N+1}{2}}, & N \text{ is odd} \\ \frac{\left(\hat{y}_{i,\frac{N}{2}} + \hat{y}_{i,\frac{N+2}{2}}\right)}{2}, & N \text{ is even} \end{cases} \tag{34}$$

$$Mean = \sum_{j=1}^{N} p_{i,j} \cdot \hat{y}_{i,j}. \tag{35}$$

4. Empirical Results

4.1. Data Sources and Preprocessing

In the current investigation, we use wind power data from the Elia Belgian power grid company website as empirical data to verify and test the validity of the proposed model. For this purpose, the data from the aggregate Belgian wind farms are chosen for a period from 1 January to 28 February 2022. Since the original data have a 15 min frequency, they are resampled to a 1 h frequency to lessen the computational effort and for the ease of recording. According to the demonstrated processed data in Figure 5, it is evident that the wind power series data are highly variable and random. As a result, the probability density prediction of wind power is necessary for quantifying the uncertainty of wind power output and providing results that would be more informative to relevant decision-makers and stakeholders. About 80% of the data, the purple solid line part (from 1 January 2022 00:00:00 to 17 February 2022 03:00:00), are chosen to be exploited as the training set, whereas the remaining 20% of the data, the brown dashed line part (from 17 February 2022 04:00:00 to 28 February 2022 23:00:00), are utilized as the test set. A sliding window of 168 periods (seven days) is employed to construct the feature variables, meaning that $y_{t-167}, y_{t-166}, \cdots, y_t$ is employed to predict the value of y_{t+1}. After the above process is completed, the 3D tensor data from both the training and test sets are normalized to prepare for the NNM fitting. Table 1 provides information on the main parameters of the NNMs used in the present work.

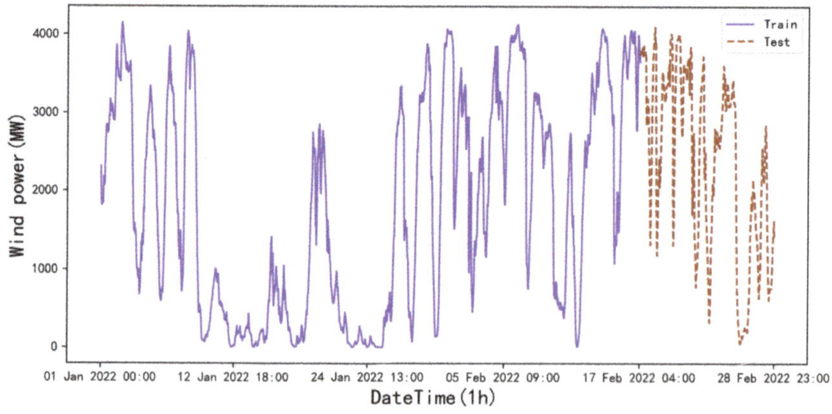

Figure 5. Wind power plot of aggregate Belgian wind farms from 1 January to 28 February 2022.

Table 1. The main parameters of the NNMs.

Parameter	Value
Number of hidden layers	2
Number of neurons in the hidden layer	[64, 32]
Batch size	32
Maximum number of iterations	50
Embedding layer dimension	32
Number of multi-head attention heads	4
Level τ	0.5

4.2. Comparison of the Model Prediction Results

Nine models are utilized for comparison in order to evaluate the prediction results of classical point estimation methods. These models are appropriately analyzed via four metrics: MAE, RMSE, MAPE, and R^2. A comparison of the point prediction results of some of the models is presented in Figure 6. The depicted results indicate that the predicted and actual values for the four models are relatively close. The models exhibit highly accurate prediction performance for intervals where the wind power data are monotonic, while more deviations for intervals are observed for the cases in which the wind power fluctuates and varies. Notably, the QRNN model predicts more dramatic fluctuations between 24 February 2022 and 27 February 2022, which could be related to its training process that utilizes an absolute value loss function. The four error metrics calculated for all models on the test set are given in Table 2.

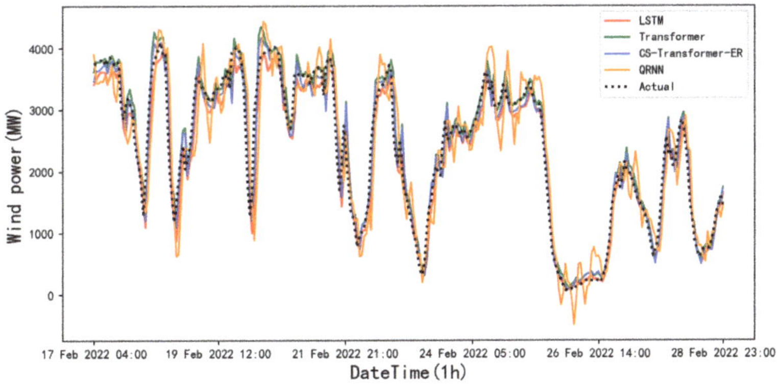

Figure 6. Plots of the point prediction results based on the partial models.

Table 2. Comparison of the point prediction results for all examined models.

Model	MAE	RMSE	MAPE	R^2
ER	428.4206	565.2859	1.8311	0.7672
QRNN	354.1195	460.9649	1.8252	0.8452
LSTM	207.8788	278.8663	1.7486	0.9434
GRU	228.8504	308.0818	**1.7363**	0.9309
MLP	361.4761	478.3939	1.8812	0.8392
RNN	239.0550	321.7016	1.7537	0.9246
Transformer	190.5266	269.8500	1.8787	0.9470
Transformer-ER	194.9061	268.8835	1.7945	0.9473
CS-Transformer-ER	**183.9616**	**252.6901**	1.8142	**0.9535**

From the results presented in Table 2, the following conclusions can be drawn:

(1) Among all the models, the transformer model has the best performance in predicting wind power data. The prediction results of the Transformer-ER model when $\tau = 0.5$ should be theoretically similar to those of the transformer model, and any minor differences between them can be attributed to the numerical calculation variations. In this type of sequential data, compared to the RNN-based model, the transformer model demonstrates superior performance in capturing the internal correlation of longer sequential data. In addition, based on the common sense of deep learning, this effect becomes more noticeable as the amount of training data rises.

(2) The CS algorithm is effective in searching for hyperparameters of NNMs. Additionally, the achieved results from the CS-Transformer-ER model, which exploits the hyperparameters found through the CS algorithm, also exhibit superior performance in all four evaluation metrics. It is crucial to mention that the low MAPE values for the GRU model could be skewed. The MAPE may not be as reliable as the other three indicators in assessing the prediction performance of the models on the test set due to the presence of intervals in the test set data that contain zero values or close to them. This may lead to the calculated MAPE values tending towards infinity, making the metric unreliable. Furthermore, further optimization of the CS algorithm with more iterations could possibly lead to even better predictions.

(3) The linear model (i.e., ER) exhibits the worst performance among the benchmark models. Although the MLP and QRNN models are essentially nonlinear, they fail in full consideration of the temporal relationship between data and thus exhibit lower prediction performance than the RNNs. Among the three recurrent neural networks, namely RNN, LSTM, and GRU, the best performance is achieved for the LSTM, which is exploited by most researchers. However, the corresponding MAE error metric of the prediction results is almost 9% higher than that of the transformer model.

4.3. The Predicted Results Based on the Various Levels of τ

The model has been trained and tested with different levels of τ. The effect of the prediction curve is presented in Figure 7, and the corresponding evaluation metrics calculated are presented in Figure 8.

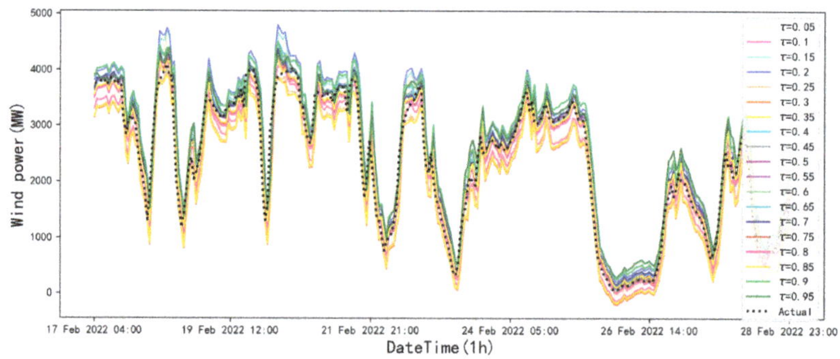

Figure 7. The graphed prediction results for various levels of τ.

Figure 7 illustrates the plotted prediction results based on different τ values. It is apparent that the prediction curves are highly similar in trend and degree of fluctuation and are superimposed to configure a confidence interval covering the actual value. It is feasible and reliable to use these predicted values to construct probability density estimation curves.

From Figure 8, it is obtainable that the prediction performance is better and more consistent with less error in the case of τ in the range of 0.4 to 0.85. When the value of τ is considered too large or too small, it leads to a strong asymmetry in the loss function, which is appropriate for describing the corresponding conditional distribution, but the overall prediction performance is poorer.

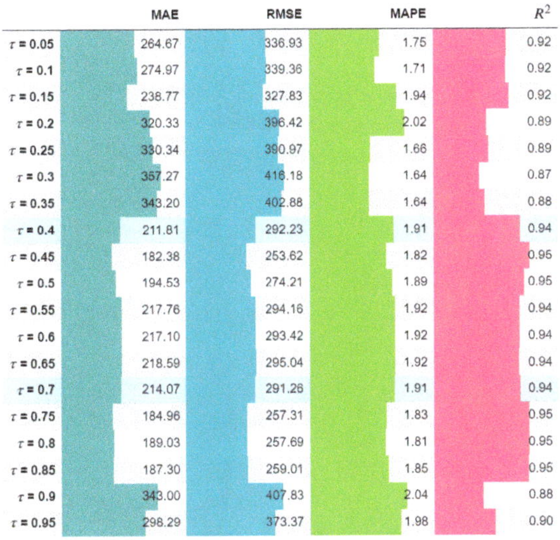

Figure 8. Prediction metrics for different levels of τ.

4.4. Probability Density Prediction Results

Before performing the kernel density estimation, the optimal bandwidth size selected is appropriately verified by a leave-one-out cross-validation for each group of bit data in the test set. The box plots of all optimal bandwidths (h) are demonstrated in Figure 9. Figure 9 clearly displays that the majority of the optimal bandwidths (h) are in the range of 40–90.

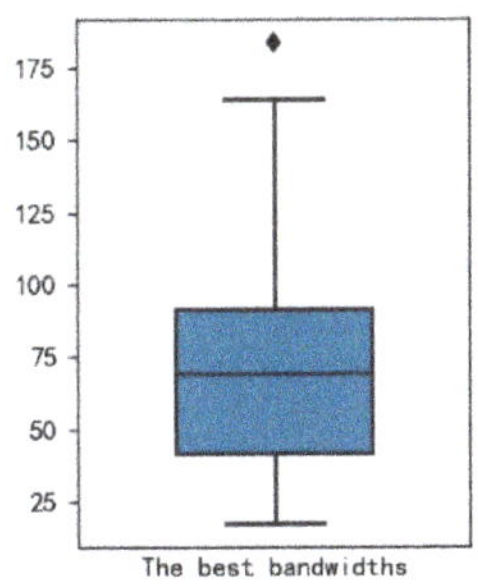

Figure 9. Optimal range of the bandwidth.

The first nine points of the test set (from 17 February 2022 04:00:00 to 17 February 2022 12:00:00) are chosen, and the actual values and probability density curves of the wind power are demonstrated in Figure 10. The blue curve and the red dashed line represent the kernel density estimation curve and the actual values of the test set, respectively. All the actual values clearly fall within the predicted probability density curve, with the majority of the values being concentrated around the peak of the estimated probability density. This indicates that the estimated probability density effectively captures the inherent uncertainty in wind power generation. The location of the estimated probability density curve peak may be the true value of the wind power data. The probability density estimation offers several advantages such as quantifying uncertainty and improving prediction accuracy, providing decision-makers with more precise information about the WPF.

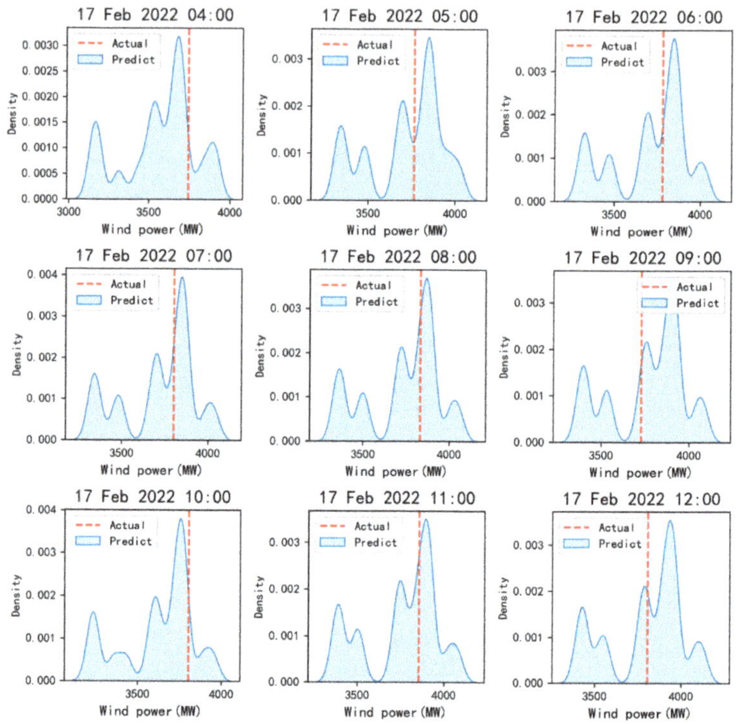

Figure 10. Probability density curves of the wind power for the partial test set.

The results of the probability density estimation for the proposed model, QRNN, and ER are compared in Table 3. The evaluation metrics for the point estimates (i.e., mode, median, and mean) constructed from the probability density estimates of each model are given in this table. Additionally, the corresponding histograms are presented in Figure 11, providing a visual representation of the performance of each model.

Table 3. The evaluated metrics for point estimation based on several approaches.

Methods	Point Estimates	MAE	RMSE	MAPE	R^2
Transformer-ER	Mode	203.1176	282.0029	1.9020	0.9421
	Median	191.4592	271.6375	1.8843	0.9463
	Mean	**187.0430**	**263.9477**	1.8611	**0.9493**
QRNN	Mode	334.9645	428.0405	1.8301	0.8665
	Median	310.3674	419.9150	**1.7790**	0.8716
	Mean	303.6449	411.5041	1.7964	0.8767
ER	Mode	428.8808	564.8208	1.8274	0.7676
	Median	430.2076	566.8995	1.8285	0.7659
	Mean	430.0491	566.1494	1.8323	0.7665

The presented results in Table 3 and Figure 11 display that the point prediction errors based on the probability density estimation of the proposed Transformer-ER model are substantially lower in comparison to those of the QRNN and linear ER models, which do not take into account temporal effects. Additionally, regardless of the model or method used, the mode, median, and mean values of the probability density predictions are relatively similar in terms of performance. The mean accuracy is slightly higher than mode and median accuracies because it takes into account all the information of the predicted data.

Among all the models and methods, the Transformer-ER model exhibits the lowest MAE and RMSE and the highest R^2 for the mean probability density, making it the best point prediction result. Its error metric is smaller than the point prediction results of almost all models in Table 2. It is worth mentioning that the exploitation of the MAPE may not be reliable due to the presence of values close to or equal to zero in the test set.

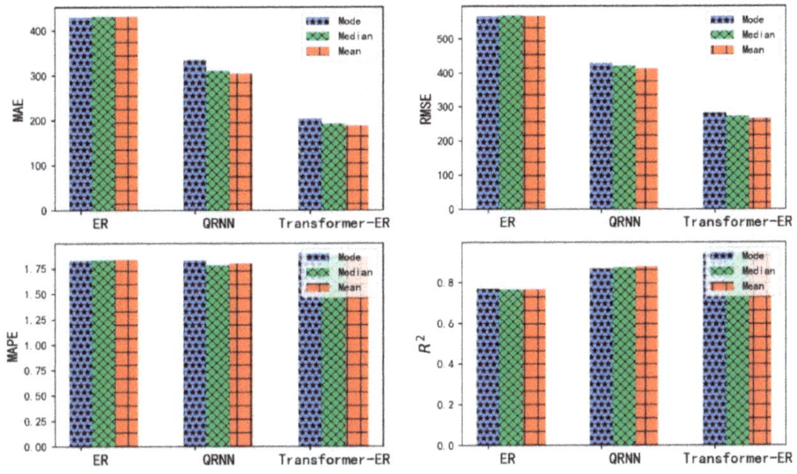

Figure 11. Evaluation metrics for point estimates constructed by the probability density prediction.

The evaluation metrics for interval estimation are provided in Table 4. The PICP values of the Transformer-ER, QRNN, and ER models are remarkably different. The QRNN model presents a high PICP, accordingly presenting a low PIEE. The ER of the linear model fails to satisfactorily fit the uncertainty of the wind power data, with a PICP value of only 42.25%. However, the higher PICP of the QRNN is derived from a larger average width of the prediction interval. This means that the QRNN gives an extensive prediction interval, which is of little significance for practical decision making. On the contrary, the Transformer-ER-based model exhibits a more moderate PICP and a smaller PINAW, and its composite index CWC has the smallest value. Therefore, the probability density prediction interval of the Transformer-ER model exhibits higher quality than that of other models.

Table 4. Evaluation metrics for the interval prediction of various approaches.

Methods	PICP	PIEE	PINAW	CWC
Transformer-ER	0.8697	0.0064	**0.1728**	**0.3510**
QRNN	**0.9824**	**0.0008**	0.5580	0.5580
ER	0.4225	0.0572	0.1781	0.4732

As can be observed from Figures 12 and 13, while the PIs obtained from the QRNN model cover a majority of the actual values of the wind power, they also exhibit a broader range compared to the PIs from the Transformer-ER model. This broader range of PIs from the QRNN model could lead to a growth of uncertainty in the prediction of wind power forecasting; thus, it could not be beneficial in power planning and decision making. The PIs of the Transformer-ER model are more precise, as they are narrower in zones where the wind power data exhibit a monotonic increase or decrease and broader in zones where the wind power is volatile and variable. This issue would be effectively helpful in capturing the uncertainty in wind power forecasting, providing decision-makers with more relevant and useful information.

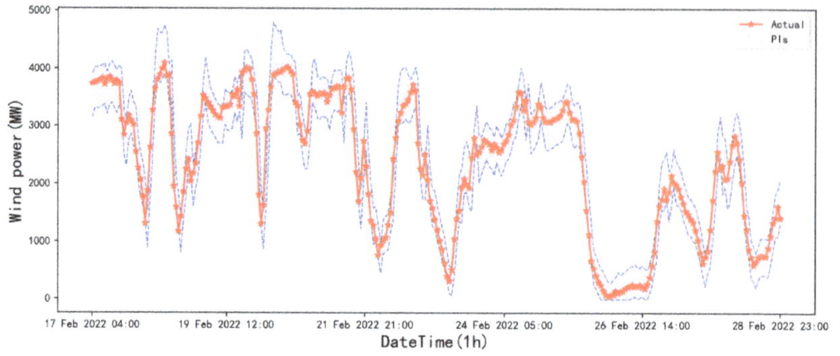

Figure 12. Plot of the prediction intervals of the ERNN-based model.

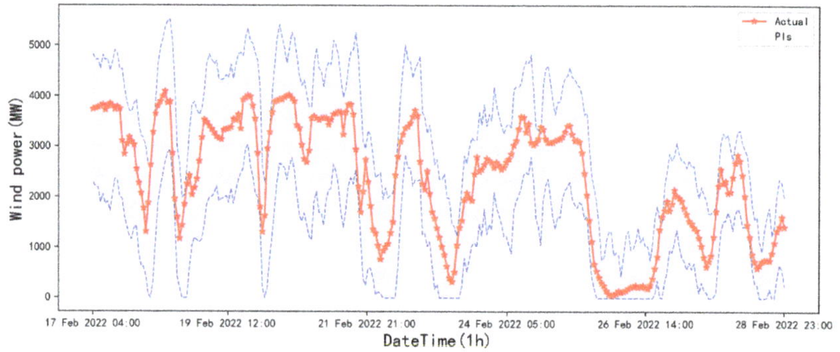

Figure 13. Plot of the prediction intervals of the QRNN-based model.

5. Conclusions

In the current investigation, a combination of the transformer network that performed best in the sequential data task and expectile regression is proposed for effective wind power prediction via an ERNN structure. The model is optimized by employing the cuckoo search algorithm. The methodology of kernel density estimation is then exploited to achieve the complete probability density curve, which is then built into the point and interval prediction. These predicted results are separately evaluated to provide comprehensive information on the uncertainty of the wind power. The proposed approach is then validated and tested based on the wind power generation data from the Belgian power grid company Elia. The major obtained conclusions are as follows: (1) The proposed model effectively addresses the volatility and stochastic nature of wind power data, provides comprehensive and accurate prediction, reduces the operational risks associated with wind power generation, and enhances the stability of power systems. (2) The transformer network, when compared to the commonly exploited recurrent neural networks, demonstrates the superior capability to capture the internal correlations and dependencies in long sequences and yields a higher level of prediction accuracy. (3) The proposed probability density prediction approach in this paper is capable of providing more comprehensive information for relevant stakeholders and decision-makers and has been proven to be more robust and accurate than point predictions. (4) The proposed ERNN-based model produces more accurate and narrow prediction intervals compared to QRNN models and thereby leads to higher quality prediction intervals in general.

Author Contributions: Conceptualization, X.H.; methodology, H.X. and X.H.; software, H.X.; validation, H.X.; formal analysis, H.X.; investigation, H.X.; resources, H.X.; data curation, H.X.; writing—original draft preparation, H.X.; writing—review and editing, H.X., X.H. and C.L.; visualization, H.X.;

supervision, X.H. and C.L.; project administration, X.H.; funding acquisition, X.H. All authors have read and agreed to the published version of the manuscript.

Funding: This research was supported by the National Natural Science Foundation of China (NSFC) (No. 11201356) and Hubei Province Key Laboratory of Systems Science in Metallurgical Process (Wuhan University of Science and Technology) (No. Y202201).

Data Availability Statement: The data presented in this study are available from the following website: https://www.elia.be/en/grid-data/power-generation/wind-power-generation (accessed on 2 January 2023).

Conflicts of Interest: The authors declare no conflict of interest.

References

1. Tian, C.; Niu, T.; Wei, W. Developing a wind power forecasting system based on deep learning with attention mechanism. *Energy* **2022**, *257*, 124750. [CrossRef]
2. He, Y.; Zhang, W. Probability density forecasting of wind power based on multi-core parallel quantile regression neural network. *Knowl.-Based Syst.* **2020**, *209*, 106431. [CrossRef]
3. Niu, D.; Sun, L.; Yu, M.; Wang, K. Point and interval forecasting of ultra-short-term wind power based on a data-driven method and hybrid deep learning model. *Energy* **2022**, 124384. [CrossRef]
4. Wang, Y.; Zou, R.; Liu, F.; Zhang, L.; Liu, Q. A review of wind speed and wind power forecasting with deep neural networks. *Appl. Energy* **2021**, *304*, 117766. [CrossRef]
5. Qiao, B.; Liu, J.; Wu, P.; Teng, Y. Wind power forecasting based on variational mode decomposition and high-order fuzzy cognitive maps. *Appl. Soft Comput.* **2022**, *129*, 109586. [CrossRef]
6. Ding, M.; Zhou, H.; Xie, H.; Wu, M.; Liu, K.-Z.; Nakanishi, Y.; Yokoyama, R. A time series model based on hybrid-kernel least-squares support vector machine for short-term wind power forecasting. *ISA Trans.* **2021**, *108*, 58–68. [CrossRef]
7. Liu, K. A random forest-based method for wind power system output power prediction. *Light Source Light.* **2022**, *07*, 165–167.
8. Zha, W.; Liu, J.; Li, Y.; Liang, Y. Ultra-short-term power forecast method for the wind farm based on feature selection and temporal convolution network. *ISA Trans.* **2022**, *129*, 405–414. [CrossRef]
9. Wang, W.; Liu, H.; Chen, Y.; Zheng, N.; Li, Z.; Ji, X.; Yu, G.; Kang, J. Wind power prediction based on LSTM recurrent neural network. *Renew. Energy* **2020**, *38*, 1187–1191.
10. Jin, Y.; Kang, J.; Chen, Y. Wind power prediction technology based on LSTM recurrent neural network algorithm. *Electron. Test.* **2022**, *36*, 49–51.
11. Cui, Y.; Chen, Z.; He, Y.; Xiong, X.; Li, F. An algorithm for forecasting day-ahead wind power via novel long short-term memory and wind power ramp events. *Energy* **2022**, *263*, 125888. [CrossRef]
12. Ahmad, T.; Zhang, D. A data-driven deep sequence-to-sequence long-short memory method along with a gated recurrent neural network for wind power forecasting. *Energy* **2022**, *239*, 122109. [CrossRef]
13. Niu, Z.; Yu, Z.; Tang, W.; Wu, Q.; Reformat, M. Wind power forecasting using attention-based gated recurrent unit network. *Energy* **2020**, *196*, 117081. [CrossRef]
14. Li, L.L.; Liu, Z.F.; Tseng, M.L.; Jantarakolica, K.; Lim, M.K. Using enhanced crow search algorithm optimization-extreme learning machine model to forecast short-term wind power. *Expert Syst. Appl.* **2021**, *184*, 115579. [CrossRef]
15. Jalali SM, J.; Ahmadian, S.; Khodayar, M.; Khosravi, A.; Shafie-khah, M.; Nahavandi, S.; Catalão, J.P.S. An advanced short-term wind power forecasting framework based on the optimized deep neural network models. *Int. J. Electr. Power Energy Syst.* **2022**, *141*, 108143. [CrossRef]
16. Jiang, X.; Xu, Y.; Song, C. A new method for short-term wind power load forecasting. *J. Beijing Norm. Univ.* **2022**, *58*, 39–46.
17. Chen, D.; Hong, W.; Zhou, X. Transformer Network for Remaining Useful Life Prediction of Lithium-Ion Batteries. *IEEE Access* **2022**, *10*, 19621–19628. [CrossRef]
18. Vaswani, A.; Shazeer, N.; Parmar, N.; Uszkoreit, J.; Jones, L.; Gomez, A.N.; Kaiser, Ł.; Polosukhin, I. Attention is all you need. *Adv. Neural Inf. Process. Syst.* **2017**, *30*, 5998–6008. [CrossRef]
19. Wang, L.; He, Y.; Liu, X.; Li, L.; Shao, K. M2TNet: Multi-modal multi-task Transformer network for ultra-short-term wind power multi-step forecasting. *Energy Rep.* **2022**, *8*, 7628–7642. [CrossRef]
20. Wu, H.; Meng, K.; Fan, D.; Zhang, Z.; Liu, Q. Multistep short-term wind speed forecasting using transformer. *Energy* **2022**, *261*, 125231. [CrossRef]
21. Qu, K.; Si, G.; Shan, Z.; Kong, X.; Yang, X. Short-term forecasting for multiple wind farms based on transformer model. *Energy Rep.* **2022**, *8*, 483–490. [CrossRef]
22. Xiong, B.; Lou, L.; Meng, X.; Wang, X.; Ma, H.; Wang, Z. Short-term wind power forecasting based on Attention Mechanism and Deep Learning. *Electr. Power Syst. Res.* **2022**, *206*, 107776. [CrossRef]
23. Zhou, X.; Liu, C.; Luo, Y.; Wu, B.; Dong, N.; Xiao, T.; Zhu, H. Wind power forecast based on variational mode decomposition and long short term memory attention network. *Energy Rep.* **2022**, *8*, 922–931. [CrossRef]

24. Zhang, J.; Yan, J.; Infield, D.; Liu, Y.; Lien, F. Short-term forecasting and uncertainty analysis of wind turbine power based on long short-term memory network and Gaussian mixture model. *Appl. Energy* **2019**, *241*, 229–244. [CrossRef]
25. Zhou, M.; Wang, B.; Guo, S.; Watada, J. Multi-objective prediction intervals for wind power forecast based on deep neural networks. *Inf. Sci.* **2021**, *550*, 207–220. [CrossRef]
26. He, Y.; Liu, R.; Li, H.; Wang, S. Short-term power load probability density forecasting method using kernel-based support vector quantile regression and Copula theory. *Appl. Energy* **2017**, *185*, 254–266.
27. He, Y.; Li, H. Probability density forecasting of wind power using quantile regression neural network and kernel density estimation. *Energy Convers. Manag.* **2018**, *164*, 374–384. [CrossRef]
28. Jiang, C.; Jiang, M.; Xu, Q.; Huang, X. Expectile regression neural network model with applications. *Neurocomputing* **2017**, *247*, 73–86.
29. Zhao, H.; Zhang, S.; Zhao, Y.; Liu, H.; Qiu, B. Short-term power load interval forecasting based on adaptive noise-complete empirical modal decomposition-sample entropy-long-term memory neural network and kernel density estimation. *Mod. Electr.* **2021**, *38*, 138–146.
30. Li, J.; Zhang, S.; Yang, Z. A wind power forecasting method based on optimized decomposition prediction and error correction. *Electr. Power Syst. Res.* **2022**, *208*, 107886.
31. Lu, W.; Duan, J.; Wang, P.; Ma, W.; Fang, S. Short-term Wind Power Forecasting Using the Hybrid Model of Improved Variational Mode Decomposition and Maximum Mixture Correntropy Long Short-term Memory Neural Network. *Int. J. Electr. Power Energy Syst.* **2023**, *144*, 108552. [CrossRef]
32. Ewees, A.A.; Al-qaness MA, A.; Abualigah, L.; Elaziz, M.A. HBO-LSTM: Optimized long short term memory with heap-based optimizer for wind power forecasting. *Energy Convers. Manag.* **2022**, *268*, 116022. [CrossRef]
33. Dong, Y.; Zhang, H.; Wang, C.; Zhou, X. Wind power forecasting based on stacking ensemble model, decomposition and intelligent optimization algorithm. *Neurocomputing* **2021**, *462*, 169–184. [CrossRef]
34. Zhang, Y.; Pi, Z.; Zhu, R.; Song, J.; Shi, J. Wind power prediction based on WOA-BiLSTM neural network. *Electrotechnology* **2022**, *10*, 28–31.
35. Kang, H.; Li, Q.; Yu, S.; Yao, S. Ultra-short-term wind power output prediction based on SA-PSO-BP neural network algorithm. *Inn. Mong. Power Technol.* **2020**, *38*, 64–68.
36. Yang, X.S.; Deb, S. Cuckoo search via Lévy flights. In Proceedings of the 2009 World Congress on Nature & Biologically Inspired Computing (NaBIC), IEEE, Coimbatore, India, 9–11 December 2009; pp. 210–214.
37. Yang, X.S. *Nature-Inspired Metaheuristic Algorithms*; Luniver Press: Coventry, UK, 2010; pp. 16–17.

sustainability

MDPI

Article

Lightweight Network-Based Surface Defect Detection Method for Steel Plates

Changqing Wang [1,2,3], Maoxuan Sun [1,2,3], Yuan Cao [1,2,3,*], Kunyu He [1,2,3], Bei Zhang [1,2,3], Zhonghao Cao [1,2,3] and Meng Wang [1,2,3]

1 College of Electronics and Electrical Engineering, Henan Normal University, Xinxiang 453007, China
2 Henan Key Laboratory of Optoelectronic Sensing Integrated Application, Xinxiang 453007, China
3 Henan Engineering Laboratory of Additive Intelligent Manufacturing, Xinxiang 453007, China
* Correspondence: xyuan_cao@163.com

Abstract: This article proposes a lightweight YOLO-ACG detection algorithm that balances accuracy and speed, which improves on the classification errors and missed detections present in existing steel plate defect detection algorithms. To highlight the key elements of the desired area of surface flaws in steel plates, a void space convolutional pyramid pooling model is applied to the backbone network. This model improves the fusion of high- and low-level semantic information by designing feature pyramid networks with embedded spatial attention. According to the experimental findings, the suggested detection algorithm enhances the mapped value by about 4% once compared to the YOLOv4-Ghost detection algorithm on the homemade data set. Additionally, the real-time detection speed reaches about 103FPS, which is about 7FPS faster than the YOLOv4-Ghost detection algorithm, and the detection capability of steel surface defects is significantly enhanced to meet the needs of real-time detection of realistic scenes in the mobile terminal.

Keywords: defect detection; lightweight; cavity spatial convolution; spatial attention

Citation: Wang, C.; Sun, M.; Cao, Y.; He, K.; Zhang, B.; Cao, Z.; Wang, M. Lightweight Network-Based Surface Defect Detection Method for Steel Plates. *Sustainability* **2023**, *15*, 3733. https://doi.org/10.3390/su15043733

Academic Editors: Luis Hernández-Callejo, Sergio Nesmachnow and Sara Gallardo Saavedra

Received: 18 November 2022
Revised: 9 February 2023
Accepted: 9 February 2023
Published: 17 February 2023

1. Introduction

With the rapid development of industrial automation technology, the study of automated [1,2] detection of defects in industrial production is receiving more and more attention. Due to the influence of various uncertainties, the surface of the steel plate in the production process will produce a variety of defects [3–7], such as scratches, deformation, welds, holes, etc. These defects [8–12] not only affect the integrity of the steel plate but also make a certain impact on the quality of the steel plate, so a more accurate detection of defects [13–16] on the surface of the steel plate is of paramount importance.

Conventional inspection methods use manual observation to detect defects, which is not only time-consuming and labor-intensive, but the results still do not meet the expected requirements. Based on the traditional industrial inspection methods proposed, the automated defect detection technology has been driven to a new level. Experts and scholars at home and abroad have conducted more profound research and practice on traditional machine vision in the detection of defects in steel plates. The enhanced BP detection algorithm was presented by Peng et al. [17] to detect flaws in steel plates. While this technique has a decent detection performance for flaws that are clear targets, it has a sluggish convergence rate and poor performance for small samples. Wang Yixin et al. [18] suggested a comparative detection approach utilizing machine vision; however, despite its high accuracy in recognizing faults in steel plates, it has a higher environmental impact and is incapable of detecting flaws in harsh conditions due to its difficulty with extracting feature images.

At this juncture, the accuracy of steel plate surface flaw detection [19] has increased due to the rapid development of deep learning technology in industrial inspection. Tian Siyang et al. investigated at timeframe instances of hot-rolled strip steel surface faults, identifying two faults, watermarks and water droplets. The one-stage identification YOLOv2 [20] algorithm was developed and tested for a wide range of surface flaws on steel sheets, as well as against several interference effects caused by false defects. Although the approach can detect surface flaws in hot-tied steel sheets with an average mAP of 92.54%, the detection speed of just 14 FPS prevents real-time detection. Xu Qian et al. used a modified YOLOv3 [21] network structure for the detection of surface defects in steel plates, reducing the model size of YOLOv3 by using a lighter MobileNet [22] network model, adding a cavity convolutional neural network [23] to the network to improve the defect detection capability of steel plates, and adding the Inceptionv3 structure to the network to make the number of layers richer.

In this paper, a defect detection technique for YOLO-ACG is proposed. First, the model's detection accuracy and speed have significantly increased thanks to the use of GhostNet as the replacement for the backbone network of CSPMakenet53 in the YOLOv4 network. Secondly, by replacing the spatial convolutional pooling pyramids at different scales in the original YOLOv4-Ghost network with more accurate spatial convolutional pooling pyramids in the null space, the focus of the model on the significant regions of the feature map target is increased and the perceptual regions of the feature map are enhanced by combining the semantic information of the context. Finally, the pyramid network structure of feature fusion with spatial attention mechanism is embedded in the network design, and the loss of information at the edges of the feature map is addressed by using the FPN structure to connect the fusion of two channels from top to bottom, which facilitates the fusion of information at different scales. The experimental results demonstrate that the YOLO-ACG algorithm can detect surface flaws in steel sheets more quickly and accurately than other lightweight methods, meeting the expectations of industrial inspection. The article will be followed by a more detailed analysis of the YOLO-ACG algorithm in terms of network structure and experimental data.

2. Methodology

2.1. The YOLOv4 Backbone Network

The YOLO (You Only Look Once) algorithm was put forth as a ONE-STAGE target detection technique by Redmon et al. [24] in 2016. The fundamental idea behind the YOLO algorithm is to approach the object recognition problem as a regression problem and utilize a convolutional neural network [25] structure to directly forecast the bounding box and category probabilities from the input image. The fourth iteration of the YOLO algorithm, YOLOv4, employs a variety of algorithmic network architectures, including feature pyramid networks and complete convolutional networks. The CSPDarknet53 backbone network, which is seen in Figure 1, replaces the YOLOv3 algorithm's Darknet53 backbone network topology. Additionally, YOLOv4 use a Mish function for the activation function, logistic regression for the categorization of images, and a feature pyramid network for multi-scale target detection, all of which maintain a high accuracy rate and ensure real-time monitoring.

The YOLOv4 algorithm's backbone network is CSPDakrnet53, which is also one of the best backbone networks. CSPDarknet53 generates three outputs, designated P3, P4, and P5, after applying convolutional layers 1×1 and 3×3. In this process, P3 and P4 are convolved once for 1×1 and then input to the enhanced feature extraction network for feature fusion. P5 is convolved three times and input to the void pyramid pooling layer, with the pooling results being input to the enhanced feature extraction network for feature fusion.

2.2. GhostNet

Yolov4-ghost will replace the current YOLOv4 backbone network with the GhostNet module [26], making the network lighter and easier to deploy on mobile terminals. The head network uses a PAN (path aggregation network) network structure, while the backbone network consists of convolution, spatial pyramid pooling (SPP), and GhostBottleneck.

	Type	Filters	Size	Output
	Convolutional	32	3×3	256×256
	Convolutional	64	3×3/2	128×128
1×	CrossStagePartial			
	Convolutional	32	1×1	
	Convolutional	64	3×3	
	Residual			128×128
	Convolutional	128	3×3/2	64×64
2×	CrossStagePartial			
	Convolutional	64	1×1	
	Convolutional	64	3×3	
	Residual			64×64
	Convolutional	256	3×3/2	32×32
8×	CrossStagePartial			
	Convolutional	128	1×1	
	Convolutional	128	3×3	
	Residual			32×32
	Convolutional	512	3×3/2	16×16
8×	CrossStagePartial			
	Convolutional	256	1×1	
	Convolutional	256	3×3	
	Residual			16×16
	Convolutional	1024	3×3/2	8×8
4×	CrossStagePartial			
	Convolutional	512	1×1	
	Convolutional	512	3×3	
	Residual			8×8

Figure 1. CSPDarknet53 backbone network.

GhostBottleneck is replaced by GhostNet as the network's hub. In place of other network modules, a plug-and-play reusable module called GhostBottleneck dramatically decreases the computational load and model volume of method models. Two Ghost modules are stacked to create a GhostBottleneck. The primary goal of the first Ghost module is to enable the addition of additional channels and dimensions to the feature extraction, usually in the form of an extension layer. The second Ghost module checks to see if the feature extraction dimension still matches its input after lowering the number of channels. The input and output of the two Ghost modules are finally connected. When the second Ghost module is utilized, the ReLU function is not. Because of the variation in the input data distribution between the front layer and the back layer following the activation function, constant matching is required, which reduces training efficiency.

2.3. Loss Function

Steel plates can have a wide variety of imperfections, so the algorithm used in the detection process needs to be extremely precise to identify the types and locations of flaws. The three components that make up the loss function are as follows: (1) confidence loss; (2) classification loss; and (3) bounding box regression loss.

$$\begin{aligned} loss_a &= \sum_{i=0}^{S^2}\sum_{j=0}^{B} W_{ij}^{obj}[\hat{C}_i^j \log(C_i^j) + (1-\hat{C}_i^j)\log(1-C_i^j)] \\ &-l_{noobj}\sum_{i=0}^{S^2}\sum_{j=0}^{B}(1-W_{ij}^{obj})[\hat{C}_i^j \log(C_i^j) + (1-\hat{C}_i^j)\log(1-C_i^j); C_i^j = P_{i,j} * IUO_{pred}^{truth} \end{aligned} \quad (1)$$

$$loss_b = \sum_{i=0}^{S^2}\sum_{j=0}^{B} W_{ij}^{obj}\sum_{c=1}^{C}[\hat{p}_i^j \log(p_i^j(c)) + (1-\hat{p}_i^j(c))\log(1-p_i^j(c))] \quad (2)$$

$$loss_c = 1 - IOU + \frac{\rho(d,d^{st})}{c^2} + \alpha v;\ v = \frac{4}{\pi}(\arctan\frac{w^{gt}}{h^{gt}} - \arctan\frac{w}{h})^2 \quad (3)$$

where W_{ij}^{obj} represents whether the jth anchor box is in the predicted ith grid, C_i^j is represented by the confidence that the jth bounding box in the ith grid has produced, $P_{i,j}$

represents the objective function's discriminant, $\hat{C}_i^j$ shows the measurement's actual value; $\hat{P}_i^j(c)$ is represented as the expected likelihood that the *j*th bounding box in I grids belongs to c; $P_i^j(c)$ is defined as the actual likelihood that *j*th bounding box in *i*th grid belongs to c.; d^{st}, d as the intersection of the true and anticipated boxes, ρ is represented by the location of the d^{st}, d centroid. In Figure 2, where the square box represents the prediction box and the rectangular box represents the real box, c is the diagonal distance between the minimum closed loops of the two boxes.

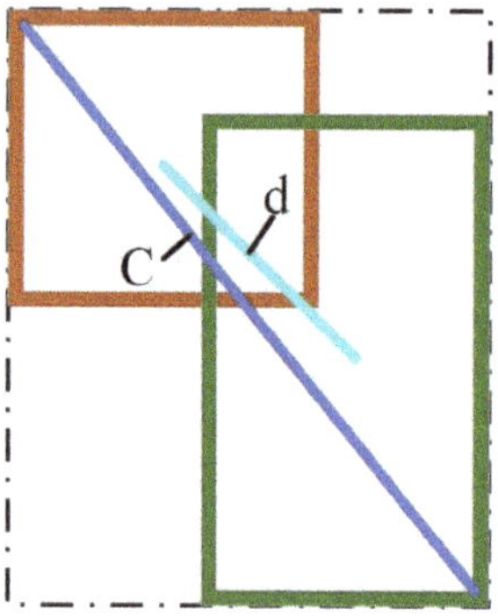

Figure 2. Anchor box.

3. Our Approach

3.1. YOLO-ACG Algorithm

This study proposes the three-part YOLO-ACG network model, which is based on the YOLOv4 algorithm and is visible in Figure 3. It includes a backbone network, a feature fusion network, and a detection head network. RGB images with a three-channel output are used as the input. The first step is to enlarge the feature scale of the image to 52 × 52, 26 × 26 and 13 × 13 for image information screening and extraction through the P3, P4, and P5 levels of the backbone network. Second, the extracted results are sent to the CBM and CA attention mechanisms, and the ASPP module is added to the output of the extracted features to enhance the effective recognition of target defect differences by merging global and local characteristics with various perceptual fields. The CA attention mechanism then reinforces the output features to improve their location correlations and interactions across latitudes. The problem of the higher-layer network losing the feature information of the lower-layer network during the information extraction process is effectively resolved by fusing the extracted three-feature layers with various semantic information, passing the feature information, and allowing the feature information to enter the feature fusion network after the CA attention mechanism is finished. Finally, the non-maximal suppression algorithm (NMS), whose thresholds are used to further filter the redundant anchor frames during the NMS processing, is combined with the center distance factor of the prediction frame to create the final prediction frame.

3.2. Ghost Module

The redundancy of feature maps in neural convolutional networks is one of the most important features. When the feature maps are output for visualization, there are many outputs with very similar features, which can be obtained by simple linear transformations without complicated operations. As shown in Figure 4, the working principle of the standard convolution and Ghost modules are presented separately.

As shown in Figure 4a for standard convolution, although ordinary convolution extracts features by using a large number of convolutions, and thus, generates a feature map, the excessive number of convolution kernels with the number of channels generates redundant information and leads to an increase in computation. Additionally, the Ghost module in Figure 4b separates the regular convolution into two parts, extracts the partial convolution to create the feature map, and then efficiently creates the whole feature map using straightforward linear operations.

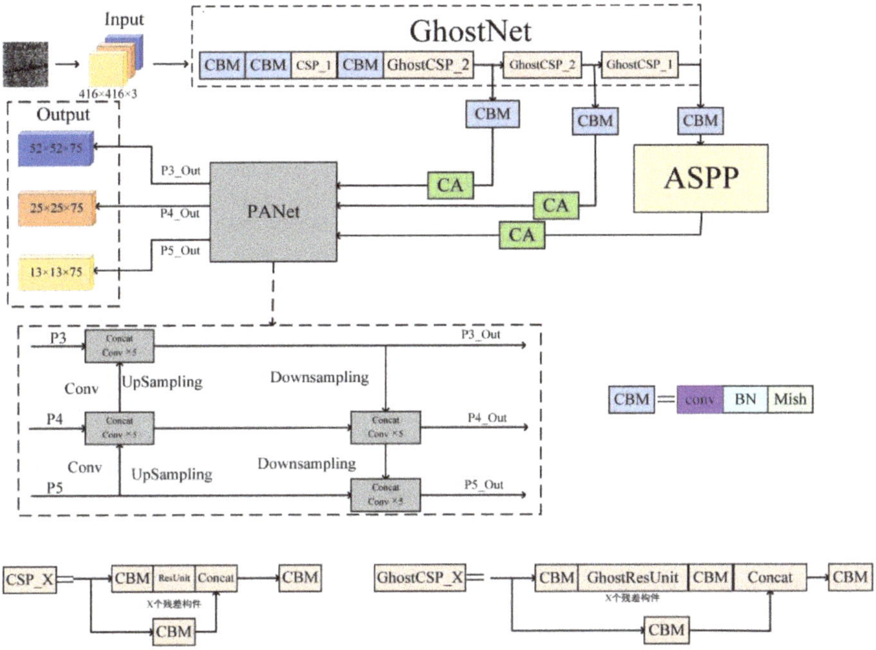

Figure 3. YOLO-ACG network model.

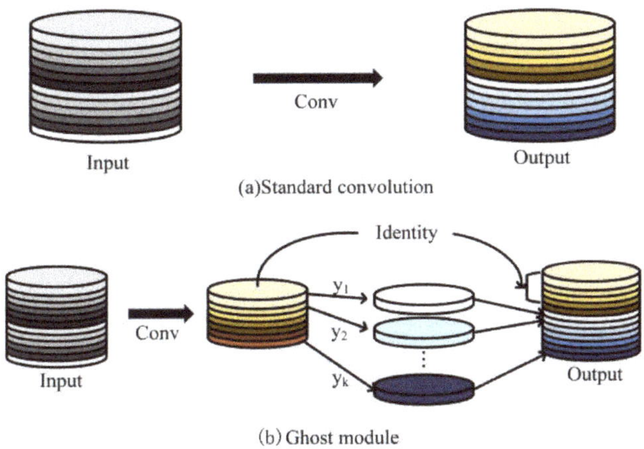

Figure 4. Standard convolution and Ghost module.

To reflect the benefits of Ghost convolution in convolutional computing, the input feature map's width, height, and channel count are all assumed to be w, h, and c, respectively. The output after one convolution is $n * h' * w'$, where k and d are the sizes of the linearly variable convolution kernel and standard convolution kernel, respectively. Equation (4) illustrates that the amount of convolution computation performed in the Ghost module serves as the denominator and the numerator of the equation, respectively.

Comparing the two convolutional calculations reveals that for the same parameters, the normal convolutional computation is s times larger than the convolutional computation in the Ghost module.

$$\begin{aligned} r_s &= \frac{n \cdot h' \cdot w' \cdot c \cdot k \cdot k}{\frac{n}{s} \cdot h' \cdot w' \cdot c \cdot k \cdot k + (s-1) \cdot \frac{n}{s} \cdot h' \cdot w' \cdot d \cdot d} \\ &= \frac{c \cdot k \cdot k}{\frac{1}{s} \cdot c \cdot k \cdot k + \frac{s-1}{s} \cdot d \cdot d} \approx \frac{s \cdot c}{s+c-1} \approx s \end{aligned} \tag{4}$$

3.3. Improved ASPP Module

The SPP [27] structure serves as the pyramidal pooling module for the whole YOLOv4 network. Large numbers of picture features must be stored in the SPP structure, and feature extraction from the feature map requires laborious multi-stage training that takes too long. The ASPP structure addresses the drawbacks of SPP by substituting the pooling process in the SPP structure with null convolution, which is typically utilized as the global feature extraction of the feature image and can be employed in the feature map with emphasis on the picture, preventing the loss of image data, although improving the semantic segmentation ability of the feature map and enhancing the perceptual field also results in the loss of information details on the edges of the feature map. The ASPP structure is a good answer to the aforementioned issue since it increases the perceptual field of the feature map without losing the finer details of the edge information.

The ASPP structure has two parts, the first part consists of an 1 × 1 convolutional layer and three 3 × 3 null convolutional layers with sampling rates of [6,12,18], respectively, whose convolutional kernels have a size of 256; the second part is a convolutional operation of 1 × 1 by global level pooling, and the same convolutional kernels also have a size of 256. Figure 5 depicts the ASPP module's structural layout. The method of null convolution and upsampling and the multiscale structure are used to realize the feature extraction of images in the environment of high resolution and perceptual field. This makes a significant improvement in the perceptual field of the feature image and the processing of the details of the edge of the feature image. The expansion rate is introduced in the convolution layer, expressed as the number of zero values in the convolution kernel.

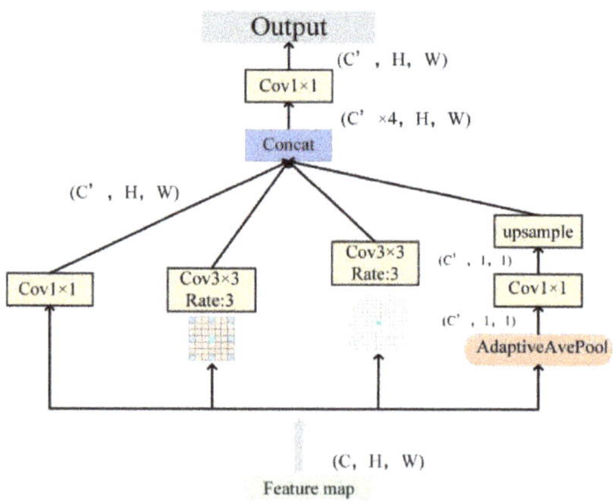

Figure 5. ASPP module.

3.4. CA Attention Mechanism Module

It was discovered throughout the target detection procedure that there was a lack of effectiveness in the detection of subtle faults. To solve this issue, the network was enhanced with the CA (coordinate attention) spatial attention mechanism [28], which increases the significance of the location relationship and cross-latitude interaction on the channel attention mechanism and improves the accuracy and sensitivity of the entire network model to the information and location of the defective targets. To solve this issue, the CA (coordinate attention) spatial attention mechanism was incorporated into the network to increase the significance of its location relationship and cross-latitude interaction on the channel attention mechanism, making the entire network model more accurate and sensitive to the information and location of the defective targets. Figure 6 depicts the structure of the added CA attention network, which involves pooling the feature maps globally to obtain feature information in both directions. Sub-associative fusion and 1×1 convolutional transforms were then used to account for feature variation; finally, the integrated feature maps were divided into two feature maps with an equal number of channels by two 1×1 convolutions through transformation before being output by the added CA attention network. The CA module encodes the feature map's precise location to produce the width and height, such as the output concatenate for feature fusion, which is represented in Equation (5); Equation (6) represents the feature transformation of two independent features to make the input's dimensionality consistent; combining g^n and g^m to create a weight matrix in Equation (7) represents the outcome.

$$f = \beta(F([z^n, z^m])) \tag{5}$$

$$g^n = \delta(F_n(f^n)), g^m = \delta(F_m(f^m)) \tag{6}$$

$$y_a(i,j) = x_a(i,j) \times g_c^n(i) \times g_c^m(j) \tag{7}$$

where f denotes the mapped feature map, β denotes the nonlinear activation function, z^n and z^m denote the horizontal and vertical position relationship, g^n and g^m denote the feature map after the sigmoid output of two identical number of channels. Finally, x_a denotes the connected jump feature information.

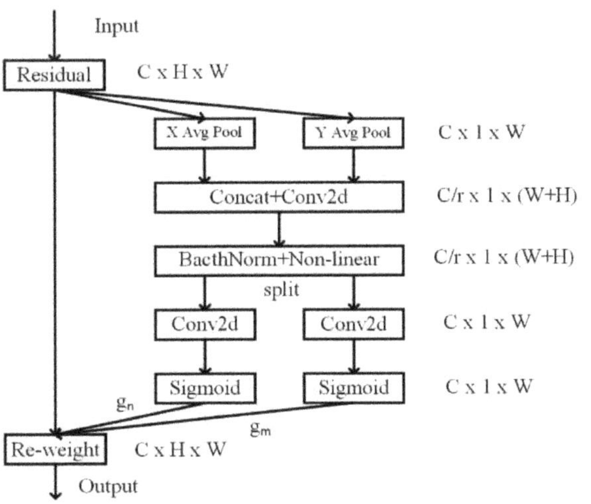

Figure 6. CA attention network.

4. Experimental Preparation

4.1. Test Environment

The experimental platform is Win10 OS, CPU is 12th Gen Intel(R) Core(TM) i7-12700KF 3.60 GHz; memory is 64 GB; GUP is NVIDIA GeForce RTX3090Ti.Pytorch 1.10.2 is used; the software runs in Anaconda 3.6; Cuda 10.0 and Cudnn 7.5 were installed to help speed up the GUP process, and Tensorflow 1.13.1, Opencv4.1, and Numpy 1.14.2 were installed in the environment. The auxiliary databases were installed to make the code run correctly.

4.2. Production of Data Set

In terms of data set, it was discovered that the three types of flaws that have the most impact on the quality of steel plate during the identification of defects in steel plate were weld, hole, and scratch. The number of feature photos offered by these three categories of steel plate defects was insufficient to meet the training needs, even though the existing data collection of the surface defects of German DAGM steel plate has a total of 10 types of steel plate defects. Therefore, the data set was expanded by combining the actual scene shooting with the public data set. A total of 4500 defect feature images were obtained from the entire self-made data set through on-site collection and selection of public data sets. Because it was discovered that the format and size of the feature map would influence the detection efficiency during the detection process, LabelImg labeling software was used to label the area of each type of image in proportion to the area to be labeled, and the length and width of the image were less than or equal to 3:1. This will help to better train the neural convolution network model. XML files should be used to store critical data from marked defect boxes in order to apply it to neural convolution network learning.

The target's proportion in the image varies slightly as a result of variations in the camera's viewing distance, and the model's capacity for adaptation is reduced by the various target sizes. The method of random scaling, clipping, and distribution of the logarithmic data set, which is more accurate for the training of the data set, was utilized to carry out the random splicing of photos in the preprocessing stage to tackle this problem. Figure 7 displays an illustration of data enhancement.

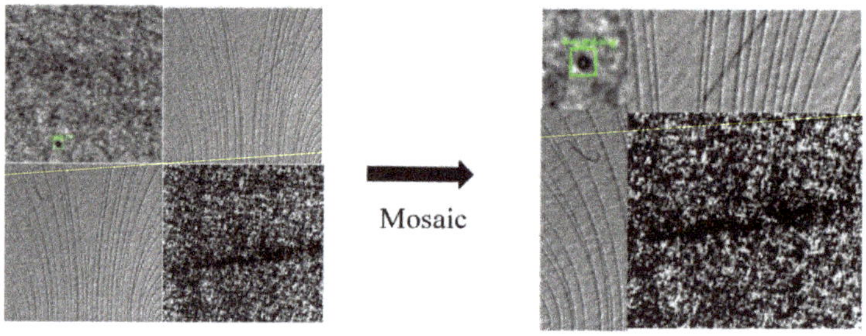

Figure 7. Example of data enhancement.

5. Results and Discussion

The self-made data set utilized in this experiment randomly separated the training set and test set at a ratio of 6:4. Several ablation experiments were then set up to assess the effects of each model improvement on the training effect and to choose the best model. The usefulness of the proposed algorithm was further confirmed by several sets of comparison experiments with the better steel defect detection methods already in use, and the superiority of the algorithm was assessed by the average detection accuracy (map) and the detection speed (FPS).

5.1. Training Model

The experiments in this paper were carried out in accordance with the predetermined parameters: the image input size was 416 × 416; the epoch was set to 300 rounds; the batch size was 128 for the first 50 rounds and 64 for the last 250 rounds; the learning rate was 1×10^{-2} for the first 50 rounds and 1×10^{-3} for the last 250 rounds; momentum was the amount of stochastic gradient descent in order to obtain a better convergence effect, momentum was set at 0.937. Momentum is the expression of the reduction in the learning efficiency of the initialization of the weights in terms of momentum. Figure 8 depicts the loss curve during the training phase.

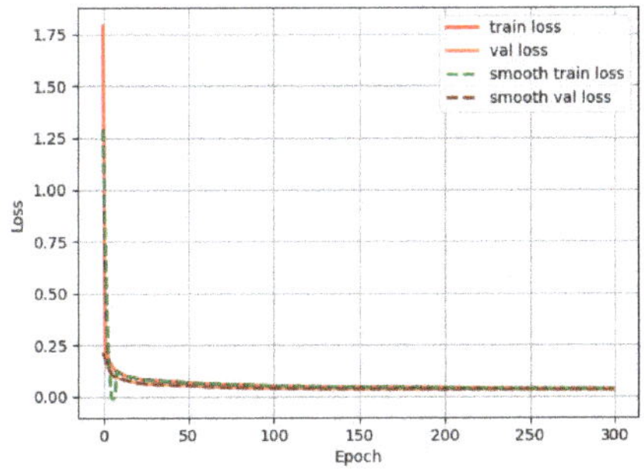

Figure 8. Loss curve during training.

It can be observed from the figure that the loss value decreases continuously as the epoch increases, and when the training proceeds to 50 rounds the loss curve is basically in a stable state without generating an overfitting situation, and the loss value of the YOLO-ACG algorithm converges to about 0.19 and 0.14 by increasing the accuracy of the recognition model, so the overall parameter setting of the algorithm is reasonable.

5.2. Comparison Experiment

The YOLO-ACG method is compared with current popular detection techniques, such as YOLOv4, YOLOv4-MobileNetv1, YOLOv4-MobileNetv2 [29], YOLOv4-MobileNetv3, YOLOv3-tiny and YOLOv4-tiny, in a self-made data set, as shown in Table 1, to confirm that the algorithm's improvement is more genuine and reliable.

Table 1. Data set to compare the experimental results.

Model	mAP	Model Size/MB	FPS
YOLOv4	96.35%	244.7	85.3
YOLOv4-MobileNetv1	88.39%	40.95	47.6
YOLOv4-MobileNetv2	89.52%	39.06	40.1
YOLOv4-MobileNetv3	89.75%	39.99	43.2
YOLO-ACG	92.49%	69.82	102.91

The chart shows that when employing a big network model, such as the YOLOv4 [30] detection method, the algorithm has a very high detection accuracy of 96.35% but the model's size is relatively large at 244.7 MB, making it difficult to deploy on mobile devices. Some lightweight models, Models 2 to 6 in Table 1, enhance the YOLO algorithm. The comparison results reveal that the lightweight model's size has been significantly decreased in comparison to YOLOv4, which is more readily implemented in mobile devices, but that the model's detection accuracy has been significantly reduced in comparison to the YOLOv4 algorithm.

In view of this, the YOLO-ACG network model proposed in this paper takes into account the computational speed of the model, the detection accuracy, and the size of the model. Its model size is about 1/4 times that of the YOLOv4 network model, although slightly higher than that of Table 2–6 models, about 1/3 times that of 2–6 network models. The suggested approach has unquestionable advantages in terms of operation speed, reaching up to 103FPS. It surpasses roughly 18FPS, as compared to YOLOv4. It exceeds by almost two times 2–6 models, realizing high-speed detection. The accuracy of the YOLOv4-ACG model is around 2% greater than that of the models in Table 2–6, although being about 4% lower than that of the YOLOv4 model. From the above three aspects, we can see that YOLO-ACG is more efficient when deployed on mobile devices.

5.3. Ablation Experiment

The ablation experiment is to improve different modules based on the YOLOv4 Ghost algorithm and use self-made data sets to conduct training and performance evaluation. Table 2 shows the comparison of evaluation results of all models.

Table 2. Training results of different algorithm models.

Experiment	SPP	ASPP	SE	ECA	CBAM	CA	mAP	FPS	Size/MB	Recall
1	√						89.17%	95.88	43.63	67.34%
2	√		√				88.49%	96.53	44.25	71.42%
3	√			√			88.37%	95.17	44.61	71.91%
4	√				√		87.84%	96.01	44.26	70.42%
5	√					√	88.61%	97.88	43.84	68.77%
6		√					91.64%	97.89	69.57	75.81%
7		√	√				91.09%	94.28	70.23	76.52%
8		√		√			90.49%	96.53	69.63	72.26%
9		√			√		89.76%	95.36	70.26	72.72%
10		√				√	92.49%	102.91	69.12	77.49%

As can be seen from the table, experiments conducted with the introduction of SPP and ASPP modules in the model reveal that the model size of the algorithm is approximately 1.5 times larger when the ASPP module is introduced than when the SPP module is introduced. In terms of accuracy, it is about 2% higher than the SPP module. In terms of recall rate, there is a significant improvement of approximately 8% compared to the SPP module. The detection speed is also improved compared with the introduction of the SPP module, which can reach about 98FPS. The comparison of the total ablation experiment reveals that, despite the model's size being only slightly larger—by about a third—than that of the method with the SPP module, it has far faster and more accurate detection. As a result, the ASPP module is elected as the algorithm's primary pooling layer. It can be shown from comparative tests 6–9 that the accuracy and speed of the algorithm in the detection algorithm with only the ASPP module are marginally improved, while the model size is slightly decreased, when compared to the introduction of ASPP module and the addition of SE [31], ECA, and CBAM. By comparing Experiments 6 to 10, it can be seen that Experiment 10's model size is only slightly smaller than Experiments 6 to 9's. In terms of recall, Experiment 10 exceeded Experiments 6 to 9 by about 2% to 5%, but experiment 10's precision is about 3% to 4% higher, and Experiment 10's speed is about 103 FPS, which is higher than Experiment 6 to 9's speed by about 6 to 9 FPS.

The aforementioned studies demonstrate that the revised YOLO-ACG algorithm approaches are efficient and increase the model's accuracy in the detection of steel plate surface flaws. In order to identify steel plate surface flaws more effectively, it possesses the qualities of quick detection speed, lightweight models, and ease of deployment in real-world situations.

6. Conclusions

This study suggests the YOLO-ACG method addresses the shortcoming of the YOLOv4 algorithm in handling flaw identification of steel plate data. From the three points below, the algorithm has been improved. First, lower the model size and substitute the present YOLOV4 method's backbone network with the lightweight Ghost module to make the algorithm simple to install on mobile devices. Then, to increase the maximum pooling efficiency of the YOLOv4 algorithm, the ASPP module is introduced to replace the maximum pooling layer. This considerably enhances the processing of the feature image's edge details and the feature image's receptive field. Finally, employing the pyramid feature fusion network CA module enables the enhancement of feature map fusion effectiveness in various scale spaces and further enhances the analysis of feature maps' edge information.

From the analysis of the experimental results, the proposed YOLO-ACG target detection algorithm applied to a homemade data set has a higher mAP of about 3% compared to the existing YOLOv4-MobileNet algorithm model. In comparison with the size of the YOLOv4 algorithm model, the proposed algorithm is about 1/4 of the YOLOv4 algorithm model. The detection speed of the YOLO-ACG algorithm can reach about 103 FPS, which is twice as fast as the existing YOLOv4-MobileNet algorithm model. Therefore, YOLO-ACG target detection has significantly improved the ability to detect defects on the surface of steel plates and meet the mobile requirements for the real-time detection of realistic scenes.

Author Contributions: Conceptualization, C.W. and M.S.; methodology, C.W.; software, M.S.; validation, M.S.; formal analysis, C.W.; investigation, M.S.; resources, C.W.; data curation, M.S.; writing—original draft preparation, M.S.; writing—review and editing, C.W., Y.C., B.Z., K.H., Z.C. and M.W. All authors have read and agreed to the published version of the manuscript.

Funding: This project was supported by the National Natural Science Foundation of China (Fund Numbered 52177004).

Institutional Review Board Statement: Not applicable.

Informed Consent Statement: Not applicable.

Data Availability Statement: The data sets used and analyzed in the current study are available from the corresponding author on reasonable request.

Conflicts of Interest: The authors declare no conflict of interest.

References

1. Chen, K.; Li, H.; Li, C.; Zhao, X.; Wu, S.; Duan, Y.; Wang, J. An Automatic Defect Detection System for Petrochemical Pipeline Based on Cycle-GAN and YOLO v5. *Sensors* **2022**, *22*, 7907. [CrossRef] [PubMed]
2. Dai, J.; Li, T.; Xuan, Z.; Feng, Z. Automated Defect Analysis System for Industrial Computerized Tomography Images of Solid Rocket Motor Grains Based on YOLO-V4 Model. *Electronics* **2022**, *11*, 3215. [CrossRef]
3. Jung, H.; Rhee, J. Application of YOLO and ResNet in Heat Staking Process Inspection. *Sustainability* **2022**, *14*, 15892. [CrossRef]
4. Zhao, Z.; Ge, Z.; Jia, M.; Yang, X.; Ding, R.; Zhou, Y. A Particleboard Surface Defect Detection Method Research Based on the Deep Learning Algorithm. *Sensors* **2022**, *22*, 7733. [CrossRef] [PubMed]
5. Zhang, Y.; Liu, X.; Guo, J.; Zhou, P. Surface Defect Detection of Strip-Steel Based on an Improved PP-YOLOE-m Detection Network. *Electronics* **2022**, *11*, 2603. [CrossRef]
6. Luo, Q.; Fang, X.; Liu, L.; Yang, C.; Sun, Y. Automated visual defect detection for flat steel surface: A survey. *IEEE Trans. Instrum. Meas.* **2020**, *69*, 626–644. [CrossRef]
7. Shi, T.; Kong, J.; Wang, X.; Liu, Z.; Zheng, G. Improved Sobel algorithm for defect detection of rail surfaces with enhanced efficiency and accuracy. *J. Cent. South Univ.* **2016**, *23*, 2867–2875. [CrossRef]
8. Thomas, B.G.; Jenkins, M.S.; Mahapatra, R.B. Investigation of strand surface defects using mould instrumentation and modelling. *Ironmak. Steelmak.* **2004**, *31*, 485–494. [CrossRef]
9. Liu, Y.; Xu, K.; Wang, D. Online surface defect identification of cold rolled strips based on local binary pattern and extreme learning machine. *Metals* **2018**, *8*, 197. [CrossRef]
10. Wang, Z.; Zhu, D. An accurate detection method for surface defects of complex components based on support vector machine and spreading algorithm. *Measurement* **2019**, *147*, 106886. [CrossRef]
11. Kang, G.W.; Liu, H.B. Surface defects inspection of cold rolled strips based on neural network. In Proceedings of the 2005 International Conference on Machine Learning and Cybernetics, Guangzhou, China, 18–21 August 2005; Volume 8, pp. 5034–5037.
12. Di, H.; Ke, X.; Peng, Z.; Zhou, D. Surface defect classification of steels with a new semi-supervised learning method. *Opt. Lasers Eng.* **2019**, *117*, 40–48. [CrossRef]
13. Schlegl, T.; Seeböck, P.; Waldstein, S.M.; Schmidt-Erfurth, U.; Langs, G. Unsupervised anomaly detection with generative adversarial networks to guide marker discovery. In *International Conference on Information Processing in Medical Imaging*; Springer: Berlin/Heidelberg, Germany, 2017; pp. 146–157.
14. Lee, S.Y.; Tama, B.A.; Moon, S.J.; Lee, S. Steel surface defect diagnostics using deep convolutional neural network and class activation map. *Appl. Sci.* **2019**, *9*, 5449. [CrossRef]
15. Tabernik, D.; Šela, S.; Skvarč, J.; Skočaj, D. Segmentation-based deep-learning approach for surface-defect detection. *J. Intell. Manuf.* **2020**, *31*, 759–776. [CrossRef]
16. Prappacher, N.; Bullmann, M.; Bohn, G.; Deinzer, F.; Linke, A. Defect detection on rolling element surface scans using neural image segmentation. *Appl. Sci.* **2020**, *10*, 3290. [CrossRef]
17. Peng, K.; Zhang, X. Classification technology for automatic surface defects detection of steel strip based on improved BP algorithm. In Proceedings of the 2009 Fifth International Conference on Natural Computation, Tianjin, China, 14–16 August 2009; pp. 110–114.

18. Wang, Y.X.; Guang-Hui, Y.U.; Qiang, X.U. A Machine Vision Based Printing Defect Detection Technology for Product Packaging. *J. Jiangsu Univ. Technol.* **2019**, *25*, 7–14.
19. Wang, Z.Y. Research on steel plate surface defects detection method based on machine vision. *Comput. Modern* **2013**, *7*, 97–117.
20. Wang, L.; Wei, C.; Li, W.; Zhang, Y. Pedestrian detection based on YOLOv2 with pyramid pooling module in underground coal mine. *Comput. Eng. Appl.* **2018**, *55*, 133–139.
21. Yang, W.; Zhou, G.L.; Gu, Z.W.; Jiang, X.D.; Lu, Z.M. Safety Helmet Wearing Detection Based on an Improved Yolov3 Scheme. *Int. J. Innov. Comput. Inf. Control.* **2022**, *18*, 973–988.
22. Liu, M.; Mao, J. Deep Face Recognition Algorithm Based on Improved Mobilenet Algorithm. *Inf. Commun. Technol.* **2019**, *1*, 41–46.
23. Cheng, S.; Zhou, B. Recognition of Characters in Aluminum Wheel Back Cavity Based on Improved Convolution Neural Network. *Comput. Eng.* **2019**, *45*, 182–186.
24. Redmon, J.; Divvala, S.; Girshick, R.; Farhadi, A. You only look once: Unified, real-time object detection. In Proceedings of the IEEE Conference on Computer Vision and Pattern Recognition, Las Vegas, NV, USA, 26 June–1 July 2016; pp. 779–788.
25. Krizhevsky, A.; Sutskever, I.; Hinton, G.E. Imagenet classification with deep convolutional neural networks. *Commun. ACM* **2017**, *60*, 84–90. [CrossRef]
26. Han, K.; Wang, Y.; Tian, Q.; Guo, J.; Xu, C.; Xu, C. Ghostnet: More features from cheap operations. In Proceedings of the IEEE/CVF Conference on Computer Vision and Pattern Recognition, Seattle, WA, USA, 14–19 June 2020; pp. 1580–1589.
27. Yang, K.; Jiao, Z.; Liang, J.; Lei, H.; Li, C.; Zhong, Z. An application case of object detection model based on Yolov3-SPP model pruning. In Proceedings of the 2022 IEEE International Conference on Artificial Intelligence and Computer Applications (ICAICA), Dalian, China, 24–26 June 2022; pp. 578–582.
28. Hou, Q.; Zhou, D.; Feng, J. Coordinate attention for efficient mobile network design. In Proceedings of the IEEE/CVF Conference on Computer Vision and Pattern Recognition, Virtual, 19–25 June 2021; pp. 13713–13722.
29. Sandler, M.; Howard, A.; Zhu, M.; Zhmoginov, A.; Chen, L.-C. Mobilenetv2: Inverted residuals and linear bottlenecks. In Proceedings of the IEEE Conference on Computer Vision and Pattern Recognition, Salt Lake City, UT, USA, 18–23 June 2018; pp. 4510–4520.
30. Bochkovskiy, A.; Wang, C.Y.; Liao HY, M. Yolov4: Optimal speed and accuracy of object detection. *arXiv* **2020**, arXiv:2004.10934.
31. Hu, J.; Shen, L.; Sun, G. Squeeze-and-excitation networks. In Proceedings of the IEEE Conference on Computer Vision and Pattern Recognition, Salt Lake City, UT, USA, 18–23 June 2018; pp. 7132–7141.

applied sciences

MDPI

Article

TCNformer Model for Photovoltaic Power Prediction

Shipeng Liu [1,2], Dejun Ning [1,*] and Jue Ma [1,2]

[1] Shanghai Advanced Research Institute, Chinese Academy of Sciences, Shanghai 200120, China
[2] University of Chinese Academy of Sciences, Beijing 100049, China
* Correspondence: ningdj@sari.ac.cn; Tel.: +86-186-1653-6063

Abstract: Despite the growing capabilities of the short-term prediction of photovoltaic power, we still face two challenges to longer time-range predictions: error accumulation and long-term time series feature extraction. In order to improve the longer time range prediction accuracy of photovoltaic power, this paper proposes a seq2seq prediction model TCNformer, which outperforms other state-of-the-art (SOTA) algorithms by introducing variable selection (VS), long- and short-term time series feature extraction (LSTFE), and one-step temporal convolutional network (TCN) decoding. A VS module employs correlation analysis and periodic analysis to separate the time series correlation information, LSTFE extracts multiple time series features from time series data, and one-step TCN decoding realizes generative predictions. We demonstrate here that TCNformer has the lowest mean squared error (MSE), mean absolute error (MAE) and mean absolute percentage error (MAPE) in contrast to the other algorithms in the field of the short-term prediction of photovoltaic power, and furthermore, the effectiveness of each module has been verified through ablation experiments.

Keywords: transformer; SkipGRU; TCN; photovoltaic power prediction; time series data prediction

Citation: Liu, S.; Ning, D.; Ma, J. TCNformer Model for Photovoltaic Power Prediction. *Appl. Sci.* **2023**, *13*, 2593. https://doi.org/10.3390/app13042593

Academic Editor: Sergio Nesmachnow

Received: 7 February 2023
Revised: 15 February 2023
Accepted: 16 February 2023
Published: 17 February 2023

1. Introduction

At present, with the rapid development of perovskite solar cell technology [1,2], the maximum efficiency [3] and stability [4] of photovoltaic power have been greatly improved. Photovoltaic power is increasingly important in the field of new energy. According to the data of the International Energy Agency (IEA), the growth rate of global photovoltaic installed capacity has reached as much as 49%. It is estimated that global photovoltaic power will reach 16% of the total power in 2050 [5]. At the same time, China is promoting the construction of a new power system with new energy as the principal part. Photovoltaic power using solar energy is an important branch of new energy and one of the important means for China to achieve the goal of carbon neutrality. After the large-scale integration of photovoltaic power stations into the energy network, the manner by which to accurately predict photovoltaic power and then accordingly dispatch the power grid has become an urgent problem to be addressed. Therefore, improving the prediction accuracy of photovoltaic power is significant for improving the operation efficiency of power stations themselves and for maintaining the stability of power grids.

Many scholars in China and abroad have carried out a lot of research on the prediction of photovoltaic power. At present, the mainstream prediction methods focus on traditional random learning and deep learning methods. In the field of traditional random learning, literature [6] uses historical weather data and historical power data as inputs of a support vector machine (SVM) to build a short-term photovoltaic power prediction model, which has a higher level of accuracy than the traditional autoregressive model (AR) or the radial basis function (RBF) models. One study [7] proposed a model based on Support Vector Regression (SVR) and achieved better prediction performance. In the field of deep learning, recurrent neural network (RNN) structures, such as long short-term memory (LSTM), gated recurrent unit (GRU), and seq2seq structural models, are widely used to analyze and predict time series data for such applications as stock price prediction [8], gold price prediction [9], traffic flow [10], voice classification [11], etc. The prediction of photovoltaic power can also be regarded as a kind of time series data prediction, so the above algorithms have been used to predict short-term global horizontal irradiance (GHI) or comprehensive solar loads [12,13]. Furthermore, in order to ensure the accuracy as much as possible and reduce the training time, the GRU network has been applied to short-term photovoltaic power prediction [14], and the multivariable GRU model [15–17] has been used to predict solar irradiance or power. Some hybrid models have been applied in the field of photovoltaic power generation prediction, such as the combination of a deep learning model and a heuristic algorithm [18,19], the combination of a deep learning model and a traditional random learning method [20,21], the combination of multiple deep learning models [22,23], etc. The seq2seq structural model represented by the Transformer series model takes the photovoltaic power prediction problem as a experimental sample of its model, such as Autoformer and Informer [24,25]. However, in these models, usually the photovoltaic power prediction data are only used for prediction; that is, the corresponding weather data is not fully used, and the time series features of the data are not fully extracted.

Compared with traditional LSTM, GRU, and other models, the Transformer series seq2seq model can avoid the problem of error accumulation and read longer input data [26], but it is still limited by the length of the input data. It is difficult for the seq2seq series model to capture longer time series features. For this problem, [27] proposes long- and short-time series network (LSTnet) models. The Skip recurrent neural network (Skip RNN) structure is used to capture more long-term time series features.

Based on the above analysis, the current research mainly focuses on the prediction of data within a few hours. When applied to predict a longer time range [28] for photovoltaic power, these methods typically suffer from two major challenges: error accumulation and long-term time series feature extraction in order to simultaneously extract multiple time series features in the historical data of photovoltaic power and weather factors, and to avoid error accumulation. Inspired by the application of LSTM, LSTnet, and Transformer series models in the field of photovoltaic power prediction, this paper proposes a long and short temporary correction network (TCNformer), and we verified the model by using the real data of a photovoltaic station in Australia. According to the experimental results, the TCNformer model greatly optimizes various indicators compared with LSTM, SkipGRU, Transformer, and Informer, improving the accuracy of photovoltaic power prediction.

The contributions of this paper include the following:

(1) According to the different impacts of various weather factors on photovoltaic power generation, a VS module was designed to screen and process data through correlation analysis and periodic analysis of data.
(2) Aiming at the challenge to extract long-term time series features due to the limitation of the traditional Transformer, a LSTFE module was designed to extract multiple time series features through LSTM and a SkipGRU network.
(3) In order to improve the temporal feature extraction and avoid error accumulation, one-step temporal convolutional network (TCN) decoding was used to realize the generative prediction.

2. Preliminary

2.1. Time Series Features of Photovoltaic Power Data

According to the literature [29,30], the current photovoltaic power prediction problem is usually defined as a time series data prediction problem. However, as the time granularity increases, the degree of the photovoltaic power data affected by external factors increases, and the self-similarity decreases. The basic photovoltaic power data studied in this paper are collected at a 15-min granularity, and they are greatly affected by external factors that have a certain regularity and contingency, so the statistical features of photovoltaic power data show certain periodicity, abruptness, and contingency.

As shown in Figure 1, the 4-day power history data of a photovoltaic station were randomly selected, showing obvious periodicity and volatility.

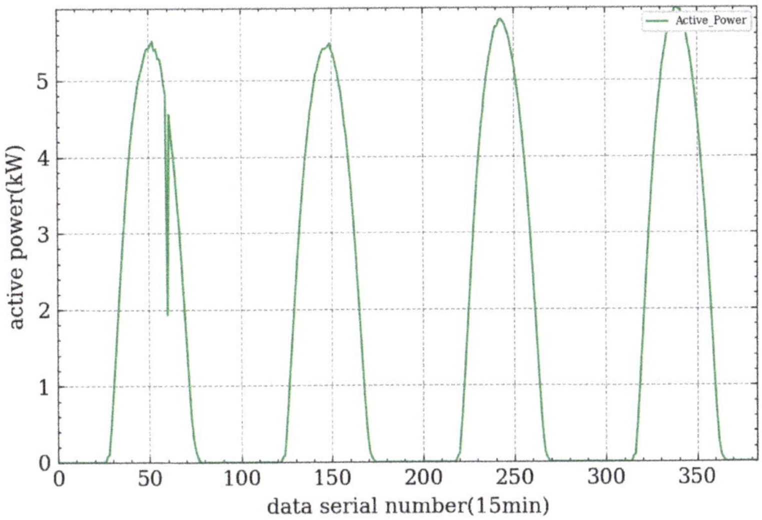

Figure 1. Graph of 15-min data from 4 continuous days.

As shown in Figure 2, in order to explore the long-term time series features of photovoltaic power data, this study employed the classic skills of a seasonal prediction model to select the historical data of a photovoltaic power station at 8:30 for 4 consecutive years. Although they show greater volatility, a certain periodicity can still be seen.

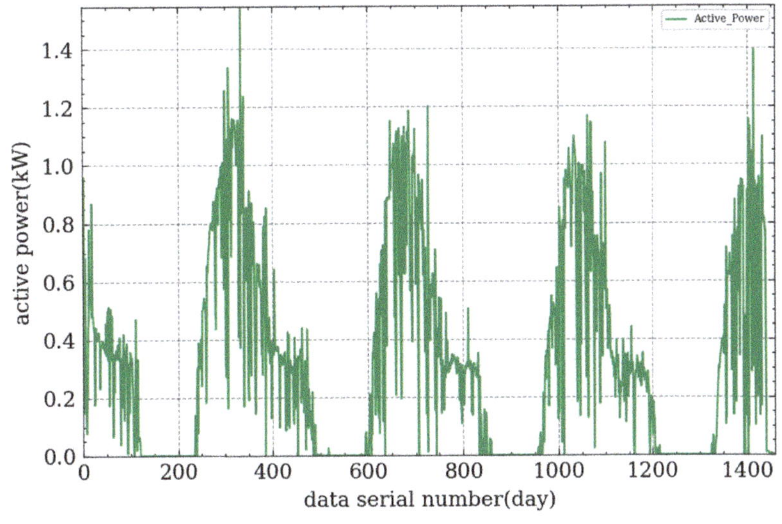

Figure 2. Graph of 8:30 data for 4 continuous years.

2.2. *LSTM and SkipGRU*

LSTM is a classic model in the field of time series prediction. In the prediction process, LSTM updates the internal state and the external state at the same time, mainly through three gates: a forgetting gate, an input gate, and an output gate.

The GRU network [31] is a variant of the LSTM network, which combines the three gates of the LSTM unit into two gates. The SkipGRU module skips the connection layer. By sampling at intervals, it can look back for a longer period of time when the length of the sampling sequence remains unchanged so as to capture the long-term features.

2.3. Self-Attention Mechanism and ProbSparse Self-Attention Module

The calculation formula of a traditional self-attention mechanism is as follows:

$$Q, K, V = XW^Q, XW^K, XW^V \quad (1)$$

$$A(Q, K, V) = softmax\left(\frac{QK^T}{\sqrt{d}}\right)V \quad (2)$$

In the formula, W^Q, W^K, and W^V are the three weight matrices. After random initialization, three vectors, Q, K, and V, are generated according to Equation (1), and then the result $A(Q, K, V)$ weighted ion mechanism is calculated according to Equation (2). The result contains the information via the attention of all of the input data.

The ProbSparse Self-Attention proposes to calculate the sparsity measurement of each query using KL divergence:

$$M(q_i, K) = Ln\sum_{j=1}^{L_K} \exp\left(\frac{q_i k_j^T}{\sqrt{d}}\right) - \frac{1}{L_K}\sum_{j=1}^{L_K} \frac{q_i k_j^T}{\sqrt{d}} \quad (3)$$

Based on the calculated sparsity metric, each key focuses on only u main queries to achieve probsparse self-attention:

$$A(Q, K, V) = softmax\left(\frac{\overline{Q}K^T}{\sqrt{d}}\right)V \quad (4)$$

In the formula, $\overline{Q}$ is a sparse matrix with the same size as q, and it only contains top-u queries under sparse metric $M(q, K)$.

2.4. Temporal Convolutional Network (TCN) Module

TCN is a variant of a convolutional neural network for processing sequence modeling tasks. It combines RNN and CNN architectures. TCN performs better than standard recursive networks on different tasks and data sets, and it demonstrates more long-term and efficient memory. The main component of the TCN network is Dilate Causal Conv. Other components are similar to the Feedforward module, which plays a role in deepening the linear features.

2.5. Problem Definition

The present study abstracts the photovoltaic power prediction problem as a multistep time series prediction problem, which can be defined as a data series with an input of $I \times n$ and an output of $O \times 1$, where I is the length of the input data, and O is the length of the output data. For example, under a 15-min sampling frequency, if the historical data of photovoltaic power in the past 30 days are used to predict the photovoltaic power data in the future 24 h, the I length is 2880, and the O length is 96.

3. Methodology

3.1. Transformer Based TCNformer Solution

For the time series features of photovoltaic power data, this paper proposed a TCNformer prediction model. The structure of the model is shown in Figure 3. Based on the traditional Transformer architecture, the TCNformer model mainly includes four modules: a variable selection (VS) module, an long- and short-time series feature extraction (LSTFE) module, an Encoder, and a Decoder.

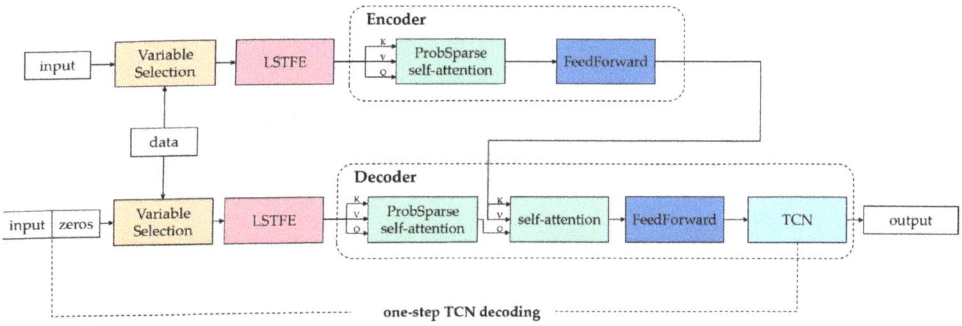

Figure 3. The structure of TCNformer model.

The overall TCNformer network design follows the traditional Transformer structure, in which the Encoder module and the Decoder module are designed with a multilayer structure.

3.2. Variable Selection (VS) Module

Combined with the information shown in Figures 1 and 2, the historical data of photovoltaic power not only have timing features in the short term, but they also have certain timing features over the long term. Considering the length of the long-term cycle (as shown in Figure 2, the cycle is close to 365 days) and the subsequent optimization problems, it is difficult for the traditional model to capture these timing features at the same time. So, we designed a VS module to divide the input sequence into three dimensions through preliminary analysis and selection of the historical data. Then, the results from the VS module are transferred to the LSTFE module for feature fusion.

$$d_l, d_s, d_t = VariableSelection(data, input) \tag{5}$$

In the formula, $data \in \mathbb{R}^{I\times n}, d_l \in \mathbb{R}^{I_l\times n_l}, d_s \in \mathbb{R}^{I_s\times n_s}, d_t \in \mathbb{R}^{I\times n_t}$ respectively represent preprocessed raw data, month-level time series data, week-level time series data, and day-level time series data. n, n_l, n_s and n_t respectively represent the number of influencing factors. $VariableSelection(\cdot)$ represents the VS module, and the specific calculation method is as follows.

Photovoltaic power data often show strong time series features. Although the volatility is strong, they still have a certain periodicity over a longer time range. In this paper, the Fourier transform decomposition curve of photovoltaic power data and its influencing factor data are selected for periodicity analysis [32] in order to obtain the fluctuation periods of different periodic curves and to provide a certain degree of reference for the analysis of photovoltaic power prediction. The formula of the Fourier transform is as follows:

$$X(k) = \sum_{n=0}^{N-1} x(n) W_N^{nK}, k = 0, 1, \ldots, N-1 \tag{6}$$

$$x(n) = (1/N) \sum_{n=0}^{N-1} X(k) W_N^{-nK}, k = 0, \ldots, N-1 \tag{7}$$

$$W_N^{nK} = e^{-j(2\pi/N)kn} \tag{8}$$

$X(k)$ represents the Fourier series, $x(n)$ represents the Fourier coefficient, W_N^{nK} represents the complex function, k represents the x coordinate in the frequency domain, and N represents the period.

Photovoltaic power is correlated with a large number of weather factors, especially the strong correlation between solar radiation intensity and photovoltaic power. In this study, the Pearson correlation coefficient was selected for correlation analysis, and the calculation formula is as follows:

$$P_{x,y} = \frac{cov(x,y)}{\sigma_x \sigma_y} = \frac{E[(x_i - \overline{x})(y_i - \overline{y})]}{\sigma_x \sigma_y} \quad (9)$$

The VS module processes the month-level time series data, week-level time series data, and day-level time series data according to the analytical results of correlation and periodicity.

3.3. *Long- and Short-Time Series Feature Extraction (LSTFE) Module*

In this study, we designed an LSTFE module, and we used it to extract time series features from each time scale. The structure of the LSTFE module is shown in Figure 4. The LSTFE mainly includes the LSTM unit, the SkipGRU unit, and the CycleEmbed unit.

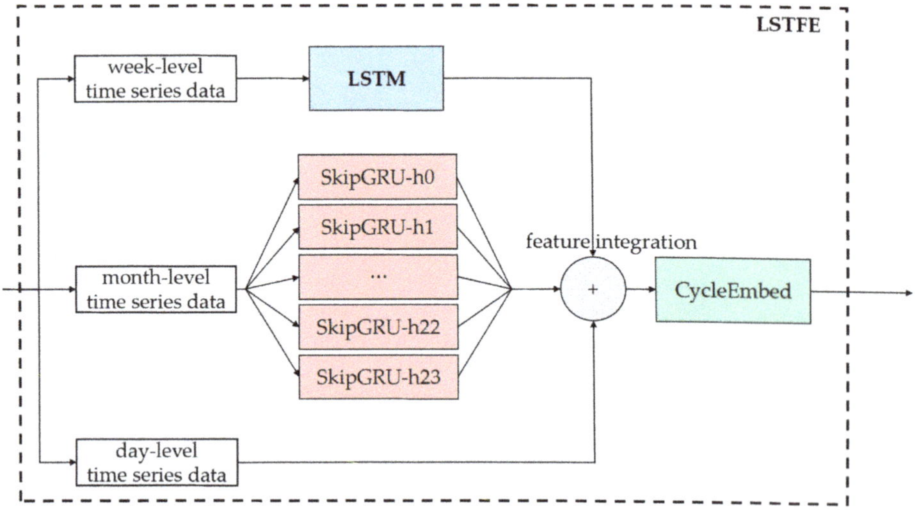

Figure 4. The structure of LSTFE module.

We transferred the week-level time-series-related data and the month-level time-series-related data to the LSTM network and the SkipGRU network in the LSTFE module for prediction. The prediction results of the LSTM network made full use of the short-term time series features, while the SkipGRU network made full use of the long-term time series features:

$$f_l = LSTM(d_l) \quad (10)$$

$$f_s = SkipGRU(d_s) \quad (11)$$

$$X = Integration(d_t, f_l, f_s) \quad (12)$$

$$X_{en}^0 = CycleEmbed(X) \quad (13)$$

In Formulas (10) and (11), $f_l \in \mathbb{R}^I$ and $f_S \in \mathbb{R}^I$ represent the month-level time series feature extraction results and the week-level time series feature extraction results in the LSTFE module, respectively. Using the excellent feature extraction capabilities of the LSTM and the SkipGRU, the extracted feature results were transformed into the input length I of the Encoder module.

Using the LSTM and the SkipGRU, the time series features at weekly and monthly levels were extracted, but we were left wondering how to extract the time series features at an annual level? To solve this problem, we designed the CycleEmbed module.

The structure of the CycleEmbed unit is shown in Figure 5, including data projection, position coding, cycle coding, and timing coding.

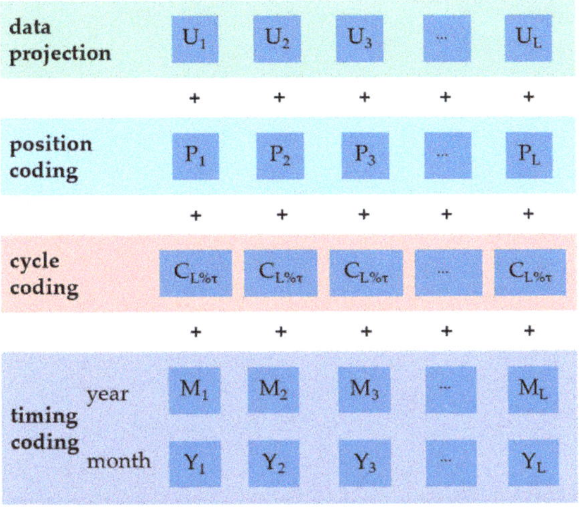

Figure 5. The structure of the CycleEmbed module.

Data projection is based on the results of correlation and periodic analysis, mapping the output data to the vector of dimension, and aligning the dimensions. The alignment tool is a one-dimensional convolution filter.

The position coding is calculated in the same way as in Transformer:

$$P(pos, 2j) = \sin\left(\frac{pos}{(2L_x)^{\frac{2j}{d_{model}}}}\right) \tag{14}$$

$$P(pos, 2j+1) = \cos\left(\frac{pos}{(2L_x)^{\frac{2j}{d_{model}}}}\right). \tag{15}$$

In Formulas (14) and (15), $j \in \left\{1, \ldots, \lfloor \frac{d_{model}}{2} \rfloor\right\}$, L_x is the input sequence length, and d_{model} is the Encoder input dimension.

Cycle coding is divided according to the results of periodic analysis and calculation. τ is the number of cycle data steps, which is determined by the results of periodic analysis T and the granularity of the data sampling time g; that is, $\tau = T/g$. Then, the cycle information of the input data is coded according to the results of τ; that is, there are τ results in cycle coding, $C_i = C_{i\%\tau}$.

Timing coding is used to add the month and year to the coding to extract the longer time series features. In this way, the annual time series features of the data are introduced into the codec along with the embedding operation.

Combining the results of the four parts, the output result of the final period embedding module is the input of the Encoder:

$$CycleEmbed_t[i] = U_i + P(L_x(t-1)+i) + C_i + M_i + Y_i \tag{16}$$

3.4. Encoder

The input of the Encoder is the output of the LSTFE module. The structure of the Encoder is a multilayer network structure. Each layer of the Encoder is mainly composed of a sparse attention unit and a composition unit.

$$S_{en}^{l,1} = ProbSelfAttention\left(X_{en}^{l-1}\right) \tag{17}$$

$$S_{en}^{l,2} = FeedForward\left(S_{en}^{l,1}\right) \tag{18}$$

$$X_{en}^{l} = S_{en}^{l,2} \tag{19}$$

In Formula (17), $S_{en}^{l,1} \in \mathbb{R}^{I \times d_{model}}$ is the calculation result of the sparse attention mechanism in the Layer 1 Encoder module, $S_{en}^{l,2} \in \mathbb{R}^{I \times d_{model}}$ is the calculation result of the Feedforward layer in the Layer 1 Encoder module, and $FeedForward(\cdot)$ is an important part of the traditional Transformer network structure which is used to deepen the linear representation and better extract the features. The Feedforward structure used in this paper is shown in Figure 6. ProbSelfAttention$(\cdot)$ is the sparse attention mechanism in the Informer model [24].

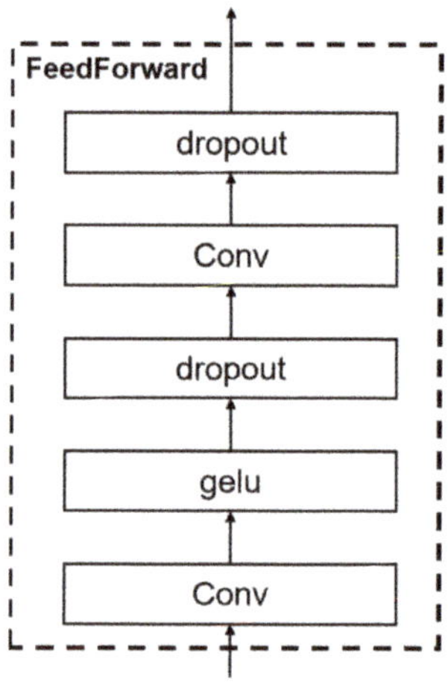

Figure 6. The structure of the Feedforward layer.

3.5. Decoder

In the Transformer model, the Encoder can be calculated in parallel, but the Decoder needs to decode step by step. As with the LSTM model, error accumulation will occur. This study introduced a one-step TCN decoding operation:

$$X_0 = Zeros[O, d] \tag{20}$$

$$X_{des} = concat(X, X_0) \tag{21}$$

$$X_{de}^0 = CycleEmbed(X_{des}) \tag{22}$$

In Formula (20), X_0 is the result of the zero-filling operation. One-step decoding divides the Decoder's input into two parts through a zero-filling operation. The first I datum is a known sequence, the last O datum is a sequence to be predicted, and $X_{de}^0 \in \mathbb{R}^{(I+O) \times d_{model}}$ is the Decoder's input data. At this time, part of the time information of the data to be predicted is also transmitted to the Decoder through the period embedding module for prediction. The prediction process of the Decoder is similar to that of the Encoder, but it has a more of a self-attention layer than does the Encoder.

$$S_{de}^{l,1} = ProbSelfAttention\left(X_{de}^{l-1}\right) \tag{23}$$

$$S_{de}^{l,2} = SelfAttention\left(S_{de}^{l,1}, X_{en}^{N}\right) \tag{24}$$

$$S_{de}^{l,3} = FeedForward\left(S_{de}^{l,2}\right) \tag{25}$$

$$X_{de}^{l} = S_{de}^{l,3} \tag{26}$$

In Formulas (23)–(25), $S_{de}^{l,1} \in \mathbb{R}^{(I+O)\times d_{model}}$ is the calculation result of the sparse attention mechanism in the Layer l Decoder module, $S_{de}^{l,2} \in \mathbb{R}^{(I+O)\times d_{model}}$ is the result of matching the sparse attention mechanism in the Layer l Decoder module with the feature map obtained in the Encoder, and $S_{de}^{l,3} \in \mathbb{R}^{(I+O)\times d_{model}}$ is the calculation result of the Feedforward layer in the Layer l Decoder module. The calculation method of FeedForward$(\cdot)$ and ProbSelfAttention$(\cdot)$ is the same as above. SelfAttention$(\cdot)$ is the self-attention mechanism. (See Section 2.3 for the calculation method.)

$$X_{pred} = TCN\left(X_{de}^{M}\right) \tag{27}$$

$X_{pred} \in \mathbb{R}^{O\times d_{model}}$ is the final prediction result of TCNfomer, which uses TCN to make generative predictions. The TCN structure used in this paper is shown in Figure 7.

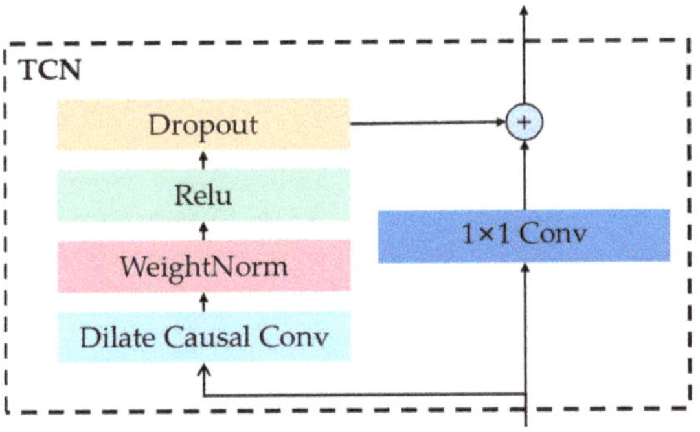

Figure 7. The structure of the TCN.

4. Experiment

4.1. Experimental Design

4.1.1. Data Preparation

The data set included an open-source data set of photovoltaic power conducted on a solar farm in Australia [33] from 2015 to 2016. The time interval is 15 min, there are 96 data points every day, and there are 70,176 samples in total. Each sample contains 13 data, including a time stamp, received active energy, average value at the current stage, active power, performance ratio, wind speed, weather temperature in Celsius, weather relative humidity, global horizontal radiation, diffuse horizontal radiation, wind direction, daily rainfall, global tilt of radiation, and diffuse tilt of radiation. The test set used data from the last 2 months of 2016.

All data for two years are shown in Figure 8. The x-axis is the number of days, the y-axis is 96 time points per day (sampling granularity is 15 min, and 24-h data processing includes 96 event points), and the z-axis is the photovoltaic power data.

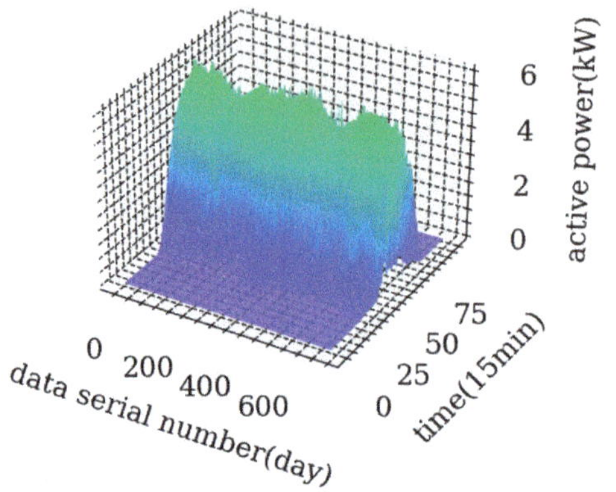

Figure 8. Historical data set of photovoltaic power.

4.1.2. Data Preprocessing

Since the dimensions between variables are not identical, linear normalization is required for prediction, and the conversion function is:

$$x_{norm} = \frac{x_i - min(x_i)}{max(x_i) - min(x_i)} \tag{28}$$

In the formula, x_{norm} is the preprocessing result of the data after linear normalization; x_i is the variable input value to be normalized; $max(x_i)$ is the maximum value of the variable in the original dataset x_i; and $min(x_i)$ is the minimum value of the variable in the original data set x_i.

4.1.3. Evaluation Index

In order to verify the prediction accuracy of the model, the root mean square error (MSE), the average absolute error (MAE) and mean absolute percentage error (MAPE) were used as the evaluation indicators of the model performance. The specific calculation formula is:

$$MSE = \frac{1}{N}\sum_{i=1}^{n}(X_i - \hat{X}_i)^2 \tag{29}$$

$$MAE = \frac{1}{N}\sum_{i=i}^{N}|X_i - \hat{X}_i| \tag{30}$$

$$MAPE = \frac{100\%}{N}\sum_{i=i}^{N}\left|\frac{X_i - \hat{X}_i}{X_i}\right| \tag{31}$$

X_i is the actual output value of the i th data point of the test set; $\hat{X}_i$ is the output prediction value of the i th data point; and N is the total number of samples in the test set.

4.1.4. Experimental Environment and Parameter Setting

The experimental environment used an Intel i7-9700 K processor and an NVIDIA GeForce RTX 3080Ti graphics card, and the algorithm model used Python 3.8 as the programming language. The model-related network was built based on the open-source machine learning framework PyTorch. The Python libraries directly used in the experiment included: pandas, numpy, matplotlib, torch, math, and time. In this study, the random search method was used to determine the final super parameter settings. The final super parameter settings are shown in Table 1.

Table 1. Model parameter setting.

Parameter	Value
LSTM hidden layers	2
SkipGRU hidden layers	2
Encoder layers	2
LSTM hidden unit	64
SkipGRU hidden unit	64
Decoder layers	1
d_{model}	512
batch-size	8
learn_rate	0.0001
epochs	2000

4.2. Variable Selection Results and Discussion

The VS module in the long- and short-sequence correction network includes correlation analysis and periodicity analysis. The results of the correlation analysis on photovoltaic power are shown in Table 2.

Table 2. Results of correlation analysis.

Variable	p
Wind speed	0.2096
Temperature	0.4246
Humidity	−0.4072
Direct radiation	0.9690
Scattered radiation	0.5183
Wind direction	−0.0444
Rainfall	−0.0244

It can be seen from Table 2 that photovoltaic power is positively correlated with direct radiation intensity, scattered radiation intensity, temperature, and wind speed, while it is negatively correlated with humidity, wind direction, and rainfall. According to their numerical values, the data were filtered by 0.1. It can be seen that the correlation between direct radiation intensity and photovoltaic power is the largest, while variables such as scattered radiation intensity, temperature, humidity, and wind speed have a certain correlation with photovoltaic power, which show that these influencing factors have a certain degree of impact on the photovoltaic power, and the impact decreases in turn. Although wind direction and rainfall are negatively related to the photovoltaic power, the value is too small to impact the output.

It can be seen from Table 3 that the cycle of photovoltaic power, humidity, direct radiation intensity, and scattered radiation intensity is 24.03 h, approximately 1 day, and the cycle of the wind speed, wind direction, and rainfall is 0.17 h, which can be regarded as a periodicity. The temperature cycle is 8760 h; that is, the temperature cycle conforms to the changes in the four seasons, and the above results basically conform to the natural logic.

Table 3. Results of periodicity analysis.

Variable	Cycle
Active power	24.03
Wind speed	0.17
Temperature	8760
Humidity	24.03
Direct radiation	24.03
Scattered radiation	24.03
Wind direction	0.17
Rainfall	0.17

By correlation analysis, five influencing factors should be selected, including direct radiation, scattered radiation, temperature, humidity, and wind speed. Three influencing factors, namely, direct radiation, scattered radiation, and humidity, were screened through periodic analysis. Finally, the time series related variables of photovoltaic power were screened through the VS module, those being direct radiation, scattered radiation, and humidity.

4.3. Prediction Results of Different Prediction Steps

In order to explore the prediction performance of each model under different prediction steps, this study selected LSTM, SkipGRU, Transformer, and Informer to compare with TCNformer.

The results are shown in Table 4. It can be seen from the results that, when the number of prediction steps is 1, the MSE errors of the five models have little difference. With the increase in the number of the prediction steps, the LSTM model demonstrated the largest error growth rate, and the error accumulation is obvious. Informer and TCNformer use the generative prediction method, so the error was relatively stable, and the error accumulation was low. The TCNformer model proposed in this paper not only had a low level of error accumulation, but it also had the lowest MSE error. In order to more intuitively observe the error accumulation in the models, the prediction results were visualized, as shown in Figure 9.

Table 4. Prediction Accuracy (MSE) Results of periodicity analysis.

	LSTM	SkipGRU	Transformer	Informer	TCNformer
1	0.1409	0.1272	0.2188	0.2273	0.0395
8	0.6859	0.3554	0.3135	0.3623	0.1080
16	0.9509	0.3852	0.3411	0.3644	0.1149
24	1.1062	0.4077	0.4562	0.5756	0.1322
32	1.3072	0.3501	0.6278	0.6356	0.1232
40	1.2668	0.4535	0.7122	0.5967	0.1300
48	1.4285	0.4546	0.6979	0.5571	0.1399
56	1.3687	0.5232	0.9031	0.6691	0.1234
64	1.4027	0.7204	0.8710	0.6438	0.1258
72	1.2783	0.7330	0.9645	0.7385	0.1377
80	1.3385	0.9499	0.8236	0.6510	0.1360
88	1.3601	1.1452	1.1139	0.7544	0.1382
96	1.4285	1.1624	1.1186	0.7455	0.1349

4.4. Prediction Performance of Different Models

In this experiment, each model was trained five times in a 24-h (96 prediction steps) scenario, and the average value was taken. The final test set prediction results are shown in Table 5.

Table 5. The 24-h scenario prediction results.

	MSE	MAE	MAPE	Train Time (s)	Run Time (ms)
LSTM	1.4285	0.7536	131.6989	139.7855	0.5945
SkipGRU	1.1624	0.4303	64.8742	88.7615	0.4832
Transformer	1.1186	0.5385	3.7577	76.1636	2.2314
Informer	0.7455	0.3779	2.9381	16.9919	1.8821
TCNformer	0.1349	0.1888	2.4987	153.4314	1.2910

As shown in Table 5, the TCNformer performs best according to the three indicators of the MSE, MAE, and MAPE. Compared with the time series prediction model Informer, the MSE, MAE, and MAPE decreased by 81.90%, 50.03%, and 14.98%, respectively. The training time (153.43 s) and running time (1.29 ms) of TCNformer are relatively long, but considering the 15-min sampling granularity and 24-h prediction scenario, the training time and running time do not affect the practical application of TCNformer.

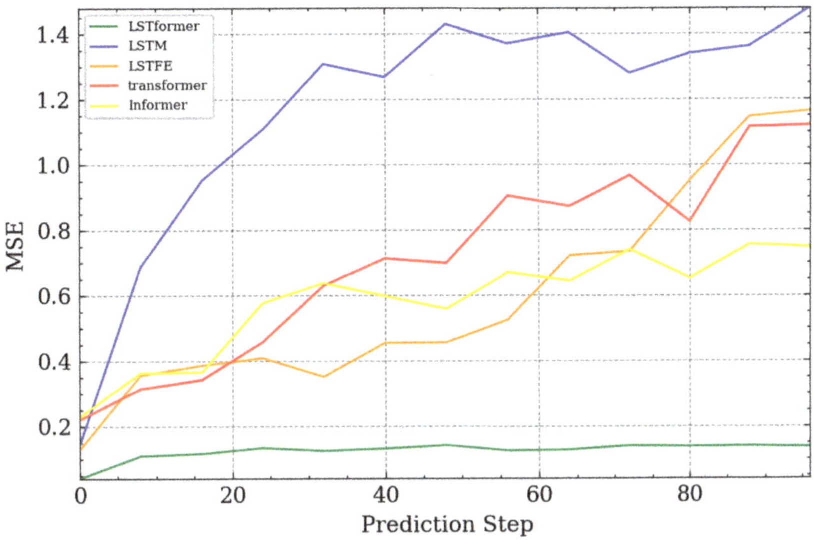

Figure 9. Prediction performance of the different numbers of prediction steps for each model.

As shown in Figure 10, this we visualized the prediction results of TCNformer using the test data set. The prediction results shown in the figure are 30 sets of 24-h prediction results, with little deviation when compared with the real data. It can be seen that TCNformer has a high level of accuracy and low number of errors.

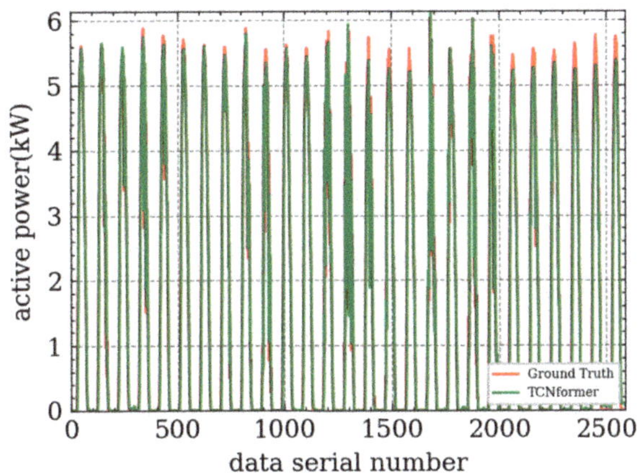

Figure 10. Results of the TCNformer model.

4.5. Error Analysis

Because the prediction of TCNformer model is a time series, we did not calculate the standard error for multiple series. Instead, error analysis was carried out through the MSE of prediction and ground truth. Figure 11 shows the standard error diagram. The error bar in the diagram represents the standard error. Table 6 shows the mean value, standard deviation (SD), and standard error (SE) of the error under different sample numbers.

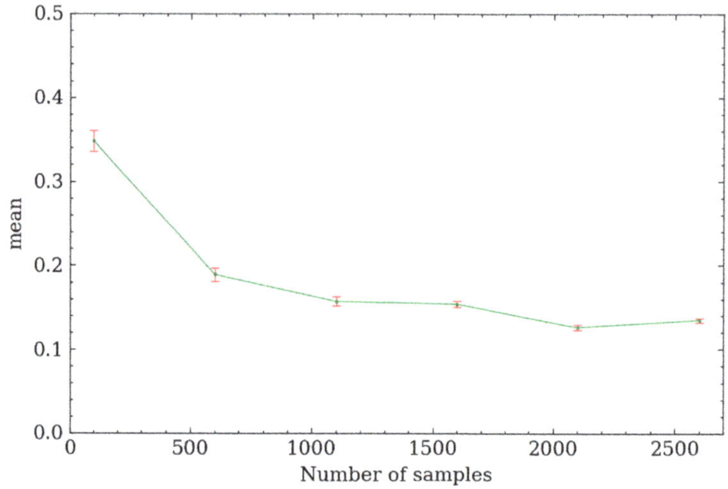

Figure 11. The standard error diagram of the MSE.

Table 6. Results of error analysis.

Sample Numbers	Mean	SD	SE
100	0.3482	0.1253	0.0125
600	0.1893	0.1977	0.0081
1100	0.1574	0.1798	0.0054
1600	0.1543	0.1584	0.0040
2100	0.1264	0.1507	0.0033
2600	0.1349	0.1394	0.0027

As shown in Table 6 and Figure 11, with the increase in the number of samples, the standard deviation and standard error gradually decreased, and the average value was closer to the average value of the overall sample. Therefore, the prediction result of the TCNformer model has a relatively stable level of error and a high level of reliability.

4.6. Ablation Experiment

In order to verify the effectiveness of each optimization module of the TCNformer model, we conducted ablation experiments, and we removed three innovative modules from the TCNformer model for comparative experiments, that is, we set them separately:

Experiment 1: Removal of the VS module.

Experiment 2: Removal of the long- and short-time series feature extraction module.

Experiment 3: Removal of the seq2seq structure, and use of the VS module + long- and short-time series feature extraction module + full connection network.

Experiment 4: Removal of one-step TCN decoding.

Experiment 5: Use of the complete TCNformer model.

As shown in Table 7, the three innovations proposed in this paper are a VS module, an LSTFE module, and the seq2seq generative model structure combined with Informer and Transformer. No matter which module was removed, the error of the model was increased. When the seq2seq model structure was not used, the error was the largest, and the VS module had the smallest impact on the overall model, but it still caused a decline in accuracy. From these data, it can be concluded that the TCNformer model proposed in this paper is effective, and its innovative modules are useful.

Table 7. Results of ablation experiment.

	MSE	MAE	MAPE
Experiment 1	0.4480	0.8385	3.5746
Experiment 2	0.6997	0.5339	3.4223
Experiment 3	0.8610	0.5288	4.2124
Experiment 4	0.3695	0.1949	2.7078
Experiment 5	0.1349	0.1888	2.4987

5. Conclusions

In this paper, a TCNformer model was proposed for photovoltaic power prediction, and we can draw the following three conclusions based on the experiment results:

1. The TCNformer model adopts the Transformer structure and introduces the sparse attention mechanism into the Informer model. The experimental results show that the photovoltaic output prediction accuracy is improved effectively.
2. The VS module, LSTFE module, and one-step TCN decoding extract more efficiently the impact of multiple time series features and other weather factors on photovoltaic power by classifying the data based on the time series, periodicity, and correlation.
3. Compared with the LSTM model and the Transformer series model, the TCNformer model has a higher level of accuracy in multistep prediction, but there is still room for optimization when the prediction range is further enlarged. In the follow-up study, we will focus on ways to solve the multistep prediction problem with a further increase in the time dimension.

Author Contributions: Methodology, S.L.; Software, S.L.; Validation, S.L.; Investigation, J.M.; Writing—original draft, S.L.; Writing—review & editing, D.N. All authors have read and agreed to the published version of the manuscript.

Funding: This research was funded by Research and development of large dispatching level data acquisition and monitoring control system grant number E212641B01 and AI assisted optimization of hybrid energy system and techno-enviro-economic analysis of green hydrogen supply chain grant number PT19797.

Institutional Review Board Statement: Not applicable.

Informed Consent Statement: Not applicable.

Data Availability Statement: Not applicable.

Conflicts of Interest: The authors declare no conflict of interest.

References

1. Naqvi, F.H.; Ko, J.-H. Structural Phase Transitions and Thermal Degradation Process of MAPbCl3 Single Crystals Studied by Raman and Brillouin Scattering. *Materials* **2022**, *15*, 8151. [CrossRef] [PubMed]
2. Yerezhep, D.; Omarova, Z.; Aldiyarov, A.; Shinbayeva, A.; Tokmoldin, N. IR Spectroscopic Degradation Study of Thin Organometal Halide Perovskite Films. *Molecules* **2023**, *28*, 1288. [CrossRef] [PubMed]
3. Omarova, Z.; Yerezhep, D.; Aldiyarov, A.; Tokmoldin, N. In Silico Investigation of the Impact of Hole-Transport Layers on the Performance of CH3NH3SnI3 Perovskite Photovoltaic Cells. *Crystals* **2022**, *12*, 699. [CrossRef]
4. Imani, S.; Seyed-Talebi, S.M.; Beheshtian, J.; Diau, E.W.G. Simulation and characterization of CH3NH3SnI3-based perovskite solar cells with different Cu-based hole transporting layers. *Appl. Phys. A* **2023**, *129*, 143. [CrossRef]
5. IEA. *Solar Photovoltaic Energy, IEA Technology Roadmaps*; OECD Publishing: Paris, France, 2015. [CrossRef]
6. Zeng, J.; Qiao, W. Short-term solar power prediction using a support vector machine. *Renew. Energy* **2013**, *52*, 118–127. [CrossRef]
7. Zagouras, A.; Pedro, H.T.; Coimbra, C.F. On the role of lagged exogenous variables and spatio–temporal correlations in improving the accuracy of solar forecasting methods. *Renew. Energy* **2015**, *78*, 203–218. [CrossRef]
8. Kumar, K.; Haider, M.T.U. Enhanced prediction of intra-day stock market using metaheuristic optimization on RNN-LSTM network. *New Gener. Comput.* **2021**, *39*, 231–272. [CrossRef]
9. Vidal, A.; Kristjanpoller, W. Gold Volatility Prediction using a CNN-LSTM approach. *Expert Syst. Appl.* **2020**, *157*, 113481. [CrossRef]

10. Shao, H.; Soong, B.H. Traffic flow prediction with Long Short-Term Memory Networks (LSTMs). In Proceedings of the TENCON 2016—2016 IEEE Region 10 Conference, Singapore, 22–25 November 2016.
11. Bae, S.H.; Choi, I.; Kim, N.S. Acoustic Scene Classification Using Parallel Combination of LSTM and CNN. In Proceedings of the Detection and Classification of Acoustic Scenes and Events 2016, Budapest, Hungary, 3 September 2016.
12. Yu, Y.; Cao, J.; Zhu, J. An LSTM Short-Term Solar Irradiance Forecasting Under Complicated Weather Conditions. *IEEE Access* **2019**, *7*, 145651–145666. [CrossRef]
13. Sethi, R.; Kleissl, J. Comparison of Short-Term Load Forecasting Techniques. In Proceedings of the 2020 IEEE Conference on Technologies for Sustainability (SusTech), Santa Ana, CA, USA, 23–25 April 2020; pp. 1–6. [CrossRef]
14. Wang, Y.; Liao, W.; Chang, Y. Gated Recurrent Unit Network-Based Short-Term Photovoltaic Forecasting. *Energies* **2018**, *11*, 2163. [CrossRef]
15. Wang, K.; Qi, X.; Liu, H. A comparison of day-ahead photovoltaic power forecasting models based on deep learning neural network. *Appl. Energy* **2019**, *251*, 113315. [CrossRef]
16. Wojtkiewicz, J.; Hosseini, M.; Gottumukkala, R.; Chambers, T.L. Hour-Ahead Solar Irradiance Forecasting Using Multivariate Gated Recurrent Units. *Energies* **2019**, *12*, 4055. [CrossRef]
17. Abdel-Nasser, M.; Mahmoud, K. Accurate photovoltaic power forecasting models using deep LSTM-RNN. *Neural Comput. Applic* **2019**, *31*, 2727–2740. [CrossRef]
18. Srinivasan, R.; Balamurugan, C.R. Deep Neural Network Based MPPT Algorithm and PR Controller Based SMO for Grid Connected PV System. *Int. J. Electron.* **2021**, *109*, 576–595. [CrossRef]
19. Gumar, A.K.; Demir, F. Solar Photovoltaic Power Estimation Using Meta-Optimized Neural Networks. *Energies* **2022**, *15*, 8669. [CrossRef]
20. Zhou, M.; Huang, Y.; Yang, X. Ultra-short-term photovoltaic power forecasting of multifeature based on hybrid deep learning. *Int. J. Energy Res.* **2022**, *46*, 1370–1386.
21. Kallio, S.; Siroux, M. Photovoltaic power prediction for solar micro-grid optimal control. *Energy Rep.* **2023**, *9* (Suppl. S1), 594–601. [CrossRef]
22. Korkmaz, D.; Akgz, H.; Yldz, C. A Novel Short-Term Photovoltaic Power Forecasting Approach based on Deep Convolutional Neural Network. *Int. J. Green Energy* **2021**, *18*, 54–62. [CrossRef]
23. Lim, S.-C.; Huh, J.-H.; Hong, S.-H.; Park, C.-Y.; Kim, J.-C. Solar Power Forecasting Using CNN-LSTM Hybrid Model. *Energies* **2022**, *15*, 8233. [CrossRef]
24. Zhou, H.; Zhang, S.; Peng, J.; Zhang, S.; Li, J.; Xiong, H.; Zhang, W. Informer: Beyond Efficient Transformer for Long Sequence Time-Series Forecasting. *arXiv* **2020**, arXiv:2012.07436. [CrossRef]
25. Wu, H.; Xu, J.; Wang, J.; Long, M. Autoformer: Decomposition Transformers with Auto-Correlation for Long-Term Series Forecasting. *arXiv* **2021**, arXiv:2106.13008.
26. Lin, T.; Wang, Y.; Liu, X.; Qiu, X. A Survey of Transformers. *arXiv* **2021**, arXiv:2106.04554. [CrossRef]
27. Lai, G.; Chang, W.-C.; Yang, Y.; Liu, H. Modeling Long- and Short-Term Temporal Patterns with Deep Neural Networks. *arXiv* **2018**, arXiv:1703.07015.
28. Meisenbacher, S.; Turowski, M.; Phipps, K.; Rätz, M.; Müller, D.; Hagenmeyer, V.; Mikut, R. Review of automated time series forecasting pipelines. *arXiv* **2022**, arXiv:2202.01712. [CrossRef]
29. *GB/T 33590.2-2017*; China Electricity Council. Technical Specification for Smart Grid Dispatching Control System-Part 2: Terminology. General Administration of Quality Supervision, Inspection and Quarantine of the People's Republic of China and China National Standardization Administration: Beijing, China, 2017.
30. Cho, K.; van Merrienboer, B.; Gulcehre, C.; Bahdanau, D.; Bougares, F.; Schwenk, H.; Bengio, Y. Learning Phrase Representations using RNN Encoder-Decoder for Statistical Machine Translation. *Comput. Sci.* **2014**, 1724–1734.
31. Khademi, M.; Moadel, M.; Khosravi, A. Power Prediction and Technoeconomic Analysis of a Solar PV Power Plant by MLP-ABC and COMFAR III, considering Cloudy Weather Conditions. *Int. J. Chem. Eng.* **2016**, *2016*, 1031943. [CrossRef]
32. Katoh, K.; Misawa, K.; Kuma, K.; Miyata, T. MAFFT: A novel method for rapid multiple sequence alignment based on fast Fourier transform (describes the FFT-NS-1, FFT-NS-2 and FFT-NS-i strategies). *Nucleic Acids Res.* **2002**, *30*, 3059–3066. [CrossRef]
33. DKA Solar Centre. Available online: http://dkasolarcentre.com (accessed on 28 September 2020).

sustainability

MDPI

Article

A Two-Stage Bilateral Matching Study of Teams-Technology Talents in New R&D Institutions Based on Prospect Theory

Lin Jiang * and Biyun Chen

School of Information Engineering, Yancheng Teachers University, Yancheng 224002, China
* Correspondence: jiangl@yctu.edu.cn

Abstract: This study considers two-stage bilateral matching of teams and scientific and technological talents in new R&D organizations and proposes a two-stage dual-objective bilateral matching method based on prospect theory. The matching of teams and scientific and technological talent in new R&D institutions is divided into two stages: elimination matching in the first stage and selection matching in the second stage. In the first stage, the evaluation index of the team to talent and the cost index of talent are constructed, the dual reference points of peer and expectation are set for evaluating talent, and the bottom-line reference points are set for talent cost. The comprehensive prospect value in the first stage is calculated based on prospect theory, and the matching in the first stage is completed based on the dual-objective optimization model with the highest evaluation value and the lowest cost value. In the second stage, using the matching results of the first stage, the team evaluates the talent again, while the talent ranks the team to obtain the satisfaction value, and completes the second stage of bilateral matching based on prospect theory and the dual-objective optimization model with the highest evaluation value and the highest satisfaction value. Finally, a case study and method comparison show that the proposed method is feasible and effective.

Keywords: prospect theory; new R&D institutions; scientific and technological talent; performance assessment; bilateral matching

Citation: Jiang, L.; Chen, B. A Two-Stage Bilateral Matching Study of Teams-Technology Talents in New R&D Institutions Based on Prospect Theory. *Sustainability* **2023**, *15*, 3494. https://doi.org/10.3390/su15043494

Academic Editors: Luis Hernández-Callejo, Sergio Nesmachnow and Sara Gallardo Saavedra

Received: 25 July 2022
Revised: 7 February 2023
Accepted: 8 February 2023
Published: 14 February 2023

1. Introduction

In China, new R&D institutions, with their impressive innovation achievements and rapid development momentum, have developed into the pioneering force for source science and technology innovation and development of strategic emerging industries in various regions, creating a new model of science and technology R&D that leaps and bounds to enhance source innovation capacity and rapidly realize industrialization. The new R&D institution is a "four different", not exactly like a university, with different culture; not exactly like a scientific research institute, with different content; not exactly like an enterprise, with different objectives; not exactly like an institution, with different mechanisms. In particular, its essential characteristics of marketization, industrialization, diversification, socialization and internationalization also foreshadow an important direction for the reform and development of scientific research institutions in China.

The development of science and technology cannot be separated from talent, and scientific and technological talent is the first resource for the development of new R&D institutions. With the development of new R&D institutions, the requirements for scientific and technological talents are becoming increasingly high-end and comprehensive. First, they should be patriotic and dedicated, transforming their love for the country and the will to strengthen the country into the act of serving the country. Second, they should be active in innovative thinking and conduct research oriented by the scientific and technological needs of the country and the market as well as "high precision and shortage" projects. At the same time, new R&D institutions should also create various conditions for all kinds of talent to settle down, feel at ease and work, respect talent and creativity, innovate talent assessment mechanisms, and improve the assessment system of scientific and technological talent oriented toward innovation ability, quality, and contribution. There are many teams within the new R&D institutions, and these institutions give the teams a lot of autonomy to choose the direction of scientific research independently, issue performance and rewards independently, and implement the science and technology management system independently. Therefore, teams of new R&D institutions also have autonomy in selecting and employing people. The two-way choice between a team of new R&D institutions and scientific and technological talents is the issue of bilateral matching. On the one hand, the team of the new R&D institution looks for scientific and technological talents according to its own needs, and each team will propose its own personalized evaluation index that classifies and selects scientific and technological talents through an evaluation based on a personalized index, which will realistically consider the benefits and costs together. On the other hand, the tech talent will also look for teams according to their own needs; there will also be a ranking of the teams.

To study the bilateral matching of new R&D institutions and talents, the author searched the SCI literature on the performance assessment of new R&D institutions, performance assessment of scientific and technological talents, and bilateral matching. Moliterno et al. [1] proposed that performance comparison is central to the behavioral theory of the firm, that is, companies assess their performance based on their own prior performance ("historical comparison") and the performance of other organizations ("social comparison") and base subsequent organizational changes on this performance feedback. Bode and Singh [2] argue that the provision of opportunities for employees to participate in social activities helps attract, motivate, and retain employee talent. Abramo et al. [3] proposed that the ultimate goal of research innovation activities is not publication, but scientific and technological progress useful to science or practice, and that there is no incentive to produce low-value papers if innovation performance is assessed and funds are allocated based on the total impact of publications, rather than on the number of publications. Yin et al. [4] proposed the use of a blend of subjective and objective methods to assess green technology innovation capabilities which should consider indicators in four areas: input elements, technological output, economic aggregates, and social effects. Sun and Cao [5] point out that Chinese academic research on innovation has paid particular attention to R&D expenditures, performance assessments, regional innovation ecosystems, the role of state-owned enterprises in innovation, and the role of the Chinese Communist Party in innovation. Mao et al. [6] show that organizational innovation climate, knowledge management capabilities, and internal collaboration networks have a significant positive impact on innovation performance, and that internal collaboration networks have a significant mediating role between them. Chomać et al. [7] showed that consumer knowledge and preferences in the field of renewable energy determine the diffusion of RES solutions in personal use, thus stimulating the progress of energy transition. In general, to evaluate the performance of new R&D institutions, first, the evaluation indicators should be as comprehensive as possible, taking into account both subjective and objective factors. Second, in addition to assessing the established results, the innovation potential should also be assessed. Third, more attention should be paid to the quality of publications rather than the quantity. Fourth, the investment of funds and scientific and technological talents should be increased; and fifth, the economic promotion benefits as well as the social effects generated should be considered.

Chamorro et al. [8] discussed three methods for assessing talent: machine-learning algorithms, social sensing technologies, and user experience. Pillai et al. [9] investigated the application of AI technologies in talent acquisition, designed technology-organization-environment (TOE) and task-technology-adaptation (TTF) frameworks and proposed a model to explore the adoption of AI technologies for talent acquisition. Jiang et al. [10] proposed the development of technological talent in line with the globalization context and to integrate talent acquisition, research and development, technological innovation, and enterprise development. Wiblen and Marler [11] propose the role of digitization in talent identification, showing how the same digital talent management techniques can produce different ways to identify talent. Chaudhuri et al. [12] argued that company management with PhDs in key roles outperforms similar company management. Agarwal et al. [13] argue that stable shared leadership is at the root of firms becoming the center of gravity of their industry, accounting for the largest share of output. In general, the performance assessment of scientific and technical talents should be conducted by considering both their results and their potential. Second, regular, immediate, and dynamic assessment of scientific and technical talents should be conducted to form a digital database of assessments. Third, the capabilities of scientific and technical talents should be integrated with the development of the company and linked to economic and social benefits; and fourth, an artificial intelligence-based approach should be provided, and input assessment data can be quickly and accurately predicted and reasoned to draw assessment conclusions.

Eirinakis et al. [14] propose a time-optimal algorithm that identifies all stable worker-firm pairs and allocations under pairwise stability, individual preferences, and max-min criteria. Wang et al. [15] studied the bilateral matching decision problem using heterogeneous information and attribute associations. Kanoria and Saban [16] introduced a dynamic bilateral search model in which strategic agents incur costs to discover their value for each potential partner and can do so nonsimultaneously. Nguyen et al. [17] developed a many-to-one matching market model in which agents with multiunit demand aim to maximize the underlying linear objective subject to a multidimensional backpack constraint. Johari et al. [18] proposed that, in a service platform, the job type is known, but the worker type is unknown and must be learned by observing the matching results. Deng et al. [19] find that buyer and supplier conformance levels, conformance types, and inconsistency directions affect project performance. Chen et al. [20] found that matching the nature of CEO human capital and the type of acquisitions they make is associated with stronger performance. Chomać et al. [21] showed that global electricity price increases can be effectively reduced by conducting feasibility and matching analyses of renewable energy sources based on consumer investment and willingness to support them. In general, research on bilateral matching should consider the characteristics of heterogeneity, uncertainty, and incompleteness of input information. Second, improve the satisfaction and efficiency of bilateral matching based on certain methods. Third, consider multistage decision models; and fourth, match dynamically, conduct dynamic assessments, tap dynamic preferences, and consider dynamic reference point values, among others.

New R&D institutions have the mission of "high precision and shortage" of science and technology innovation, the mission of diversified and flexible reform of the science and technology system and mechanism, the mission to respond to market demand and generate economic and social benefits, and the mission of gathering high-end science and technology talents. It is particularly important and meaningful to study the matching of new R&D institutions with scientific and technological talents. Based on the fact that there is particularly little literature on the bilateral matching of new R&D institution teams and scientific and technological talents, this study applies the idea of bilateral matching. First, the new R&D organization team and technology talent are divided into two phases: the elimination matching phase and the selection matching phase. In the first stage of matching, since the number of talents is greater than the number of teams and a team can only match one talent, there are bound to be talents that are rounded off and cannot be matched, so we call it elimination matching. In the second stage of matching, the number of talents is equal to the number of teams, so we call it selection matching. Second, elimination matching considers the team's assessment of talent and the cost of talent introduction, and selection matching considers the mutual assessment of team and talent. Third, elimination matching considers the interval grey number to characterize the uncertain assessment and cost values, and selection matching considers the mean value of expert assessment and the talent's preference order value. Fourth, consider the psychological factors of decision makers, set a historical and desired double reference point for assessment value, set a bottom-line reference point for cost value and preference order value, and calculate the prospect value based on the reference point. Fifth, consider double objective matching, eliminate matching considering the double objective of maximum assessment value and minimum cost value, select matching considering the double objective of maximum assessment value and maximum satisfaction, reduce the double objective to a single objective, and construct a 0–1 integer programming optimal matching model. In this paper, the dual objectives are linear, and for simplicity of calculation, the dual objectives plus the weight information of the objectives can be simplified to a single objective. In order to make the matching results optimal, this paper uses 0–1 integer programming for solving the optimal solution.

2. Phase I Elimination Match between the Team of New R&D Institutions and Scientific and Technological Talents

The matching of new R&D organization teams and technology talents is divided into two stages. The first stage is the elimination of bilateral matching, that is, each team can match technology talents, the number of talents is greater than the number of teams, so not all technology talents can match the team, limited by the number of teams and the number of matches, there are always technology talents are eliminated in this matching process. In this study, there are n teams of new R&D institutions, and each team can only match one technological talent, and there are m technological talents, m > n; thus, so there will be m-n technological talents are eliminated. Therefore, this stage is referred to as elimination matching. Elimination matching must consider many factors, because eventually the new R&D organization has to explain the elimination reason for the eliminated tech talents, so the method of matching is very demanding.

In the first stage of matching, the highest value of the team's assessment of talent is considered, and the lowest value of the introduction cost of scientific and technological talent is also considered, historical reference points and desired reference points are set for the assessment value, bottom-line reference points are set for the cost value, the psychological factors of the team in the new R&D organization are fully considered, and scientific and technological talent is better compared, and finally, using the "0–1" integer planning model for bilateral matching. In the first phase of elimination matching, the total number of talents was greater than the number of R&D teams. Red smiling faces represent successfully matched tech talents, and green smiling faces represent unmatched tech talents that will be eliminated, as shown in Figure 1.

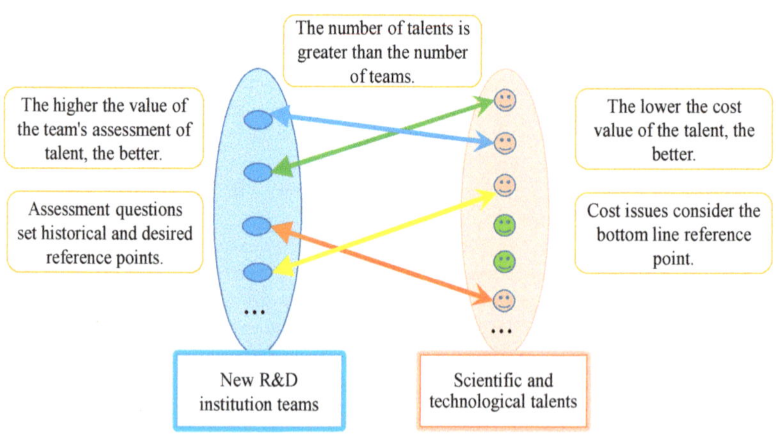

Figure 1. Phase I elimination matching chart for teams and talents in new R&D institutions.

2.1. *Constructing Indicators for the Assessment of Scientific and Technological Talents by the Teams of New R&D Institutions*

The assessment of scientific and technological talent has new requirements in this new era. Since new R&D institutions focus on breaking through core technologies, solving "neck" problems and transforming results, the development process of new R&D institutions requires scientific and technological talents to pay more attention to patriotism, technical potential and transformation ability. In Chinese history, many noble men and women had a strong sense of concern for the country and the people, and they took the affairs of the country as their responsibility, and they defended the motherland and cared for people's livelihood. No matter what environment we are in, we should love our country and take the affairs of our country as our responsibility. For example, in 2020, when we were fighting against the new coronavirus, many outstanding scientific and technological talents emerged in our country, creating Chinese speed in the field of nucleic acid detection, vaccines, saving critically ill patients, building hospitals, etc. They are not only a demonstration of scientific and technological ability, but more importantly, they have a patriotic heart and passion to serve the country. The technical potential of scientific and technological talent depends on their experience and personal drive. Graduating from a prestigious university can help him take fewer detours, critical innovation can keep him challenging new heights, cross-discipline can give him more inspiration for innovation, and so on. Transformation ability should first have good communication and collaboration ability, because the projects of new R&D institutions are usually completed by teams; second, there should be transformed results that generate economic and social benefits, as well as talent benefits, etc.

The author visited 10 new R&D institutions, researched scientific and technological talents, reviewed 50 papers on skills and literature on R&D institution development, and summarized and refined 15 new R&D institutions' benefit assessment indicators for scientific and technological talents, as shown in Table 1. Ten new R&D institutions were located in the Science and Technology Park of Yancheng City, Jiangsu Province. Yancheng is a coastal city with the largest mudflat wetlands and the largest wind energy. Ten new R&D institutions belong to different industries: two in the saline rice industry, two in the cable industry, two in the machinery industry, two in the wind power industry, and two in the energy industry. I visited the human resources departments of these new R&D institutions and communicated with the person in charge of their talent introduction, who generally reflected the high cost of talent introduction and insufficient satisfaction. They also proposed modifications to our assessment index framework, such as serving the motherland, graduating from prestigious universities, cross-discipline, and providing scientific and technological insights.

During my visit, I also found that different teams of new R&D institutions have different indicators for assessing scientific and technological talent because their requirements are different. For example, new R&D organization A has a strict confidentiality sign posted from the entrance, and there are confidentiality signs everywhere on the stairs and restrooms, indicating that the team has a particularly high requirement for confidentiality. The high-end nature of the technology of the new R&D organization makes the technology accessible to a certain extent. Any knowledge is public to a certain extent and confidential to a certain extent. Intellectual property, for example, may have many public methods, but there are still some details that are kept confidential. Secrecy sometimes represents importance, sophistication, and high-end, so that a few talented people can still learn and improve, but just not be known to the general public. Another example is new R&D organization B. Their recruitment website shows that they are looking for PhD students who have graduated from "985," which means that they require graduates from prestigious universities. Therefore, in this study, we consider the assessment of individualized indicators of scientific and technological talent by teams of new R&D institutions. The first team selected its own personalized indicators from the 15 indicators, and the second team selected its own personalized indicators, k_{1j} from the 15 indicators. The second team selected its own desired indicators k_{2j} from the 15 indicators until all n teams selected their desired indicators k_{nj} from the 15 indicators. The teams assign indicator weights as they select the indicators ω_{ij}.

Table 1. Indicators for assessing the benefits of new R&D organization teams for talents.

T-S Indicators	Assessment Indicators	Content of the Assessment
The team's evaluation indicators for talent	1. Serving the motherland	Love the country and love the Party, personal ideals into the great national development, and take pride in this, strong country has my, sincere enthusiasm to serve the country only grows.
	2. Professional ethics	The code of conduct to be followed in professional activities, patriotism and respect for work, honesty and friendliness, compliance with laws and regulations, etc.
	3. Critical innovation	It is important to think critically and creatively, to break down old traditions, to build new ideas, new methods, new processes, etc.
	4. Innovative results	Scientific and technical papers, scientific and technical projects, scientific and technical patents and other achievements.
	5. Entrepreneurial outcomes	The breadth and depth of application and transformation of scientific and technological achievements, generating economic benefits, social benefits, talent benefits, etc.
	6. Graduated from a prestigious school	Graduated from a prestigious school or conducted research under the guidance of a renowned teacher.
	7. Communication and collaboration	Good communication skills, an academic background in teamwork, and regular participation in various academic networking events where they actively contribute or share.
	8. Cross-cutting disciplines	Cross-disciplines can break down limitations, stand higher, plan farther, grasp more accurately, and generate more energy.
	9. Technology Insights	Accurate judgment of the situation and clear positioning for career development.
	10. Design and programming	Must have some design skills, have some programming skills and be able to maintain data and other information.
	11. Keeping secrets	Keep the contents, methods and processes of research and development strictly confidential.
	12. Self-control and self-discipline	Aspire to be a person of value, value time, pace yourself, spend your energy on things of value, read a lot of literature, do a lot of science experiments, and regular paper writing.
	13. Self-reliance and self-improvement	Good at independent thinking and inquiry questioning, adhere to independent innovation, always on the road of struggle and exploration.
	14. Physical fitness	Having a healthy body allows you to focus on the big things.
	15. Psychological quality	Resilient to pressure and work under pressure with the team's goals and mission in mind.

Because there are differences in the process of assessment of scientific and technological talents by the team of new R&D institutions due to the assessment experts' own learning, preferences, reference points, information asymmetry, etc., and the assessment results have a certain degree of uncertainty, this study uses interval grey numbers to characterize the uncertain assessment values. The interval grey number is an uncertain number that takes values in a certain interval or in a general set of numbers, and only the range of information values is known without knowing the exact information values, which is usually denoted by the symbol $\otimes$ to denote it. A grey number that has both a lower and an upper bound is called an interval grey number and is denoted as $\otimes \in [\underline{a}, \bar{a}]$ with $\underline{a}$

denotes the lower bound and $\bar{a}$ denotes the upper bound. Using the interval grey number to characterize the team's assessment value of talent, we obtain the n-decision matrix $A^k = \left[a_{ij}^k\right], i = 1, 2, \cdots m, j = 1, 2, \cdots p, k = 1, 2, \cdots n$, i denotes that there are m scientific and technical talents, j denotes that there are p assessment metrics, and k denotes that there are n teams.

$$A^k = \left[a_{ij}^k(\otimes)\right] = \begin{bmatrix} \underline{a}_{11}^k, \bar{a}_{11}^k & \underline{a}_{12}^k, \bar{a}_{12}^k & \cdots & \underline{a}_{1p}^k, \bar{a}_{1p}^k \\ \underline{a}_{21}^k, \bar{a}_{21}^k & \underline{a}_{22}^k, \bar{a}_{22}^k & \cdots & \underline{a}_{2p}^k, \bar{a}_{2p}^k \\ \cdots & \cdots & \cdots & \cdots \\ \underline{a}_{m1}^k, \bar{a}_{m1}^k & \underline{a}_{m2}^k, \bar{a}_{m2}^k & \cdots & \underline{a}_{mp}^k, \bar{a}_{mp}^k \end{bmatrix} \tag{1}$$

2.2. Constructing Cost Indicators for the Introduction of Scientific and Technological Talent

New R&D institutions must pay a certain price when introducing scientific and technological talent. We take the introduction of three years as an example: the cost in three years generally includes the settlement fee, science and technology start-up fee, salary, insurance, provident fund, performance incentives, project dividends, and other incentives. Some items under the difference are also relatively large, and we choose more representative indicators as the matching indicators, as shown in Table 2.

Table 2. Cost assessment indicators of new R&D organization teams for scientific and technological talents.

S-T Indicators	Assessment Indicators	Content of the Assessment
Cost metrics for talent	1. Wage insurance	The portion of institutional payments for salaries, insurance, etc., paid on an annual or monthly basis.
	2. Subsidy for house purchase	There is a lump sum settlement fee, annual or monthly provident fund contributions for home purchase, plus some allowance for home purchase, etc.
	3. Performance incentives	Performance is generally divided into four grades, excellent, good, pass, and fail, and the award is taken at whichever grade is met after the performance appraisal, which is estimated in this paper as the good grade.
	4. Project dividends	Estimated by dividing the total project amount by the number of people in the team on average.
	5. Other awards	There are some additional incentives when scientific and technological achievements are transformed into economic, social and talent benefits, and there are also institutions that implement training and incentives for particularly outstanding scientific and technological people, which are all costed and calculated on an average basis.

Some scientific and technological talents can receive high rewards in teams of new R&D institutions, while others receive very little reward. Since the mechanism of new R&D institutions is very flexible and financially autonomous, there are many uncertain cost factors, and sometimes the rewards can far exceed salary income. In general, the cost of bringing in scientific and technological talent with higher performance evaluation values is relatively high. It is a real problem to match the teams of new R&D institutions with scientific and technical talents with high assessment values and low costs. In this study, the team of the new R&D organization and the technology talent are communicated and based on the cost assessment index in this study, both parties negotiate and finally determine an acceptable cost range. Using the interval grey number to characterize the cost value of talent introduction, we obtain a decision matrix of $1B = \left[b_{ij}\right], i = 1, 2, \cdots m, j = 1, 2, \cdots q$, with i denotes the number of m a scientific and technical talent, and j denotes that there are q evaluation index.

$$B = \left[b_{ij}(\otimes)\right] = \begin{bmatrix} \underline{b}_{11}, \overline{b}_{11} & \underline{b}_{12}, \overline{b}_{12} & \cdots & \underline{b}_{1q}, \overline{b}_{1q} \\ \underline{b}_{21}, \overline{b}_{21} & \underline{b}_{22}, \overline{b}_{22} & \cdots & \underline{b}_{2q}, \overline{b}_{2q} \\ \cdots & \cdots & \cdots & \cdots \\ \underline{b}_{m1}, \overline{b}_{m1} & \underline{b}_{m2}, \overline{b}_{m2} & \cdots & \underline{b}_{mq}, \overline{b}_{mq} \end{bmatrix} \quad (2)$$

2.3. *Phase I Elimination Matching Results Based on Prospect Theory and Bilateral Matching Models*

According to prospect theory and grey target theory, this study sets a double reference point for the appraisal value of the historical reference point and desired reference point. This study sets a bottom-line reference point for the cost value, uses the reference point as the bull's eye, and applies prospect theory to the data set.

In the first stage of matching the team of a new R&D institution with scientific and technological talents, although the assessment value of scientific and technological talents is given, the assessment value does not reflect whether the scientific and technological talents are good, how good they are, and whether the team is satisfied. Therefore, it is necessary to find a reference point for comparison, and the data for the reference point must be easily accessible.

2.3.1. Setting Reference Points for Assessment Values—Historical Reference Points and Desired Reference Points

The historical reference point is the assessment of the scientific and technological talents already introduced in the history of the new R&D institution team and the comparison with the assessment of the talents in the past, which can tell whether the batch of scientific and technological talents is better than the past or not as good as the historical scientific and technological talents. In general, the new R&D organization team hopes that the technology talent brought in is improving. If the assessment value is higher than the historical reference point, the new R&D institution team is satisfied. If the assessment value is lower than the historical reference point, the new R&D institution team is not satisfied.

The expectation reference point is the goal that the new R&D organization team expects the tech talent to achieve, and comparing it with the expectation in the decision maker's mind will determine whether the batch exceeds or falls short of expectations. In general, the new R&D organization team expects the technological talent to meet expectations, but expectations are usually not too low. If the assessment value is higher than the expected reference point value, the new R&D organization team is satisfied. If the assessment value is lower than the expected reference point value, the new R&D organization team is not satisfied.

There are many teams within the new R&D organization, and each team sends different evaluation experts; therefore, the reference point values are different for each team. The historical reference point values can be obtained by collating historical data, and the expected reference point values can be given by the evaluation experts, so that the historical reference point vector C_1^k and the desired reference point vector C_2^k The historical reference point vector and the desired reference point vector can be obtained.

$$C_1^k = \left[c_{1j}^k(\otimes)\right] = \left\{\left[\underline{c}_{11}^k, \overline{c}_{11}^k\right], \left[\underline{c}_{12}^k, \overline{c}_{12}^k\right], \cdots \left[\underline{c}_{1p}^k, \overline{c}_{1p}^k\right]\right\} \quad (3)$$

$$C_2^k = \left[c_{2j}^k(\otimes)\right] = \left\{\left[\underline{d}_{21}^k, \overline{d}_{21}^k\right], \left[\underline{d}_{22}^k, \overline{d}_{22}^k\right], \cdots \left[\underline{d}_{2p}^k, \overline{d}_{2p}^k\right]\right\} \quad (4)$$

2.3.2. Set the Reference Point for the Cost Value—Bottom Line Reference Point

New R&D institutions provide cost value for the introduction of individual scientific and technical talent, but it is not known whether the institution is satisfied. Therefore, a bottom-line reference point was set, and the cost value was compared with the bottom-line reference point. The bottom-line reference point is the bottom-line cost value that the institution can afford, and if the cost value is lower than the bottom-line reference point value, the new R&D institution feels a gain. If the cost value is higher than the bottom-line reference point value, the new R&D institution feels a loss. The bottom-line reference point data are jointly provided by the evaluation experts of the new R&D organization team, and because the team pays about the same cost to the scientific and technological talents, the bottom-line reference point represents the bottom-line cost of the entire new R&D organization. After the evaluation experts' determination, we obtained the bottom-line reference point value vector C_3.

$$C_3 = \left[c_{3j}\left(\bigotimes\right)\right] = \left\{[\underline{c}_{31}, \bar{c}_{31}], [\underline{c}_{32}, \bar{c}_{32}], \cdots \left[\underline{c}_{3q}, \bar{c}_{3q}\right]\right\} \tag{5}$$

2.3.3. Calculate the Distance from the Appraised Value and the Cost Value to the Reference Point Separately

The reference point is the bullseye of the gray target decision. If the assembled data is greater than zero, the target is hit. If the assembled data are less than 0, it is considered off-target. The values of the new R&D organization team's assessment of technological talent and the cost of introducing technological talent are uncertain and are characterized by interval grey numbers. Historical reference point values, desired reference point values, and bottom-line reference point values also have uncertainty and are characterized by interval grey numbers. To calculate the distance from the appraised value and cost value to the reference point, the formula for calculating the distance between the interval grey number and the interval grey number is used. In this study, we consider the kernel and half-interval length of the interval grey number to calculate the distance between the interval grey numbers.

Definition 1. *Let two grey numbers* $\otimes_1 \in [\underline{a}, \bar{a}]$, $\otimes_2 \in [\underline{c}, \bar{c}]$, *and define the kernel of the two interval grey numbers as*

$$\widehat{\bigotimes}_1 = \frac{1}{2}(\underline{a} + \bar{a}),\ \widehat{\bigotimes}_2 = \frac{1}{2}(\underline{c} + \bar{c}) \tag{6}$$

If $\widehat{\otimes}_1 > \widehat{\otimes}_2$, *then* $\otimes_1 > \otimes_2$.

Definition 2. *Let two grey numbers* $\otimes_1 \in [\underline{a}, \bar{a}]$, $\otimes_2 \in [\underline{c}, \bar{c}]$, *and define the length of the two interval grey numbers as*

$$1(\bigotimes_1) = \frac{1}{2}(\bar{a} - \underline{a}), 1(\bigotimes_2) = \frac{1}{2}(\bar{c} - \underline{c}) \tag{7}$$

Definition 3. *Let two grey numbers* $\otimes_1 \in [\underline{a}, \bar{a}]$, $\otimes_2 \in \left[\underline{b}, \bar{b}\right]$, *and define the distance between the two interval grey numbers as*

$$d(\bigotimes_1, \bigotimes_2) = \left|\widehat{\bigotimes}_1 - \widehat{\bigotimes}_2\right| + \frac{1}{2}|1(\bigotimes_1) - 1(\bigotimes_2)| \tag{8}$$

We calculated the assessed values separately according to the algorithm of interval grey numbers A^k *and the historical reference point value* C_1^k *and the distance between the evaluated value* A^k *from the desired reference point value* C_2^k *and the distance of the cost value B and the bottom-line reference point value* C_3 *to obtain the distance matrix* $d\left(A^kC_1^k\right), d\left(A^kC_2^k\right), d(BC_3)$.

2.3.4. Calculation of Prospective Values of Assessed and Cost Values Based on Distance

We compare the distances of the assessed and cost values calculated above to the reference point values by kernels of interval grey numbers. We substitute them into the prospect theory equation to obtain k matrix of prospective values of assessed values based on historical reference points, the k prospect value matrix of appraisal values based on desired reference points, and 1 prospect value matrix of cost values based on bottom-line reference points.

$$V_{ij(1)}^{kc_1} = \begin{cases} \left(d\left(A^k C_1^k\right)\right)^{\alpha} & \widehat{\otimes}_{a_{ij}^k(\otimes)} > \widehat{\otimes}_{c_{1j}^k(\otimes)} \\ -\theta * \left(d\left(A^k C_1^k\right)\right)^{\beta} & \widehat{\otimes}_{a_{ij}^k(\otimes)} < \widehat{\otimes}_{c_{1j}^k(\otimes)} \end{cases} \tag{9}$$

$$V_{ij(1)}^{kc_2} = \begin{cases} \left(d\left(A^k C_2^k\right)\right)^{\alpha} & \widehat{\otimes}_{a_{ij}^k(\otimes)} > \widehat{\otimes}_{c_{2j}^k(\otimes)} \\ -\theta * \left(d\left(A^k C_2^k\right)\right)^{\beta} & \widehat{\otimes}_{a_{ij}^k(\otimes)} < \widehat{\otimes}_{c_{2j}^k(\otimes)} \end{cases} \tag{10}$$

$$V_{ij(1)}^{c_3} = \begin{cases} (d(BC_3))^{\alpha} & \widehat{\otimes}_{b_{ij}(\otimes)} < \widehat{\otimes}_{c_{3j}(\otimes)} \\ -\theta * (d(BC_3))^{\beta} & \widehat{\otimes}_{b_{ij}(\otimes)} > \widehat{\otimes}_{c_{3j}(\otimes)} \end{cases} \tag{11}$$

As the assessed value is a benefit type of data, the larger the value, the better, with gains above the reference point and losses below it. Cost values, on the other hand, are cost-based data; the smaller the value, the better with gains below the reference point and losses above the reference point.

Formula (9) in $V_{ij(1)}^{kc_1}$ denotes the prospective value of the first stage team's assessment of talent based on the historical reference point, as determined by the assessment value $a_{ij}^k(\otimes)$ and the reference point value $c_{1j}^k(\otimes)$ calculated as a power function of the distance between the assessment value and the reference point value. Equation (10) in $V_{ij(1)}^{kc_2}$ denotes the prospect value of the assessment value based on the desired reference point, calculated as a power function of the distance between the assessment value $a_{ij}^k(\otimes)$ and the value of the reference point $c_{2j}^k(\otimes)$ is calculated as a power function of the distance between the assessed value and the reference point value. Equation (11) in $V_{ij(1)}^{c_3}$ indicates that the assessment value is based on the prospective value of the historical reference point, obtained by a power function of the distance between the assessment value $b_{ij}(\otimes)$ and the value of the reference point $c_{3j}(\otimes)$ is calculated as a power function of the distance between the assessment value and the reference point value. The prospect value is determined by the subjective perception of the decision maker, and the gains and losses are relative to the reference point.α and β denote the decision-maker's risk attitude coefficients in the gain and loss regions, respectively. $\alpha, \beta < 1$ denote that the decision-maker's sensitivity is decreasing, and in this study, $\alpha = 0.88$, $\beta = 0.88$. θ is the coefficient of the decision-maker's loss perception, and because decision-makers are risk-averse in the face of gains and risk-averse in the face of losses. When $\theta > 1$, decision-makers are steeper and more sensitive in the loss region than in the gain region. In this study, $\theta = 2.25$.

2.3.5. Normalization of Assessed Value Prospect and Cost Value Prospect and Aggregation

The foreground values we obtained according to foreground theory do not meet the requirements of normalization and need to be normalized so that their values fall on $[-1, 1]$. We used the maximum value method for normalization, which is easy and convenient to operate and retains the characteristics of the original data.

$$M_{ij(1)}^{kc_1} = \frac{V_{ij(1)}^{kc_1}}{\max\limits_i \max\limits_j V_{ij(1)}^{kc_1}} \cdot M_{ij(1)}^{kc_2} = \frac{V_{ij(1)}^{kc_2}}{\max\limits_i \max\limits_j V_{ij(1)}^{kc_2}} \cdot M_{ij(1)}^{c_3} = \frac{V_{ij(1)}^{c_3}}{\max\limits_i \max\limits_j V_{ij(1)}^{c_3}} \tag{12}$$

Let the kth team be given an assessed value indicator weight of ω_j^k and the weight of the cost value indicator is μ_j then, the matrix of prospective values of assessed values based on historical reference points $N_i^{kc_1}$, the assessed value prospect value matrix based on the desired reference point $N_i^{kc_2}$, and the cost value prospect values based on the bottom-line reference point $N_i^{c_3}$ are.

$$N_{i(1)}^{kc_1} = \sum_{j=1}^{p} \omega_j^k M_{ij(1)}^{kc_1},\ N_{i(1)}^{kc_2} = \sum_{j=1}^{p} \omega_j^k M_{ij(1)}^{kc_2},\ N_{i(1)}^{c_3} = \sum_{j=1}^{q} \mu_j M_{ij(1)}^{c_3} \tag{13}$$

For $N_{i(1)}^{kc_1}$, $N_{i(1)}^{kc_2}$, and $N_{i(1)}^{c_3}$ transpose processing.

$$TS_{ts(1)}^{c_1} = \left[N_i^{kc_1T}\right],\ TS_{ts(1)}^{c_2} = \left[N_i^{kc_2T}\right],\ TS_{ts(1)}^{c_3} = \left[N_i^{c_3T}\right] \tag{14}$$

Let the historical reference point weights of this study be θ and the desired reference point weight is $1-\theta$. First, the assessment value prospect value based on the historical reference point and the desired reference point is set to obtain the assessment value composite prospect value. Then, the assessment value composite prospect value is set with the cost value prospect value, and the weight of the assessment value composite prospect value is δ and the weight of the cost-value prospect value is $1-\delta$. After agglomeration, we obtained the first-stage matched composite prospect value, $S_{ij(1)}$.

$$TS_{ts(1)} = \delta \times \left(\theta \times \left(TS_{ts(1)}^{c_1}\right) + (1-\theta) \times \left(TS_{ts(1)}^{c_2}\right)\right) + (1-\delta) \times \left(TS_{ts(1)}^{c_3}\right) \tag{15}$$

2.3.6. Construction of a Bilateral Matching Model to Derive the Results of the First Phase of Elimination Matching

Professor Gale of Brown University and the renowned economist Professor Sharply pioneered the theory of bilateral matching decisions in 1962 with their article, "College Admissions and Stable Marriages". Let $\mu : P \cup Q \to Q \cup P$ be a one-to-one mapping, if $\forall P_i \in P$. $\forall Q_j \in Q$ satisfies $\mu P_i = Q_j \in Q$, and $\mu Q_j = P_i \in P$, then we call μ is a two-way matching. $\mu(P_i) = Q_j$ denote P_i with Q_j in μ in the match, and $\mu Q_j = Q_j$ indicates that Q_j in μ does not match.

For the bilateral matching decision problem between teams and scientific and technological talent in new R&D institutions, forming a reasonable and effective bilateral matching scheme is the common demand of teams and talent, and constructing a bilateral matching model and proposing a solution algorithm is the most crucial step. The traditional Gale—Sharply algorithm performs research in preference order, and this paper proposes bilateral matching based on uncertain interval grey numbers and prospect theory based on reference points. In the first stage, the new R&D organization team and the scientific and technological talents are many-to-many matching. The new R&D organization achieves complete matching, the scientific and technological talents are incomplete matching, and the scientific and technological talents without successful matching are eliminated. In this study, we constructed the first-stage elimination matching model M-1.

$$\text{MAX } Z = \sum_{t=1}^{n} \sum_{s=1}^{m} \pi_{ts} TS_{ts(1)} \tag{16}$$

$$\text{s.t.} \begin{cases} \pi_{ts} \in [0 \text{ or } 1] \\ \sum\limits_{t=1}^{n} \pi_{ts} = 1 \\ \sum\limits_{s=1}^{m} \pi_{ts} \leq 1 \end{cases} \tag{17}$$

The objective of the M-1 model is to transform the dual objective of maximizing the prospective value of the assessed value and prospective value of the cost value into a single objective. The constraints are as follows: the weight vector π_{ts} can only be 0 or 1; each new R&D organization team must be matched to a scientific and technological talent, and only one talent can be matched; each scientific and technological talent may not be matched to a new R&D organization team.

Theorem 1. *M-1 The model must have an optimal solution.*

Note: According to the optimality existence theorem, any single-objective programming with bounded feasible domain must be optimal over its feasible domain. M-1 The model is a single-objective programming problem, and the feasible domain of the model exists and is bounded. Therefore, the M-1 model must have an optimal solution. M-1 It can be solved using LINGO software, and the solution of $\pi_{ts} = 1$ means that the tth new R&D institution team and the sth technological talent are successfully matched. $\pi_{ts} = 0$ means that the tth new R&D institution team and the sth technological talent do not match successfully, and the m-n technological talents are eliminated.

3. New R&D Organization Team and Science and Technology Talent Second Stage selection Match

After the first phase of elimination matching, the number of teams and talents is equal, and the matching is still many-to-many. After one year, when the trial period expires, the team re-evaluates the talent and determines the team they will eventually work for. In the second stage of matching, the highest evaluation value of the team to the talent is considered, and the highest satisfaction of the feedback of the ranking value of the team by the tech talent is also considered, setting historical reference points and expectation reference points for the evaluation value and bottom-line reference points for the satisfaction, fully considering the psychological factors of the new R&D organization team and the tech talent in both directions to better realize the two-way selection, and finally using the 0–1 The integer planning model is used for bilateral matching. In the second stage of selection matching, because the number of scientific and technical talents is the same as the number of R&D teams, the pressure on decision makers will be much less, there will be no more unmatched situations, and teams and talents achieve complete matching, as shown in Figure 2.

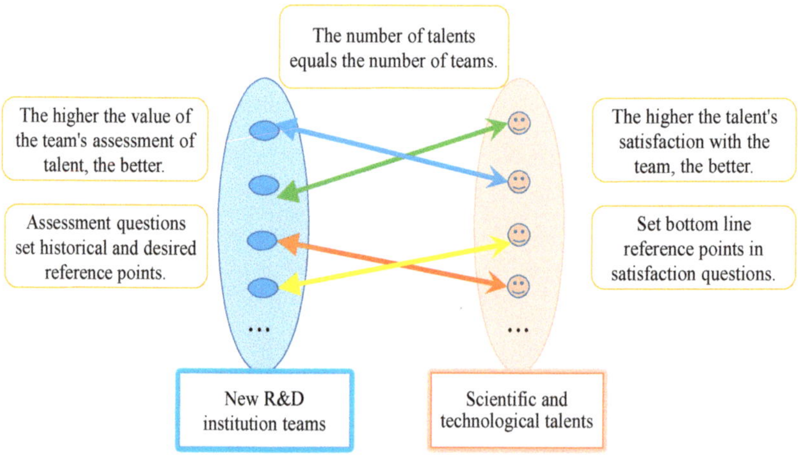

Figure 2. Matching diagram of the second stage of selection of teams and talents in new R&D institutions.

3.1. Collecting the Team's Assessment Value of Scientific and Technical Talents and the Ranking Value of Talents to the Team

The first phase of elimination matching is dominated by new R&D institutions and does not consider the dominant weight of talents. Therefore, the second phase of selection matching considers the assessment of scientific and technological talents by the team of new R&D institutions and also considers the ranking of scientific and technological talents to new R&D institutions.

We collected the assessed values of the second stage of the new R&D organization team for scientific and technological talents, characterized by the average value of the expert assessment. The reference point remains unchanged, and the data of the historical reference point and the desired reference point of the first stage are still used, since the expert assessments in this study are all scored on a percentage scale. In order to achieve the standardization requirements of the data, all the data are averaged and then divided by 100, so that all the assessed values, the reference point values lie between 0 and 1. We obtain the matrix of assessment values $E^k = \left[e_{ij}^k\right]$, $i = 1, 2, \cdots m$, $j = 1, 2, \cdots p, k = 1, 2, \cdots n$.

We consider the ranking of teams in new R&D institutions by scientific and technical talents to obtain information on the preference order of scientific and technical talents for teams $F = [f_{ts}], t = 1, 2, \cdots n, s = 1, 2, \cdots m$. To facilitate the assembly of data, the preference order was transformed into satisfaction according to certain rules. The transformation rules are listed in Table 3. According to the transformation rules, it is transformed into satisfaction information $G = [g_{ts}], t = 1, 2, \cdots n, s = 1, 2, \cdots m$.

Table 3. Rules for converting the preference order of talents to satisfaction with the teams.

Preference Order	Satisfaction
1	1
2	1-1/n
3	1-2/n
4	1-3/n

3.2. Matching Integrated Prospect Values Based on Historical, Desired, and Bottom-Line Reference Point Sets for the Second Stage

Based on the historical reference point value and the desired reference point value, the distance from the team's assessment of talent to the historical and desired reference points is calculated. Because it is a real number, the distance is directly subtracted and calculated by taking the absolute value. Let the bottom-line reference point of scientific and technological talents' satisfaction with the team be 0.5. Calculate the distance from the satisfaction to the bottom-line reference point, and because it is a real number, the distance is directly subtracted and calculated by taking the absolute value. The prospect value is calculated according to prospect theory.

$$V_{ij(2)}^{kc_1} = \begin{cases} \left|e_{ij}^k - e_j^{kc_1}\right|^\alpha & e_{ij}^k > e_j^{kc_1} \\ -\theta * \left|e_{ij}^k - e_j^{kc_1}\right|^\beta & e_{ij}^k < e_j^{kc_1} \end{cases} \tag{18}$$

$$V_{ij(2)}^{kc_2} = \begin{cases} \left|e_{ij}^k - e_j^{kc_2}\right|^\alpha & e_{ij}^k > e_j^{kc_2} \\ -\theta * \left|e_{ij}^k - e_j^{kc_2}\right|^\beta & e_{ij}^k < e_j^{kc_2} \end{cases} \tag{19}$$

$$V_{ts(2)}^{c_3} = \begin{cases} \left|g_{ts} - 0.5\right|^\alpha & f_{ts} > 0.5 \\ -\theta * \left|g_{ts} - 0.5\right|^\beta & f_{ts} < 0.5 \end{cases} \tag{20}$$

Combining the prospect value of the team's assessment of talent based on the historical reference point $V_{ij}^{kc_1}$ and the prospect value of the assessed value based on the desired reference point $V_{ij}^{kc_1}$, combining the indicator weights of the assessed values given by the team as ω_j^k,

$$N_{i(2)}^{kc_1} = \sum_{j=1}^{p} \omega_j^k M_{ij(1)}^{kc_1} . N_{i(2)}^{kc_2} = \sum_{j=1}^{p} \omega_j^k M_{ij(1)}^{kc_2} \tag{21}$$

For $N_{i(1)}^{kc_1}$, $N_{i(1)}^{kc_2}$ to transpose the treatment.

$$TS_{ts(2)}^{c_1} = \left[N_i^{kc_1 T}\right], TS_{ts(2)}^{c_2} = \left[N_i^{kc_2 T}\right] \tag{22}$$

point weight be $1-\theta$. The assessment value prospect values based on the historical reference point and the desired reference point are first pooled to obtain the assessment value composite prospect value, and then the assessment value composite prospect value is pooled with the satisfaction prospect value, and the weight of the assessment value composite prospect value is σ and the weight of the satisfaction prospect value is $1-\sigma$. After agglomeration, we obtain the second-stage matched composite prospect value, $S_{ij(2)}$.

$$TS_{ts(2)} = \sigma \times \left(\theta \times \left(TS^{c_1}_{ts(2)}\right) + (1-\theta) \times \left(TS^{c_2}_{ts(2)}\right)\right) + (1-\sigma) \times \left(V^{c_3}_{ts(2)}\right) \quad (23)$$

3.3. Construction of a Bilateral Matching Model to Derive the Second Stage Selection Matching Results

In the second stage, the new R&D organization team and the scientific and technological talents are many-to-many matched, the new R&D organization achieves complete matching, the scientific and technological talents are incompletely matched, and the scientific and technological talents without successful matching are eliminated. In this study, we constructed the first-stage elimination matching model, M-2.

$$\text{MAX } Z = \sum_{t=1}^{n}\sum_{s=1}^{m} \varphi_{ts} TS_{ts} \quad (24)$$

$$\text{s.t.} \begin{cases} \varphi_{ts} \in [0 \text{ or } 1] \\ \sum\limits_{t=1}^{n} \varphi_{ts} = 1 \\ \sum\limits_{s=1}^{m} \varphi_{ts} = 1 \end{cases} \quad (25)$$

The objective of the M-1 model is to transform the dual objective of maximizing the prospect value of the team's assessment of talent and the prospect value of talent's satisfaction with the team's sorting transformation into a single objective. The constraints are as follows: the weight vector φ_{ts} can only be 0 or 1; each new R&D organization team must be matched to a technology talent, and only one talent can be matched; each technology talent must be matched to a new R&D organization team; and only one team can be matched.

Theorem 2. *M-2 The model must have an optimal solution.*

Note: According to the optimality existence theorem, any single-objective programming problem with a bounded feasible domain must be optimal over its feasible domain. The M-2 model is a single-objective programming problem, and the feasible domain of the model exists and is bounded. Thus, the M-2 model must have an optimal solution. The M-2 model can be solved using LINGO software, and the solution of $\varphi_{ts} = 1$ means that the tth new R&D institution team and sth technological talent are successfully matched. $\varphi_{ts} = 0$ means that the tth new R&D institution team and the sth technological talent are not successfully matched.

In the second stage, the number of teams of new R&D institutions and scientific and technological talent are equal, and one team matches one scientific and technological talent, which is a many-to-many exact match, and a two-way selection match due to the consideration of two-way assessment.

3.4. Methodological Steps of This Paper

The steps of this study's prospect theory-based two-stage bilateral matching method for teams and scientific and technological talents in new R&D institutions are as follows.

In the first step, based on the first stage of elimination matching of new R&D institution teams and scientific and technological talents, a general framework of assessment indicators for new R&D institution teams, personalized assessment indicators, and indicators for the cost of introducing scientific and technological talents are constructed.

In the second step, the team-to-talent assessment data of new R&D institutions based on personalized assessment indicators in the first stage, historical reference point data and desired reference point data, cost of introducing technology talent data, and bottom-line reference point data, all characterized by interval grey numbers, were collected to calculate the benefit prospect value and cost prospect value of team-to-talent based on prospect theory and the given indicator weights, respectively.

In the third step, based on the benefit-cost weights, the integrated prospect values of the first stage are integrated, substituted into the M-1 model, and solved using LINGO software to obtain the elimination matching results of the first stage.

In the fourth step, based on the second stage of selection matching between the team of the new R&D institution and the scientific and technological talents, the assessment data of the team of the new R&D institution to the scientific and technological talents are collected, characterized by the mean value, following the historical reference point and desired reference point data in the first stage. The data were divided by 100 for normalization. While the ranking value of the scientific and technological talents to the team is collected, the ranking value is converted into satisfaction and given the satisfaction bottom-line reference point value. Based on the prospect theory and the given index weights, the evaluation prospect values of both team and talent are calculated.

In the fifth step, the prospect values of the second stage are integrated based on the team-talent weights, substituted into the M-2 model, and solved using LINGO software to obtain the selection matching results of the second stage.

4. Case Studies

4.1. Background of the New R&D Agency Team

The new R&D institution T was established in 2019 by the talent team of the University of D, the government, and social capital, of which the talent team holds 70% of the shares. The new R&D institution T aims at the frontier of science and technology and market demand and implements the specific measures of General Secretary Xi Jinping's "three firsts," reflecting the organic unity of superior disciplines, innovation clusters, and high-quality development. Currently, institutions are mainly engaged in new energy and intelligent distributed power generation, new high-efficiency refrigeration and heat pumps and equipment, key technologies for building environment and air quality, and new energy system optimization and monitoring research, actively building a national research platform for building intelligent environmental energy. As a leading domestic intelligent energy development and innovation enterprise, the institute adheres to the business philosophy of integrity, professionalism, aggressiveness, and cooperation and has accumulated rich experience in research and development, manufacturing, construction, operation, and maintenance in the fields of building energy conservation, distributed photovoltaic power generation, and comprehensive energy utilization, which is dedicated to providing customers with high-quality, fast, and efficient services. Few energy companies do energy extraction independently, and generally they work in partnership with many other companies. Now that many new energy companies are emerging, and they are eager to find clean alternative energy sources, there is a greater need for collaboration among scientific and technical talents in multiple fields. The need to recruit more scientific and technical talents and make their research and development direction match the team.

The new R&D institution T currently has four teams: the production team of high-efficiency building energy-saving equipment, the R&D team of intelligent environmental energy, the team of power engineering design and construction, and the team of distributed energy management. Since its establishment, the new R&D organization T has introduced 36 high-end talents from home and abroad, who have PhD degrees, achieved remarkable results in the industry, won national awards, and earned considerable economic income. The new R&D organization T has always attached importance to the introduction and training of scientific and technological talent. Scientific and technological talent development is good—constantly pioneering and innovative— and has won more than 20 national projects, the National Technical Invention Award, more than 20 national projects, one national technical invention award, more than 160 invention patents authorized, and four incubated enterprises.

The new R&D organization T is always looking for scientific and technical talents in environmental engineering and electronic information, requiring solid knowledge of environmental monitoring, theoretical knowledge of electronic information technology, relevant papers published in foreign professional academic journals, familiarity with laboratory-related technical experimental processes, a PhD degree, 1–3 years of relevant work experience, strong work responsibility, excellent language skills, and communication and coordination skills. Other requirements include the ability to handle complex problems and critical incidents independently, a strong work motivation, a sense of proactive service, etc. An annual salary of ¥100,000–¥1,000,000 is offered working in the Yangtze River Delta.

The new R&D institution T—in August this year, four teams put forward plans to recruit talents, the personnel department after screening, preliminary tests, practice, and other links. Finally, identified were six scientific and technical talents, the decision-making power to the team, and the team sent experts to conduct the first stage of elimination matching and the second stage of selection matching. Based on the results of the first-stage matching, the scientific and technical talents will enter four teams for a one-year probationary period and work in each team for three months. Science and technology talents will finalize which team to work on based on the second stage of selection matching. According to the two-way assessment index system proposed in this paper for teams and tech talents in new R&D institutions, both teams and talents have completed the assessment, and the data and bilateral matching are presented in detail next.

4.2. Phase I Phase-Out Matching of Teams and Scientific and Technological Talents in New R&D Institutions

We obtained the assessed values of the T1-T4 team for six scientific and technical talents from the HR director of the new R&D organization T; the values are characterized by interval gray numbers, as shown in Tables 4–7. There are 22 science and technology teams in the new R&D organization T. Four of the teams had similar specialties, and each team sent three experts to assess the science and technology talents under each indicator. Twelve experts from each of the four teams filled in the assessment values under the percentage system. In this study, the uncertainty of the team's assessment value of talent is fully considered; therefore, the interval grey number is used to characterize the assessment value.

Table 4. Assessed values of talents S by team T1 in new R&D institutions.

T1-S	Render Service to One's Country	Critical Innovation	Entrepreneurial Achievements	Interdisciplinary Subject (in Science)	Self-Control and Self-Discipline
S1	[78, 88]	[68, 75]	[80, 81]	[69, 79]	[70, 73]
S2	[77, 84]	[70, 70]	[55, 76]	[74, 75]	[82, 88]
S3	[70, 79]	[75, 81]	[50, 66]	[68, 75]	[77, 81]
S4	[82, 88]	[65, 80]	[68, 70]	[68, 71]	[65, 69]
S5	[80, 83]	[76, 78]	[73, 88]	[72, 90]	[68, 73]
S6	[76, 78]	[66, 88]	[74, 78]	[70, 73]	[68, 74]
Historical reference point	[70, 80]	[60, 90]	[50, 80]	[60, 75]	[70, 85]
Desired reference point	[75, 85]	[70, 80]	[70, 85]	[70, 85]	[70, 80]

Table 5. The assessed values of the new R&D organization team T2 for the talents S.

T2-S	Professional Ethics	Innovative Results	Entrepreneurial Achievements	Communication and Collaboration	Keep a Secret
S1	[70, 73]	[68, 75]	[76, 78]	[75, 90]	[83, 88]
S2	[75, 80]	[67, 88]	[85, 88]	[73, 85]	[65, 78]
S3	[77, 86]	[62, 78]	[75, 86]	[65, 71]	[72, 85]
S4	[73, 82]	[65, 75]	[67, 69]	[72, 73]	[66, 88]
S5	[68, 92]	[68, 81]	[62, 70]	[72, 75]	[65, 82]
S6	[66, 77]	[68, 74]	[65, 69]	[70, 76]	[72, 75]
Historical reference points	[70, 80]	[60, 80]	[60, 85]	[55, 80]	[60, 90]
Desired reference point	[70, 90]	[70, 90]	[70, 90]	[60, 80]	[70, 85]

Table 6. Assessed values of the new R&D organization team T3 for talents S.

T3-S	Render Service to One's Country	Entrepreneurial Achievements	Top Students	Technology INSIGHTS	Physical Quality (in Ideological Education)
S1	[73, 77]	[78, 88]	[75, 87]	[70, 76]	[68, 78]
S2	[65, 74]	[77, 84]	[72, 76]	[70, 84]	[80, 90]
S3	[73, 78]	[65, 79]	[66, 88]	[72, 90]	[71, 83]
S4	[68, 85]	[62, 88]	[72, 74]	[79, 88]	[80, 91]
S5	[82, 83]	[61, 83]	[70, 82]	[61, 77]	[78, 85]
S6	[75, 78]	[66, 71]	[66, 85]	[72, 78]	[70, 84]
Historical reference points	[70, 90]	[60, 85]	[60, 80]	[65, 85]	[70, 90]
Desired reference point	[70, 90]	[70, 90]	[70, 85]	[70, 90]	[75, 90]

Table 7. Assessed values of the new R&D organization team T4 for the talents S.

T4-S	Professional Ethics	Innovative Results	Communication and Collaboration	Design and Programming	Psychological Quality (in Ideological Education)
S1	[70, 81]	[56, 83]	[67, 91]	[68, 75]	[67, 76]
S2	[65, 76]	[75, 80]	[72, 82]	[73, 87]	[73, 82]
S3	[65, 86]	[68, 86]	[70, 92]	[68, 79]	[67, 86]
S4	[76, 90]	[74, 85]	[50, 68]	[68, 82]	[61, 85]
S5	[77, 86]	[50, 64]	[80, 83]	[88, 92]	[77, 86]
S6	[80, 82]	[70, 77]	[66, 88]	[77, 78]	[68, 70]
Historical reference points	[60, 85]	[60, 85]	[70, 85]	[50, 80]	[60, 80]
Desired reference point	[70, 90]	[70, 90]	[70, 90]	[70, 85]	[70, 85]

As can be seen from the above table:

(1) Each of the four teams of new R&D institutions selected their own personalized assessment indicators and assessed their talent based on personalized assessment indicators.

(2) Assessment data have a certain level of uncertainty, characterized by interval grey numbers, which are different for each assessment.

(3) The four teams of new R&D institutions provide both historical and desired reference point data, which are also uncertain and characterized by interval grey numbers.

The data in the above table do not show a bilateral match between the NRA team and scientific and technological talent, and a suitable method of integration is needed to facilitate comparison and analysis. Since the teams of the new R&D institutions rely on the new R&D institutions to select and hire scientific and technological talents, but the mechanism is very flexible, the new R&D institutions give the teams great hiring decisions and autonomy, and the funds are also allocated in a lump sum way, so the team leaders were invited to use the Delphi method to determine the indicator weights, and after several confirmations, it was finally determined that the personalized indicator weights of the four teams of the new R&D institutions were $\omega_{1j} = (0.2, 0.1, 0.3, 0.2, 0.2)$, and $\omega_{2j} = (0.1, 0.3, 0.3, 0.2, 0.1)$, and $\omega_{3j} = (0.1, 0.4, 0.2, 0.2, 0.1)$, and $\omega_{4j} = (0.15, 0.3, 0.15, 0.25, 0.15)$. In this study, a historical reference point and an expected reference point were set, and assuming that both reference points are equally important, $\theta = 0.5$.

Based on the interval grey number distance formula and foreground theory formula, indicator weights, and reference point weights, we integrate to obtain the matrix of foreground values based on the dual reference points of historical reference points and desired reference points, as shown in Table 8.

Table 8. Prospect values for the assessment of talents S by the new R&D organization teams T.

T-S	S1	S2	S3	S4	S5	S6
T1	0.0632	−0.0103	−0.3488	−0.1857	0.2396	−0.1200
T2	0.0641	0.2749	−0.0912	−0.4754	−0.1589	−0.3928
T3	−0.1087	0.0490	−0.2983	0.0006	−0.1589	−0.6959
T4	−0.2245	0.0500	−0.0045	−0.0671	0.1089	−0.1271

As can be seen from the table above:

(1) The results of the assessment of scientific and technological talents by the four teams of the new R&D institutions are different: some are satisfied, some are not, and the degree of satisfaction and dissatisfaction are also different and ranked differently.

(2) Team T1 of the new R&D institution is satisfied with S1 and S5, and dissatisfied with S2, S3, S4 and S6, with a satisfaction rate of 33.33%, and the ranking of scientific and technological talents is S5 $\succ$ S1 $\succ$ S2 $\succ$ S6 $\succ$ S4 $\succ$ S3.

(3) Team T2 of the new R&D organization is satisfied with S1 and S2, and is dissatisfied with S3, S4, S5 and S6, with a satisfaction rate of 33.33%, and ranks the scientific and technological talents as S2 $\succ$ S1 $\succ$ S3 $\succ$ S5 $\succ$ S6 $\succ$ S4.

(4) Team T3 of the new R&D organization is satisfied with S2 and S4, and is dissatisfied with S1, S3, S5 and S6, with a satisfaction rate of 33.33%, and ranks the scientific and technological talents as S2 $\succ$ S4 $\succ$ S1 $\succ$ S5 $\succ$ S3 $\succ$ S6.

(5) Team T4 of the new R&D organization is satisfied with S2 and S5, and is dissatisfied with S1, S3, S5 and S6, with a satisfaction rate of 33.33%, and ranks the scientific and technological talents as S5 $\succ$ S2 $\succ$ S3 $\succ$ S4 $\succ$ S6 $\succ$ S1.

If we let the new R&D organization team T and tech talent S match directly through the satisfaction of prospect value representation, then it appears that teams T1 and T5 select technology talent S5, and teams T3 and T4 select tech talent S2 so that only two tech talents are selected. The phenomenon of internal talent grabbing occurs, which does not achieve the overall optimum of the new R&D organization. In the first stage, we select the technology talent that can match the applicable stage and eliminate the two tech talents. Some may also say that removing the least satisfactory ones is necessary. Our analysis found that T1 team eliminated S3, T2 team eliminated S4, T3 team eliminated S6, and T4 team eliminated S1. This eliminates too many again because we do not know how dominant the team is in the process.

In the first stage of matching the team of the new R&D organization and the scientific and technological talent belonging to the elimination stage, we must consider the high satisfaction of the team of the new R&D organization with the assessed value of the scientific and technological talent relative to the reference point, in addition to the low cost of introducing scientific and technological talent. We collected data on the cost of introducing these six scientific and technological talents into different teams, as shown in Table 9.

Table 9. Costs of bringing in scientific and technological talents S.

S-T	Wage Insurance	Home Purchase Subsidy	Performance Incentives	Project Dividends	Other Awards
S1	[25, 30]	[45, 65]	[20, 26]	[10, 25]	[5, 16]
S2	[24, 28]	[50, 60]	[18, 20]	[10, 12]	[4, 18]
S3	[30, 40]	[20, 50]	[15, 30]	[8, 16]	[5, 12]
S4	[23, 28]	[40, 50]	[16, 28]	[6, 20]	[3, 18]
S5	[30, 32]	[55, 60]	[25, 28]	[12, 15]	[6, 14]
S6	[28, 35]	[38, 45]	[14, 30]	[7, 20]	[5, 10]
Bottom line reference point	[32, 32]	[50, 50]	[25, 25]	[15, 15]	[15, 15]

As can be seen from the table above:

(1) The cost of introducing scientific and technical talent S is different and uncertain and can only be estimated as an approximate range within which any value taken is possible, which we characterize as an interval grey number.

(2) The cost of bringing in scientific and technological talent under different projects varies, with the cost of housing subsidies generally being higher, and other incentives lower.

(3) The new R&D organization gives a bottom-line reference point, and feels a loss when the cost of bringing in technology talent exceeds the bottom-line reference point, and a gain when the cost of bringing in technology talent is less than the bottom-line reference point.

The introduction cost alone does not indicate the situation of scientific and technological talents, and we need to use the bottom-line reference point to measure whether the new R&D institutions pay the introduction cost as a gain or a loss. In this paper, we calculate the introduction cost of scientific and technological talents based on the bottom-line reference point, and set the index weight of the introduction cost of scientific and technological talents as $\omega_{5j} = (0.25, 0.25, 0.2, 0.2, 0.1)$ to get the prospective value of the introduction cost of six scientific and technological talents.ST_{1j} = $(-0.2509, 0.0188, 0.2156, 0.3149, -0.0846, 0.2919)$. From the prospective values of the introduction cost of scientific and technological talent, it can be seen that:

(1) Scientific and technological talents S1 and S5 are beyond the bottom-line reference point, and new R&D institutions would feel a loss if they brought them in.

(2) Scientific and technological talents S2, S3, S4, and S6 do not exceed the bottom-line reference point, and new R&D institutions will feel the benefits of bringing them in.

(3) The highest prospective value of the introduction cost of S4 for scientific and technological talents indicates that S4 is the least expensive, but this does not mean that the introduction of S4 is appropriate and has to be considered in conjunction with the assessment of S4 by the team of the new R&D institution.

We cannot consider the selection and recruitment of scientific and technological talent from the perspective of cost alone, but should be combined with the possible benefits of scientific and technological talent for comprehensive consideration. In this study, we consider dual-objective bilateral matching between the maximum team satisfaction and the minimum cost of introducing scientific and technological talents in the new R&D organization, and simplify the dual objective into a single objective to construct a matching model. To achieve the overall optimum of the new R&D organization, we apply the M-1 bilateral matching model in this paper for matching and set the weights of the benefit assessment of the team and the scientific and technological talents as $\delta = 0.5$, and the weight of cost assessment is $1 - \delta = 0.5$. This is because if technology talent can create more value, the more promising technology talent cannot be missed because of the low cost of the new R&D organization. The comprehensive prospect value is based on the combination of historical reference points and desired reference points, and the introduction cost, which fully considers the psychological and cost factors of decision makers. The higher the comprehensive prospect value, the higher the satisfaction of decision-makers, and the lower the comprehensive prospect value, the lower the satisfaction of decision-makers. Based on the prospect value of the assessment value, the prospect value of the introduction cost, and the weights of the two prospect values δ, we obtain the matrix of integrated prospect values for the first-stage matching of new R&D organization team T and scientific and technological talent S, as shown in Table 10.

Table 10. Combined Phase I outlook values for new R&D organization teams T and talents S.

T-S	S1	S2	S3	S4	S5	S6
T1	−0.0939	0.0042	−0.0666	0.0646	0.0775	0.0859
T2	−0.0934	0.1468	0.0622	−0.0802	−0.1218	−0.0505
T3	−0.1798	0.0339	−0.0413	0.1578	−0.1218	−0.2020
T4	−0.2377	0.0344	0.1056	0.1239	0.0121	0.0824

As can be seen from the table above:

(1) After incorporating the cost of introduction, the satisfaction rate and ranking of scientific and technological talents by the team of new R&D institutions changed.

(2) Team T1 of the new R&D organization is satisfied with S2, S4, S5, and S6, and is dissatisfied with S1 and S3, with a satisfaction rate of 66.67%, and ranks scientific and technological talents as S6 $\succ$ S5 $\succ$ S4 $\succ$ S2 $\succ$ S3 $\succ$ S1.

(3) Team T2 of the new R&D organization is satisfied with S2 and S3, and is dissatisfied with S1, S4, S5 and S6, with a satisfaction rate of 33.33%, and ranks the scientific and technological talents as S2 $\succ$ S3 $\succ$ S6 $\succ$ S4 $\succ$ S1 $\succ$ S5.

(4) Team T3 of the new R&D organization is satisfied with S2 and S4, and is dissatisfied with S1, S3, S5 and S6, with a satisfaction rate of 33.33%, and ranks the scientific and technological talents as S4 $\succ$ S2 $\succ$ S3 $\succ$ S5 $\succ$ S1 $\succ$ S6.

(5) Team T4 of the new R&D organization is satisfied with S2, S3, S4, S5 and S6, and is dissatisfied with S1, with a satisfaction rate of 83.33%, and ranks the scientific and technological talents as S4 $\succ$ S3 $\succ$ S6 $\succ$ S2 $\succ$ S5 $\succ$ S1.

The combined prospect values of the first-stage matching of new R&D organization team T and technology talent S were substituted into the M-1 bilateral matching model and solved using LINGO software to obtain the first stage team-talent matching results, as shown in Table 11.

Table 11. Results of the first stage matching of new R&D organization teams T and talents S.

Matching Teams	Phase I Matching Technology Talent	Phase I Combined Prospect Value
T1	S6	0.0859
T2	S2	0.1468
T3	S4	0.1578
T4	S3	0.1056

From the above table we can see that:

(1) New R&D organization teams T1, T2, and T3 were matched with the highest overall prospect value of science and technology talent, and T4 was more satisfied with S3.

(2) Tech talents S1 and S5 are eliminated during the first matching phase.

The match between the team of the new R&D institution and the scientific and technological talents in the first stage considers the team's assessment of the talents and the cost of introducing talent, which is more in line with the actual situation. In the process of the team's assessment of talent, the team's personalized indicators, historical reference points, and desired reference points are taken into account so that the team's psychological gain and loss can be obtained, and whether the team is satisfied, and the degree of satisfaction can be seen from the assessment prospect value. In the process of negotiating the introduction cost between the team and the talent, the bottom-line reference points of the new R&D organization are taken into account so that the gain and loss of the new R&D organization as a whole on the cost can be obtained, whether it is satisfied, and the degree of satisfaction from the cost prospect value. The prospect values of the combined benefits and costs of the team and talent were substituted into the M-1 model for bilateral matching and the matching results obtained were more convincing.

4.3. Second-Stage Trial Matching of Teams and Scientific and Technical Talents in New R&D Institutions

After the first phase of elimination matching, S2, S3, S4, and S6 enter the second phase of probationary matching. Each science and technology talent enters the new R&D institution to work and learn for a one-year probationary period, which is applied in four teams on a rotating basis for a three-month probationary period. After a year of probation in team T of the new R&D organization, the new R&D organization team again evaluated the scientific and technical talents to determine which team the scientific and technical talents would eventually work inside. In this study, we collect the values of the second stage of the assessment of scientific talent by the new R&D organization team, which is characterized by the mean value, with the historical and desired reference points also using the data given in the first stage. All data are divided by 100 and normalized so that they fall between [0,1], as shown in Tables 12–15. For ease of calculation, the scientific and technical talents were still matched with their original numbers.

Table 12. Assessment values of talents S by the new R&D organization team T1 phase 2.

T1-S	Render Service to One's Country	Critical Innovation	Entrepreneurial Achievements	Interdisciplinary Subject (in Science)	Self-Control and Self-Discipline
S2	0.8475	0.7605	0.7325	0.7500	0.7775
S3	0.8025	0.7900	0.8025	0.7975	0.8475
S4	0.8475	0.8125	0.6975	0.7375	0.7150
S6	0.7850	0.7050	0.7325	0.8075	0.7800
Historical reference points	0.7500	0.7500	0.6500	0.6750	0.7750
Desired reference point	0.8000	0.7500	0.7750	0.7750	0.7500

Table 13. Assessment values of talents S by the new R&D organization team T2 phase 2.

T2-S	Professional Ethics	Innovative Results	Entrepreneurial Achievements	Communication and Collaboration	Keep a Secret
S2	0.8575	0.8300	0.7200	0.8175	0.7850
S3	0.7900	0.7625	0.8000	0.7600	0.8150
S4	0.8650	0.8225	0.7750	0.8075	0.7275
S6	0.8025	0.8125	0.8125	0.7650	0.7775
Historical reference points	0.7500	0.7000	0.7250	0.6750	0.7500
Desired reference point	0.8000	0.8000	0.8000	0.7000	0.7750

Table 14. Assessment values of talents S by the new R&D organization team T3 phase 2.

T3-S	Render Service to One's Country	Entrepreneurial Achievements	Top Students	Technology INSIGHTS	Physical Quality (in Ideological Education)
S2	0.8275	0.7300	0.7350	0.7700	0.8175
S3	0.7900	0.8500	0.8200	0.7450	0.7200
S4	0.8575	0.6975	0.8325	0.7500	0.7675
S6	0.8125	0.7950	0.7775	0.7950	0.8575
Historical reference points	0.8000	0.7250	0.7000	0.7500	0.8000
Desired reference point	0.8000	0.8000	0.7750	0.8000	0.8250

Table 15. Assessment values of talents S by the new R&D organization team T4 phase 2.

T4-S	Professional Ethics	Innovative Results	Communication and Collaboration	Design and Programming	Psychological Quality (in Ideological Education)
S2	0.8175	0.7825	0.7475	0.8200	0.7325
S3	0.7375	0.6800	0.7700	0.7725	0.7575
S4	0.8350	0.7700	0.8375	0.7400	0.7900
S6	0.7950	0.8725	0.8725	0.7750	0.8400
Historical reference points	0.7250	0.7250	0.7750	0.6500	0.7000
Desired reference point	0.8000	0.8000	0.8000	0.7750	0.7750

As can be seen from the above table:

(1) The assessed values of the four teams of the new R&D organization on the S of the scientific and technological talents have changed compared to the assessed values in the first stage. Because scientific and technological talents have been on trial for a period of time, the new R&D organization teams are not familiar with the scientific and technological talents to the same extent and have a new understanding of the scientific and technological talents. The scientific and technological talents are also influenced by the team culture and values in the new working environment. For example, if everyone in the team is determined

to serve the motherland, and meetings are often held with ideological education, then scientific and technical talents will also strengthen their love for the motherland in this atmosphere, and these changes are reflected in the changes in assessment values.

(2) The reason for taking the average value in this assessment is that the match in the second stage is a fit-for-post match and no more talent will be eliminated, so the matching pressure will be much less. Eventually, the teams will all be matched with tech talent, and the tech talent will all be matched with teams. For simpler and easier calculations, the uncertainty is ignored in the second-stage match.

(3) Reference points were set and teams of new R&D institutions were asked to provide the corresponding reference point data. The reference point data were not updated in the second stage, also for the convenience of calculation, and because the time difference between the first and second stages is only one year, the time of scientific and technological talents in each team is only three months, and the change in the reference points will not be very large, so the dynamics of the reference points were not considered in this study.

In the second stage, the reference point values are the average of the reference point values in the first stage divided by 100 normalized data, and the indicator weights and reference point weights are the same as those in the first stage. Based on the assessed values, reference point values, indicator weights, and reference point weights of the second stage of the new R&D organization team T to the scientific and technological talent S, we obtained the prospect value matrix of the second-stage match between the new R&D organization team T and the scientific and technological talent S, as shown in Table 16.

Table 16. Prospect values for the second stage assessment of talents S by the new R&D organization teams T.

T-S	S2	S3	S4	S6
T1	0.0235	0.0910	−0.0313	0.0080
T2	0.0398	0.0379	0.0578	0.0733
T3	−0.0524	0.0172	−0.0773	0.0391
T4	0.0218	−0.0772	0.0309	0.1056

As can be seen from the table above:

(1) After the trial of scientific and technical talents, both the satisfaction and satisfaction rates of the new R&D organization team with scientific and technical talents have changed.

(2) Team T1 of the new R&D organization is satisfied with S2, S3, and S6, but is not satisfied with S4, with a satisfaction rate of 75%, and ranks the scientific and technological talents as S3 ≻ S2 ≻ S6 ≻ S4.

(3) Team T2 of the new R&D organization is satisfied with all scientific and technological talents, with a satisfaction rate of 100%, and the ranking of scientific and technological talents is S6 ≻ S4 ≻ S2 ≻ S3.

(4) Team T3 of the new R&D organization is satisfied with S3 and S6, and is dissatisfied with S2 and S4, with a satisfaction rate of 50%, and ranks the scientific and technological talents as S6 ≻ S3 ≻ S2 ≻ S4.

(5) Team T4 of the new R&D organization is satisfied with S2, S4, S6, and is not satisfied with S3, with a satisfaction rate of 75%, and ranks the scientific and technological talents as S6 ≻ S4 ≻ S2 ≻ S3.

After the probationary period, the tech talent will be called a full staff member in the new R&D organization team; in the second stage of matching, the new R&D organization team is 50% dominant, and we need to consider the choice of technology talent. In this study, scientific and technical talents ranked the four teams, as shown in Table 17.

Table 17. Ranking of talents S on the second stage of the new R&D institution teams T.

S-T	S2	S3	S4	S6
T1	4	3	3	2
T2	3	1	2	1
T3	1	2	1	4
T4	2	4	4	3

Since the ordinal values cannot be matched, ordinal values are converted into satisfaction values: ordinal number 1 indicates satisfaction of 1, ordinal number 2 indicates satisfaction of 0.75, ordinal number 3 indicates satisfaction of 0.5, and ordinal number 4 indicates satisfaction of 0.25. Let satisfaction of 0.5 be the reference point of the scientific and technical talents. According to prospect theory, we obtain the prospect value of the scientific and technical talents' satisfaction with the team, as shown in Table 18.

Table 18. Ranking prospect values of talents S for the second stage of the new R&D organization team T.

S-T	S2	S3	S4	S6
T1	−0.6643	0.0000	0.0000	0.2952
T2	0.0000	0.5434	0.2952	0.5434
T3	0.5434	0.2952	0.5434	−0.6643
T4	0.2952	−0.6643	−0.6643	0.0000

We weighted the prospect values of the team-to-talent assessment and the prospect values of the talent-to-team assessment in the second stage to obtain the combined prospect values of the second stage match, as shown in Table 19.

Table 19. New R&D organization teams T and talents S Phase II composite prospect values.

Combined Prospect Value	S2	S3	S4	S6
T1	−0.3204	0.0455	−0.0157	0.1516
T2	0.0199	0.2907	0.1765	0.3083
T3	0.2455	0.1562	0.2330	−0.3126
T4	0.1585	−0.3707	−0.3167	0.0528

As can be seen from the table above:

(1) After combining the rankings of talent degree teams, the satisfaction rate, satisfaction rate, and ranking of the bilateral match between teams of new R&D institutions and scientific and technological talents changed.

(2) Team T1 of the new R&D organization is satisfied with S3 and S6, and is dissatisfied with S2 and S4, with a satisfaction rate of 50%, and ranks the scientific and technological talents as S6 $\succ$ S4 $\succ$ S4 $\succ$ S2.

(3) Team T2 of the new R&D organization is satisfied with S2 and S3, and is dissatisfied with S1, S4, S5, and S6, with a satisfaction rate of 33.33%, and ranks the scientific and technological talents as S6 $\succ$ S4 $\succ$ S4 $\succ$ S2.

(4) Team T3 of the new R&D organization is satisfied with S2 and S4, and is dissatisfied with S1, S3, S5 and S6, with a satisfaction rate of 33.33%, and ranks the scientific and technological talents as S2 $\succ$ S4 $\succ$ S3 $\succ$ S6.

(5) Team T4 of the new R&D organization is satisfied with S2, S3, S4, S5 and S6, and is dissatisfied with S1, with a satisfaction rate of 83.33%, and ranks the scientific and technological talents as S2 $\succ$ S6 $\succ$ S4 $\succ$ S3.

The prospective values of the team's second-stage assessment of talent were substituted into the M-2 model for bilateral matching, and the results of second-stage matching were obtained, as shown in Table 20. The overall assessment satisfaction for this match was 0.2478, and the match was stable.

Table 20. Results of the second stage matching of the new R&D organization teams T and the talents S.

Matching Teams	Phase 2 Matches Tech Talent	Phase II Composite Prospect Value
T1	S6	0.1516
T2	S3	0.2907
T3	S4	0.2330
T4	S2	0.1585

From the above table we can see that:

(1) The results of the matching between the new R&D organization team T and the scientific and technological talent S have been completely different from the previous matching results, and the indicators of each team's assessment of the scientific and technological talent have not changed, indicating that after a period of understanding, the team's understanding of the scientific and technological talent has changed. Therefore, bilateral matching should be based on dynamic ideas so that the matching results can be more accurate.

(2) In terms of the composite prospect values, the data on the composite prospect values for Phase 2 also validated the Phase 1 match, with all reaching satisfaction, but with plenty of room for satisfaction improvement. The combined team and talent satisfaction is more balanced, with the highest satisfaction rate for tech talent S3.

4.4. Comparison of Methods

The first stage of this paper is elimination matching, where different methods eliminate talents with different results. The second stage of selection matching only evaluates and ranks the scientific and technical talents matched in the first stage; therefore, other methods cannot obtain all the data of the second stage. This paper conducts a method comparison on the first stage of matching.

Method A: The approach in this study, considering a dual historical and aspirational reference point, is based on prospect theory, considering the cost of bringing in scientific and technical talent, and considering two stages.

Method B: Based on the approach presented in this paper, only historical reference points were considered for one phase.

Method C: Based on the method presented in this paper, only the desired reference point was considered for one stage.

Method D: on the data in this study, a phase was considered without considering the cost of introducing scientific and technological talent.

Method E: on the data in this study, a stage was considered based on regret theory.

Method F: on the data in this study, a phase was considered based on the grey correlation.

After calculation, the results are shown in Table 21.

Table 21. Bilateral Matching Results of Teams and Talents in New R&D Institutions under Different Methods.

Approach	First Phase Matching Results	Phase I Elimination of Talent	Reference Point	Consider the Cost of Introduction	Consider Psychological Factors
A	[T1,S6] [T2,S2] [T3,S4] [T4,S3]	S1, S5	Historical reference point Desired reference point	yes	Yes
B	[T1,S6] [T2,S1] [T3,S4] [T4,S2]	S3, S5	Historical reference point	yes	Yes
C	[T1,S1] [T2,S2] [T3,S4] [T4,S5]	S3, S6	Desired reference point	yes	yes
D	[T1,S5] [T2,S2] [T3,S4] [T4,S3]	S1, S6	Historical reference point Desired reference point	no	yes
E	[T1,S5] [T2,S2] [T3,S1] [T4,S6]	S3, S4	Historical reference point Desired reference point	yes	yes
F	[T1,S5] [T2,S2] [T3,S4] [T4,S3]	S1, S6	Historical reference point Desired reference point	yes	no

As can be seen from the table above:

(1) Under different approaches, some consider historical and desired dual reference points, some consider historical or desired single reference points, some consider introduction costs, some do not, some consider psychological factors, and some do not; thus, the results of bilateral matching are different.

(2) For the first stage of elimination matching, the scientific and technical talent eliminated is different under different methods, so the program still has a greater impact on the outcome of the elimination.

(3) Among the six different methods, the probability of being eliminated for scientific and technological talents S1, S3, and S6 was 50%, the probability of being eliminated for scientific and technological talents S5 was 33.33%, the probability of being eliminated for scientific and technological talents S5 was 33.33%, and the probability of being eliminated for scientific and technological talents S4 was 16.67%. Therefore, for scientific and technological talents, the higher the probability of being eliminated, the higher the risk, and it is necessary to find the direction of improvement from the personalized assessment index of the team of new R&D institutions, find the deficiency, and make efforts to improve so as to enhance their scientific and technological capabilities.

5. Conclusions

In summary, the following conclusions can be drawn.

First, this study characterizes the uncertain new R&D organization team's assessed value of scientific and technological talents and the cost of introducing scientific and technological talents by interval grey numbers, which shows the characteristics of multi-expert group decision-making of new R&D organization teams, inconsistent expert backgrounds and preferences, and the existence of an adjustment interval of talent policy, which is closer to reality.

Second, this study considers personalized assessment indicators for teams in new R&D institutions, because there are different teams in new R&D institutions, and different teams do not have exactly the same requirements for scientific and technological talents. Although the teams belong to the same new R&D institution, the focus of work is not quite the same, some focus more on market research, some focus more on R&D innovation, and some focus more on communication and collaboration. Therefore, this study provides a general framework of assessment metrics and allows teams to select individualized assessment metrics under the general framework, that is, some metrics will be selected and some will not, but the selected metrics must come from this metric framework. Different teams may choose different assessment metrics that are both relevant and easy to use.

Third, this study sets a dual reference point of historical reference point and desired reference point in the team assessment of talent, and a bottom-line reference point in the agreed cost of introduction of the team and talent, applying prospect theory based on the reference point to turn the assessed value and cost value into prospect value, fully considering people's psychology, and no longer comparing the absolute value of assessed value or cost value but comparing the relative value based on the reference point.

Fourth, the bilateral matching of teams and scientific and technological talents in new R&D institutions is divided into two stages, the first stage is elimination matching and the second stage is selection matching, both stages are many-to-many matching, but in the elimination matching of the first stage, all teams are matched and not all scientific and technological talents are matched, more scientific and technological talents than the number of teams will be eliminated. In the second stage of selection matching, there are equal numbers of teams and talents. Considering these two stages shows the idea of a dynamic assessment.

Fifth, in the first stage of elimination matching, both the team's assessed value of talent and the cost value of scientific and technological talent are considered. The larger the assessed value, the better the lower the cost value, so that the scientific and technological talent matched out by forming dual objectives is in line with the principle of maximizing benefits and will not be too costly, and the assessment factor and cost factor can be adjusted by weighting.

Sixth, in the second stage of selection matching, both the team's assessment value of the talent and the talent's ranking of the team are considered. The larger the assessment value, the better, and the smaller the ranking, the better, so as to form a double target matching out of the scientific and technological talent. The satisfaction of both sides is not too low, fully considering the psychological factors of both the team and the talent matching, and the assessment factor and the ranking factor are adjustable by weighting.

In summary, this study considers two stages of team and talent matching in new R&D organizations. The first considers the team's assessment of talent and the cost of talent introduction; the assessment considers historical reference points and desired reference points, the cost considers bottom-line reference points, and combines prospect theory with high assessment value and low cost to optimize the solution. The second stage considers the team's evaluation of the talent and the talent's satisfaction with the team. The evaluation still considers the historical reference point and the desired reference point, which considers the bottom-line reference point. This optimizes the solution based on high evaluation value and high satisfaction in combination with prospect theory, fully considering the psychological factors of decision makers. In real life, assessment values, ranking values, reference point values, etc., are decided by multiple participants, so these values mostly have uncertainty. The interval grey number possesses an independent algorithm that characterizes uncertainty but is simple to calculate. The method in this paper solves the problems of high cost and low satisfaction of talent introduction raised by HR directors of new R&D organizations and greatly improves the efficiency of team and talent matching. New R&D organization teams should comply with market changes, update personalized assessment indicators, dynamically assess scientific and technical talent, continuously build a reasonable talent ladder, identify talent deficiencies and train them, and adjust talent policies to promote output. Scientific and technological talent should strive to cope with the matching situation of the team of new R&D institutions, adjust their efforts according to the team's personalized assessment indicators in a timely manner, conduct scientific and technological work with the team's goals as the focus, create more scientific and technological achievements, and realize their self-worth. Since there is not much data acquisition, the author has selected only one new R&D institution for method calculation. The sample size is still small; therefore, it has some limitations. The current study only considers factors such as team's evaluation of talent, talent's cost, and talent's satisfaction with the team, which has some limitations. Later on, we can combine government factors, leadership factors of teams, self-motivation factors of talents, and project-based management factors with this study. The current study only considered static reference points. In the future, the author will also conduct research on bilateral matching of team and scientific and technological talents in new R&D institutions based on government supervision, re-research on bilateral matching of team and scientific and technological talents in new R&D institutions based on dynamic reference points, and research on trilateral matching of team-project-talent in new R&D institutions based on reference points.

Author Contributions: Conceptualization, L.J.; Methodology, L.J.; Writing—original draft, L.J.; Writing—review & editing, B.C. All authors have read and agreed to the published version of the manuscript.

Funding: 1. The National Natural Science Foundation of China (NSFC) project: A method and application of dynamic emergency group decision-making driven by ternary group intelligence information interaction for public emergencies, Project No. 72071106. 2. Jiangsu Province Education Reform and Development Strategic and Policy Research Major Project: Jiangsu high-level teacher team construction research, Project No. 202000206. 3. Project of Jiangsu Provincial Education Department: Research on talent management innovation of universities based on big data in the context of "double first-class," Project No. 2019SJA1734.

Institutional Review Board Statement: There are no human ethical aspects to this study.

Informed Consent Statement: Informed consent was obtained from all subjects involved in the study.

Data Availability Statement: The author obtained data from HR managers of the new R&D organization T for two phases of two-way team and talent assessment. For the new R&D organization T, these data are not publicly available and are not published in a publicly available dataset. These data are presented in the article. The focus of this paper is to demonstrate the validity of the method by simulating the data, and the data from the computation process are shown in detail in the paper.

Conflicts of Interest: The authors declare no conflict of interest.

References

1. Moliterno, T.; Beck, N.; Beckman, C.; Meyer, M. Knowing Your Place: Social Performance Feedback in Good Times and Bad Times. *Organ. Sci.* **2014**, *25*, 1684–1702. [CrossRef]
2. Bode, C.; Singh, J. Taking a hit to save the world? Employee participation in a corporate social initiative. *Strateg. Manag. J.* **2018**, *39*, 1003–1030. [CrossRef]
3. Abramo, G.; Angelo, C.; Costa, F. When research assessment exercises leave room for opportunistic behavior by the subjects under evaluation. *J. Informetr.* **2019**, *13*, 830–840. [CrossRef]
4. Yin, S.; Zhang, N.; Li, B. Enhancing the competitiveness of multi-agent cooperation for green manufacturing in China: An empirical study of the measure of green technology innovation capabilities and their influencing factors. *Sustain. Prod. Consump.* **2020**, *23*, 63–76. [CrossRef]
5. Sun, Y.; Cao, C. The dynamics of the studies of China's science, technology and innovation (STI): A bibliometric analysis of an emerging field. *Scientometrics* **2020**, *124*, 1335–1365. [CrossRef]
6. Mao, Y.; Kang, X.; Fang, Y. Research on the impact of innovation climate and knowledge management on innovation performance. *Stud. Sci. Sci.* **2021**, *39*, 519–529.
7. Chomać, E.; Sobczak, A.; Urbańczyk, E. RES market. RES market development and public awareness of the economic and environmental dimension of the energy transition in Poland. Dimension of the Energy Transformation in Poland and Lithuania. *Energies* **2022**, *15*, 5461. [CrossRef]
8. Chamorro, T.; Akhtar, R.; Winsborough, D.; Sherman, R. The datafication of talent: How technology is advancing the science of human potential at work. *Curr. Opin. Behav. Sci.* **2017**, *18*, 13–16. [CrossRef]
9. Pillai, R.; Sivathanu, B. Adoption of artificial intelligence (AI) for talent acquisition in IT/ITeS organizations. *Benchmark Int. J.* **2020**, *27*, 2599–2629. [CrossRef]
10. Jiang, J.; Tan, W.; Zhu, X.; Liu, J.; Liu, T. Innovative Models and Practices for Engineering and Technology Talent Cultivation in Transnational Higher Education. *Transform. Bus Econ.* **2021**, *20*, 37–52.
11. Wiblen, S.; Marler, J. Digitalised talent management and automated talent decisions: The implications for HR professionals. *Int. J. Hum. Resour. Man.* **2021**, *32*, 2592–2621. [CrossRef]
12. Chaudhuri, R.; Ivkovic, Z.; Pollet, J.; Trzcinka, C. A Tangled Tale of Training and Talent: PhDs in Institutional Asset Management. *Manag. Sci.* **2020**, *66*, 5623–5647. [CrossRef]
13. Agarwal, R.; Braguinsky, S.; Ohyama, A. Centers of gravity: The effect of stable shared leadership in top management teams on firm growth and industry evolution. *Strateg. Manag. J.* **2020**, *41*, 467–498. [CrossRef]
14. Eirinakis, P.; Magos, D.; Mourtos, I.; Miliotis, P. Finding All Stable Pairs and Solutions to the Many-to-Many Stable Matching Problem. *INFORMS J. Comput.* **2012**, *24*, 245–259. [CrossRef]
15. Wang, R.; Li, D.; Yu, G. A bilateral matching decision method considering attribute association in a heterogeneous information environment. *J. Intell. Fuzzy Syst.* **2020**, *38*, 4779–4792. [CrossRef]
16. Kanoria, Y.; Saban, D. Facilitating the Search for Partners on Matching Platforms. *Manag. Sci.* **2021**, *67*, 5990–6029. [CrossRef]
17. Nguyen, H.; Nguyen, T.; Teytelboym, A. Stability in Matching Markets with Complex Constraints. *Manag. Sci.* **2021**, *67*, 7438–7454. [CrossRef]
18. Johari, R.; Kamble, V.; Kanoria, Y. Matching While Learning. *Oper. Res.* **2021**, *69*, 655–681. [CrossRef]
19. Deng, C.; Yan, T.; Mao, J.; Yin, S. Thinking alike or not: Performance implications of transactional psychological contract congruence in buyer-supplier new product development projects. *J. Oper. Manag.* **2021**, *68*, 4–32. [CrossRef]
20. Chen, G.; Huang, S.; Meyer, P.; Mindruta, D. Generalist versus specialist CEOs and acquisitions: Two-sided matching and the impact of CEO characteristics on firm outcomes. *Strateg. Manag. J.* **2021**, *42*, 1184–1214. [CrossRef]
21. Chomać, E.; Sobczak, A.; Soboń, D. Wind Energy Market in Poland against the background of the Baltic countries in the era of COVID-19. Era of the COVID-19 Pandemic. *Energies* **2022**, *15*, 2470. [CrossRef]

electronics

Article

Comparative Estimation of Electrical Characteristics of a Photovoltaic Module Using Regression and Artificial Neural Network Models

Jonghwan Lee and Yongwoo Kim *

Department of System Semiconductor Engineering, Sangmyung University, Cheonan 31066, Republic of Korea
* Correspondence: yongwoo.kim@smu.ac.kr

Abstract: Accurate modeling of photovoltaic (PV) modules under outdoor conditions is essential to facilitate the optimal design and assessment of PV systems. As an alternative model to the translation equations based on regression methods, various data-driven models have been adopted to estimate the current–voltage (I–V) characteristics of a photovoltaic module under varying operation conditions. In this paper, artificial neural network (ANN) models are compared with the regression models for five parameters of a single diode solar cell. In the configuration of the proposed PV models, the five parameters are predicted by regression and neural network models, and these parameters are put into an explicit expression such as the Lambert W function. The multivariate regression parameters are determined by using the least square method (LSM). The ANN model is constructed by using a four-layer, feed-forward neural network, in which the inputs are temperature and solar irradiance, and the outputs are the five parameters. By training an experimental dataset, the ANN model is built and utilized to predict the five parameters by reading the temperature and solar irradiance. The performance of the regression and ANN models is evaluated by using root mean squared error (RMSE) and mean absolute percentage error (MAPE). A comparative study of the regression and ANN models shows that the performance of the ANN models is better than the regression models.

Keywords: regression; artificial neural network; I–V characteristics; photovoltaic module

Citation: Lee, J.; Kim, Y. Comparative Estimation of Electrical Characteristics of a Photovoltaic Module Using Regression and Artificial Neural Network Models. *Electronics* **2022**, *11*, 4228. https://doi.org/10.3390/electronics11244228

Academic Editors: Luis Hernández-Callejo, Sergio Nesmachnow and Sara Gallardo Saavedra

Received: 21 November 2022
Accepted: 16 December 2022
Published: 19 December 2022

1. Introduction

The output power of photovoltaic (PV) systems is strongly affected under arbitrary operating conditions such as temperature and solar irradiance of PV modules [1,2]. However, highly predictive and efficient models across different temperatures and irradiances have not been established [3–6]. In addition, their nonlinear characteristics make highly predictive modeling even more difficult [7–13]. The single-diode model (SDM) with five parameters is widely utilized to reproduce the current–voltage (I–V) characteristics [5–8]. Owing to the inherent implicit expression for the electrical equivalent circuit of the SDM, analytical and explicit I–V models have been proposed to calculate the I–V relationship [1,14–16]. The explicit I–V model based on the Lambert W function is simple and efficient, while the implicit model requires more computational time [14–16]. Although optimization methods have been proposed to obtain the five parameters at the standard test condition (STC), significant extraction efforts are required to consider the dependence of unknown parameters on temperature and solar irradiance [3–8,17–21]. For arbitrary operating conditions, the performance of the parameter translation model is greatly limited by the chosen translation equation and correction factors [13,17–21]. In order to construct a complete PV model for climatic conditions, the translational formula should be further modified [17,19] and new parameters may need to be taken into account [18,20,21]. Moreover, the accuracy of the translational formula varies significantly at low irradiance levels [3–5]. However, artificial neural network (ANN) models provide parameter identification, I–V prediction with higher accuracy directly from the measured data [22,23], and fault detection and diagnosis for photovoltaic systems [24–26]. To accurately estimate the performance of PV modules under varying operating conditions, it is necessary to establish a data-driven model for the five parameters with change in irradiance and temperature [27–37]. In recent years, improved ANN models have been proposed by adding more variables [38] and utilizing the efficient training schemes and processing of neural networks [39–42]. This paper compares the performance of regression and ANN models for the five parameters in predicting the I–V relationship of a PV module based on an explicit expression. The results show that the ANN model provides better performance than the regression model. The novelty of the proposed approach lies primarily in the successful integration of comparative models into an analytical and explicit Lambert W function, in contrast with the previous practice for the electrical equivalent circuits. (1) In this new framework, temperature and solar irradiance serve as inputs to establish the regression and ANN models for I–V prediction under arbitrary operating conditions. (2) An advanced ANN model for the five parameters is developed by determining an optimum ANN architecture to improve the estimation of the model. The ANN model developed can provide an efficient method with higher accuracy in predicting I–V characteristics, compared to the regression model.

In this work, the modeling process begins in Section 2 with the theoretical formulation of an explicit I–V model and translation equations for the five parameters. Section 3 describes the regression and ANN models, followed in Section 4 by a comparative validation of both the models against the experimental data for a PV module. In Section 5, the main conclusions are drawn.

2. Theoretical Models

2.1. Explicit and Analytical I–V Model

The PV-equivalent circuit of a single diode with two resistors is shown in Figure 1. The I–V relationship of a PV module can be expressed with a single diode as [1,2,5–8]:

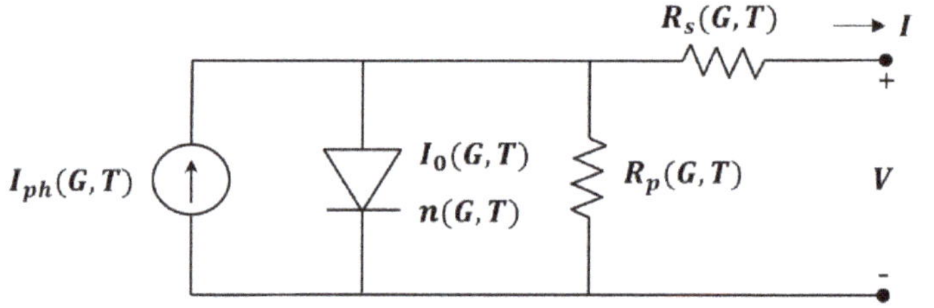

Figure 1. PV-equivalent circuit of a single diode with series and parallel resistance at arbitrary irradiance (G) and temperature (T).

$$I = I_{ph} - I_0\left[\exp\left(\frac{V + IR_s}{nV_t}\right) - 1\right] - \frac{V + IR_s}{R_p} \quad (1)$$

where I_{ph} is the photogenerated current, I_0 is the diode reverse saturation current, n is the ideality factor, and R_s and R_p is the series and parallel resistance, respectively. The thermal voltage is given by $V_t = N_s k_b T/q$, where N_s is the number of series-connected cells, k_b is the Boltzmann constant, and T is the temperature. The explicit solution of the PV module transcendent Equation (1) is given as a function of the Lambert W function [1,14–16]:

$$I = \frac{R_p\left(I_{ph} + I_0\right) - V}{R_s + R_p} - \frac{nV_t}{R_s} W(\alpha(V)) \quad (2)$$

$$\alpha(V) = \frac{R_s R_p I_0}{n V_t (R_s + R_p)} e^{\frac{R_p(R_s I_{ph} + R_s I_0 + V)}{n V_t (R_s + R_p)}} \tag{3}$$

2.2. Five Parameters as a Function of Temperature and Solar Irradiance

Environmental conditions such as temperature (T) and solar irradiance (G) have strong effects on the I–V characteristics of the PV module. In order to extract accurate estimates of the model parameters under arbitrary T and G, mathematical expressions for the five parameters are reformulated by using the advantage of the previous formula [1,2,5–8]. The I–V curve translation for desired solar irradiance and temperature (G, T) from STC (G_0,T_0) is obtained by using the short circuit current (I_{sc}) and the open circuit voltage (V_{oc}). Assuming the condition $I_{sc} \approx I_{ph}$, $I_{sc}(G,T)$ and $V_{oc}(G,T)$ are determined as [1,2,5–8]:

$$I_{sc}(G,T) \approx I_{ph}(G,T) = \left(\frac{G}{G_0}\right)\left[I_{ph0} + \alpha_i (T - T_0)\right] \tag{4}$$

$$V_{oc}(G,T) = V_{oc0}\left[1 + \alpha_v (T - T_0) + \beta_v V_t \ln\left(\frac{G}{G_0}\right)\right] \tag{5}$$

where I_{ph0} and V_{oc0} are the photogenerated current and open circuit voltage at standard test conditions, respectively; α_i and α_v are temperature coefficients and β_v is an irradiance coefficient. From the relationship $n = n_0(V_{oc}/V_{oc0})$ and $R_{s,p} = R_{s0,p0}(V_{oc}/V_{oc0})(I_{sc0}/I_{sc})$, the values of the translated parameters $n(G,T)$, $R_s(G,T)$, and $R_p(G,T)$ are calculated as follows [1,2,7]:

$$n(G,T) = n_0\left[1 + \alpha_n (T - T_0) + \beta_n V_t \ln\left(\frac{G}{G_0}\right)\right] \tag{6}$$

$$R_s(G,T) = R_{s0}\frac{1 + \alpha_{R_s}(T - T_0) + \beta_{R_s} V_t \ln\left(\frac{G}{G_0}\right)}{\left(\frac{G}{G_0}\right)\left[1 + \alpha^*_{R_s}(T - T_0)\right]} \tag{7}$$

$$R_p(G,T) = R_{p0}\frac{1 + \alpha_{R_p}(T - T_0) + \beta_{R_p} V_t \ln\left(\frac{G}{G_0}\right)}{\left(\frac{G}{G_0}\right)\left[1 + \alpha^*_{R_p}(T - T_0)\right]} \tag{8}$$

where n_0, R_{s0}, and R_{p0} are the ideality factor, series resistance, and parallel resistance at standard test conditions, respectively; α_n, α_{R_s}, $\alpha^*_{R_s}$, α_{R_p}, and $\alpha^*_{R_p}$ are temperature coefficients and β_n, β_{R_s}, and β_{R_p} are irradiance coefficients. By setting $I = 0$ in (1), the translation expression for the reverse saturation current $I_0(G,T)$ as a function of $I_{ph}(G,T)$, $V_{oc}(G,T)$, $n(G,T)$, and $R_{s,p}(G,T)$ is obtained as [1,2,7]:

$$I_0(G,T) = \frac{I_{ph}(G,T) - \frac{V_{oc}(G,T)}{R_p(G,T)}}{\exp\left(\frac{V_{oc}(G,T)}{n(G,T)\,V_t}\right) - 1} \tag{9}$$

3. Parameter Identification Approaches

3.1. Multiple Regression

The multivariate regression analysis is employed to model the statistical relationship between inputs (G, T) and outputs (I_{ph},n,R_s,R_p,I_0). The parametric regression equation for a linear or nonlinear function f is expressed as [43–45]:

$$y = f(X_1, \ldots, X_n; \theta_1, \ldots, \theta_m) + \varepsilon \tag{10}$$

where y is an $n \times 1$ vector of dependent variable, $X_1, \ldots, X_n$ are an $n \times m$ matrix of independent variables, $\theta_1, \ldots, \theta_m$ are an $m \times 1$ vector of regression parameters, and ε is an $n \times 1$ vector of random error. The regression parameters $\theta_1, \ldots, \theta_m$ are usually determined using the least square method (LSM) to minimize $\sum_{i=1}^{k} (y_i - f(X_{1i}, \ldots, X_{ni}; \theta_1, \ldots, \theta_m))^2$ with different sample points k. Based on the estimated regression parameters, the optimum regression model is chosen for prediction.

3.2. Artificial Neural Network (ANN)

The artificial neural network is specified in modeling complex systems, especially nonlinear or random variable systems. The multilayer perceptron (MLP), known as the fully connected feed-forward network for supervised learning, is the most common and successful for modeling nonlinear systems [22,23,31]. The MLP network configuration has an input layer, two hidden layers, and an output layer. The input layer consists of two neurons (G, T), and the output layer contains five neurons (I_{ph},n,R_s,R_p,I_0). Every neuron in one layer is fully connected to every neuron in the next layer. By using the activation function of a hyperbolic tangent sigmoid function for N neurons, the output $h_i^{(k)}$ of the ith neuron in the kth hidden layer is computed as follows [22,23]:

$$h_i^{(1)} = \tanh\left(\sum_{j=1}^{2} w_{ij}^{(1)} x_j + b_i^{(1)}\right) \tag{11}$$

$$h_i^{(2)} = \tanh\left(\sum_{j=1}^{N} w_{ij}^{(2)} h_j^{(1)} + b_i^{(2)}\right) \tag{12}$$

where x_j is the jth input to the neuron, $w_{ij}^{(k)}$ is the weight for the ith neuron and jth input in the kth hidden layer, and $b_i^{(k)}$ is the bias for the ith neuron in the kth hidden layer. With the use of a linear activation function for neurons in the output layer, the network's output can be written as

$$y_i = \sum_{j=1}^{N} w_{ij}^{(3)} h_j^{(2)} + b_i^{(3)} \tag{13}$$

In the vectorized form with a weight matrix $\boldsymbol{W}^{(k)}$, an activation vector $\boldsymbol{h}^{(k)}$, and a bias vector $\boldsymbol{b}^{(k)}$, the network's computations are given by [22,23]:

$$\boldsymbol{h}^{(1)} = \tanh\left(\boldsymbol{W}^{(1)}\boldsymbol{x} + \boldsymbol{b}^{(1)}\right) \tag{14}$$

$$\boldsymbol{h}^{(2)} = \tanh\left(\boldsymbol{W}^{(2)}\boldsymbol{h}^{(1)} + \boldsymbol{b}^{(2)}\right) \tag{15}$$

$$\boldsymbol{y} = \boldsymbol{W}^{(3)}\boldsymbol{h}^{(2)} + \boldsymbol{b}^{(3)} \tag{16}$$

where $\boldsymbol{x} = [G\ T]^T$ is the transpose of the input vector $\boldsymbol{x}$. The neural network is trained by using the Levenberg–Marquardt algorithm, a method used extensively for learning a feed-forward network, to realize the rapid correction of network weights and biases. Since the initial value of weight and bias affects the training results, the neural network can be retrained several times to obtain a neural network with excellent universality. The configuration of the proposed model is summarized in Figure 2. The five parameters are predicted by a MLP neural network, and these parameters are put into the Lambert W function.

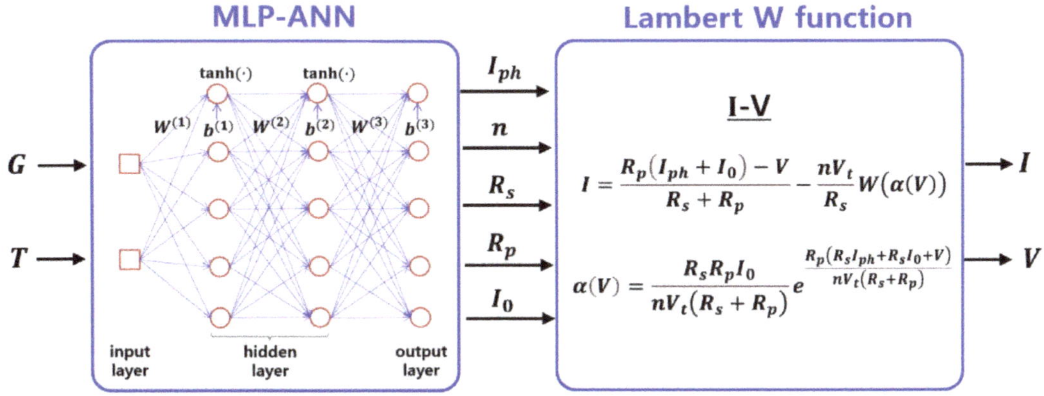

Figure 2. Configuration of the proposed photovoltaic model.

4. Model Verification

The validation of the regression and ANN models for the five parameters (I_{ph},n,R_s,R_p,I_0) is assessed by using the experimental data of the monocrystalline SM55 PV panel ($N_s = 36$) [1]. The five parameters are extracted by using the quality factor variation method from the manufacturer's data sheet [1,17]. After determining the parameters at standard test conditions ($G_0 = 1000$ W/m^2, $T_0 = 298$ K), the procedure is applied to estimate variations in the five parameters for different temperature levels ($T = 298 \sim 343$ K) and solar irradiance levels ($G = 200 \sim 1000$ W/m^2). The database obtained from the procedure is used to develop the multiple regression models for $I_{ph}(G,T)$, $n(G,T)$, $R_s(G,T)$, $R_s(G,T)$, and $I_0(G,T)$.

Table 1 shows the regression models and coefficient of determination R^2 estimated by using Equations (4)–(9). As shown in Table 1, the regression models have a high coefficient of determination ($R^2 = 0.9692 \sim 1.000$), and $I_0(G,T)$ is obtained by other parameters with a high coefficient of determination.

Table 1. Regression model and R^2 for parameters estimated from experimental data.

Parameters	Regression Models and R^2
$I_{ph}(G,T)$ (A)	$\left(\frac{G}{G_0}\right)\left[3.457 + 1.407 \times 10^{-3}(T - T_0)\right]$, $R^2 = 1.000$
$V_{oc}(G,T)$ (V)	$21.63\left[1 - 3.434 \times 10^{-3}(T - T_0) + 1.752 \times 10^{-4} V_t \ln\left(\frac{G}{G_0}\right)\right]$, $R^2 = 0.9985$
$n(G,T)$	$1.084\left[1 - 8.455 \times 10^{-4}(T - T_0) + 2.749 \times 10^{-4} V_t \ln\left(\frac{G}{G_0}\right)\right]$, $R^2 = 0.9692$
$R_s(G,T)$ (Ω)	$0.4724 \frac{1+1.405\times10^{-2}+6.854\times10^{-4}V_t \ln\left(\frac{G}{G_0}\right)}{\left(\frac{G}{G_0}\right)\left[1+3.488\times10^{-2}(T-T_0)\right]}$, $R^2 = 0.9938$
$R_p(G,T)$ (Ω)	$222 \frac{1+1.890\times10^{-2}+7.246\times10^{-4}V_t \ln\left(\frac{G}{G_0}\right)}{\left(\frac{G}{G_0}\right)\left[1+2.515\times10^{-2}(T-T_0)\right]}$, $R^2 = 0.9926$
$I_0(G,T)$ (A)	$\left[I_{ph}(G,T) - \frac{V_{oc}(G,T)}{R_p(G,T)}\right] / \left[\exp\left(\frac{V_{oc}(G,T)}{n(G,T)V_t}\right) - 1\right]$

The best ANN model has an input layer with two variables, two hidden layers with five neurons in each layer, and an output layer with five variables (2-5-5-5 topology). Logarithm data preprocessing is used to improve the ANN model accuracy for the reverse saturation current $I_0(G,T)$. For training the ANN model, the offline method is utilized to generate the dataset for the five parameters of the PV panel, which is extracted from manufacturer's data sheet. Figure 3 shows the dependence of five parameters on temperature and solar irradiance for the regression model and ANN model. The correlation coefficients are employed to evaluate the accuracy of the ANN model for the five parameters, including the training, validation, and testing phases. As a result, the correlation coefficients with values greater than 99.85% were observed between the predicted and measured data in all network phases.

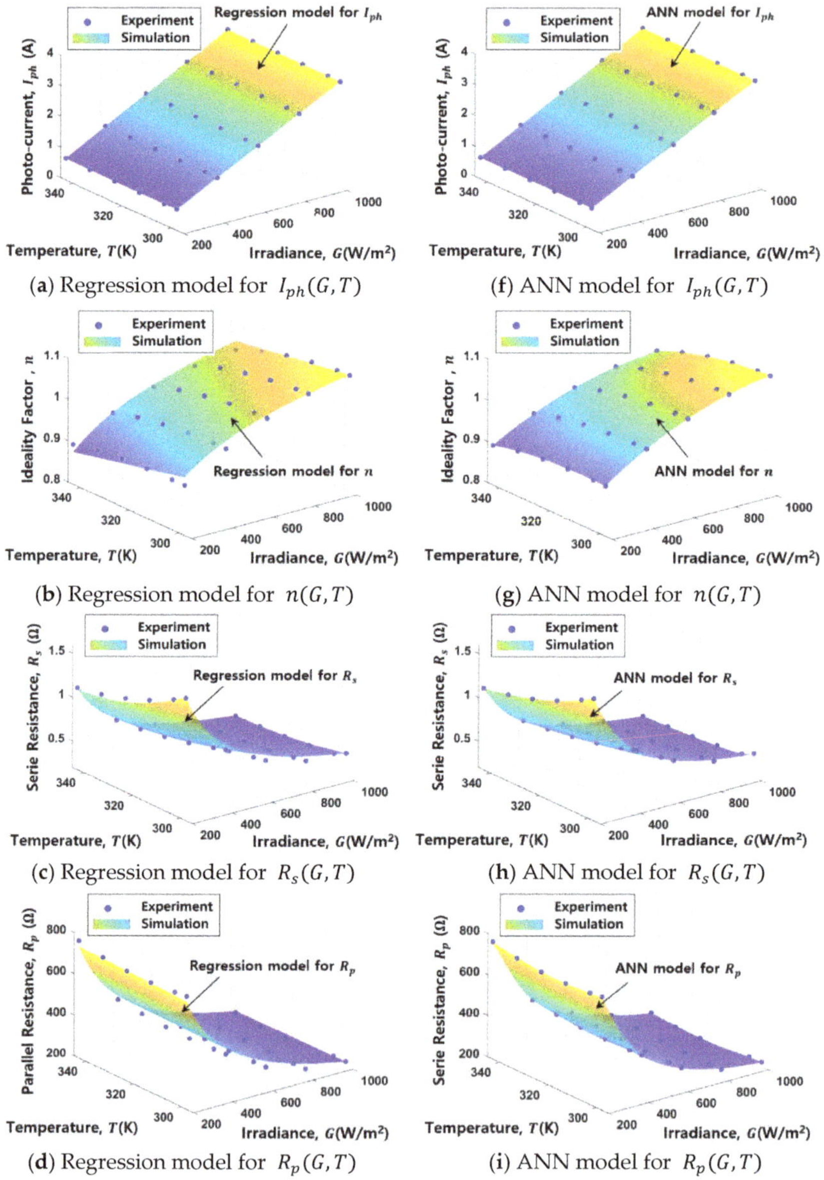

(a) Regression model for $I_{ph}(G,T)$ (f) ANN model for $I_{ph}(G,T)$

(b) Regression model for $n(G,T)$ (g) ANN model for $n(G,T)$

(c) Regression model for $R_s(G,T)$ (h) ANN model for $R_s(G,T)$

(d) Regression model for $R_p(G,T)$ (i) ANN model for $R_p(G,T)$

Figure 3. *Cont.*

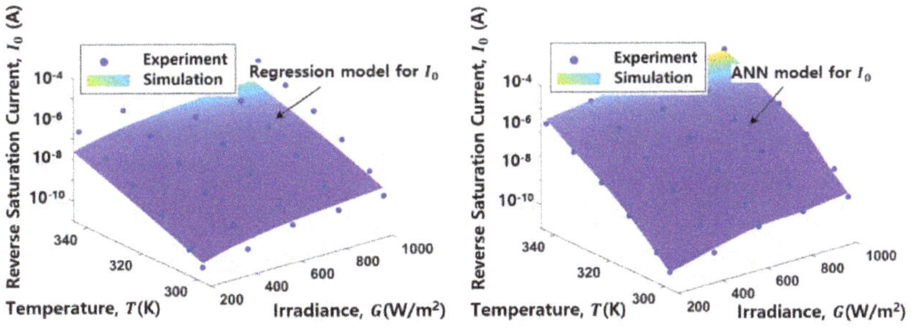

(e) Regression model for $I_0(G,T)$ (j) ANN model for $I_0(G,T)$

Figure 3. Dependence of the five parameters on temperature and solar irradiance for the regression model (**a–e**) and ANN model (**f–j**).

As plotted in Figure 4, two different statistical metrics based on the measured and estimated five parameters are employed to compare the accuracy of the ANN model with the regression model, including root mean squared error (RMSE) and mean absolute percentage error (MAPE), as follows [14,15]:

$$RMSE = \sqrt{\frac{1}{n}\sum_{i=1}^{n}(y_{m,i} - y_{e,i})^2} \tag{17}$$

$$MAPE = \frac{1}{n}\sum_{i=1}^{n}\left|\frac{y_{m,i} - y_{e,i}}{y_{m,i}}\right| \times 100\ (\%) \tag{18}$$

where $y_{m,i}$ and $y_{e,i}$ are the measured and estimated values, respectively. Both RMSE and MAPE values of the ANN models are lower than those of the regression models. This is attributed to the strong capability of MLP–ANN models to learn the nonlinear relationship between the inputs and the outputs, whereas the regression models may be limited to a specific condition. Figure 5a,b show I–V characteristics and the absolute error by simulated and experimental data at different irradiances and temperatures, respectively. It is found that the absolute error values of the ANN models are lower than the regression models, resulting from the better performance of the ANN models for the five parameters shown in Figure 3. Table 2 shows the comparison of maximum absolute errors of the I–V curves estimated from the proposed ANN model and different models for the SM55 PV panel. As can be seen, the maximum absolute errors of the proposed ANN model are much lower than the other models for different irradiances and temperatures. These results justify the higher accuracy of the ANN models, compared with other works [1,4,46].

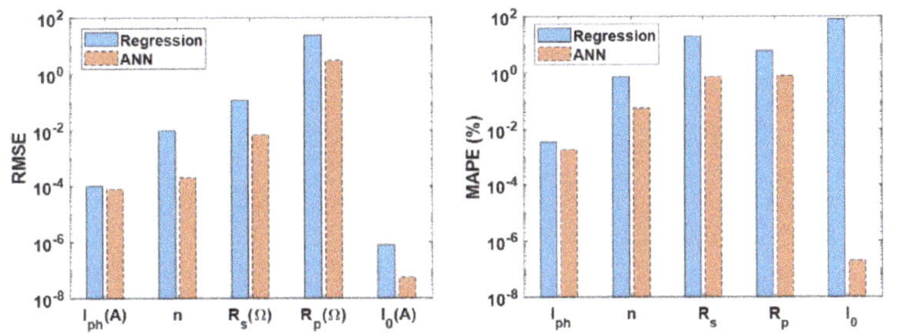

(**a**) (**b**)

Figure 4. Statistical metrics for the five parameters in evaluating the accuracy of the regression model and ANN model: (**a**) RMSE and (**b**) MAPE.

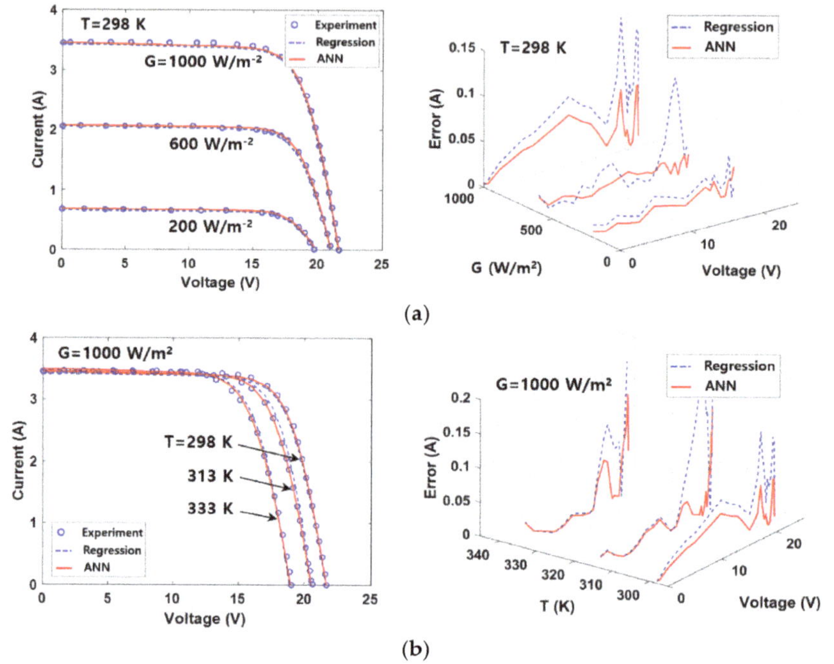

Figure 5. I–V curves and absolute error by simulated and experimental data (**a**) at different irradiances and (**b**) at different temperatures.

Table 2. Comparison of maximum absolute errors estimated from different models.

Temperature *T* (K)	Irradiance *G* (W/m^2)	Maximum Absolute Errors (A) of I–V Curves for SM55 Panel			
		Proposed ANN Model	Model 1 [1]	Model 2 [4]	Model 3 [46]
298	200	0.02	0.03	0.03	0.08
	600	0.01	0.04	0.06	0.05
	1000	0.06	0.08	0.06	0.06
298	1000	0.04	0.07	0.06 (293 K)	0.10 (293 K)
313		0.09	0.16	0.15	0.09
333		0.09	0.12	0.10	0.20

5. Conclusions

The electrical characteristics of a PV module under arbitrary operating conditions have been estimated by using the regression and ANN models. The models are utilized to predict the five parameters of a single diode solar cell, and the parameters are combined with an explicit equation for I–V characteristics. The inputs of the regression and ANN models are temperature and solar irradiance, while the outputs are the five parameters. The dataset needed for the five parameters was extracted from manufacturer's data sheet and used to construct the regression and ANN models. The best neural network architecture had a 2-5-5-5 topology for the five parameters, leading to correlation coefficients with values greater than 99.85%. Both the RMSE and MAPE values of the ANN models were found to be lower than those of the regression models. With comparative results, the ANN models show better performance than the regression models in predicting I–V characteristics under varying temperature and solar irradiance. It is applicable to extend the higher capability of ANN models to the prediction of electrical characteristics for diverse solar cells under actual weather conditions.

Author Contributions: Conceptualization, J.L. and Y.K.; methodology, J.L.; software, Y.K.; validation, J.L. and Y.K.; writing—original draft preparation, J.L.; writing—review and editing, J.L. and Y.K.; visualization, J.L.; supervision, Y.K.; project administration, Y.K.; funding acquisition, Y.K. All authors have read and agreed to the published version of the manuscript.

Funding: This work is funded by a 2020 research grant from Sangmyung University.

Conflicts of Interest: The authors declare no conflict of interest.

References

1. Zaimi, M.; Achouby, E.; Ibral, A.; Assaid, E.M. Determining combined effects of solar radiation and panel junction temperature on all model-parameters to forecast peak power and photovoltaic yield of solar panel under non-standard conditions. *Sol. Energy* **2019**, *191*, 341–359. [CrossRef]
2. Ibrahim, H.; Anani, N. Variations of PV module parameters with irradiance and temperature. *Energy Procedia* **2017**, *134*, 276–285. [CrossRef]
3. Teh, C.T.Q.; Drieberg, M.; Soeung, S.; Ahmad, R. Simple PV modeling under variable operating conditions. *IEEE Access* **2021**, *9*, 96546–96558. [CrossRef]
4. Dongue, S.B.; Njomo, D.; Ebengai, L. An improved nonlinear five-point model for photovoltaic modules. *Int. J. Photoenergy* **2013**, *2013*, 680213-1-11.
5. Shinong, W.; Qianlong, M.; Jie, X.; Yuan, G.; Shilin, L. An improved mathematical model of photovoltaic cells based on datasheet information. *Sol. Energy* **2020**, *199*, 437–446. [CrossRef]
6. Brano, V.L.; Orioli, A.; Ciulla, G.; Gangi, A.D. An improved five parameter model for photovoltaic modules. *Sol. Energy Mater. Sol. Cells* **2010**, *94*, 1358–1370. [CrossRef]
7. Anani, N.; Ibrahim, H. Adjusting the single-diode model parameters of a photovoltaic module with irradiance and temperature. *Energies* **2020**, *13*, 3326. [CrossRef]
8. Hao, P.; Zhang, Y.; Lu, H.; Lang, Z. A novel method for parameter identification and performance estimation of PV module under varying operation conditions. *Energy Convers. Manag.* **2021**, *247*, 114689. [CrossRef]
9. Deng, J.; Wang, Y. Research on MPPT of photovoltaic system based on PSO with time-varying compression factor. *IEICE Electron. Express* **2022**, *19*, 20220165. [CrossRef]
10. Wang, Y.; Zhang, B.; Yang, Y.; Wen, H.; Zhang, Y.; Chen, X. A new optimized control system architecture for solar photovoltaic energy storage application. *IEICE Electron. Express* **2021**, *18*, 20200404. [CrossRef]
11. Carrasco, J.A.; de Quiros, F.G.; Alaves, H.; Navalon, M. A PWM multiplier for maximum power point estimation of a solar panel. *IEICE Electron. Express* **2018**, *15*, 20180496. [CrossRef]
12. Liu, H.D.; Lin, C.H.; Lu, S.D. A novel MPPT algorithm considering solar photovoltaic modules and load characteristics for a single stage standalone solar photovoltaic system. *IEICE Electron. Express* **2020**, *17*, 20200099. [CrossRef]
13. Picault, D.; Raison, B.; Bacha, S.; de la Casa, J.; Aguilera, J. Forecasting photovoltaic array power production subject to mismatch losses. *Sol. Energy* **2010**, *84*, 1301–1309. [CrossRef]
14. Jain, A.; Sharma, S.; Kapoor, A. Solar cell array parameters using Lambert W-function. *Sol. Energy Mater. Sol. Cells* **2006**, *90*, 25–31. [CrossRef]
15. Fathabadi, H. Lambert W function-based technique for tracking the maximum power point of PV modules connected in various configurations. *Renew. Energy* **2015**, *74*, 214–226. [CrossRef]
16. Hmamou, D.B.; Elyaqouti, M.; Arjdal, E.; Ibrahim, A.; Abdul-Ghaffar, H.I.; Aboelsaud, R.; Obukhov, S.; Diab, A.A.Z. Parameters identification and optimization of photovoltaic panels under real conditions using Lambert W-function. *Energy Rep.* **2021**, *7*, 9035–9045. [CrossRef]
17. Achouby, E.; Zaimi, M.; Ibral, A.; Assaid, E.M. New analytical approach for modelling effects of temperature and irradiance on physical parameters of photovoltaic solar module. *Energy Convers. Manag.* **2018**, *177*, 258–271. [CrossRef]
18. Silva, E.A.; Bradaschia, F.; Cavalcanti, M.C.; Nascimento, A.J.; Michels, L., Jr.; Pietta, L.P. An eight-parameter adaptive model for the single diode equivalent circuit based on the photovoltaic module's physics. *IEEE J. Photovolt.* **2017**, *7*, 1115–1123. [CrossRef]
19. Zhang, C.; Zhang, Y.; Su, J.; Gu, T.; Yang, M. Modeling and prediction of PV module performance under different operating conditions based on power-law I-V model. *IEEE J. Photovolt.* **2020**, *10*, 1816–1827. [CrossRef]
20. Hejri, M.; Mokhtari, H. On the comprehensive parameterization of the photovoltaic cells and modules. *IEEE J. Photovolt.* **2017**, *7*, 250–258. [CrossRef]
21. Silva, A.; Bradaschia, F.; Cavalcanti, M.C.; Nascimento, A.J. Parameter estimation method to improve the accuracy of photovoltaic electrical model. *IEEE J. Photovolt.* **2016**, *6*, 278–285. [CrossRef]
22. Mittal, M.; Bora, B.; Saxena, S.; Gaur, A.M. Performance prediction of PV module using electrical model and artificial neural network. *Sol. Energy* **2018**, *176*, 104–117. [CrossRef]

23. Karamirad, M.; Omid, M.; Alimardani, R.; Mousazadeh, H.; Heidari, S.N. ANN based simulation and experimental verification of analytical four- and five-parameters models of PV modules. *Simul. Model. Pract. Theory* **2013**, *34*, 86–98. [CrossRef]
24. Zhu, H.; Lu, L.; Yao, J.; Dai, S.; Hu, Y. Fault diagnosis approach for photovoltaic arrays based on unsupervised sample clustering and probabilistic neural network model. *Sol. Energy* **2018**, *176*, 395–405. [CrossRef]
25. Chine, W.; Mellit, A.; Lughi, V.; Malek, A.; Sulligoi, G.; Massi Pavan, A. A novel fault diagnosis technique for photovoltaic systems based on artificial neural networks. *Renew. Energy* **2016**, *90*, 501–512. [CrossRef]
26. Li, B.; Delpha, C.; Diallo, D.; Migan-Dubois, A. Application of artificial neural networks to photovoltaic fault detection and diagnosis: A review. *Renew. Sustain. Energy Rev.* **2021**, *138*, 110512. [CrossRef]
27. Zecevic, Z.; Rolevski, M. Neural network approach to MPPT control and irradiance estimation. *Appl. Sci* **2020**, *10*, 5051. [CrossRef]
28. Karatepe, E.; Boztepe, M.; Colak, M. Neural network based solar cell model. *Energy Convers. Manage.* **2006**, *47*, 1159–1178. [CrossRef]
29. Xu, E.; Zhang, X.; Huang, Z.; Xie, S.; Gu, W.; Wang, X.; Zhang, L.; Zhang, Z. Current characteristics estimation of Si PV modules based on artificial neural network modeling. *Materials* **2019**, *12*, 3037. [CrossRef]
30. Chikh, A.; Chandra, A. Adaptive neuro-fuzzy based solar cell model. *IET Renew. Power Gener.* **2014**, *8*, 679–686. [CrossRef]
31. Rizzo, S.A.; Scelba, G. ANN based MPPT method for rapidly variable shading conditions. *Appl. Energy* **2015**, *145*, 124–132. [CrossRef]
32. Cortes, B.; Sanchez, R.T.; Flores, J.J. Characterization of polycrystalline photovoltaic cell using artificial neural network. *Sol. Energy* **2020**, *196*, 157–167. [CrossRef]
33. Elsheikh, A.H.; Sharshir, S.W.; Elaziz, M.A.; Kabeel, A.E.; Guilan, W.; Haiou, Z. Modeling of solar energy systems using artificial neural network: A comprehensive review. *Sol. Energy* **2019**, *180*, 622–639. [CrossRef]
34. Kalogirou, S.A.; Mathioulakis, E.; Belessiotis, V. Artificial neural network for the performance prediction of large solar systems. *Renew. Energy* **2014**, *63*, 90–97. [CrossRef]
35. Castro, R. Data-driven PV modules modelling: Comparison between equivalent electric circuit and artificial intelligence based models. *Sustain. Energy Technol. Assess.* **2018**, *30*, 230–238. [CrossRef]
36. Celik, A.N. Artificial neural network modelling and experimental verification of the operating current of mono-crystalline photovoltaic modules. *Sol. Energy* **2011**, *85*, 2507–2517. [CrossRef]
37. Almonacid, F.; Rus, C.; Hontoria, L.; Munoz, F.J. Characterization of PV CIS module by artificial neural networks. A comparative study with other methods. *Renew. Energy* **2010**, *35*, 973–980. [CrossRef]
38. Huang, C.; Bensoussan, A.; Edesess, M.; Tsui, K.L. Improvement in artificial neural network-based estimation of grid connected photovoltaic power output. *Renew. Energy* **2016**, *97*, 838–848. [CrossRef]
39. Bonanno, F.; Capizzi, G.; Graditi, G.; Napoli, C.; Tina, G.M. A radial basis function neural network based approach for the electrical characteristics estimation of a photovoltaic module. *Appl. Energy* **2012**, *97*, 956–961. [CrossRef]
40. Zhang, L.; Bai, Y.F. Genetic algorithm-trained radial basis function neural networks for modelling photovoltaic panels. *Eng. Appl. Artif. Intell.* **2005**, *18*, 833–844. [CrossRef]
41. Chen, Z.; Yu, H.; Luo, L.; Wu, L.; Zheng, Q.; Wu, Z.; Cheng, S.; Lin, P. Rapid and accurate modeling of PV modules based on extreme learning machine and large datasets of I–V curves. *Appl. Energy* **2021**, *292*, 116929. [CrossRef]
42. Liu, Y.H.; Liu, C.L.; Huang, J.W.; Chen, J.H. Neural-network-based maximum power point tracking methods for photovoltaic systems operating under fast changing environments. *Sol. Energy* **2013**, *89*, 42–53. [CrossRef]
43. Barhmi, S.; Elfatni, O.; Belhaj, I. Forecasting of wind speed using multiple linear regression and artificial neural networks. *Energy Syst.* **2020**, *11*, 935–946. [CrossRef]
44. Asilturl, I. Predicting surface roughness of hardened AISI 1040 based on cutting parameters using neural networks and multiple regression. *Int. J. Adv. Manuf. Technol.* **2012**, *63*, 249–257. [CrossRef]
45. Wang, Y.M.; Elhag, T.M.S. A comparison of neural network, evidential reasoning and multiple regression analysis in modelling bridge risks. *Expert Syst. Appl.* **2007**, *32*, 336–348. [CrossRef]
46. Khalid, C.; Mohammed. F.; Mohcine, M. An accurate modelling of PV modules based on two-diode model. *Renew. Energy* **2021**, *167*, 294–305.

energies

Article

Prediction of Solid Conversion Process in Direct Reduction Iron Oxide Using Machine Learning

Masih Hosseinzadeh [1], Hossein Mashhadimoslem [1,2], Farid Maleki [3] and Ali Elkamel [2,4,*]

1 Department of Chemical Engineering, Iran University of Science and Technology (IUST), Narmak, Tehran 16846, Iran
2 Department of Chemical Engineering, University of Waterloo, Waterloo, ON N2L 3G1, Canada
3 Department of Polymer Engineering & Color Technology, Amirkabir University of Technology, Tehran 15916, Iran
4 Department of Chemical Engineering, Khalifa University, Abu Dhabi P.O. Box. 59911, United Arab Emirates
* Correspondence: aelkamel@uwaterloo.ca

Abstract: The direct reduction process has been developed and investigated in recent years due to less pollution than other methods. In this work, the first direct reduction iron oxide (DRI) modeling has been developed using artificial neural networks (ANN) algorithms such as the multilayer perceptron (MLP) and radial basis function (RBF) models. A DRI operation takes place inside the shaft furnace. A shaft furnace reactor is a gas-solid reactor that transforms iron oxide particles into sponge iron. Because of its low environmental pollution, the MIDREX process, one of the DRI procedures, has received much attention in recent years. The main purpose of the shaft furnace is to achieve the desired percentage of solid conversion output from the furnace. The network parameters were optimized, and an algorithm was developed to achieve an optimum NN model. The results showed that the MLP network has a minimum squared error (MSE) of 8.95×10^{-6}, which is the lowest error compared to the RBF network model. The purpose of the study was to identify the shaft furnace solid conversion using machine learning methods without solving nonlinear equations. Another advantage of this research is that the running speed is 3.5 times the speed of mathematical modeling.

Keywords: direct reduction; MIDREX; neural network; optimization; algorithm; modeling

Citation: Hosseinzadeh, M.; Mashhadimoslem, H.; Maleki, F.; Elkamel, A. Prediction of Solid Conversion Process in Direct Reduction Iron Oxide Using Machine Learning. *Energies* **2022**, *15*, 9276. https://doi.org/10.3390/en15249276

Academic Editors: Luis Hernández-Callejo, Sergio Nesmachnow and Sara Gallardo Saavedra

Received: 7 November 2022
Accepted: 4 December 2022
Published: 7 December 2022

1. Introduction

Direct reduction of iron oxide (DRI) is one of the most important non-catalytic gas-solid reactions in industry, and it continues to be an important field of study in chemical engineering [1,2]. The MIDREX process, which is one of the direct-reduction technologies, has received a lot of interest because it is a great technology for considerably reducing carbon dioxide (CO_2) emissions from steel plants [3,4]. This is primarily accomplished by using natural gas instead of coke or coal [5]. Several approaches were used to develop these solutions, whose overview is provided in Figure 1.

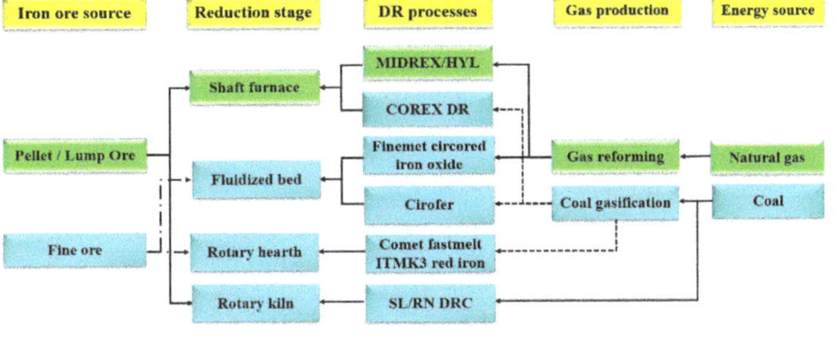

Figure 1. Direct reduction methods have been developed extensively [6].

Despite the global COVID-19 pandemic, global DRI output in 2020 is 104.4 million tonnes which has a 3.4% decrease compared with the previous year's record of 108.1 million tonnes. India and Iran produced about half of the world's DRI [7].

The shaft furnace, reformer, and recuperator are the three main parts of the Midrex process, of which the shaft furnace is the most important. Within the shaft furnace, reduction processes take place, and iron oxide turns into sponge iron. Researchers have recently worked to regenerate hydrogen and develop the new MIDREX process design. Pimm et al. improved the MIDREX process to use renewable energies to satisfy the energy needs of the revised MIDREX process and the hydrogen-based MIDREX unit. According to Rechberger et al.'s research, the carbon footprint of the power used to manufacture hydrogen has a significant impact on the amount of potential that the hydrogen-based pathway offers for environmentally friendly steelmaking [8,9].

Figure 2 indicates direct reduction processes for the production of sponge iron which uses natural gas as the major reducing agent. Today these processes provide for more than 70% of the overall production of DRI and hot briquetted iron (HBI). Natural gas is transformed into reducing agents, mostly carbon monoxide and hydrogen, which operate as iron oxide reducers [6]. The shaft furnace is divided into three main parts: (i) reduction zone, (ii) transition zone, and (iii) cooling zone. The most fundamental part of the shaft furnace is the place where reduction occurs. Therefore, most of the modeling has been conducted around this area. The unreacted shrinking core model (USCM) is an assumption adopted by the majority of prior simulations at the pellet scale [10–12]. Furthermore, some modeled direct reduction reactors in industrial units use this model and achieved desirable results [13,14]. Nevertheless, the grain model can be better than the USCM at predicting plant data [15].

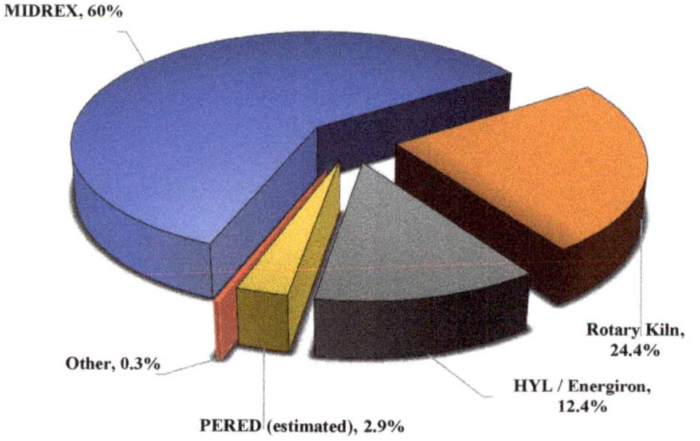

Figure 2. World DRI Production report (2020) using different technologies [7].

Hamadeh et al. assumed that the shaft furnace had pellets made of grains and crystals [4]. In some reactor models, only one reductant gas is used, such as pure H_2 gas [12,16–18], pure CO gas [19,20], and H_2 and CO mixtures [21,22]. In a real shaft furnace, the reducing gas is a combination of H_2, CO, H_2O, CH_4, and CO_2 [4]. Modeling and simulation of industrial direct reduction furnaces have been performed by numerical solutions and computational fluid dynamics (CFD), which are summarized in Table 1. Additionally, some notable non-industrial modeling is given in Table 2.

It is employed at research centers today to comprehend the uses of machine learning (ML) in both the present and the future of energy systems. One of the most effective techniques employed in a variety of industrial fields is the deployment of ML algorithms. Because hybrid approaches take advantage of two or more ways to make an accurate forecast, they sometimes yield greater results than a single method. In light of this, it is advised to use hybrid ML strategies in the future [23–25].

Table 1. Summarized simulation works were validated using industrial plants.

Authors	Modeling/Simulation	Shaft Furnace Name	Remark	Reference
Parisi and Laborde	USCM/Numerical (Rung–Kutta and Dormand–Prince)	Gilmore Siderca	The model fits the data from two MIDREX plants successfully. Furthermore, they enable investigation of the reactor's behavior under various operating situations. In terms of metallization, it has been determined that if it increases to 100% (a 6% increase in metallization), the output must reduce to 70% (A 30% loss in production).	[13]
Valipour and Saboohi	USCM/CFD (FVM)	Gilmore	They concluded that hematite was reduced entirely to magnetite during the reduction of the haematite pellets in a shaft furnace. However, wustite was transformed into iron in the lower section of the bed. Additionally, magnetite was changed into wustite in the center of the bed. Because there was more hydrogen in syngas than carbon monoxide, endothermic reactions were avoided, and as a result, the temperature along the bed was reduced.	[26,27]
Nouri et al.	USCM and Grain model with product layer resistance/Numerical Rung–Kutta	Mobarakeh	The impact of reducing gas parameters and pellet properties on the degree of reduction has been examined. They observed that the grain model predicted plant data better than the shrinking unreacted core model. Their calculations demonstrate that even at the lowest flow rate, the Sherwood number is large enough to eliminate the mass transfer resistance effectively.	[15]
Alamsari et al.	USCM/CFD (FEM)	Krakatau	The increase in H_2 composition is predicted, whereas the attenuation of CO results in a higher metallization degree. The degree of metallization increases as the gas inlet temperature rises. It was discovered that lowering the gas temperature below 973 °C (1246 K) is not suggested because sticky iron production would occur. They also investigated the impact of lowering the iron's temperature on the generation of total carbon in the cooling zone and isobaric iron reactor.	[28]
Khalid Alhumaizi et al.	USCM/Numerical (Developed computer code)	Saudi Arabia	The simulation of a complete MIDREX unit (shaft furnace, reformer, and recuperator) and the study of complex mass–energy interactions between the reformer and the reduction furnace were investigated in an iron plant based on MIDREX technology. According to their parametric study, the scrubber's exit temperature might be lowered in order to create sufficient water vapor to avoid carbon formation, subject to the limitations placed on preventing carbon deposition in the reformer tubes. The same constraint should apply to the optimization of the recycle ratio. The ratio of natural gas to iron ore is optimal. If we want to improve the natural gas flow rate, it is best to do so at the transition zone, where methane breaks down into carbon and hydrogen and boosts metalization and carburization.	[29]

Table 1. *Cont.*

Authors	Modeling/Simulation	Shaft Furnace Name	Remark	Reference
Shams and Moazeni	USCM/Numerical (Rung–Kutta and Gill)	Gilmore	Complete shaft furnace simulation (reduction zone, cooling zone, transition zone) and validation using Gilmore unit data. The pressure drop for the reduction zone's 9.75 m length was demonstrated to be approximately 0.87 bar. The length of the reduction zone is the only length that has any bearing on pressure drop because of the upward gas flow. Methane causes the production of carbon. The final product has a carbon content of 1.4%. A small portion of the cooling gas that helps produce carbon flows upward and mixes with the reducing gas to cool the solid to about 56 °C.	[30]
Ghandi et al.	USCM/CFD (FVM)	Mobarakeh Gilmore	It was discovered that using the twin gas injection approach increases the radial average hydrogen concentration. As a result of the addition of a new gas entrance that injects hot and fresh reducing gas, the overall degree of reduction increases when operating a reactor with a dual injection system.	[31]
Mirzajani et al.	Three-interface USCM/Numerical (Rung-Kutta-Dormand–Prince)	Khorasan	According to this study, the two factors that impact the conversion of iron oxide are particle size and gas flow rate. This study has several limitations because most industrial data is only available for the top and bottom of the reactor, not all along the reactor.	[14]
Hamadeh et al.	Grain model with Crystallite/CFD (FVM)	Contrecoeur Gilmore	A multi-scale approach accurately describes the principal physical, chemical, and thermal phenomena. The moving bed is assumed to be composed of pellets with grain and crystallites. Eight heterogeneous chemical reactions and two homogeneous chemical reactions are also considered. One of their important findings is that they found a central area with a lower temperature and conversion.	[4]
Béchara et al.	Grain model/Aspen Plus model (FDM)	Contrecoeur Gilmore	They built the aspen plus model of the DR shaft derived from the REDUCTOR. Reduced CO_2 emissions were investigated. Computer-aided optimization was used to adjust a set of ten operating parameters at the same time. The results revealed a 15% improvement over the original emissions for comparable output values.	[3,32]
This Study	ML	Gilmore Siderca Khorasan Mobarakeh	The first modeling was performed using ML, SGD, Adam, and LBFG optimization methods and optimization functions, and the best network for modeling was found.	

Table 2. Summarized simulation works using non-industrial modeling DRI.

Authors	Modeling/Simulation	Shaft Furnace Name	Remark	Ref.
Hara et al.	Three-interface USCM/Numerical	Pilot Plant	This model is used for the modeling of the variations in the degree of reduction in an experimental shaft furnace with a diameter of 0.1 m and a height of 4.0 m. It has been discovered that the model can accurately simulate the reduction behavior in the furnace and that the rate constants obtained from the simulation calculation of the experimental data are very close to those found in the previous literature on the fundamental research of iron ore pellet reduction.	[33]
Szekely and Evanse	Grain model and Pore model/Numerical	Pilot Plant	Part I and II series articles "a structural model for gas-solid reactions with a moving boundary" introduce two porous models for non-catalytic solid and gas reactions. They investigate the effects of porosity, grain size, and temperature on the reduction.	[34,35]
Takenaka et al.	Grain model/Numerical (Rung–Kutta and Gill)	Pilot Plant (Commercial-scale)	They concluded that by increasing the temperature of the reducing gas, the gas ratio required to produce the desired degree of reduction of product is decreased, thereby improving the gas economy. When the H/CO ratio of the reducing gas is high, the endothermic process due to reduction with hydrogen occurs more strongly, decreasing the temperature within the furnace and so retarding the reduction. In this case, either the temperature of the reducing gas needs to be raised, or the gas ratio needs to be raised.	[36]
Negri et al.	Three-interface USCM/Numerical	Pilot Plant Yanagiya [37] and Takenaka [36].	The consumption of hydrogen is greater than that of carbon monoxide; this is consistent with the higher reduction rate expected by the given kinetic data for the first reactant.	[22]
Usai	Zone and Grain model/Experimental and Numerical	Pilot Plant	Based on the grain model, they explored the isothermal reduction of wustite pellet with hydrogen under pseudo-steady state and unsteady state circumstances. They said that the pseudo-steady state solution of their model performed well when compared to the unsteady state solution.	[18]
Kang et al.	USCM/Experimental and Analytical	Pilot Plant	Using Ishida–Mathematical Wen's model, the effect of the size and form of iron oxide samples during reduction with CO–CO_2 mixtures at 800 and 900 °C were discussed.	[38,39]
Bonalde et al.	Grain model/Experimental and Numerical	Pilot Plant	Using the grain model, the kinetics of reducing hematite pellets using hydrogen–carbon monoxide mixtures as the reducing agent was characterized. The experimental results were compared to the model's predictions.	[40]
Rahimi and Niksiar	Grain Model/Numerical	Pilot Plant Takenaka [36].	Their findings indicate that the feed temperature appears to have little effect on reactor performance. On the other hand, it is expected that the incoming gas flow rate will have a significant influence. It has also been explained that the significant interconnection of the reactor's different zones may prevent general expression.	[41]

Table 2. *Cont.*

Authors	Modeling/Simulation	Shaft Furnace Name	Remark	Reference
Ranzani da Costa et al.	Grain model with Crystallite/CFD (FVM)	Pilot Plant	They simulated a two-dimensional model of hydrogen-based DRI. Their results show that the use of hydrogen accelerates the reduction in comparison to the CO reaction, and CO_2 emissions would be reduced by more than 80%. They made the REDUCTOR model, which is the most accurate shaft model (2D, 3 zones, 10 reactions).	[42]
Ponugoti et al.	USCM/Experimental and Numerical	Pilot Plant	They et al. took a new strategy, solving physical governing equations for the solid phase at the same time as gas phase transport equations. The shrinking core model is used to determine the reaction rate, and the genetic method is used to estimate the parameters for the kinetic term.	[43]
Yongliang Yan et al.	Experimental	National Energy Technology Laboratory	This was determined using the iron Fe_3O_4-to-Fe and FeO-to-Fe, H_2 kinetic rate constants (for green hydrogen). The response rate of magnetite to iron was demonstrated. They discovered that the Fe_3O_4 to Fe reduction is a 1D growth with a slower nucleation rate, but the FeO to Fe reduction is a 1D to 2D development process that occurs more rapidly.	[44]

The reduction zone is a part of the top of the furnace. The followings are the main chemical reactions that take place in the reduction zone [13,15]:

$$Fe_2O_3 + 3H_2 \rightarrow 2Fe + 3H_2O \quad (1)$$

$$Fe_2O_3 + 3CO \rightarrow 2Fe + 3CO_2 \quad (2)$$

Before recirculation to another usage, a wet scrubber cleans and cools the gas discharged from the top of the furnace shaft. A compressor pressurizes the top gas, which contains CO_2 and H_2O, before mixing it with natural gas, preheating, and feeding it into a reformer furnace. Hundreds of reformer tubes that are filled with a nickel catalyst are installed in the reformer furnace. The mixture of the top gas and natural gas is reformed in these tubes to produce reductant gas, which consists of carbon monoxide and hydrogen. The following is the reaction that takes place in the reformer tubes [45]:

$$CH_4 + H_2O \rightarrow CO + 3H_2 \quad (3)$$

$$CH_4 + CO_2 \rightarrow 2CO + 2H_2 \quad (4)$$

After ongoing investigations on modeling, Parisi et al. developed a network for stable solid-state heterogeneous reactors by using neural networks for steam reformers that were able to respond 20 times faster than the numerical model [46]. In another research, a tubular reactor with a fixed bed full of porous pellets was developed isothermally by using an unsupervised grid with an accuracy of approximately 10^{-9} [47]. The use of ML approaches in the simulation of the shaft furnace to estimate the conversion rate of pellets for making sponge iron can overcome the challenges of using nonlinear modeling and is one of the most important achievements of this research. Due to the complexity of pellet behavior, the difficulty of modeling, and the precise prediction of pellet behavior, a new model has been proposed in this study by using an artificial neural network (ANN). Consequently, Figure 3 presents four fundamental models for the DRI process. The low error value in the ANN method, and the complexity of mathematical modeling caused by the ANN technique, make it interesting.

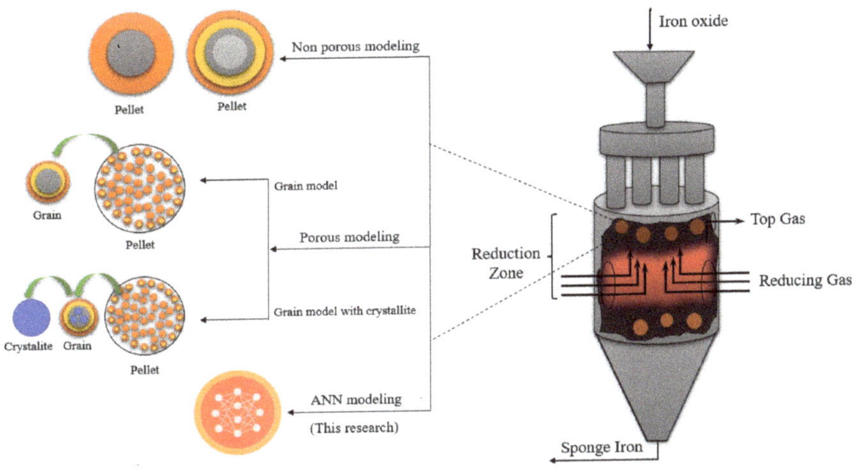

Figure 3. Models developed for pellets in shaft furnaces [3,17–19,30–34,39].

The purpose of this research is to investigate industrial units. Since various industrial data are either non-existent or limited, it has been tried to construct the network according to the simulation results that were in high conformity with the industrial data. In this investigation, modeling was conducted using the four industrial units' real data, and

MLP and RBF networks were built [13–15]. Based on the MLP network structure, Several optimization techniques tuned it after determining the optimal number of neurons in the hidden layers.

2. Numerical Modeling

As shown in Figures 3 and 4, to understand the DRI modeling process, a control volume would be positioned in the shaft furnace's cylindrical point for modeling the reduction zone, heat transfer, and mass transfer equations.

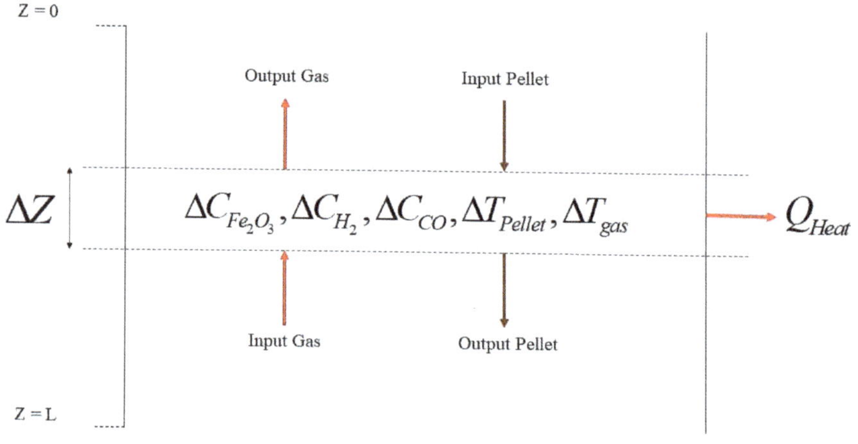

Figure 4. Schema of control volume around the reduction zone.

Firstly, the extent of the reaction should be defined to derive the balances around the element:

$$X_1 = C^{\circ}_{H_2} - C_{H_2} \tag{5}$$

$$X_2 = C^{\circ}_{CO} - C_{CO} \tag{6}$$

$$X_{rs} = X_1 + X_2 = 3(C^{\circ}_{Fe_2O_3} - C_{Fe_2O_3}) = X_3 \tag{7}$$

This value for the solid and gas phases has been concluded based on the stochiometric coefficients, mass, and energy balance as follows [13,15]:

$$u_g \frac{dX_1}{dz} + R_{r1}(X_1, X_3) = 0 \tag{8}$$

$$u_g \frac{dX_2}{dz} + R_{r2}(X_2, X_3) = 0 \tag{9}$$

$$u_s \frac{dX_3}{dz} + (R_{r1}(X_1, X_3) + R_{r2}(X_2, X_3)) = 0 \tag{10}$$

$$\frac{dT_g}{dz} - \frac{n_p A_p h(T_s - T_g)}{G_{mg} C_{pg}(X_1, X_2, T_g)} = 0 \tag{11}$$

$$\frac{dT_s}{dz} - \frac{\left[n_p A_p h(T_g - T_s) - \sum_{i=1}^{2} \Delta H_i(T_s) R_{ri}(X_i, X_3, T_s)\right]}{G_{ms}(X_1) C_{ps}(X_3, T_s)} = 0 \tag{12}$$

where u_g is gas velocity, u_s is solid velocity, R_{r1} and R_{r2} are first and second reaction rate, n_p is the quantity of pellets per unit of bed volume, A_P is pellet external area, T_g and T_s are gas and solid temperature, G_{mg} is gas molar flow, C_{pg} is heat capacity gas, h is heat transfer coefficient, and ΔH is reaction enthalpy.

The aforementioned mathematical modeling of the moving bed direct reduction reactor yields a set of nonlinear ordinary differential equations that can be solved by numerical methods such as Runge–Kutta and the shooting technique [13,15]. Several researchers have made the assumption that, since the radius of the reactor is 200–250 times greater than the pellet diameter, porosity fluctuations of the bed are disregarded [48–51].

Case Studies

Gilmore, Siderca, Mobarakeh, and Khorasan are industrial units with a comprehensive dataset on network inputs, as shown in Table 3. Using these four industrial units, 200 samples were extracted. Due to insufficient parameters, the dataset of other shaft furnaces could not be used. Developing a new general model using mentioned research dataset from Figure 5 has been considered dimensionless. We employed simulation results from previous research projects as a result of mathematical modeling for feed data. The data in Table 3, in which the error value of mathematical modeling is provided from previous research efforts (Relative Error).

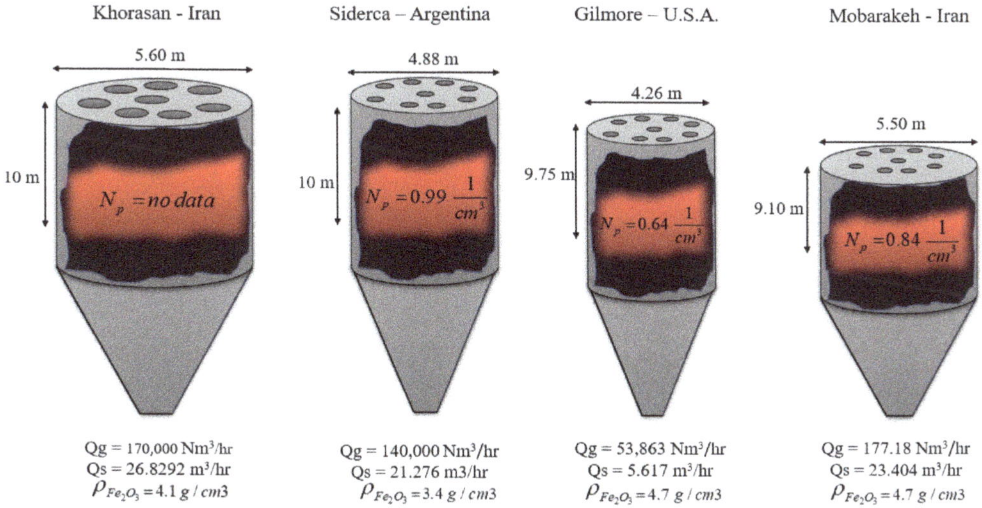

Figure 5. Different parameters in shaft furnaces at different plants.

The following variables have been selected as network input parameters based on the effective parameters such as dependent and independent in direct reduction simulation to achieve the percentage of X_3:

(i) dimensionless temperature of the gas and solid, (ii) percentage of gas entering the furnace, (iii) length-to-diameter ratio of the furnace. The network output is also investigated as a percentage of X_3, which is practical for the calculation of the degree of metallization (MD) shown in Equation (13) as a key output parameter [30].

$$MD(\%) = \frac{Fe}{Total\ Fe(Fe + FeO)} \times 100 \quad (13)$$

Table 3. Model parameters range for the current study.

Shaft Furnace Name	(L/D_{total}) (–)	X_{H_2} (–)	X_{CO} (–)	X_{H_2O} (–)	X_{CO_2} (–)	T^*_g (–)	T^*_s (–)	Q_g/Q_s (–)	X_{3final} (–)	Relative Error X_{3final} (%)	Ref.
Gilmore	0–2.2887	0.5268–0.37	0.2997–0.189	0.0465–0.212	0.048–0.143	0.71–1.08	0.33–1.15	9589.24	93	0.215	[13]
Siderca	0–2.0491	0.529–0.49	0.347–0.236	0.0517–0.124	0.0247–0.213	0.8–1.01	0.36–1.04	6580	93.7	0.106	[13]
Mobarakeh	0–1.6545	0.5357–3292	0.3425–2332	0.0583–0.2638	0.021–0.1373	0.7–1.04	0.32–1.09	7570.41	94.8	0.949	[15]
Khorasan	0–1.7857	0.53–0.4	0.345–0.197	0.048–0.180	0.022–0.171	0.52–1.08	0.32–1.11	6336.36	95.86	0.469	[14]

3. Artificial Neural Network (ANN)

This work aims to examine government machine learning (ML) strategies for addressing existing difficulties in DRI performance. Additionally, the optimization of machine learning (ML) is a promising method that quickly gains attraction in a variety of fields, such as medicine [52] and engineering [53–56].

Supervised learning is generally the task of machine learning and learning a function that maps input to output data from sample input–output pairs [57]. During the training process, information is added to the network validation, and the result data performs the network testing process. Training for the network is concluded when generalizations have improved. A variety of active functions have been used to determine the best one. Logistics, Relu, and identity were used to obtain the optimal function of the MLP network. During network training, the predicted network error should be maintained to a minimum for each step of the mean square error (MSE) in each iteration in order to determine the precise network parameter values. The MSE, square of the correlation coefficient (R^2), and root mean square error (RMSE) are used as assessment metrics to relate the model outputs to the validation dataset. MSE, RMSE, and R^2 are calculated as follows [58–60]:

$$\mathrm{MSE} = \frac{1}{n}\sum_{i=1}^{n}\left(Y_{predicted} - Y_{actual}\right)^2 \tag{14}$$

$$\mathrm{RMSE} = \sqrt{\frac{1}{n}\sum_{i=1}^{n}\left(Y_{predicted} - Y_{actual}\right)^2} \tag{15}$$

$$R^2 = \frac{\sum_{i=1}^{n}\left(Y_{predicted} - Y_{actual}\right)^2}{\sum_{i=1}^{n}\left(Y_{predicted} - Y_{mean}\right)^2} \tag{16}$$

Three types of algorithms, stochastic gradient descent (*SGD*) Equation (17) [61], adaptive moment estimation (*Adam*) Equation (18) [62], and Broyden–Fletcher–Goldfarb–Shanno (*BFGS*) Equation (19) [63], were applied to find the best ANN algorithm which has the lowest MSE, RMSE, and highest R^2. The basic concept of the mentioned algorithms is as follows:

$$w(k+1) = w(k) - \eta\frac{\partial E(k)}{\partial w(k)} + m\cdot(w(k) - w(k-1)) \tag{17}$$

where w represents the weight factor, k represents the position vector, η is the learning rate, E represents the cost function, and m represents the first moment.

$$w(k+1) = w(k) - \eta\left(\frac{m}{\sqrt{V^{(k)}}+e}\right) \tag{18}$$

where V is the second moment, and e is a small scalar that is used to prevent division by zero.

$$H_e(k+1) = H_e(k) + \frac{y_k y_k^T}{y_k^T\Delta x_k} - \frac{H_{ek}\Delta x_k \Delta x_k^T H_{ek}}{\Delta x_k^T H_{ek}\Delta x_k} \tag{19}$$

$$y_k = \nabla f(x_{k+1}) - \nabla f(x_k) \tag{20}$$

$$\Delta x_k = \psi_k p_k \tag{21}$$

where H_e is Hussein matrix, ψ is step size, and P is Direction of search. The flowchart, as shown in Figure 6, is developed to select the best model.

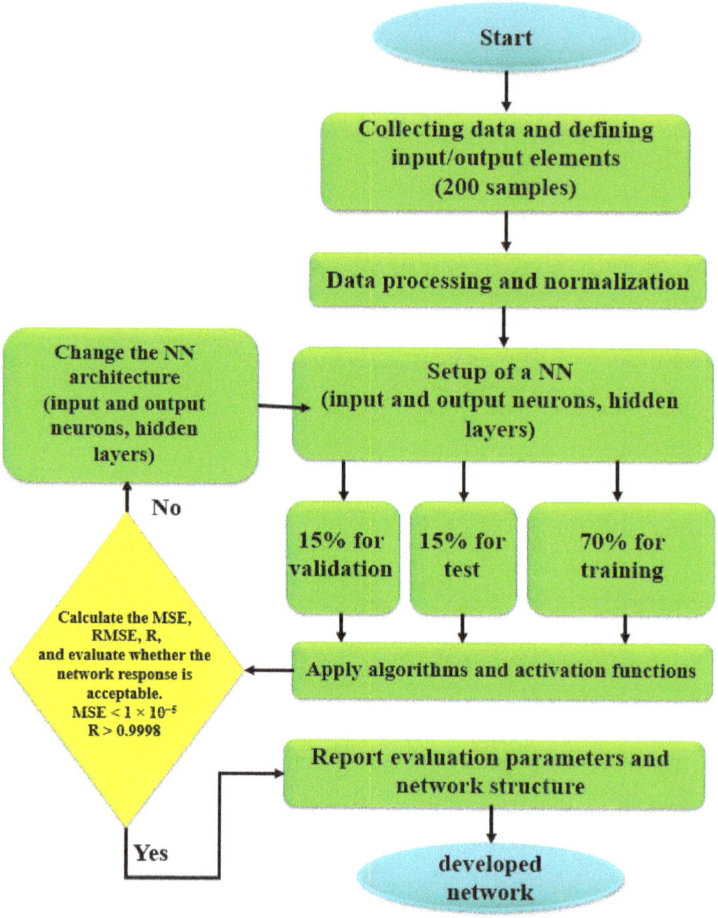

Figure 6. Flowchart of optimization for the ANN approach to finding the best network.

4. Results and Discussion

Firstly, the MLP network has been constructed by practicing all three optimization techniques and the Relu activation function. The parameters of the algorithms are optimized to achieve the optimal network. In the SGD algorithm, there are three parameters: batch size, learning rate, and momentum. The batch size is the number of training samples used in one iteration. The learning rate is the amount of each iteration's step while approaching a minimal loss function. The momentum considers the gradient of previous steps rather than depending just on the current step gradient to control the process. The selected algorithm changes the learning rate and batch size to 0.02 and 20, respectively. A high learning rate enables the model to learn more rapidly but at the expense of a less-than-optimal final weight set. A slower learning rate may enable the model to discover a more optimal or even globally optimal combination of weights, but training will take significantly longer. Concerning deep learning, most practitioners set the value of momentum to 0.9. The optimized SGD algorithm parameters are shown in Figure 7.

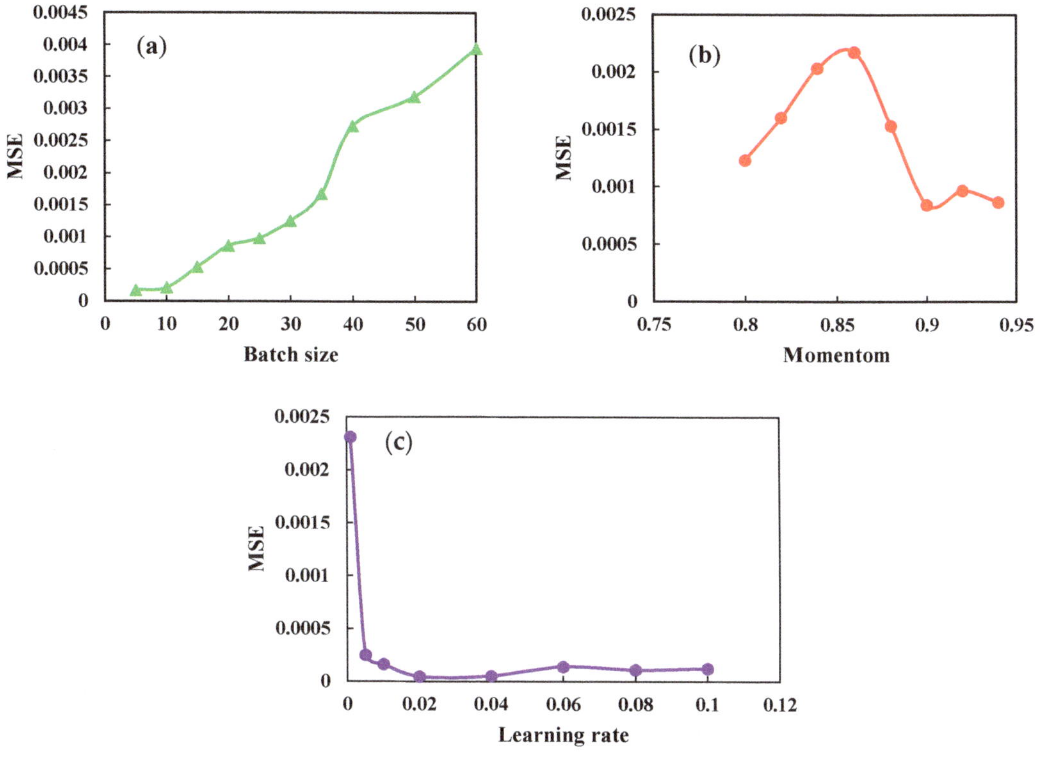

Figure 7. Optimization Parameters SGD algorithm. (**a**) Study the effect of batch size on the MSE. (**b**) Study the effect of momentum on the MSE (**c**) Study the effective learning rate on the MSE.

To examine the RBF network and compare it with the MLP network, RBF network parameters should be optimized. Therefore, according to Figure 8, the spread parameter should be optimized by network data.

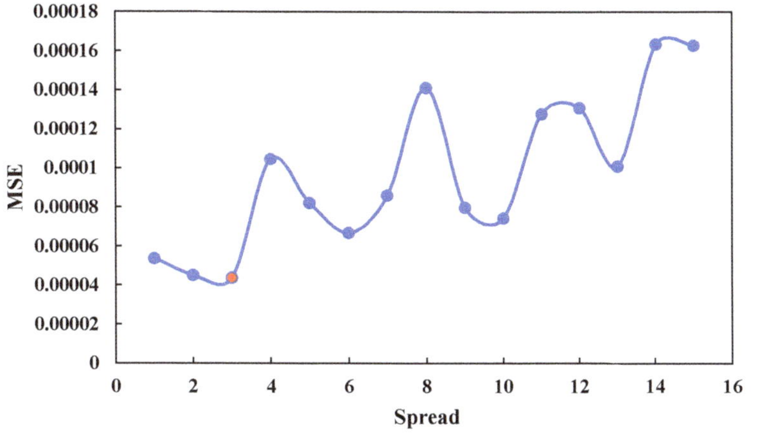

Figure 8. Optimization Spread parameter in RBF network.

4.1. Comparative Analysis of ANN Models

In order to design the structure of the MLP network, it should be optimized for the effective components of the network. One of the most crucial components is the activation function used in the network. According to Figure 9, different activation functions in the MLP network were evaluated according to the number of hidden layer neurons and were found to be the best activation functions. Although the Relu function provides an unsatisfactory output when the number of neurons is smaller than 20, it makes fewer errors when the number of neurons is greater than 20.

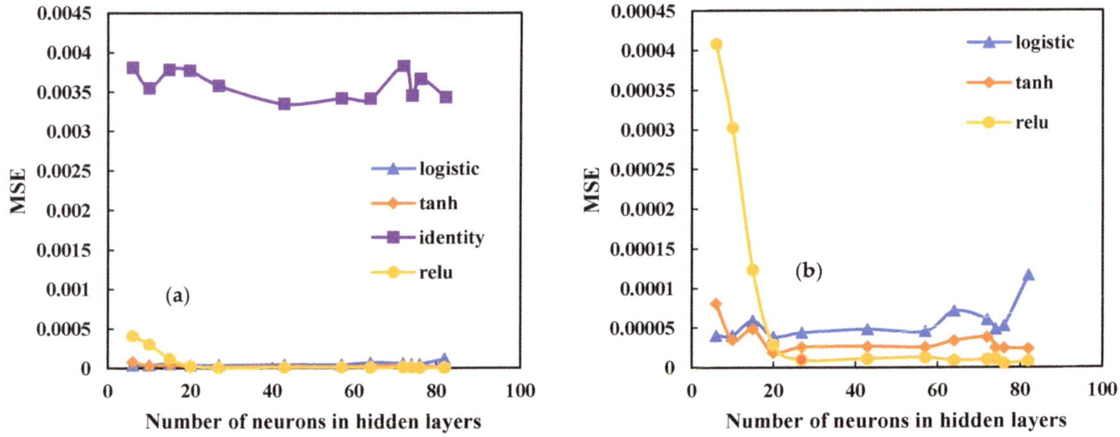

Figure 9. Comparison between different activation functions to find best function in MLP network (**a**) comparison of four activation functions; (**b**) comparison of three activation function.

The MSE, RMSE, and R parameters for the MLP and RBF networks were determined in Tables 4 and 5. The LBFGS approach is preferable based on the evaluator parameters. In the MLP network, several activation functions and optimization algorithms were also evaluated. As illustrated in Figure 9, the Relu activation function provided a more accurate result.

Figure 10 presents the R of MLP network data with the Relu activation function and different optimization methods in various hidden layer neurons for each network optimization method, which means that the BFGS approach is desirable.

As shown in Table 4, the BFGS optimization algorithm method was selected as the optimization method because it has the lowest error considering the number of hidden layer neurons. According to network comparison, the MLP network with the 27 neurons in the hidden layer has the best result and an MSE value of 8.95×10^{-6}. The structures of other networks can cause minor errors when they have more neurons. In Figure 11, three types of different optimization algorithms that have been used for the mlp network are compared with the RBF network. The comparison of different algorithms in Figure 11 shows that the lowest number of neurons concerning the mean squared of error has the LBFG optimization method.

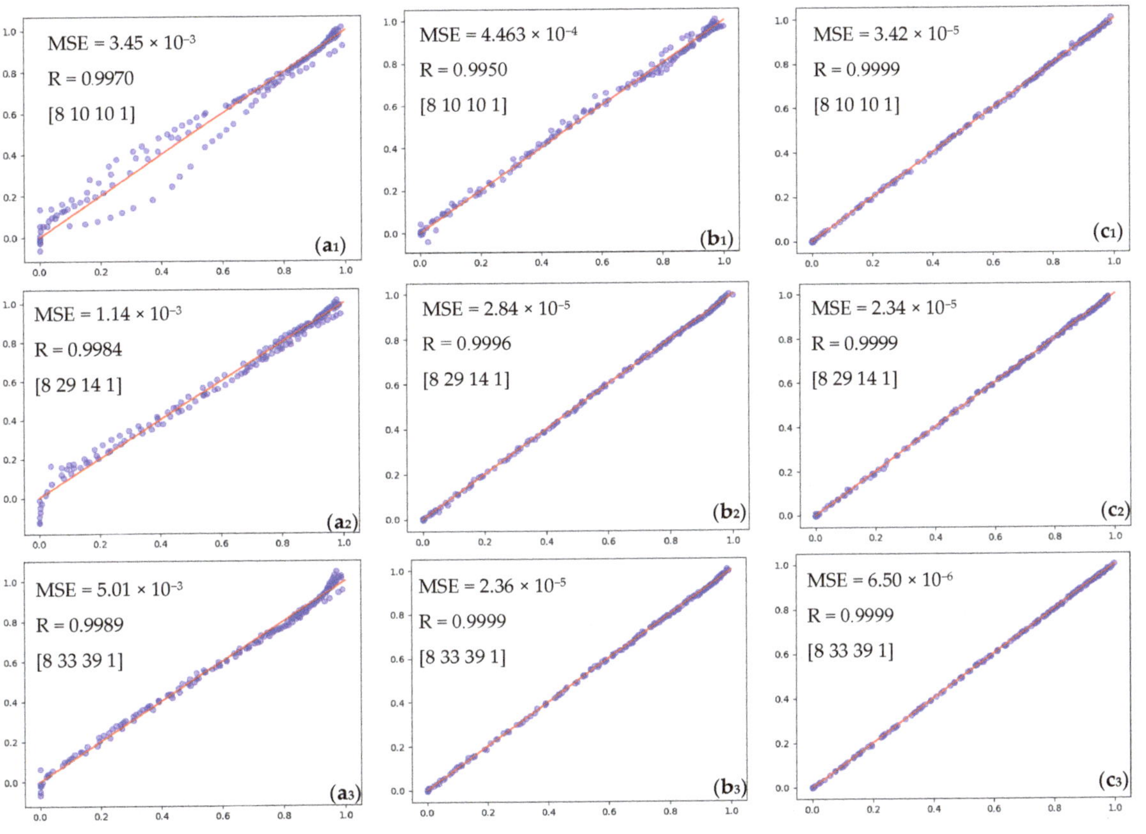

Figure 10. Comparison accuracy performance of MLP network using the Relu activation function. (**a$_1$**–**a$_3$**) SGD method, (**b$_1$**–**b$_3$**) Adam method, and (**c$_1$**–**c$_3$**) BFGS methods (blue dotes are data and the red line is the fit line).

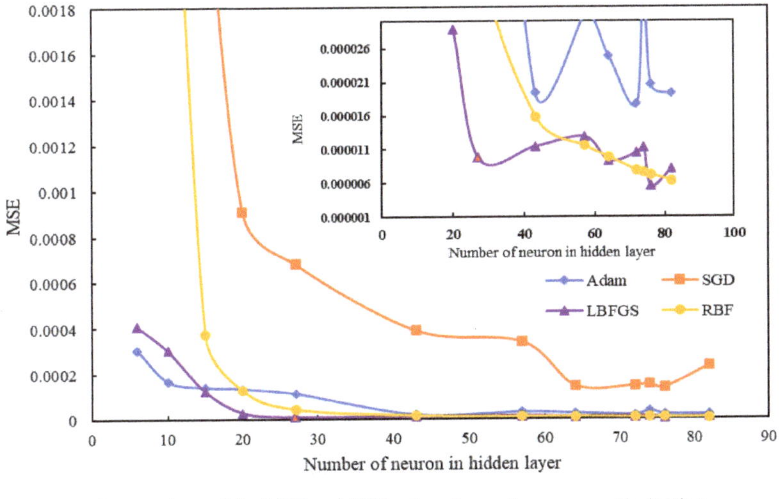

Figure 11. Comparison of the MLP and RBF networks performance optimization.

Table 4. Accuracy prediction of DRI ANN structures using MLP model algorithms with learning rate = 0.02, momentum = 0.9, and batch size = 20.

Optimization Algorithms	Training Function	*DRI−ANN* Structure									
		MSE (−)	RMSE (−)	R_{train} (%)	R (%)	Neurons in Hidden Layers	MSE (−)	RMSE (−)	R_{train} (%)	R (%)	Neurons in Hidden Layers
Adam	Relu	3.05×10^{-4}	1.74×10^{-2}	0.9985	0.9986	6	8.67×10^{-5}	9.3×10^{-3}	0.9995	0.9996	57
		1.67×10^{-4}	1.29×10^{-2}	0.9992	0.9993	10	2.48×10^{-5}	4.9×10^{-3}	0.9998	0.9998	64
		1.41×10^{-4}	1.18×10^{-2}	0.9993	0.9995	15	1.77×10^{-5}	4.2×10^{-3}	0.9999	0.9999	72
		1.35×10^{-4}	1.16×10^{-2}	0.9993	0.9994	20	3.76×10^{-5}	6.1×10^{-3}	0.9998	0.9998	74
		1.13×10^{-4}	1.06×10^{-2}	0.9994	0.9994	27	9.61×10^{-5}	9.8×10^{-3}	0.9995	0.9996	76
		7.04×10^{-5}	8.3×10^{-3}	0.9996	0.9997	43	1.94×10^{-5}	4.4×10^{-3}	0.9999	0.9999	82
Broyden, Fletcher, Goldfarb, and Shanno	Relu	4.09×10^{-4}	2.02×10^{-2}	0.9980	0.9983	6	1.29×10^{-5}	3.6×10^{-3}	0.9999	0.9999	57
		3.03×10^{-4}	1.74×10^{-2}	0.9985	0.9987	10	1.7×10^{-5}	4.13×10^{-3}	0.9999	0.9999	64
		1.24×10^{-4}	1.11×10^{-2}	0.9994	0.9996	15	1.05×10^{-5}	3.25×10^{-3}	0.9999	0.9999	72
		2.88×10^{-5}	5.37×10^{-3}	0.9998	0.9998	20	1.78×10^{-5}	4.21×10^{-3}	0.9999	0.9999	74
		$\mathbf{8.95 \times 10^{-6}}$	$\mathbf{2.99 \times 10^{-3}}$	**0.9999**	**0.9999**	**27**	5.63×10^{-6}	2.37×10^{-3}	0.9999	0.9999	76
		1.15×10^{-5}	3.39×10^{-3}	0.9999	0.9999	43	8.15×10^{-6}	2.85×10^{-3}	0.9999	0.9999	82
Scaled Conjugate Gradient	Relu	6.37×10^{-3}	3.94×10^{-2}	0.9693	0.9695	6	3.4×10^{-4}	1.84×10^{-2}	0.9984	0.9985	57
		5.2×10^{-3}	3.31×10^{-2}	0.9734	0.9735	10	1.46×10^{-4}	1.21×10^{-2}	0.9992	0.9992	64
		2.59×10^{-3}	4.19×10^{-2}	0.9877	0.9878	15	3.39×10^{-4}	1.84×10^{-2}	0.9983	0.9984	72
		9.1×10^{-4}	3.02×10^{-2}	0.9952	0.9955	20	1.53×10^{-4}	1.24×10^{-2}	0.9992	0.9993	74
		6.81×10^{-4}	2.61×10^{-2}	0.9966	0.9967	27	1.38×10^{-4}	1.18×10^{-2}	0.9993	0.9994	76
		3.9×10^{-4}	1.97×10^{-2}	0.9981	0.9984	43	2.34×10^{-4}	1.53×10^{-2}	0.9988	0.9990	82
Broyden, Fletcher, Goldfarb, and Shanno	tanh	8.08×10^{-5}	8.99×10^{-3}	0.9996	0.9997	6	2.59×10^{-5}	5.09×10^{-3}	0.9998	0.9998	57
		3.53×10^{-5}	5.94×10^{-3}	0.9998	0.9998	10	3.41×10^{-5}	5.84×10^{-3}	0.9998	0.9998	64
		4.86×10^{-5}	6.97×10^{-3}	0.9997	0.9997	15	3.82×10^{-5}	6.18×10^{-3}	0.9998	0.9998	72
		1.87×10^{-5}	4.33×10^{-3}	0.9999	0.9999	20	2.53×10^{-5}	5.03×10^{-3}	0.9998	0.9998	74
		2.56×10^{-5}	5.06×10^{-3}	0.9998	0.9998	27	2.45×10^{-5}	4.95×10^{-3}	0.9998	0.9998	76
		2.69×10^{-5}	5.18×10^{-3}	0.9998	0.9998	43	2.39×10^{-5}	4.89×10^{-3}	0.9998	0.9999	82

Table 4. *Cont.*

Optimization Algorithms	Training Function	*DRI−ANN* Structure									
		MSE (−)	RMSE (−)	R_{train} (%)	R (%)	Neurons in Hidden Layers	MSE (−)	RMSE (−)	R_{train} (%)	R (%)	Neurons in Hidden Layers
Broyden, Fletcher, Goldfarb, and Shanno	identity	3.81×10^{-3}	6.17×10^{-2}	0.9823	0.9825	6	3.42×10^{-3}	5.85×10^{-2}	0.9837	0.9839	57
		3.55×10^{-3}	5.96×10^{-2}	0.9831	0.9833	10	3.41×10^{-3}	5.84×10^{-2}	0.9841	0.9843	64
		3.79×10^{-3}	6.15×10^{-2}	0.9815	0.9818	15	3.83×10^{-3}	6.19×10^{-2}	0.9817	0.9817	72
		3.78×10^{-3}	6.14×10^{-2}	0.9828	0.9830	20	3.45×10^{-3}	5.87×10^{-2}	0.9835	0.9836	74
		3.58×10^{-3}	5.99×10^{-2}	0.9829	0.9831	27	3.67×10^{-3}	6.06×10^{-2}	0.9810	0.9812	76
		3.35×10^{-3}	5.79×10^{-2}	0.9841	0.9844	43	3.43×10^{-3}	5.86×10^{-2}	0.9841	0.9842	82
Broyden, Fletcher, Goldfarb, and Shanno	logistic	4.05×10^{-5}	6.37×10^{-3}	0.9997	0.9998	6	4.54×10^{-5}	6.74×10^{-3}	0.9997	0.9997	57
		4.03×10^{-5}	6.35×10^{-3}	0.9998	0.9998	10	7.12×10^{-5}	8.44×10^{-3}	0.9996	0.9996	64
		5.92×10^{-5}	7.7×10^{-3}	0.9997	0.9998	15	6.02×10^{-5}	7.76×10^{-3}	0.9997	0.9997	72
		3.89×10^{-5}	6.23×10^{-3}	0.9998	0.9998	20	4.84×10^{-5}	6.96×10^{-3}	0.9997	0.9997	74
		4.39×10^{-5}	6.62×10^{-3}	0.9997	0.9998	27	5.25×10^{-5}	7.25×10^{-3}	0.9997	0.9997	76
		4.8×10^{-5}	6.93×10^{-3}	0.9997	0.9997	43	1.16×10^{-4}	1.08×10^{-2}	0.9994	0.9994	82

Table 5. Accuracy prediction of DRI ANN structures using RBF model algorithms, according to the number of neurons studied and the optimal MLP network.

Optimization Algorithms	Training Function	MSE (−)	RMSE (−)	R_{train} (%)	R (%)	Neurons in Hidden Layers	MSE (−)	RMSE (−)	R_{train} (%)	R (%)	Neurons in Hidden Layers
RBF	Gaussian function	7.57×10^{-3}	8.7×10^{-2}	0.9908	0.9908	6	1.15×10^{-5}	3.39×10^{-3}	0.9999	0.9999	57
		3.66×10^{-3}	6.6×10^{-2}	0.9956	0.9956	10	9.8×10^{-6}	3.14×10^{-3}	0.9999	0.9999	64
		3.73×10^{-4}	1.93×10^{-2}	0.9995	0.9995	15	7.8×10^{-6}	2.79×10^{-3}	0.9999	0.9999	72
		1.29×10^{-4}	1.13×10^{-2}	0.9998	0.9998	20	7.6×10^{-6}	2.75×10^{-3}	0.9999	0.9999	74
		4.38×10^{-5}	6.61×10^{-3}	0.9999	0.9999	27	7.2×10^{-6}	2.68×10^{-3}	0.9999	0.9999	76
		1.58×10^{-5}	3.97×10^{-3}	0.9999	0.9999	43	6.3×10^{-6}	2.5×10^{-3}	0.9999	0.9999	82

4.2. Optimum ANN Results for Prediction of Solid Conversion for DRI Process

The accuracy of network results and the comparison of real data with the prediction amount are quite acceptable. The neural network has a low error rate, and it can calculate the percentage of X_3 from the shaft furnace based on input variables such as diameter, length, and input flow to the shaft furnace. The best result was shown using the LBFGS optimization method and the Relu activation function in Figure 12.

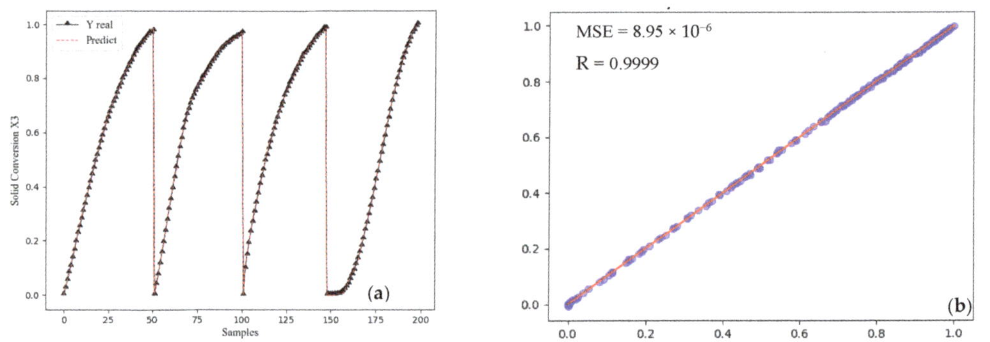

Figure 12. The best network-developed MLP with two hidden layers (**a**) Compare predicted using real data (**b**) strong positive correlation.

As a result, the most functional network consists of two hidden layers, with 13 neurons in the first layer and 14 neurons in the second layer. According to Figure 13, these neurons are fully connected using weights. Furthermore, each neuron has a bias that is listed in Appendix A, along with its weight. The matrix created using the MLP algorithm simulated for the prediction of the MIDREX process is shown in Appendix A.

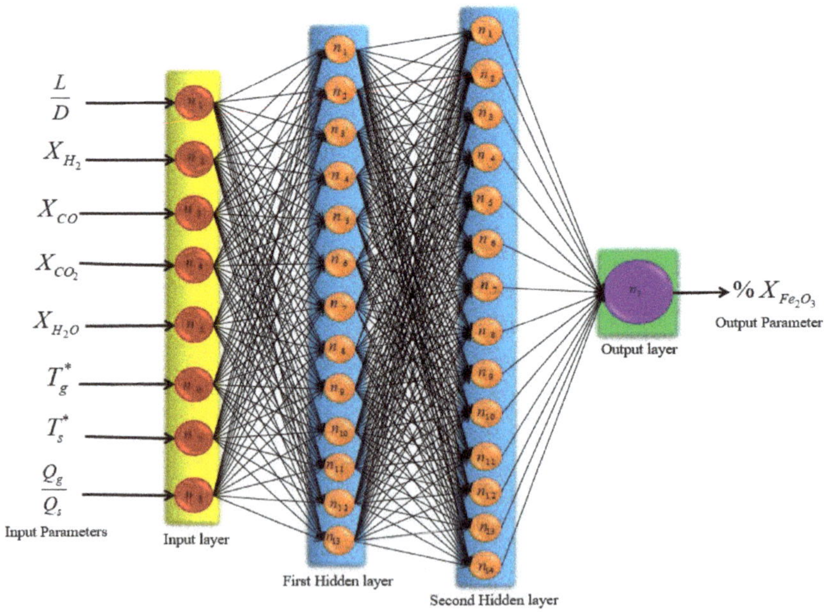

Figure 13. Schematic of final MLP structure.

4.2.1. Effect of Dimensions on Pellet Conversion Rate

Heatmap and pair plot have been used to show the effects of different parameters on the sponge iron produced, as shown in Figures 14 and 15. In the heatmap chart, Pearson's correlation coefficients show the relationship between various parameters [64]. In Figure 14, the effect of different parameters of the network is shown as a linear and nonlinear spectrum, with the interpretation that the farther from the unit value (1 and −1), the more nonlinear it is, and the closer to the unit value, the effect of two parameters is proportional and are more linear. The closer value is to one, the more linear relationship between the two parameters. If this value is positive, it indicates a direct relationship between those the two parameters, while if this value is negative, it means that those two parameters are opposite of each other (they increase or decrease in opposite directions). According to Figure 14, the correlation coefficients between the real data of the network and the prediction value of the network are equal to one, which shows that it can completely predict the real data. The correlation coefficient for the effect of dimensions on the conversion rate is 0.91, which shows that the relationship between the two is linear and direct. Based on Figure 15, the direct relationship between these parameters and the degree of linearity is clear. This direct relationship is because of the longer reactor, the longer pellets in the regeneration zone, and the higher conversion rate [15].

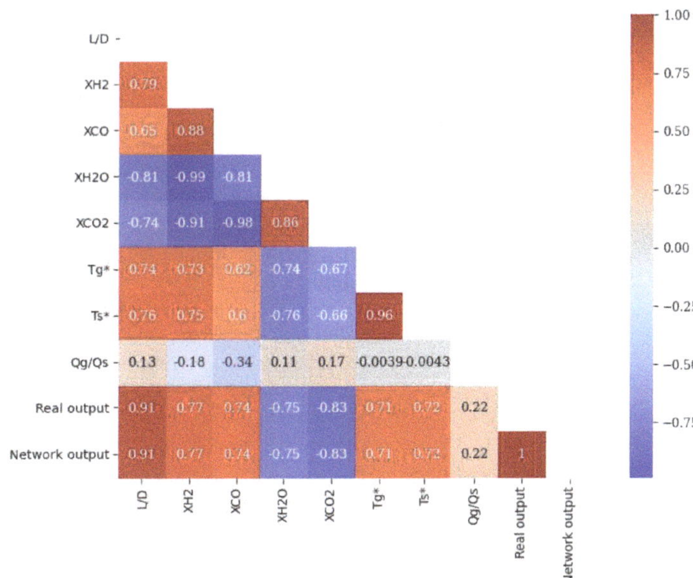

Figure 14. Heatmap of Pearson correlation coefficient matrix for network MLP.

4.2.2. The Effect of Gaseous Compounds on Pellet Conversion

According to Figure 14, the correlation coefficient between X_{CO2} and X_{H2O} with pellet conversion rate is −0.83 and −075, indicating that there is a nonlinear link between them and the direction of their changes is the opposite of each other. These values for the relationship between X_{CO} and X_{H2} with pellet conversion are equal to 0.74 and 0.77, which proves that the relationship between them is direct and nonlinear. The connection is formed when H_2 and CO gases enter the furnace from the bottom of the reduction zone. After interaction with the iron oxide pellets, they turn into CO_2 and H_2O and exit from the top of the reduction zone. Hence, changes in X_{CO2} and X_{H2O} have an inverse relationship with changes in pellet conversion rate. From the top of the regeneration zone to the bottom, the amount of pellet conversion increases, and the amount of X_{CO2} and X_{H2O} decreases. This relationship can be seen in the performance of all stimulations shown in Figure 15 for four shaft furnaces.

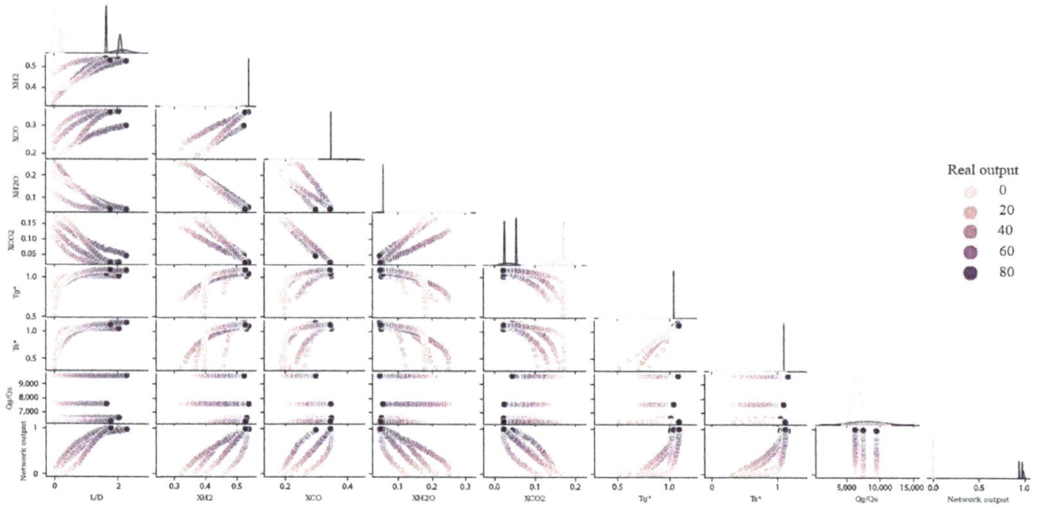

Figure 15. Comparing the effects of different parameters on each other and on the amount of solid conversion (based on dimensionless data).

4.2.3. The Effect of Flow Rate on Pellet Conversion Rate

According to Figure 14, the correlation coefficient between the ratio of gas to solid flow rate is 0.22. This value proves several facts: firstly, the relationship between the solid flow rate and the pellet conversion rate is opposite of each other because the solid flow rate is in the denominator of the dimensionless flow parameter; secondly, the relationship between the gas flow rate and the conversion rate is because the gas flow rate in the case of the dimensionless flow parameter is direct. In addition, this relationship is nonlinear since the correlation coefficient is near zero. By examination of other parameters in Figure 14, it can be seen that the most nonlinear parameter is the ratio of flow rates. This sign of nonlinearity demonstrates that the impact of this parameter on iron ore recovery is greater than other parameters. This conclusion has been proved by solving the governing equations of the problem [14].

The relationship between solid flow rate and conversion rate is very similar to the relationship between dimensions and conversion rate (both are due to residence time). Furthermore, because of the lower solid flow rate and lower speed of pellets, the pellets and the regeneration gas are more in contact with each other. Consequently, the conversion rate grows as the residence time of the pellets in the regeneration area increases. Contrary to the relationship between solid flow rate and pellet conversion, the pellet conversion rate increases with increasing gas flow rate. This increase is not due to the reduction of external mass transfer resistance from gas to solid. However, simulations reveal that even at the lowest flow rate, the Sherwood number is sufficiently large to render the mass transfer resistance insignificant [15]. Consequently, as the flow rate increases, so does the concentration of reducing gases in the top portion of the reactor.

4.2.4. The Effect of Temperature on the Pellet Conversion Rate

The lowest part of the reduction zone, where the reduction gas enters, has the highest temperature [4]. Furthermore, in this area, the pellet conversion is maximum, proving that increasing the gas and the solid temperature raises the pellet conversion rate. The correlation coefficient between the gas and solid temperature with the solid conversion rate shows the value of 0.71 and 0.72. These values show the direct and nonlinear relationship between the parameters. The smaller the particle, the higher the mass transfer coefficient, which means the film resistance is lower. Thus, the film resistance is maximized if the larger particle is chosen. To explain, if we can remove this film resistance at speed, all smaller particles can be removed at the same speed as the film resistance [50].

Considering the environmental issues, air pollution has been one of the most challenging issues for humans in recent years. Since 7% of carbon dioxide production is derived from the steel industry [9], it can be modeled in future research. The production of Midrex with green hydrogen was aimed at its feasibility from thermal and economic points of view. Another vital issue is the analysis of energy consumption in the MIDREX unit. Recently Salimi et al. investigated the technical and economic study of energy harvesting from the waste heat of the Midrex process by the Kalina cycle in the process of direct iron reduction [65].

5. Conclusions

In this study, two networks were constructed for the percentage of X_3 output parameters from the shaft furnace. Two networks were investigated using different optimization methods. The Adam and LBFGS optimization algorithm methods were faster and more accurate, delivering results for MSEs in order 8.95×10^{-6}. Furthermore, various activation functions were practiced to improve the network by the Relu, which can cause the least error due to the number of hidden layer neurons. After optimization of both networks, RBF and MLP, they were compared with the same number of hidden layer neurons. The MLP network was able to generate 8.95×10^{-6} errors and better predict the conversion percentage of the pellet as an output parameter. This network could be used to improve the performance of the shaft furnace and make modeling to predict the conversion percentage of iron oxide output of the shaft furnace straightforward. The predictive advantage of the shaft furnace can be determined using the network results, such as weights and biases, in the shortest time and with the highest accuracy. Finally, for a better analysis of the effect of different parameters, using a heat map, it was shown that the coefficient of each of the network inputs can communicate with the output and other input parameters of the network. This research could be the beginning of the utilization of ML in the direct reduction process, which is due to the complex behavior of pellets in the shaft furnace and its complex reactions (six heterogeneous reactions of iron oxide phases, methane reforming reactions, water gas shift reactions, and other side reactions) can help the pellet behavior with high speed and accuracy through ML. This modeling is superior to earlier approaches because it is more precise and quick than earlier numerical methods. Another application for ML is in unit control; for instance, it could be used to optimize the temperature of the bustling gas in the shaft furnace. This alone could make for some very fascinating research in the future. ML could be used in future studies for control purposes, such as the MIDREX process.

Author Contributions: Conceptualization, H.M., M.H. and A.E.; Methodology, H.M., M.H. and F.M.; Software, M.H.; Validation, M.H.; Formal analysis, M.H., H.M. and F.M.; Data curation, M.H.; Writing—original draft, M.H., H.M.; Writing—review & editing, H.M. and F.M.; Supervision, A.E.; Project administration, H.M.; Funding acquisition, A.E. All authors have read and agreed to the published version of the manuscript.

Funding: This research received no external funding.

Conflicts of Interest: The authors declare no conflict of interest.

Nomenclature

A_p	Pellet external area [cm^2]
b	Bias [−]
C_{ps},C_{pg}	The heat capacity of solids and gases [J/Kg K]
D	Reduction zone diameter [cm]
E	Cost function
F_k	Nonlinear functions of activation transmission
G	Function of Gaussian
f	Function
G_{mg}	gas molar flow [mol/cm^2s]
h	heat transfer coefficient [W /cm^2 K]
H	Reaction enthalpy [cal/mol]
H_e	Approximate of the Hussein matrix
i	Number of neurons in the hidden layer
k	Position vector
L	Reduction zone length [cm]
m	first moment (the mean) [−]
N	Number of data sets for training [−]
N_p	number of pellets per unit volume of the bed [$1/cm^3$]
n	Neurons
P	Direction of search
Q	Flow-rate [Nm^3/h]
R_r	Reaction rate per pellet [mol/s]
R,R^2	Correlation coefficient [%]
s	Direction vector
T^*	Dimensionless temperature [−]
T_g^*	Dimensionless gas temperature [−]
T_s^*	Dimensionless solid temperature [−]
u	Velocity [m/s]
V	Second moment (the uncentered variance)
W	Factor of weight [−]
W_{ij}	Weight-related to each hidden neuron [−]
x_k	Input variable [−]
X_1	The extent of reaction/extent of reactant conversion for H_2 [mol/cm^3]
X_2	The extent of reaction/extent of reactant conversion for CO [mol/cm^3]
X_3	Extent of reaction/extent of reactant conversion for Fe_2O_3 [mol/cm^3]
Y	Output vector [−]
y_j	Target output
Greek symbols	
α	Constants in turbulence models [−]
β_{jk}	Weight of bias for neuron j in layer k
γ_{jk}	The output of neuron j from the layer of k
ψ_k	Step size
θ	Threshold limit [−]
σ	Radial Basis Function (RBF) kernel width [−]
σ_i	Gaussian function distribution [−]
Subscripts	
c	Characteristic
rs	Reactive solid (Fe_2O_3)
g	Gas
n	Gaseous reactant
p	Pellet
s	Solid

Acronyms

ANN	Artificial Neural Networks
MSE	Mean Square Error

Terminology

Neurons	Neurons are the fundamental components of the vast neural network.
Bias	Bias is a constant that allows the model to be optimally fitted to the available data.
Activation function	This function is a mathematical function between the input that feeds the current neuron and its output that travels to the next layer.
Weight	demonstrates the significance and ability of the characteristic/input to the neurons.
Epoch	In the training process, each training step's inputs produce an output that is compared to the goal in order to determine an error. Weights and biases are calculated and adjusted in each epoch using this method.
Batch size	For stochastic optimizers, the size of mini-batches is important. (The classifier will not use minibatch if the solver is lbfgs.)
Momentum	Update on gradient descending momentum. Only when the solver is sgd.
Learning rate	For weight updates, we have to learn a rate schedule (used for SGD).

Appendix A

Table A1. Characteristic weights and biases of the DRI − ANN − MLP model with the best algorithm.

Neuron		1	2	3	4	5	6	7	8	9	10	11	12	13	14	15
First hidden layer	w_i	−0.5261	0.3122	0.3003	−0.3842	−0.1168	−0.4217	−0.4679	0.5963	−0.5029	0.2167	0.1923	0.2747	−0.0355	-	-
		0.1414	−0.6600	0.3918	−0.4975	−0.1184	−0.0795	−0.2839	0.2968	−0.1454	−0.3495	0.1552	0.2744	−0.5067		
		−0.4716	0.1534	0.1572	0.0499	−0.1170	−0.5092	0.0669	0.3481	0.3992	−0.3800	−0.0519	0.2840	0.3226	-	-
		−0.4195	0.2306	0.4402	0.3055	−0.3375	−0.2342	0.4835	0.5569	−0.0585	−0.0243	0.3713	0.2479	−0.4035	-	-
		−0.3103	−0.4300	−0.4428	−0.3570	−0.1607	−0.3180	0.3841	0.3178	0.4336	0.2486	−0.0613	−0.1156	0.1871	-	-
		0.1264	0.2072	−0.1894	0.3279	−0.4434	−0.3503	0.2591	−0.1250	−0.0045	−0.2883	−0.1393	0.5487	0.1746	-	-
		−0.0399	0.4145	−0.1454	0.4606	0.4224	−0.4450	0.4078	0.1576	−0.2562	−0.3226	−0.4084	0.4207	−0.4346	-	-
		−0.2834	0.6741	−0.3418	0.3441	0.1340	0.1127	−0.0791	−0.1540	−0.3525	−0.2026	−0.2172	−0.3762	−0.0161	-	-
	b	−0.1834	−0.1580	−0.0837	−0.2541	0.1330	0.4629	−0.1553	−0.2419	−0.4652	−0.1960	−0.3441	−0.1922	−0.1461	-	-
Second hidden layer	w_i	0.0312	0.0689	0.4262	0.2687	−0.0318	0.1865	−0.2655	−0.0196	0.1368	−0.3179	−0.3118	−0.0888	−0.2162	0.3304	-
		0.0045	0.2990	−0.0313	0.3231	0.0983	0.2352	0.2577	0.3021	−0.5072	0.3468	−0.0712	−0.6618	0.0328	0.3322	-
		−0.1573	0.1815	0.1161	−0.0604	0.0848	−0.4078	0.2779	0.1774	−0.1519	−0.2078	−0.4528	−0.3158	0.3392	−0.0341	-
		0.5107	0.2276	−0.0847	0.3726	−0.4047	0.0170	0.2782	−0.3967	0.0856	0.2555	−0.0656	0.1127	−0.0742	−0.1451	-
		−0.0711	−0.4243	0.4544	0.4096	0.1945	0.2685	−0.1691	0.2535	0.1275	−0.1207	0.0698	0.2671	−0.0571	0.3847	-
		0.4588	−0.1143	0.1844	0.4392	0.0310	0.0494	0.2749	−0.2550	−0.3050	0.2182	−0.1881	−0.0108	−0.3532	0.3005	-
		0.2885	0.2154	0.1225	0.0524	0.3704	−0.3282	0.3586	−0.4109	0.3636	−0.3674	−0.0109	−0.2198	0.3213	0.4504	-
		−0.0564	0.2155	−0.1700	−0.1756	−0.0896	−0.1456	0.1221	−0.1117	−0.3548	−0.1415	0.1060	−0.0106	0.2727	−0.5645	-
		0.0639	−0.1984	−0.1939	0.3111	−0.3191	−0.3805	0.2158	0.0610	−0.2565	0.2141	0.4456	−0.2957	−0.0531	−0.1954	-
		−0.2476	0.0253	−0.0674	−0.0173	−0.4643	0.1050	0.1627	−0.4444	−0.3693	0.3759	−0.2169	0.2445	0.3126	0.1138	-
		−0.2510	0.3462	−0.4332	0.3669	−0.4641	0.3092	0.4015	0.0386	0.0139	0.2427	−0.3794	−0.1944	−0.2762	−0.1957	-
		0.34032	0.1627	−0.0866	0.4185	−0.2898	−0.3095	0.2176	0.1026	0.1285	0.4677	0.24473	0.0041	0.14404	0.0999	-
		−0.4088	0.1685	−0.1742	0.4011	−0.0112	0.4490	0.1211	−0.1369	−0.0108	0.2971	−0.2613	0.4224	0.2733	−0.1892	-
	b	0.0027	0.3390	−0.1864	−0.3606	−0.3528	−0.0716	0.2954	−0.3802	0.2887	0.0372	0.3559	0.3273	0.0601	0.2710	-
Output layer	w_l	0.1855														
		0.5334														
		−0.1828														
		0.1919														
		0.2326														
		−0.3029														
		−0.0028														
		−0.2619														
		−0.6629														
	b	0.0341	-	-	-	-	-	-	-	-	-	-	-	-	-	-

References

1. Berg, K.; Olsen, S. Kinetics of manganese ore reduction by carbon monoxide. *Metall. Mater. Trans. B* **2000**, *31*, 477–490. [CrossRef]
2. Szekely, J.; Lin, C.; Sohn, H. A structural model for gas—Solid reactions with a moving boundary—V an experimental study of the reduction of porous nickel-oxide pellets with hydrogen. *Chem. Eng. Sci.* **1973**, *28*, 1975–1989. [CrossRef]
3. Béchara, R.; Hamadeh, H.; Mirgaux, O.; Patisson, F. Carbon impact mitigation of the iron ore direct reduction process through computer-aided optimization and design changes. *Metals* **2020**, *10*, 367. [CrossRef]
4. Hamadeh, H.; Mirgaux, O.; Patisson, F. Detailed modeling of the direct reduction of iron ore in a shaft furnace. *Materials* **2018**, *11*, 1865. [CrossRef]
5. Cavaliere, P. Clean ironmaking and steelmaking processes: Efficient technologies for greenhouse emissions abatement. In *Clean Ironmaking and Steelmaking Processes*; Springer: Berlin/Heidelberg, Germany, 2019; pp. 1–37.

6. Cavaliere, P.; Perrone, A.; Silvello, A.; Stagnoli, P.; Duarte, P. Integration of Open Slag Bath Furnace with Direct Reduction Reactors for New-Generation Steelmaking. *Metals* **2022**, *12*, 203. [CrossRef]
7. Statistics, W.D.R. *Midrex the World Leader in Direct Reduction Technology*; © 2021 Midrex Technologies, Inc.: Charlotte, NC, USA, 2020; pp. 1–15.
8. Rechberger, K.; Spanlang, A.; Sasiain Conde, A.; Wolfmeir, H.; Harris, C. Green hydrogen-based direct reduction for low-carbon steelmaking. *Steel Res. Int.* **2020**, *91*, 2000110. [CrossRef]
9. Pimm, A.J.; Cockerill, T.T.; Gale, W.F. Energy system requirements of fossil-free steelmaking using hydrogen direct reduction. *J. Clean. Prod.* **2021**, *312*, 127665. [CrossRef]
10. Szekely, J.; El-Tawil, Y. The reduction of hematite pellets with carbon monoxide-hydrogen mixtures. *Metall. Trans. B* **1976**, *7*, 490–492. [CrossRef]
11. Towhidi, N. Reduction kinetics of commercial low-silica hematite pellets with CO-H_2 mixtures over temperature range 600–1234 °C. *Ironmak. Steelmak.* **1981**, *6*, 237–249.
12. Turkdogan, E.; Vinters, J. Gaseous reduction of iron oxides: Part I. Reduction of hematite in hydrogen. *Metall. Mater. Trans. B* **1971**, *2*, 3175–3188. [CrossRef]
13. Parisi, D.R.; Laborde, M.A. Modeling of counter current moving bed gas-solid reactor used in direct reduction of iron ore. *Chem. Eng. J.* **2004**, *104*, 35–43. [CrossRef]
14. Mirzajani, A.; Ebrahim, H.; Nouri, S. Simulation of a direct reduction moving bed reactor using a three interface model. *Braz. J. Chem. Eng.* **2018**, *35*, 1019–1028. [CrossRef]
15. Nouri, S.; Ebrahim, H.A.; Jamshidi, E. Simulation of direct reduction reactor by the grain model. *Chem. Eng. J.* **2011**, *166*, 704–709. [CrossRef]
16. Kazemi, M.; Pour, M.S.; Sichen, D. Experimental and modeling study on reduction of hematite pellets by hydrogen gas. *Metall. Mater. Trans. B* **2017**, *48*, 1114–1122. [CrossRef]
17. McKewan, W. Reduction kinetics of hematite in hydrogen-water vapor-nitrogen mixtures. *Trans. Metal Soc. AIME* **1962**, *224*, 2–5.
18. Usui, T.; Ohmi, M.; Yamamura, E. Analysis of rate of hydrogen reduction of porous wustite pellets basing on zone-reaction models. *ISIJ Int.* **1990**, *30*, 347–355. [CrossRef]
19. Mondal, K.; Lorethova, H.; Hippo, E.; Wiltowski, T.; Lalvani, S. Reduction of iron oxide in carbon monoxide atmosphere—Reaction controlled kinetics. *Fuel Process. Technol.* **2004**, *86*, 33–47. [CrossRef]
20. Tien, R.; Turkdogan, E. Gaseous reduction of iron oxides: Part IV. Mathematical analysis of partial internal reduction-diffusion control. *Metall. Trans.* **1972**, *3*, 2039–2048. [CrossRef]
21. Kam, E. *A Model for the Direct Reduction of Iron Ore by Mixtures of Hydrogen and Carbon Monoxide in a Moving Bed*; Pascal and Francis: Lawrence, KS, USA, 1981.
22. Negri, E.D.; Alfano, O.M.; Chiovetta, M.G. Direct reduction of hematite in a moving-bed reactor. Analysis of the water gas shift reaction effects on the reactor behavior. *Ind. Eng. Chem. Res.* **1991**, *30*, 474–482. [CrossRef]
23. Abualigah, L.; Abu Zitar, R.; Almotairi, K.H.; Hussein, A.M.; Elaziz, M.A.; Nikoo, M.R.; Gandomi, A.H. Wind, solar, and photovoltaic renewable energy systems with and without energy storage optimization: A survey of advanced machine learning and deep learning techniques. *Energies* **2022**, *15*, 578. [CrossRef]
24. Said, Z.; Sharma, P.; Aslfattahi, N.; Ghodbane, M. Experimental analysis of novel ionic liquid-MXene hybrid nanofluid's energy storage properties: Model-prediction using modern ensemble machine learning methods. *J. Energy Storage* **2022**, *52*, 104858. [CrossRef]
25. Forootan, M.M.; Larki, I.; Zahedi, R.; Ahmadi, A. Machine Learning and Deep Learning in Energy Systems: A Review. *Sustainability* **2022**, *14*, 4832. [CrossRef]
26. Valipour, M.S.; Saboohi, Y. Numerical investigation of nonisothermal reduction of hematite using Syngas: The shaft scale study. *Model. Simul. Mater. Sci. Eng.* **2007**, *15*, 487. [CrossRef]
27. Valipour, M.S.; Saboohi, Y. Modeling of multiple noncatalytic gas–solid reactions in a moving bed of porous pellets based on finite volume method. *Heat Mass Transf.* **2007**, *43*, 881–894. [CrossRef]
28. Alamsari, B.; Torii, S.; Trianto, A.; Bindar, Y. Heat and mass transfer in reduction zone of sponge iron reactor. *Int. Sch. Res. Not.* **2011**, *2011*, 324659. [CrossRef]
29. Alhumaizi, K.; Ajbar, A.; Soliman, M. Modelling the complex interactions between reformer and reduction furnace in a midrex-based iron plant. *Can. J. Chem. Eng.* **2012**, *90*, 1120–1141. [CrossRef]
30. Shams, A.; Moazeni, F. Modeling and simulation of the MIDREX shaft furnace: Reduction, transition and cooling Zones. *JOM* **2015**, *67*, 2681–2689. [CrossRef]
31. Ghadi, A.Z.; Valipour, M.S.; Biglari, M. CFD simulation of two-phase gas-particle flow in the Midrex shaft furnace: The effect of twin gas injection system on the performance of the reactor. *Int. J. Hydrogen Energy* **2017**, *42*, 103–118. [CrossRef]
32. Béchara, R.; Hamadeh, H.; Mirgaux, O.; Patisson, F. Optimization of the iron ore direct reduction process through multiscale process modeling. *Materials* **2018**, *11*, 1094. [CrossRef]
33. Hara, Y.; Sakawa, M.; Kondo, S.-I. Mathematical model of the shaft furnace for reduction of iron-ore pellet. *Tetsu-Hagané* **1976**, *62*, 315–323. [CrossRef]
34. Szekely, J.; Evans, J. A structural model for gas-solid reactions with a moving boundary-II: The effect of grain size, porosity and temperature on the reaction of porous pellets. *Chem. Eng. Sci.* **1971**, *26*, 1901–1913. [CrossRef]

35. Szekely, J.; Evans, J. A structural model for gas—Solid reactions with a moving boundary. *Chem. Eng. Sci.* **1970**, *25*, 1091–1107. [CrossRef]
36. Takenaka, Y.; Kimura, Y.; Narita, K.; Kaneko, D. Mathematical model of direct reduction shaft furnace and its application to actual operations of a model plant. *Comput. Chem. Eng.* **1986**, *10*, 67–75. [CrossRef]
37. Yanagiya, T.; Yagi, J.; Omori, Y. Reduction of iron oxide pellets in moving bed. *Ironmak. Steelmak.* **1979**, *6*, 93–100.
38. Kang, H.W.; Chung, W.S.; Murayama, T.; Ono, Y. Effect of iron ore shape on gaseous reduction rate. *ISIJ Int.* **1998**, *38*, 1194–1200. [CrossRef]
39. Kang, H.W.; Chung, W.S.; Murayama, T. Effect of iron ore size on kinetics of gaseous reduction. *ISIJ Int.* **1998**, *38*, 109–115. [CrossRef]
40. Bonalde, A.; Henriquez, A.; Manrique, M. Kinetic analysis of the iron oxide reduction using hydrogen-carbon monoxide mixtures as reducing agent. *ISIJ Int.* **2005**, *45*, 1255–1260. [CrossRef]
41. Rahimi, A.; Niksiar, A. A general model for moving-bed reactors with multiple chemical reactions part I: Model formulation. *Int. J. Miner. Process.* **2013**, *124*, 58–66. [CrossRef]
42. Da Costa, A.R.; Wagner, D.; Patisson, F. Modelling a new, low CO_2 emissions, hydrogen steelmaking process. *J. Clean. Prod.* **2013**, *46*, 27–35. [CrossRef]
43. Ponugoti, P.V.; Garg, P.; Geddam, S.N.; Nag, S.; Janardhanan, V.M. Kinetics of iron oxide reduction using CO: Experiments and Modeling. *Chem. Eng. J.* **2022**, *434*, 134384. [CrossRef]
44. Tang, Q.; Huang, K. Determining the kinetic rate constants of Fe_3O_4-to-Fe and FeO-to-Fe reduction by H_2. *Chem. Eng. J.* **2022**, *434*, 134771. [CrossRef]
45. Ajbar, A.; Alhumaizi, K.; Soliman, M. Modeling and simulations of a reformer used in direct reduction of iron. *Korean J. Chem. Eng.* **2011**, *28*, 2242–2249. [CrossRef]
46. Parisi, D.R.; Laborde, M.A. Modeling steady-state heterogeneous gas–solid reactors using feedforward neural networks. *Comput. Chem. Eng.* **2001**, *25*, 1241–1250. [CrossRef]
47. Parisi, D.R.; Mariani, M.a.C.; Laborde, M.A. Solving differential equations with unsupervised neural networks. *Chem. Eng. Process. Process Intensif.* **2003**, *42*, 715–721. [CrossRef]
48. Wu, S.; Xu, J.; Yang, S.; Zhou, Q.; Zhang, L. Basic characteristics of the shaft furnace of COREX® smelting reduction process based on iron oxides reduction simulation. *ISIJ Int.* **2010**, *50*, 1032–1039. [CrossRef]
49. Xu, J.; Wu, S.; Kou, M.; Du, K. Numerical analysis of the characteristics inside pre-reduction shaft furnace and its operation parameters optimization by using a three-dimensional full scale mathematical model. *ISIJ Int.* **2013**, *53*, 576–582. [CrossRef]
50. Wu, S.; Xu, J.; Yagi, J.-I.; Guo, X.; Zhang, L. Prediction of pre-reduction shaft furnace with top gas recycling technology aiming to cut down CO_2 emission. *ISIJ Int.* **2011**, *51*, 1344–1352. [CrossRef]
51. Nield, D.; Bejan, A. *Convection in Porous Media*, 5th ed.; Springer: New York, NY, USA, 2016.
52. Yusufaly, T.I.; Kallis, K.; Simon, A.; Mayadev, J.; Yashar, C.M.; Einck, J.P.; Mell, L.K.; Brown, D.; Scanderbeg, D.; Hild, S.J.; et al. A knowledge-based organ dose prediction tool for brachytherapy treatment planning of patients with cervical cancer. *Brachytherapy* **2020**, *19*, 624–634. [CrossRef]
53. Zhong, S.; Zhang, K.; Bagheri, M.; Burken, J.G.; Gu, A.; Li, B.; Ma, X.; Marrone, B.L.; Ren, Z.J.; Schrier, J.; et al. Machine learning: New ideas and tools in environmental science and engineering. *Environ. Sci. Technol.* **2021**, *55*, 12741–12754. [CrossRef]
54. Lee, J.H.; Shin, J.; Realff, M.J. Machine learning: Overview of the recent progresses and implications for the process systems engineering field. *Comput. Chem. Eng.* **2018**, *114*, 111–121. [CrossRef]
55. Hegde, J.; Rokseth, B. Applications of machine learning methods for engineering risk assessment–A review. *Saf. Sci.* **2020**, *122*, 104492. [CrossRef]
56. Otchere, D.A.; Ganat, T.O.A.; Gholami, R.; Ridha, S. Application of supervised machine learning paradigms in the prediction of petroleum reservoir properties: Comparative analysis of ANN and SVM models. *J. Pet. Sci. Eng.* **2021**, *200*, 108182. [CrossRef]
57. Han, J.; Pei, J.; Kamber, M. *Data Mining: Concepts and Techniques Elsevier Science*; Elsevier: Amsterdam, The Netherlands, 2011.
58. Mashhadimoslem, H.; Vafaeinia, M.; Safarzadeh, M.; Ghaemi, A.; Fathalian, F.; Maleki, A. Development of predictive models for activated carbon synthesis from different biomass for CO2 adsorption using artificial neural networks. *Ind. Eng. Chem. Res.* **2021**, *60*, 13950–13966. [CrossRef]
59. Kolbadinejad, S.; Mashhadimoslem, H.; Ghaemi, A.; Bastos-Neto, M. Deep learning analysis of Ar, Xe, Kr, and O2 adsorption on Activated Carbon and Zeolites using ANN approach. *Chem. Eng. Process.-Process Intensif.* **2022**, *170*, 108662. [CrossRef]
60. Mashhadimoslem, H.; Ghaemi, A. Machine learning analysis and prediction of N_2, N_2O, and O_2 adsorption on activated carbon and carbon molecular sieve. *Environ. Sci. Pollut. Res.* **2022**, 1–21. [CrossRef]
61. Robbins, H.; Monro, S. A stochastic approximation method. *Ann. Math. Stat.* **1951**, *22*, 400–407. [CrossRef]
62. Kingma, D.P.; Ba, J. Adam: A method for stochastic optimization. *arXiv* **2014**, arXiv:1412.6980.
63. Fletcher, R. *Practical Methods of Optimization*; John Wiley & Sons: Hoboken, NJ, USA, 2013.
64. Egghe, L.; Leydesdorff, L. The relation between Pearson's correlation coefficient r and Salton's cosine measure. *J. Am. Soc. Inf. Sci. Technol.* **2009**, *60*, 1027–1036. [CrossRef]
65. Salemi, S.; Torabi, M.; Haghparast, A.K. Technoeconomical investigation of energy harvesting from MIDREX® process waste heat using Kalina cycle in direct reduction iron process. *Energy* **2022**, *239*, 122322. [CrossRef]

sustainability

MDPI

Article

Artificial Intelligence (AI)-Based Occupant-Centric Heating Ventilation and Air Conditioning (HVAC) Control System for Multi-Zone Commercial Buildings

Alperen Yayla [1], Kübra Sultan Świerczewska [2], Mahmut Kaya [3], Bahadır Karaca [4], Yusuf Arayici [5], Yunus Emre Ayözen [6] and Onur Behzat Tokdemir [7,*]

1 Department of Civil and Environmental Engineering, Imperial College London, Skempton Building, London SW7 2AZ, UK
2 Cundall Polska, 00-582 Warszawa, Poland
3 KPD Engineering & Consultancy, Bursa 16090, Türkiye
4 Nuclear Islands Department, Akkuyu Nuclear Power Plant, Mersin 33715, Türkiye
5 Department of Architecture and Built Environment, Northumbria University, Newcastle upon Tyne NE1 8ST, UK
6 Strategy Development Department, Ministry of Transport and Infrastructure, Ankara 06338, Türkiye
7 Department of Civil Engineering, Istanbul Technical University, Istanbul 34467, Türkiye
* Correspondence: otokdemir@itu.edu.tr

Abstract: Buildings are responsible for almost half of the world's energy consumption, and approximately 40% of total building energy is consumed by the heating ventilation and air conditioning (HVAC) system. The inability of traditional HVAC controllers to respond to sudden changes in occupancy and environmental conditions makes them energy inefficient. Despite the oversimplified building thermal response models and inexact occupancy sensors of traditional building automation systems, investigations into a more efficient and effective sensor-free control mechanism have remained entirely inadequate. This study aims to develop an artificial intelligence (AI)-based occupant-centric HVAC control mechanism for cooling that continually improves its knowledge to increase energy efficiency in a multi-zone commercial building. The study is carried out using two-year occupancy and environmental conditions data of a shopping mall in Istanbul, Turkey. The research model consists of three steps: prediction of hourly occupancy, development of a new HVAC control mechanism, and comparison of the traditional and AI-based control systems via simulation. After determining the attributions for occupancy in the mall, hourly occupancy prediction is made using real data and an artificial neural network (ANN). A sensor-free HVAC control algorithm is developed with the help of occupancy data obtained from the previous stage, building characteristics, and real-time weather forecast information. Finally, a comparison of traditional and AI-based HVAC control mechanisms is performed using IDA Indoor Climate and Energy (ICE) simulation software. The results show that applying AI for HVAC operation achieves savings of a minimum of 10% energy consumption while providing a better thermal comfort level to occupants. The findings of this study demonstrate that the proposed approach can be a very advantageous tool for sustainable development and also used as a standalone control mechanism as it improves.

Keywords: artificial intelligence (AI); automatic HVAC control; occupant behavior; model predictive control; energy efficiency

Citation: Yayla, A.; Świerczewska, K.S.; Kaya, M.; Karaca, B.; Arayici, Y.; Ayözen, Y.E.; Tokdemir, O.B. Artificial Intelligence (AI)-Based Occupant-Centric Heating Ventilation and Air Conditioning (HVAC) Control System for Multi-Zone Commercial Buildings. *Sustainability* **2022**, *14*, 16107. https://doi.org/10.3390/su142316107

Academic Editors: Luis Hernández-Callejo, Sergio Nesmachnow and Sara Gallardo Saavedra

Received: 11 October 2022
Accepted: 12 November 2022
Published: 2 December 2022

Publisher's Note: MDPI stays neutral with regard to jurisdictional claims in published maps and institutional affiliations.

1. Introduction

Due to high demand and the need for an increasing energy supply, energy efficiency becomes crucial. Restricted energy markets have wide effects in areas ranging from household budgets to international relations. Thus, due to high energy consumption, buildings are on the front line of energy efficiency research. Buildings compose approximately 40%

of the total global energy consumption (International Energy Agency, 2019 [1]), and almost 40% of this goes towards heating, ventilation, and air conditioning (HVAC) systems (Yang et al., 2014 [2]). Clearly, the development and implementation of efficient building energy control systems is essential for economic and environmental sustainability. The HVAC system is a commonly used tool to maintain thermal comfort in buildings. It also serves as an essential demand-response source for peak load reduction and system-wide activity stabilization through effective demand-side energy management strategies. Until today, this energy demand in the buildings has been measured with sensors. Since heating and cooling in large masses do not occur rapidly, the inability to respond to sudden changes in occupancy and environmental conditions makes traditional HVAC control systems energy-inefficient, especially in large commercial buildings.

An HVAC is a dynamic mechanism that includes multiple input and output variables and is subject to various fluctuations and uncertainties, including occupant behavior, external air temperature, humidity, air volume, and regulated air temperature (Alcalá et al., 2003 [3]; Mirinejad et al., 2008 [4]). These specific features and characteristics all need to be taken into consideration to operate the HVAC system effectively. Thus, the research question of this paper is "how HVAC systems can be made efficient in meeting the sudden changes in demand responses in large commercial buildings by taking into consideration the occupancy patterns and prediction?" The following section provides a critical review of the literature with related studies in energy management with HVAC control systems to establish the setting for the research.

2. Related Studies

2.1. *Traditional and Advanced Control Strategies*

HVAC control strategies can be examined in general terms under two headings: traditional control strategies (TCSs) and advanced control strategies (ACSs). This section presents a review of related studies focusing on ACSs. Different control mechanisms are examined, and then a limited number of occupancy-based control approaches are discussed.

TCSs generally include sequencing, on-off, process, and proportional-integral-derivative (PID) controls. Their simple structure, quick response, easy implementation, and low initial costs are the main advantages of TCSs. They also have many disadvantages, such as low accuracy, quality, and performance, and (thus) energy efficiency. Furthermore, they do not interact with the external environment or regulate and adapt to the input variables accurately, in terms of their setpoints, schedules, and working modes, among others (Gholamzadehmir et al., 2020 [5]). Thus, the diversity and complexity of variables make it impossible to create accurate and reliable mathematical HVAC models for TCSs.

ACTs effectively obtain superior results in HVAC applications. These can be divided into four categories: (i) soft-computing, (ii) hard-computing, (iii) hybrid, and (iv) adaptive-predictive control strategies.

2.1.1. Soft Computing Strategies

Reinforcement learning (RL), artificial neural network (ANN)-based deep learning, fuzzy logic (FL), and agent-based controls together comprise the soft-computing control strategies. As a control mechanism, this enables solutions to more complex problems by generating more accurate and statistical responses for unclear and uncertain inputs. The key benefit of fuzzy logic controllers is that no mathematical simulation is needed for controller design (Mizumoto, 1995 [6]; Mirinejad et al., 2008 [4]; Soyguder et al., 2009 [7]). The knowledge-based methodology is the fundamental aspect of a fuzzy controller. This consists of if-then rules, membership functions, and scaling factors constructed based on expert experience or learning and self-organization methods that do not involve the system's mathematical model forms.

Since the human sensation of thermal comfort is subjective, and self-reporting can vary among occupants and over time, linguistic rules, on which fuzzy logic is based, are well suited to characterize HVAC systems and thus ideal for increasing thermal comfort

(Chiou and Lan, 2005 [8]; Mirinejad et al., 2008 [4]). There are two different approaches to the automation of rule-based construction in fuzzy systems, which can be used for optimizing the fuzzy system parameters (Mirinejad et al., 2012 [9]): one involves evolutionary techniques and the other soft-computing methods and technologies, such as ANNs.

Soft-computing methods with ANNs can integrate the learning ability of neural networks with the knowledge representation of fuzzy logic. They are frequently used when the aim is to decrease the error between the fuzzy system output and the target value, as characterized by the general term "neurofuzzy system" (Mirinejad et al., 2012 [9]). ANNs can also be applied to optimize the fuzzy database, including membership functions and scaling factors in a fuzzy system (Egilegor et al., 1997 [10]; Kruse et al., 1997 [11]; Wu et al., 2011 [12]). Some studies have utilized advanced fuzzy methods to optimize the function of existing, traditional PID controllers (Malki et al., 1994 [13]; Ying, 1994 [14]; Wu et al., 1996 [15]; Patel and Mohan, 2002 [16]; Li et al., 2005 [17]), while others have used them more directly in the development of new HVAC control mechanisms (Fanger, 1972 [18]; Alcalá et al., 2003 [3]; Liang and Ru, 2008 [19]; Gacto et al., 2011 [20]; Nowak and Urbaniak, 2011 [21]).

Together with model predictive control (MPC) algorithms, fuzzy control algorithms are implemented in the hierarchical framework for HVAC device control (Nowak and Urbaniak, 2011 [21]). Wei et al. (2017) [22] presented a deep reinforced learning (RL) method to develop an HVAC system that they found to be energy-efficient compared with the traditional rule-based approach. Du et al. (2021) [23] presented a model-free deep RL framework for an optimized control approach for a multi-zone residential building. This proposed RL model was reported to provide substantial energy savings and 98% less comfort violation than a rule-based HVAC control strategy

2.1.2. Hard Computing Strategies

Hard-computing control strategies, which include auto-tuning PID control, gain-scheduling control, self-tuning control, supervisory/optimal control, MPC, and robust control, benefit from a mathematical/analytical model that needs real input variables to respond accurately and rapidly. Some important hard-computing control strategy examples are summarized below, with a focus on MPC applications as these are more important here.

Pasgianos et al. (2003) [24] applied a non-linear feedback approach for climate control in greenhouses, especially for ventilation, cooling, and moisturizing. A non-linear multi-input and multi-output model has been used for an air-handling unit (AHU) control (Moradi et al., 2010 [25]). Robust control was applied to control the temperature in a multi-zone HVAC mechanism (Al-Assadi et al., 2004 [26]) and to supply air temperature (Anderson et al., 2008 [27]). Optimal control strategies were used to manage both single zone heating in buildings (Dong, 2010 [28]) and a multi-zone air conditioning system (Mossolly et al., 2009 [29]). An adaptive optimal control approach was also employed to optimize HVAC system control using a genetic algorithm (Yan et al., 2008 [30]).

MPC is an optimization technique that involves the construction of an objective function and an input sequence considering both specified and forced constraints. Serale et al. (2018) [31] aimed to describe the problem formulation, applications, and advantages of an MPC framework for improving building and HVAC energy efficiency. MPC has four functions in buildings, related to weather, user behavior, grid, and thermal mass. Kusiak et al. (2011) [32] created a predictive model with a data-mining approach to optimize HVAC mechanisms using information gathered from an experiment performed at a research facility. Kusiak et al. (2014) [33] presented an HVAC optimization approach with data-driven models and an interior-point method. The Poisson and uniform distributions modeled the uncertainty of occupant behavior, and the internal heating gain was measured with the stochastic mechanism of the building's occupancy. The results showed that the future performance of HVAC was estimated precisely.

Another data-driven approach for optimizing HVAC energy consumption was proposed by Wei et al. (2015) [34]. For this, a quad-objective optimization problem was built to balance energy usage and occupancy comfort and solve a modified particle swarm optimization algorithm. A substantial amount of energy savings was obtained. Biyik et al. (2015) [35] and Kelman et al. (2013) [36] suggested an MPC solution in a standard commercial building for two traditional HVAC setups to maximize energy efficiency and increase occupant comfort by using weather forecasting data. The effect of occupants on internal load prediction and learning from occupant activity is one of the key features of MPC, which can have a major impact on energy efficiency (Serale et al., 2018 [31]).

2.1.3. Hybrid Strategies

Huang et al. (2015a) [37] carried out a study that proposed a hybrid MPC framework. This integrated a classical MPC with a neural network feedback linearization method to reduce the cost and energy of HVAC in commercial buildings. The results indicated that a significant level of energy-saving could be achieved without compromising thermal comfort. Garnier et al. (2015) [38] implemented predictive control for a multi-zone HVAC mechanism in non-residential buildings using EnergyPLus software for the building model and ANN-based models for the controller's internal models. This took the predicted mean vote index as a measure of thermal comfort. Basic scheduling techniques and the proposed HVAC system using a genetic algorithm for optimization were compared, and the importance of the predictive approach demonstrated. Barzin et al. (2016) [39] carried out an experimental study using weather prediction and a price-based control system for passive solar buildings, with up to 90% energy savings achieved.

Alibabaei et al. (2016) [40] explored a Matlab-TRNSYS co-simulator development for control of the TRNSYS software, which was previously designed and balanced based on a real case-study building and used an advanced predictive controller. This study is important here in terms of the co-simulation application. For various other studies, Afram and Sharifi (2013) [41] supplied a detailed literature review including control techniques that focused on the theory, and implementation of MPC approaches for the HVAC mechanism; Afram et al. (2017) [42] presented another comprehensive MPC review focusing on artificial neural network applications with a case study involving ANN models built and calibrated with the on-site data of a residential house. Trčka and Hensen (2010) [43] and Afroz et al. (2017) [44] presented a critical review of the latest simulation and modeling techniques, used in HVAC, focusing on their benefits, limitations, implementations, and efficiency.

2.1.4. Adaptive-Predictive Control Strategies

The APCS (adaptive-predicted control strategy) method can be adapted to a controlled system with time-dependent variables through online variation of its control gains. Huang et al. (2015b) [45] presented an ANN model-based system identification approach to model multi-zone buildings. This showed the thermal interactions between the zones to be well captured by the ANN model, incorporating the energy input from mechanical cooling, ventilation, changes in the weather, and the convective heat transfer between adjacent zones. Thus, more precise outcomes are obtained than a single-zone model. Javed et al. (2017) [46] introduced a random neural network (RNN)-based controller on an Internet of Things (IoT) platform combined with cloud computing to carry out RNN that estimated the number of occupants inside the area and sent information to the central RNN-based occupancy calculator placed in the sensor node.

Cardoso et al. (2018) [47] introduced a study of HVAC power-demand forecasting based on occupant activity. This influences our study in terms of the use of real data from a research building for estimation. Estimation of HVAC demand plays a vital role in developing a more efficient HVAC system. Yang et al. (2019) [48] proposed an adaptive, robust MPC and compared its performance with predictive model controllers. This study showed that adaptive modeling and robust optimization minimize unsuitable indoor conditions because of uncertainties. Zhou et al. (2019) [49] developed a non-linear MPC by MATLAB using production control systems and weather forecasts and reported a substantial decrease in energy consumption. Finally, Gholamzadehmir et al. (2020) [5]

presented a review of the adaptive-predictive control strategy for HVAC systems in smart buildings focusing on advanced control approaches and their effect on buildings according to energy consumption and cost. This study indicated that although adaptive control strategies eliminate the shortcomings of model predictive approaches, such as uncertainty and unpredictable data, a high degree of inconsistency is observed in the literature.

2.2. Occupancy Related Studies

Since the primary focus of our study is the occupancy pattern and prediction, the following paragraphs look at occupancy-related studies. Erickson et al. (2009) [50] indicate that a 14% reduction in HVAC energy usage can be provided with occupancy prediction and usage patterns. They created a wireless camera sensor network for occupancy data and estimated occupancy with an accuracy of 80%. Erickson and Cerpa (2010) [51] proposed a strategy for HVAC systems using real-time occupancy monitoring and estimation of occupancy with a sensor network of cameras, indicating energy savings of up to 20%. Oldewurtel et al. (2013) [52] developed an MPC framework using occupancy information to investigate the effect of occupancy patterns to achieve a more energy-efficient HVAC mechanism. Furthermore, an RFID-based occupancy detection was presented by Li et al. (2012) [53] to decrease the consumption of the HVAC. The study shows how demand-driven HVAC operation is efficient by integrating an occupancy detection system.

A clustering-based iterative evaluation algorithm for eliminating when and how occupants occupy a building was introduced by Yang et al. (2016) [54], who evaluated energy implications at the building level with building information modeling that provided the building geometries, HVAC system configurations, and spatial information as inputs for the computation of possible energy consequences. Capozzoli et al. (2017) [55] applied an occupancy-related HVAC operation schedule that focused on shifting groups of occupants with similar activity in the same thermal zone. As a result of the new schedule approach, HVAC-related energy use decreased by almost 14%.

Another occupant-centric model, the predictive control approach, was developed by Aftab et al. (2017) [56], who created and applied an occupancy-predictive HVAC mechanism using real-time occupancy recognition, predicting user activity, and building thermal simulation. Aftab et al. (2017) [56] focused on a single-zone mosque area whereas the research in this paper focuses on multi-zone commercial buildings and adopts the use of AI for the prediction of occupancy activity. With these advancements, the research in the paper differs from the one by Aftab et al. (2017) [56].

Shi et al. (2017) [57] used a change-point logistic regression model for precise occupancy estimation to create an occupant-centric model predictive algorithm. Their findings indicated that an HVAC control strategy with real-time occupancy estimation provides energy-saving and increases building occupant comfort. Peng et al. (2018) [58] found that 52% energy saving is possible with occupancy prediction-based cooling control using machine learning in office buildings. A demand-responsive method was developed based on energy-related occupant activity. Nikdel et al. (2018) [59] estimated the benefits of occupancy centric HVAC controls in small office buildings based on programmable thermostats; when compared with no thermostat control, their proposed HVAC control approach reduced electricity and natural gas use by up to 50% and 87%.

Ahmadi-Karvigh et al. (2019) [60] presented an automation system that continually learns occupant behavior to help service system control by determining the set of rules according to the user's preferences and behaviors. Adaptive automation gave better results than inquisitive automation in terms of benefits and occupant satisfaction. Pang et al. (2020) [61] determined the energy efficiency potential of the new HVAC system combined with occupancy sensing methods. Their study involved an energy simulation with three different occupancy scenarios, with occupancy presence sensor and occupant counting sensor providing energy savings in office buildings.

Azuatalam et al. (2020) [62] developed a reinforced learning (RL) framework to optimize and control the HVAC for a whole commercial building. Simulations showed that, compared to a handcrafted baseline controller, an energy saving of up to 22% could be reached. Deng and Chen (2020) [63] developed a smart HVAC control mechanism for multi-occupant offices using the physiological signals of occupants. They applied an ANN model to predict indoor conditions and physiological signals, such as clothing level, (wrist) skin temperature, relative skin humidity, and heart rate. The heating and cooling loads in interior offices were reduced by 90% and 30%, respectively, following coupling with the occupancy-based control through lighting sensors and wristband Bluetooth. This study was vital for our research in terms of its development of occupancy-related HVAC and direct measurement of occupant comfort level. Jung and Jazizadeh (2019) [64] presented a structured literature review examining the user-centric operations and human dynamics of HVAC systems. This study focused on occupancy, comfort, and energy savings aspects. Finally, Jazaeri et al. (2019) [65] analyzed the complex relationships among local climates, building characteristics, and occupancy patterns with the annual and peak HVAC demand of residential buildings. These studies are important for us in terms of occupancy, but as mentioned before there is no study that predicts occupancy without real-time detection tools.

2.3. *IDA Indoor Climate and Energy (ICE) Software Background*

The IDA Indoor Climate and Energy (ICE) simulation software is one of the four primary building-energy simulation tools used in research (Ryan and Sanquist, 2012 [66]) and one of the twenty main building-energy simulation software packages (Crawley et al., 2008 [67]). As with many other simulation software packages, this uses building geometry as the foundation for accurate measurements of solar radiation distribution in and between spaces. The program dynamically measures energy balances while considering climatic changes and a changing time-step. Heat balance equations are solved by the program using building geometry, design, HVAC conditions, and internal heat loads. The effectiveness and validness of the IDA-ICE software are proved in several studies over recent years (Bring et al., 1999 [68]; Achermann and Zweifel, 2003 [69]; ISO, 2003 [70]; Karlsson et al., 2007 [71]; Loutzenhiser et al., 2009 [72]; Hilliaho et al., 2015 [73]; Salvalai, 2012 [74]; Mazzeo et al., 2015 [75]; Milić et al., 2018 [76]).

3. Aim of the Research

The advanced prediction ability of AI methods can be employed with sensors to determine occupant behavior, which offers an excellent opportunity to minimize the weakness of the traditional HVAC systems. The aim of the paper is to develop an AI-based, occupant-centric HVAC control mechanism that uses actual weather predictions and continually improves its knowledge to increase energy efficiency in a commercial building. Since the cooling problem has gained importance in recent years, the focus will be on the cooling function of the HVAC systems.

The novelty of the work is twofold. Firstly, a new HVAC control algorithm is proposed, based on forecasted weather and occupancy information to establish a sensor-free mechanism. Secondly, an artificial intelligence-based occupancy forecast system is presented, which considers all parameters (weather information, time indicators, social situations) and provides year-round usage with accurate prediction. Although there are limited examples for real-time occupancy detection in multi-zone buildings, any research study involving occupancy prediction without a camera or sensors has not been performed yet. There is also no other study that constructs the relationship between occupancy prediction, real-time weather, and indoor temperature to manage HVAC control via an algorithm. While a sensor-free algorithm allows both low installation cost and high energy efficiency, AI-based occupancy forecasting provides a system that improves itself as these data increase to use control mechanism standalone and obtain better energy savings.

4. Research Methodology

Paper adopts the design science research (DSR) methodology that facilitates the development of innovative solutions for industry and organizations driven by information (Vaishnavi, et al., 2019 [77]). Its characteristics involve iterative design processes leading to development of innovative solutions in the problem domain (Wieringa, 2014 [78]). The DSR methodology integrates both social context and knowledge base technical capability to achieve the aim of the research (Markus et al., 2002 [79]). Wieringa (2014) [78] described that there are two types of DSR: These include "problem-oriented research—evaluation research", and "solution-oriented—technical research". The problem-oriented research looks at what causes/effects a problem has, or how to solve a problem, whereas the solution-oriented research design and validate a system, or a requirement (Peffers et al., 2006) [80].

With DSR, this paper promotes the adoption of AI-based and occupant-centric HVAC control systems in commercial buildings to address the research problem around inefficient energy management of the existing HVAC systems. The DSR features with social context that are relevant to the paper are given in Table 1 and the overall research methodology is illustrated in Figure 1.

Table 1. Design science research (DSR) features.

Design Science Research (DSR) Features	
AI-based occupant-centric HVAC control system in commercial buildings	Design science research
Remove inefficiencies in energy management via HVAC systems Enhance the prediction ability with AI for the determination of occupant behavior for effective energy management in commercial buildings	Design an artifact: Development of AI algorithm for occupant-centric HVAC control system Research instruments-tools: Iterative design and development process for knowledge capture and development
Development of an innovative artifact to achieve and enhance the social context.	Answering the knowledge questions: How does the artifact adopt smart heritage project principles?

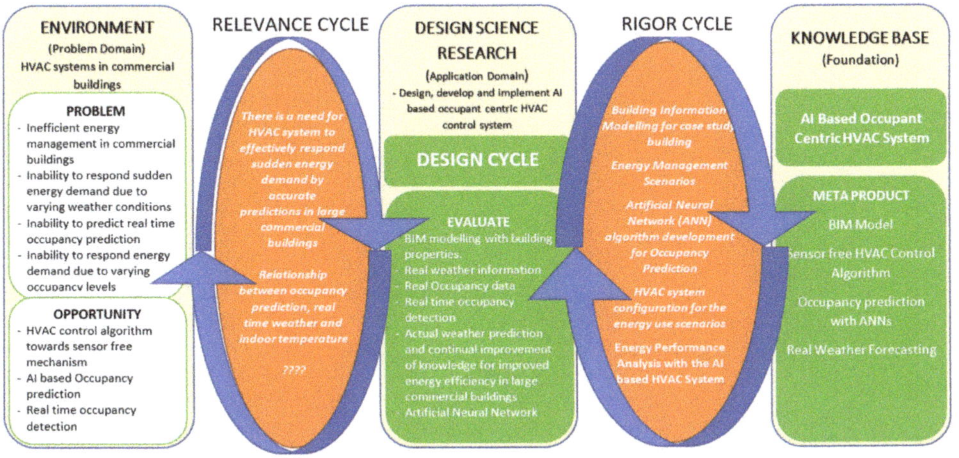

Figure 1. The flowchart of the design science research (DSR) methodology.

The design science research (DSR) methodology enabled developing the HVAC control system for accurate prediction of energy supply in commercial buildings that will serve to meet human needs for energy demand. The system developed via DSR is innovative with the key being the embedment of artificial intelligence that processes the occupancy and these weather data without the use of sensors. This paper with DSR implementation brings not only the novelty but also provides important solution to the practice for energy management in commercial buildings.

In the information systems (IS) science, the DSR methodology is highly preferred to solve identified organizational problems by developing information technologies. The paper is designed in accordance with the DSR methodology. The research problem domain and opportunities are elaborated by means of literature review through the relevance cycle. The design cycle is evaluated in Section 5 regarding the development of the artifact (AI-based occupant-centric HVAC control system) that is extended with the test and demonstration of the proposed artifact in Section 6 through the rigor cycle. This then leads to the accumulation of findings into the new knowledge base, articulated in Sections 6 and 7.

5. Design and Development of the AI-Based Occupant-Centric HVAC Control System

The artifact, which includes novelty about the research problem, is created. This artifact in this paper is the AI-based occupant-centric HVAC control system. Since the purpose of the study is to reveal energy efficiency potential of the proposed HVAC control mechanism, energy analysis according to different scenarios constitute the central part of this section. The research focuses on a specific site to obtain realistic results, using two-year occupancy and environmental conditions data of a shopping mall in Istanbul. Figure 2 shows the architecture of the system, consisting of three steps: predicting hourly occupancy, a new HVAC control mechanism, and comparison of the traditional and AI-based control systems via simulation according to different scenarios.

In the first step, building properties and real occupancy information are collected. In the second step, after determining the attributions for occupancy in the mall, hourly occupancy predictions are made using real data and ANNs, and a sensor-free HVAC control algorithm is developed with the help of occupancy data obtained from the previous stage, building characteristics, and real-time weather forecast information.

ANN is considered one of the traditional and most used artificial intelligence methods, and is still one of the most accurate and effective. This enables traditional and AI-based sensor-free HVAC control mechanism comparison to be performed in the final step, using IDA-ICE 4.8 software developed by EQUA Simulation AB based in Stockholm, Sweden.

5.1. Building Properties, Occupancy, and Environmental Information

According to the Association of Real Estate and Real Investment Companies of Turkey (2019) [81], there are currently 454 shopping malls in Turkey; across Europe, there are more than 9500 malls, with over 1000 in France and more than 1500 in the UK (STATISTA, 2021 [82]). Worldwide, there is a huge number of shopping malls, which makes them a significant target for energy savings and important in the development of sustainable energy policy. These buildings tend not to have good energy efficiency strategies because they are mostly constructed for consumption and entertainment purposes. It is commonplace for them to use varied and excessive lighting to attract people and make them feel good inside the building.

Poor heating and cooling settings disrupt the comfort area for people as well as causing energy inefficiencies. As a complicating factor and unlike office buildings, shopping malls do not have certain daily occupancy distributions, thus accurately gauging correct heating and cooling settings is not easy. It is for these reasons that this study takes a shopping mall as its case study. For more accurate energy analysis, a realistic model of the building is used that incorporates the real properties of the building elements. Furthermore, the solar radiation and weather data for the building location are obtained automatically from IDA-ICE software for energy simulations. Figure 3 shows the model of the building story; Table 2 shows the properties of the building elements.

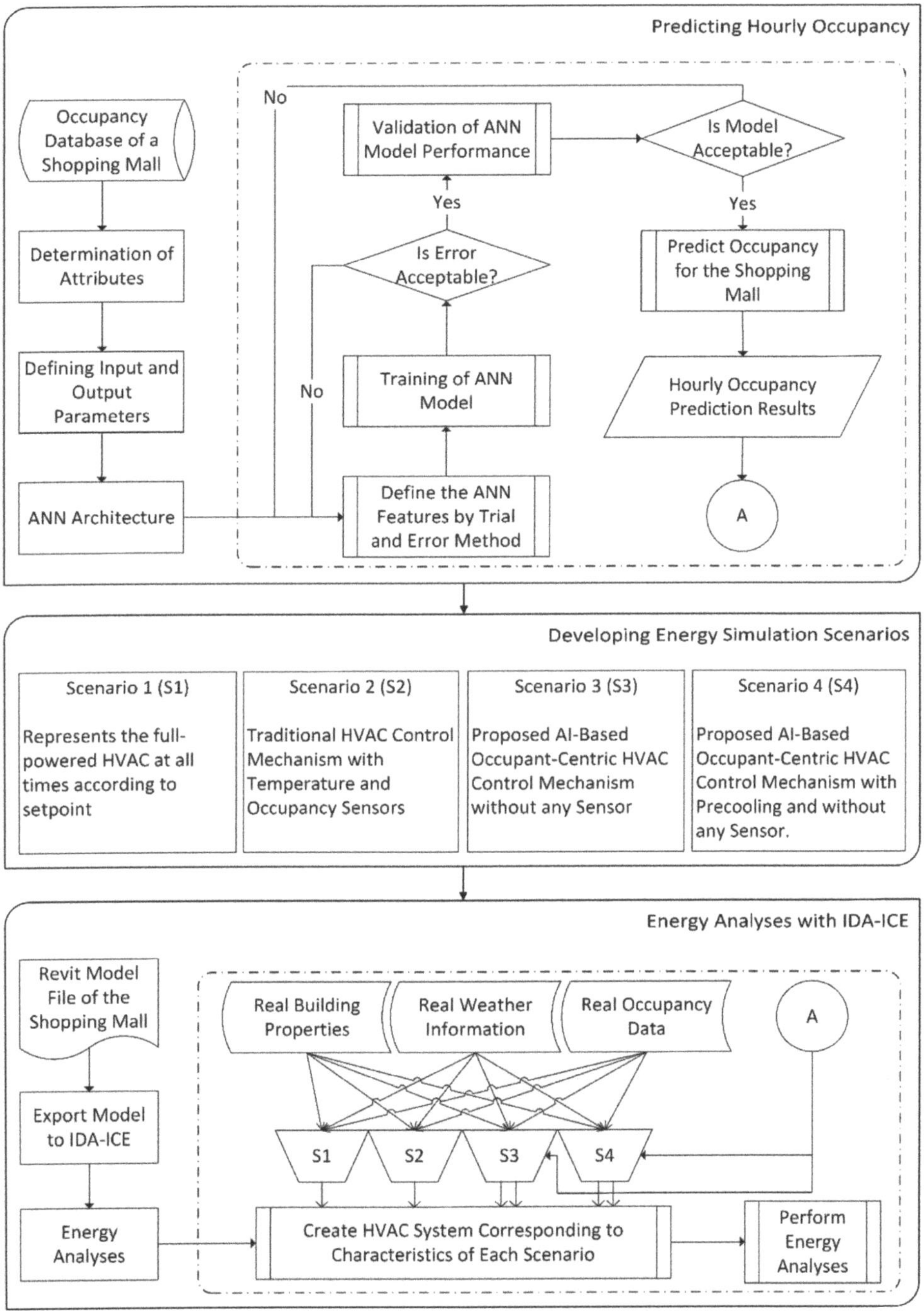

Figure 2. AI-based occupant-centric HVAC control system design.

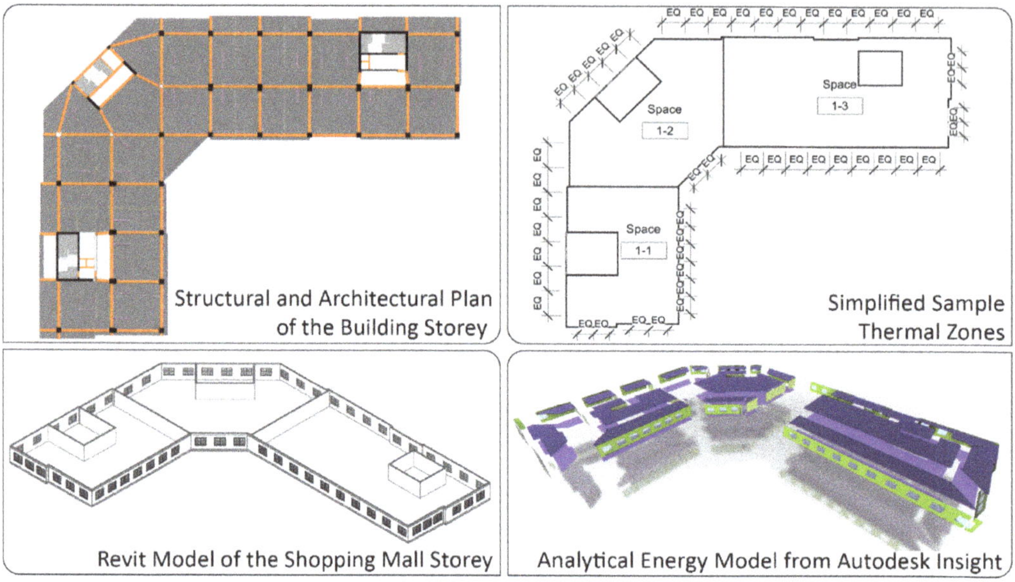

Figure 3. Sample 2D drawings and 3D models of the building story.

Table 2. Building components.

Construction	Material Layers (from Outside to Inside)
External wall	Press brick and supporters—0.088 m | Gypsum board—0.02 m | Light steel wall and XPS Insulation—0.15 m | Gypsum board—0.02 m | Plywood wall panel and supporters—0.07 m
Ground floor	Reinforced concrete—1.5 m | Concrete—0.05 m | XPS Thermal Insulation—0.06 m Concrete—0.03 m | Screed—0.005 m | Floor Covering—0.008 m
Roof	Standing seam roof sheet metal | OSB sheet—0.015 m | Corrugated steel sheet | Steel roof supporters and XPS insulation—0.12 m
Window	Low-e glass double—4 mm + 12 mm argon + 4 mm

5.2. Occupancy Prediction with ANNs

5.2.1. ANN Parameters

Many factors affect the occupancy numbers and distribution of a shopping mall. They can be divided into two categories: social and environmental. When the collected real entrance data are examined, temperature, humidity, and weather conditions along with type and time of a day come to the forefront as significant parameters. Determined as attributes in the ANN calculation, these parameters are thus:

1. Temperature: This is one of the most critical factors affecting the number of people; there are fewer visitors to the shopping mall in winter than in summer days;
2. Humidity: This affects the temperature feel; when the humidity in the air is high, warm moisture stays on people's skin longer and makes them feel hotter;
3. Weather condition: This also affects the occupant number significantly; on rainy or snowy days, shopping malls attract fewer visitors;
4. Time indicators: Days are also significant for shopping mall occupancy; on non-working days, the number of visitors is higher than on working days. In our study, the days are not separated into working and non-working days, as in some studies, but each day of the week is included in the calculation separately; furthermore, month and year information are considered as separate parameters since they are essential variables in the long-term use of the shopping mall;

5. Special days: Public (state) and religious holidays significantly affect occupancy; the number of visitors increases on national holidays and decreases considerably on religious holidays; furthermore, the first day of religious holidays is separately considered because there are far fewer visitors on these days than others;
6. Time of day: This is the most critical factor for sudden changes in the number of people visiting; for example, the occupancy number increases rapidly at the start of the lunch break and decreases rapidly when it finishes.

Table 3 illustrates the detailed categories, variables, and unit/index of attributes used in the ANN model; Table 4 shows a sample of the actual data. Furthermore, the histograms of the temperature, humidity, weather conditions, and occupancy variables are presented in Figure 4 to show the distribution of the collected data.

Table 3. Detailed category, variable, and unit/index information of attributes.

Category	Variables	Unit/Index
Environmental	Temperature	°F
	Humidity	%
	Weather Conditions	1: Fair ∣ 2: Partly cloudy ∣ 3: Mostly cloudy ∣ 4: Light rain 5: Rain ∣ 6: Heavy rain ∣ 7: Fog ∣ 8: Snow ∣ 9: Thunder
Social and Time Indicators	Weekday	1–7 (1: Monday ⋯ 7: Sunday)
	Month	1–12 (1: January ⋯ 12: December)
	Day	1–31
	Year	2017–2018–2019
	Time	10–21 (10: 10:00 a.m. ⋯ 21: 09:00 p.m.)
	Day Type	0: Normal day ∣ 1: Public holiday 2: First day of religious holidays 3: Other days of religious holidays

Table 4. Real data for ANN (sample).

Day of Week	Month	Day	Year	Time	Day Type	Temp. °F	Hum. %	Weather Conditions	Occupancy (Number of People)
Sunday	8	18	2019	10^{00}–11^{00}	Normal	70	83	Partly cloudy	661
Sunday	8	18	2019	11^{00}–12^{00}	Normal	68	88	Light rain	1346
Sunday	8	18	2019	12^{00}–13^{00}	Normal	73	78	Partly cloudy	1448
Sunday	8	18	2019	13^{00}–14^{00}	Normal	72	78	Mostly cloudy	2547
Sunday	8	18	2019	14^{00}–15^{00}	Normal	77	50	Partly cloudy	2921
Sunday	8	18	2019	15^{00}–16^{00}	Normal	79	47	Partly cloudy	3353
Sunday	8	18	2019	16^{00}–17^{00}	Normal	79	47	Partly cloudy	3181
Sunday	8	18	2019	17^{00}–18^{00}	Normal	77	47	Partly cloudy	2455
Sunday	8	18	2019	18^{00}–19^{00}	Normal	77	50	Partly cloudy	2339
Sunday	8	18	2019	19^{00}–20^{00}	Normal	75	50	Partly cloudy	2126
Sunday	8	18	2019	20^{00}–21^{00}	Normal	72	60	Partly cloudy	1644
Sunday	8	18	2019	21^{00}–22^{00}	Normal	70	68	Fair	777
Monday	8	19	2019	10^{00}–11^{00}	Normal	77	54	Mostly cloudy	463
Monday	8	19	2019	11^{00}–12^{00}	Normal	75	57	Mostly cloudy	906
Monday	8	19	2019	12^{00}–13^{00}	Normal	77	54	Mostly cloudy	1418
Monday	8	19	2019	13^{00}–14^{00}	Normal	81	48	Partly cloudy	1690
Monday	8	19	2019	14^{00}–15^{00}	Normal	81	45	Partly cloudy	1643
Monday	8	19	2019	15^{00}–16^{00}	Normal	79	51	Partly cloudy	1379
Monday	8	19	2019	16^{00}–17^{00}	Normal	77	50	Partly cloudy	1547
Monday	8	19	2019	17^{00}–18^{00}	Normal	79	51	Partly cloudy	1494
Monday	8	19	2019	18^{00}–19^{00}	Normal	77	54	Partly cloudy	1907
Monday	8	19	2019	19^{00}–20^{00}	Normal	75	61	Partly cloudy	1806
Monday	8	19	2019	20^{00}–21^{00}	Normal	72	73	Fair	1496
Monday	8	19	2019	21^{00}–22^{00}	Normal	70	78	Fair	727

Table 4. *Cont.*

Day of Week	Month	Day	Year	Time	Day Type	Temp. °F	Hum. %	Weather Conditions	Occupancy (Number of People)
⋮	⋮	⋮	⋮	⋮	⋮	⋮	⋮	⋮	⋮
Friday	8	30	2019	10^{00}–11^{00}	Pub. Hol.	77	69	Partly cloudy	1269
Friday	8	30	2019	11^{00}–12^{00}	Pub. Hol.	81	54	Partly cloudy	1406
Friday	8	30	2019	12^{00}–13^{00}	Pub. Hol.	81	48	Partly cloudy	1738
Friday	8	30	2019	13^{00}–14^{00}	Pub. Hol.	82	48	Partly cloudy	2562
Friday	8	30	2019	14^{00}–15^{00}	Pub. Hol.	82	48	Partly cloudy	2601
Friday	8	30	2019	15^{00}–16^{00}	Pub. Hol.	81	54	Partly cloudy	2990
Friday	8	30	2019	16^{00}–17^{00}	Pub. Hol.	81	51	Partly cloudy	2518
Friday	8	30	2019	17^{00}–18^{00}	Pub. Hol.	79	54	Partly cloudy	2428
Friday	8	30	2019	18^{00}–19^{00}	Pub. Hol.	77	54	Partly cloudy	2701
Friday	8	30	2019	19^{00}–20^{00}	Pub. Hol.	75	65	Partly cloudy	2262
Friday	8	30	2019	20^{00}–21^{00}	Pub. Hol.	73	65	Fair	1805
Friday	8	30	2019	21^{00}–22^{00}	Pub. Hol.	72	69	Fair	818

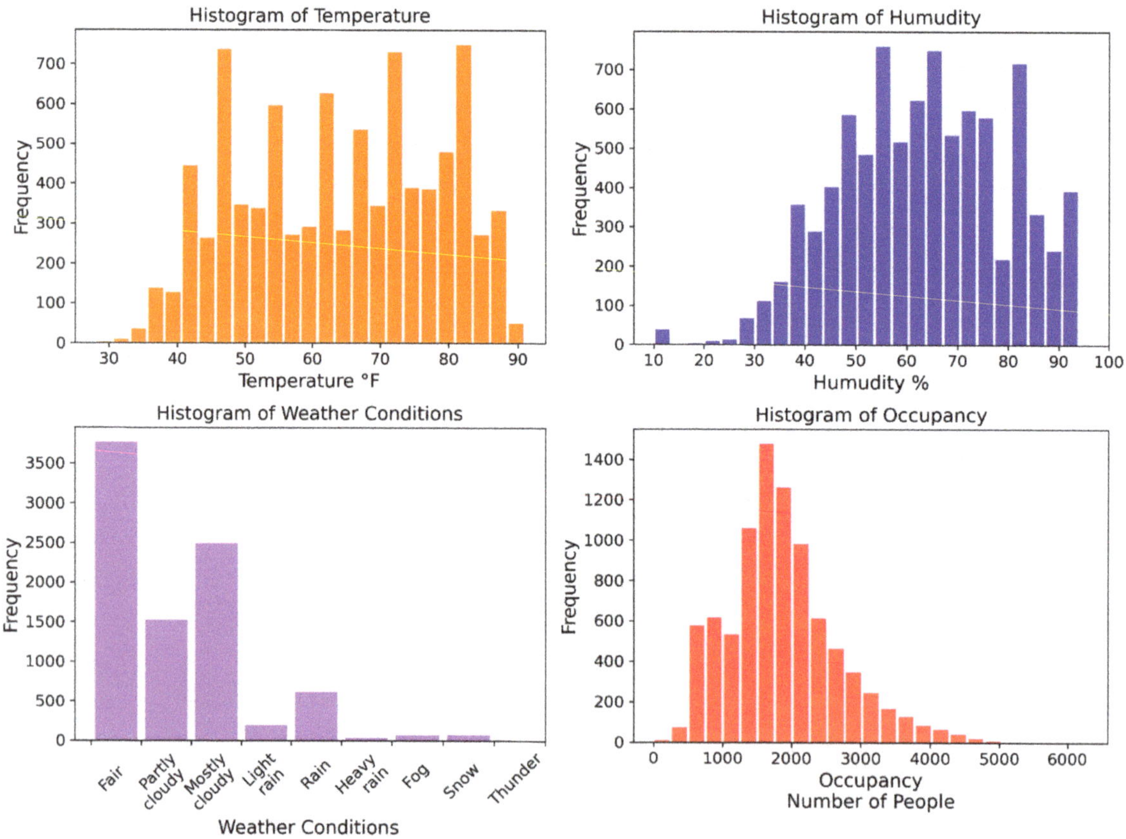

Figure 4. Histograms of the temperature, humidity, weather conditions, and occupancy variables.

5.2.2. ANN Models

Due to their strong logic, error tolerance, versatility, and generalization capabilities, AI methods are used in various applications. The ANN, a mathematical model that imitates the biological nervous system, is one of the most widely used types of AI and has been implemented to solve a variety of practical challenges in many fields of study.

The fundamental biological unit of the nervous system is the neuron, a fundamental processing factor that receives and integrates signals from other neurons through dendrite input paths. The neuron generates an output signal along the axon that links to the dendrites of several other neurons if the combined input signal is sufficiently high. An attempt to model the behavior of biological neural systems was made that led to the development of ANNs, in which artificial neurons model the components of a real neuron. An ANN is thus a set of independently linked processing units that function as parallel-distributed computing networks.

Unlike traditional computers, which are programmed to perform particular tasks, ANNs may learn from examples and eliminate the need for complicated mathematical formulas or costly physical models by acting as (human) brain-like mathematical models. They are fault-tolerant and can work with noisy data, allowing for quick generalization of unknown inputs (Wijayasekara et al., 2011 [83]). They also have specific adaptation abilities that enable them to solve highly non-linear problems in which finding analytical formulations that relate the input data to the output data is especially challenging (Hagan et al., 2014 [84]). Unlike other statistical or parametric approaches, ANNs can extract non-explicit relationships from a massive volume of correlated data using the high computational capabilities of current computers; thus, ANNs have become a prevalent problem-solving strategy in a diverse range of study areas.

The architecture of ANN models is formed by layers with complete or random connections between them. There is a connection between each neuron, and information exchange is performed. The network receives data from the input layer. The nodes in this layer do not have any weights or activation functions, thus it is not a neural computing layer. The hidden layer or intermediate layer includes data processing and computing steps and the final response to a given input, which is called the output layer (see Figure 5). The ANN model is developed using TensorFlow's Keras API and the Adam algorithm is used to train the model. Five-fold cross-validation was applied for splitting the data into two subsets, namely, training and testing. Ninety percent of cases were used for training in each trivial, and the remaining were utilized to test the model accuracy. All equations are adopted from the book *Artificial Neural Networks* by Springer US (2021) [85].

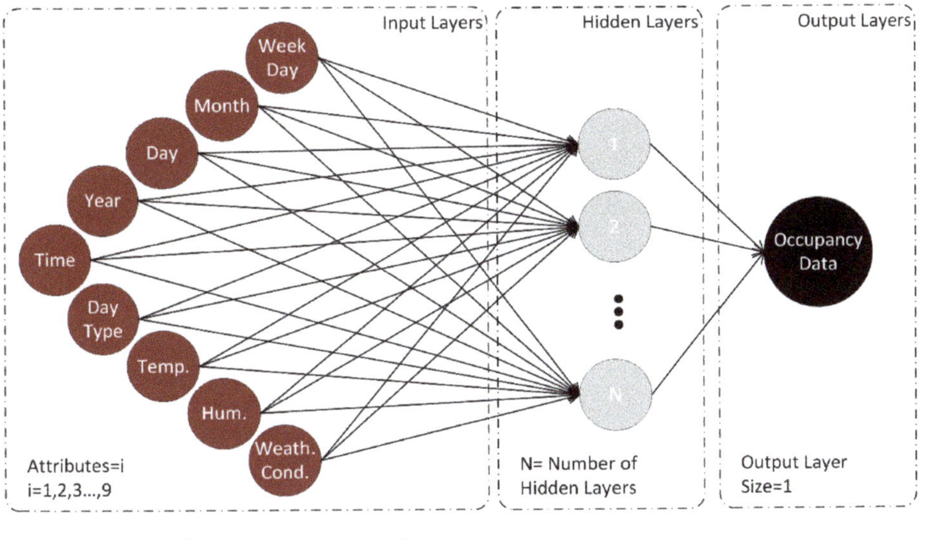

Figure 5. ANN structure for the study.

Generally, the net input of a neuron—activation potential A_i—is equivalent to the product $w_{ij}x_j$, where w_{ij} is the weight of the corresponding connection on the i-th post-synaptic neuron and x_j is the input signal (Equation (1)). Connection weights can be considered as storage of the knowledge that underlies the processing. Thus,

$$A_i = \sum_j w_{ij}x_j - a_i \tag{1}$$

where a_i is the threshold activation constant of the neuron. An output can only be obtained by propagating through a specific activation function. After the signal has been thus propagated, an output can be found thus:

$$y_i = \varphi(A_i) = \varphi\left(\sum_j w_{ij}x_j - a_i\right) \tag{2}$$

where y_i is the output of a layer and $\varphi(\bullet)$ is the transfer function.

The sigmoid activation function has been a common activation function for neural networks for a long time. Its input is converted to a value of between 0.0 and 1.0, with inputs that are significantly greater than 1.0 being converted to 1.0, and inputs that are significantly smaller than 0.0 snapped to 0.0. However, due to the vanishing gradient problem, usage of the sigmoid and hyperbolic tangent activation functions in networks with many layers is not true. This problem can be overcome by using the rectified linear activation function, allowing ANN structures to learn faster and increase performance. The formula of the rectifier or rectified linear unit (ReLU) is as follows:

$$f(x) = x^+ = max(x) \tag{3}$$

where x is the input to a neuron. This is also known as a "ramp function" and is analogous to half-wave rectification in electrical engineering. Connection weights are modified by the ANN model using a suitable learning method during the training phase. The network uses a learning mode to obtain the desired output by adjusting the weights. This is executed by introducing input and desired output to the network. The difference between the expected output and the network's output is then used to determine the error value. In the training phase, recalculations are carried out to decrease the error to an acceptable value. Due to zero occupancy on some days, mean absolute error (MAE) is used to calculate the error value, thus:

$$MAE = |y_i - \hat{y}_i| \tag{4}$$

where $\hat{y}_i$ is the corresponding desired output value. An error of close to zero shows that the ANN output values match the expected values very well and the network is well-trained. Backpropagation training is accomplished by assigning random weights to all nodes. Equation (5) is used to measure the variation quantity of the connection weights:

$$\Delta w_{ij}(t) = \lambda\delta_i - y_i + \alpha\Delta w_{ij}(t-1) \tag{5}$$

where the training rate is λ, the momentum coefficient is α, and the error of the i-th output layer is δ_i, which is calculated thus:

$$\delta_i = y_i(1 - y_i)MAE_i \tag{6}$$

MAE and mean absolute percentage error ($MAPE$) are also calculated as indices to evaluate the performance of the ANN model, thus:

$$MAPE = \frac{1}{N}\sum_{i=1}^{N}\left|\frac{y_i - \hat{y}_i}{y_i}\right| = \frac{1}{N}\sum_{i=1}^{N}|Relative\ Error_i| \tag{7}$$

where N is the total number of data sequences.

5.2.3. HVAC Control Scenarios for Energy Simulation

The primary goal of establishing HVAC control scenarios in terms of the level of development here is to measure the amount of energy to be saved with the proposed AI-based control approach. Although great progress has been made in air conditioning systems, a large proportion of commercial buildings have the most traditional type of control system, which is one that is operated manually by an attendant (janitor or similar) responsible for turning the system on and off. The most common HVAC control is based on the measurements of environmental conditions via sensors, generally temperature, humidity, and pressure sensors. The most serious deficiency of sensors in terms of energy consumption is the failure to facilitate a quickly responsive control system.

Many shopping malls serve as a lunch-places for people working near the building, which causes short-term occupancy densities during the lunch-break period. The rise in temperature due to sudden increase of people density is a slower process and, by the time this reaches the sensor, the control system responds, and the appropriate ambient temperature is provided, most people will already have left the building to return to work. Moreover, traditional building automation systems depend on quite imperfect occupancy sensors, which retards system responsiveness. Passive infrared and ultrasonic occupancy sensors, for example, are low-functioning devices for this usage since they are unable to accurately assess the occupancy condition, especially when people are stationary for an extended period and have a limited range, which especially affects their effectiveness in large areas.

AI prediction technology offers significantly more accurate occupancy information and improved energy efficiency than traditional building automation systems. Accordingly, our HVAC control mechanism takes predicted occupancy information and the maximum number of people per day and adjusts its power according to occupancy rate over time. Furthermore, new schedule algorithms are developed based on occupancy information, and weather forecasts for the scenarios (S3 and S4) explained below. The on-off status of the HVAC is determined according to these setpoint schedule algorithms. The maximum setpoint value is determined as 24 °C for all scenarios since we focus on the summer period in this study. Finally, four different scenarios—showing a level of development (from traditional to advanced)—are determined, as follows:

1. S1: The S1 scenario represents the full-powered HVAC at all times.
2. S2: The S2 scenario represents the most common traditional HVAC control mechanism based on temperature and occupancy sensors, where the HVAC control system is automatically (de)activated according to the temperature setpoints and temperature measurements from the sensors. In this scenario, occupancy is measured by CO_2 sensors, which record the level of CO_2 in the air. If the number of people in a space exceeds the amount of CO_2 allowed, the sensor triggers the HVAC mechanism to turn on. This type of sensor is more accurate than a standard motion sensor for the measurement of occupancy.
3. S3: The S3 scenario represents the proposed AI-based HVAC control system, which uses predicted occupancy numbers as produced by the ANN model. In this scenario, the HVAC system responds automatically to changes in the occupancy with no lag time, contrary to sensor-based systems. The control algorithm provides an HVAC setpoint schedule to control the system according to real weather conditions (as supplied by weather prediction services) and predicted occupant numbers. The existing sensors can still be used to monitor the real-time indoor temperature, humidity, and amount of CO_2. If the actual thermal comfort parameters exceed the desired values, the control system adjusts itself according to the sensors until thermal comfort is provided. In the energy simulations performed by IDA-ICE, the effect of the real-time sensors is not used to examine the no-sensor control mechanism. For this reason, in the illustrations and graphs for S3 and S4, dashed lines are used to show this potential.

4. S4: The S4 scenario represents the HVAC control system in the S3 scenario with a pre-cooling ability along with a quick response. The control algorithm provides pre-cooling time to control the system according to predicted weather conditions and occupant numbers. All other features are the same as for S3.

Figure 6 provides basic illustrations for the four energy simulation scenarios.

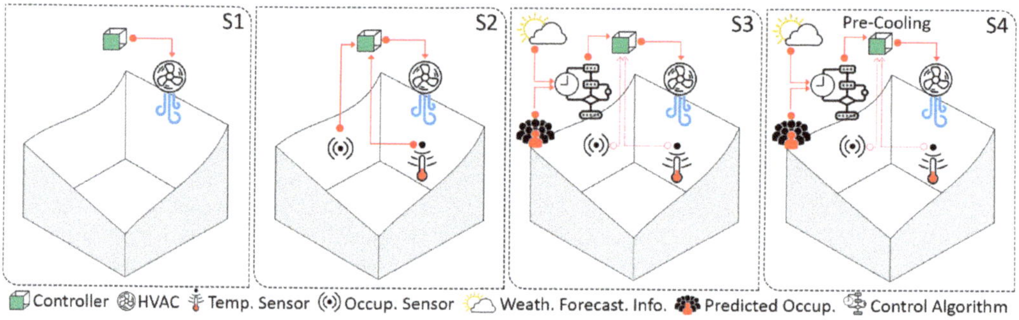

Figure 6. HVAC control scenarios for energy simulations.

Algorithm 1 and Algorithm 2, shown in below, explain the proposed HVAC control schedule algorithms in terms of cooling for S3 and S4. The control algorithm takes occupancy prediction results from the ANN analysis and real weather forecast information from provider websites as inputs (see lines 1–3). In the time intervals when the occupancy volume increases, the HVAC control activates according to the maximum setpoint (lines 4–6); otherwise, the algorithm checks the forecasted temperature and compares it with the maximum setpoint value.

Algorithm 1 HVAC Schedule Algorithm of S3 for Cooling

1 **train** ANN model
2 **make** day-ahead prediction for occupancy
3 **take** day-ahead local weather forecast information
4 **if** $occupancy_t < occupancy_{t+1}$
5   $setpoint_{max}$ ® T_{target}
6   **set** HVAC setpoint to T_{target}
7   **end**
8 **else**
9   **if** weather forecast temp.$_t$ > $setpoint_{max}$
10     $setpoint_{max}$ ® T_{target}
11   **set** HVAC setpoint to T_{target}
12   **else**
13   **deactivate** the cooling   • deactivation
14   **end if**
15 **end if**

If the weather forecast temperature for time t is greater than the maximum, the HVAC control uses the maximum setpoint (lines 8–10); if not, while the HVAC deactivates cooling automatically for S3 (lines 12 in Algorithm 1), the algorithm checks the occupancy trend of one hour later for S4. If there is a sudden increase (determined at 250 visitors), it activates the pre-cooling 30 min before the upward trend begins for S4 (line 15 in Algorithm 2).

Due to fluctuations in the occupancy numbers, sudden changes can cause comfort limit values to be exceeded, especially in situations where the number of visitors will increase too much one or two hours later, even if the occupancy trend is downward for the current time. To prevent this, S4 presents a 30-min pre-cooling. If there is no such increase, the HVAC control deactivates cooling (see line 17 in Algorithm 2), just as for the S3 algorithm.

Algorithm 2 HVAC Schedule Algorithm of S4 for Cooling

1 **train** ANN model
2 **make** day-ahead prediction for occupancy
3 **take** day-ahead local weather forecast information
4 **if** $occupancy_t < occupancy_{t+1}$
5 $setpoint_{max}$ ®T_{target}
6 **set** HVAC setpoint to T_{target}
7 **end**
8 **else**
9 **if** weather forecast $temp._t > setpoint_{max}$
10 $setpoint_{max}$ ®T_{target}
11 **set** HVAC setpoint to T_{target}
12 **else**
13 **if** $occupancy_{t+2}$ - $occupanc_{t+1} > 250$
14 **deactivate** the cooling for first t/2 • deactivation
15 $setpoint_{max}$ ®T_{target}
16 **set** HVAC setpoint to T_{target} for last t/2 • start pre-cooling
17 **else**
18 **deactivate** the cooling • deactivation
19 **end if**
20 **end if**
21 **end if**

6. Demonstration and Evaluation of the AI-Based Occupant-Centric HVAC Control System

In this stage, the designed and developed system is tested in relation to the scenarios for energy analysis. According to design science research, research can exploit experimentation, simulation, case study, proof, or other activities to demonstrate the proposed solution to the research problem. Hence, experimentation of the scenarios via simulation using IDA-ICE 4.8 software is performed for the computational energy analysis. The model of the shopping mall was created using Revit software and imported to IDA-ICE in IFC format. Four different energy analyses are carried out according to the four scenarios. For each scenario, the HVAC control system corresponding to the characteristics of the scenario is created in the simulation software using macros. Figure 7 shows the MPC algorithm framework. Simulations are carried out daily, with the results for two days explained in detail in the results section.

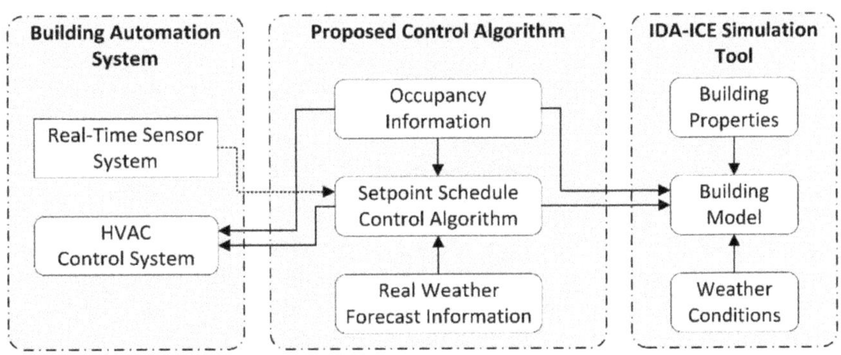

Figure 7. Model predictive control (MPC) algorithm framework.

The weather data of the software are used in the energy analysis. It is seen that the day-ahead predictions give almost the same values as real values. Therefore, the forecasted weather data for the algorithm are not used (to avoid repeating the results and graphics). In scenarios S4 and S5, both the estimated occupancy numbers found as a result of ANN calculation and the real occupancy numbers are used in different simulations to show how the small difference between the real data and ANN prediction affects the energy simulation.

The research artifact, the AI-based occupant-centric HVAC control system, is assessed, and evaluated to conceive how well the developed and demonstrated artifact is considered as a solution to the research problem. At this stage, research can benefit from surveys, feedback, and simulations. If the solution rate, which corresponds to the research problem, or functionality of the solution is not at an acceptable level, the iterative process is performed by turning stage 2 and 3.

6.1. ANN Results

In addition to the initial network settings (attributes, layer design, training algorithm, etc.), ANN parameters (hidden layer size, number of neurons in the hidden layer, batch size, which refers to the number of training examples utilized in one iteration, number of epochs, the number of complete passes through the training dataset, etc.) have a highly significant influence on the network output during the training and prediction phases. While a model with too few neurons has poor predictive performance because it cannot handle a complex model structure, if too many are selected, weak prediction performance follows as overfit too easily results from a minor fluctuation in the data.

Therefore, it is crucial to test the model's output with different design parameters. Different ANN models were trained for this study; as a result of the trials grid search methodology using the number of neurons in each layer and the number of epochs as variables by keeping the number of hidden layers constant at 8. Figure 8 shows the MAPE and R-squared results of the ANN models created using a grid search. Since the computational times are not too long and do not change much between them, they are not considered as parameters. The ANN models with 8 neurons in each hidden layer and with 500 epochs (ANN-1), with 8 neurons in each hidden layer and with 750 epochs (ANN-2), with 16 neurons in each hidden layer and with 250 epochs (ANN-3), and with 16 neurons in each hidden layer and with 500 epochs (ANN-4) give the best results with overall MAPE values of 0.1323, 0.1344, 0.1315, and 0.1335, respectively.

Figure 9 shows the learning curves of the different ANN models. It is clear from the loss curves that training and validation loss values for ANN-1 and ANN-3 (Figures 9a and 10b, respectively) are in the ideal range for model complexity. However, the distance between the training loss line and validation loss line gradually increases after a certain point because of overfitting in the ANN model with 64 neurons in each hidden layer and with 1000 epochs (Figure 9d).

The comparison between actual occupancy numbers and predicted occupancy values as given by ANN-1 and ANN-2 is illustrated in Figure 10. Although the prediction values are naturally far from the real values at some peak points, the prediction trend follows the real numbers in a general fashion. Thus, as the quantity of data increases in the future, more accurate results will be obtained.

Furthermore, to examine the results in more detail, four days (18, 19, 20, and 30 August 2019) were removed from the training data set and used as prediction values. Prediction values for these days were obtained using the ANN-1 features because of the model accuracy and processing time; a comparison with actual occupancy numbers is shown as a list (Table 5) and graphically (Figure 11).

Additionally, 18 August was a Sunday, while 30 August is a national holiday. These days are important in examining the ANN algorithm for weekdays, weekends, and special days. More people are expected to visit the shopping center on weekends and national holidays than on weekdays.

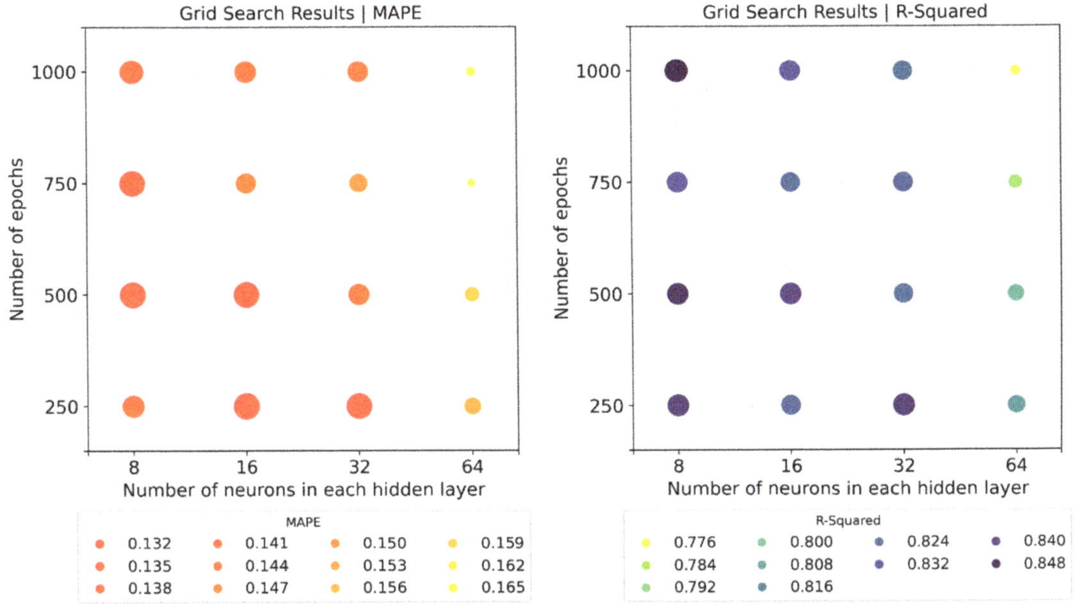

Figure 8. Grid search results of the ANN models according to MAPE and R-squared values.

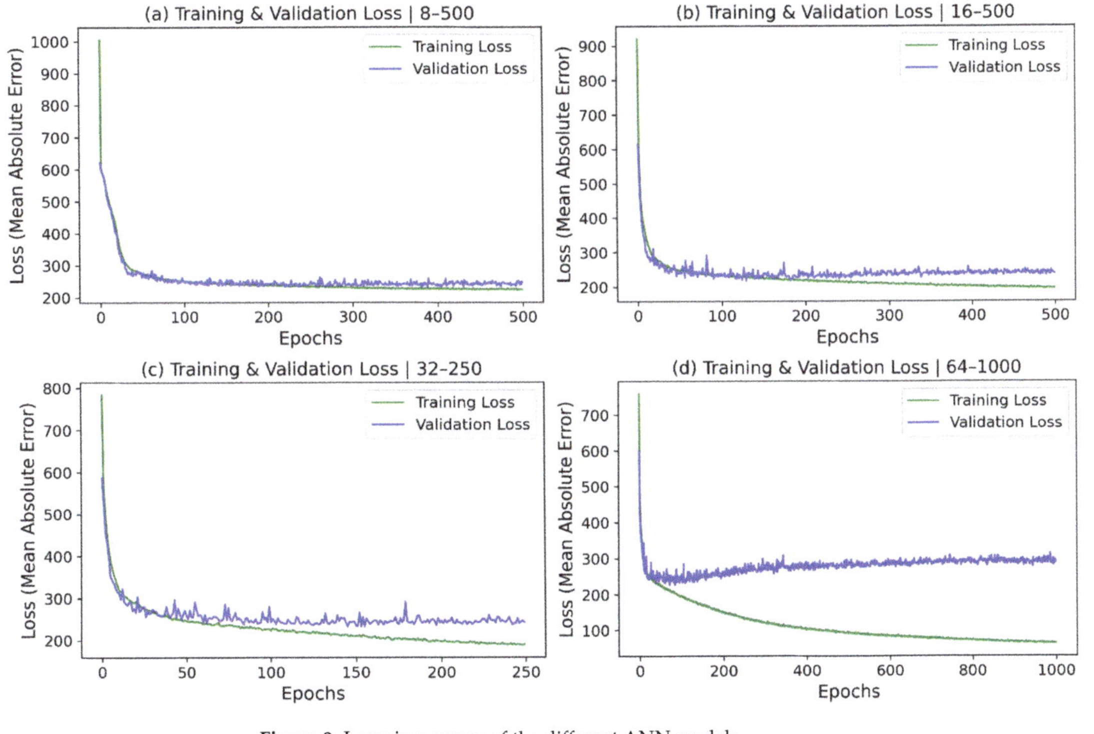

Figure 9. Learning curves of the different ANN models.

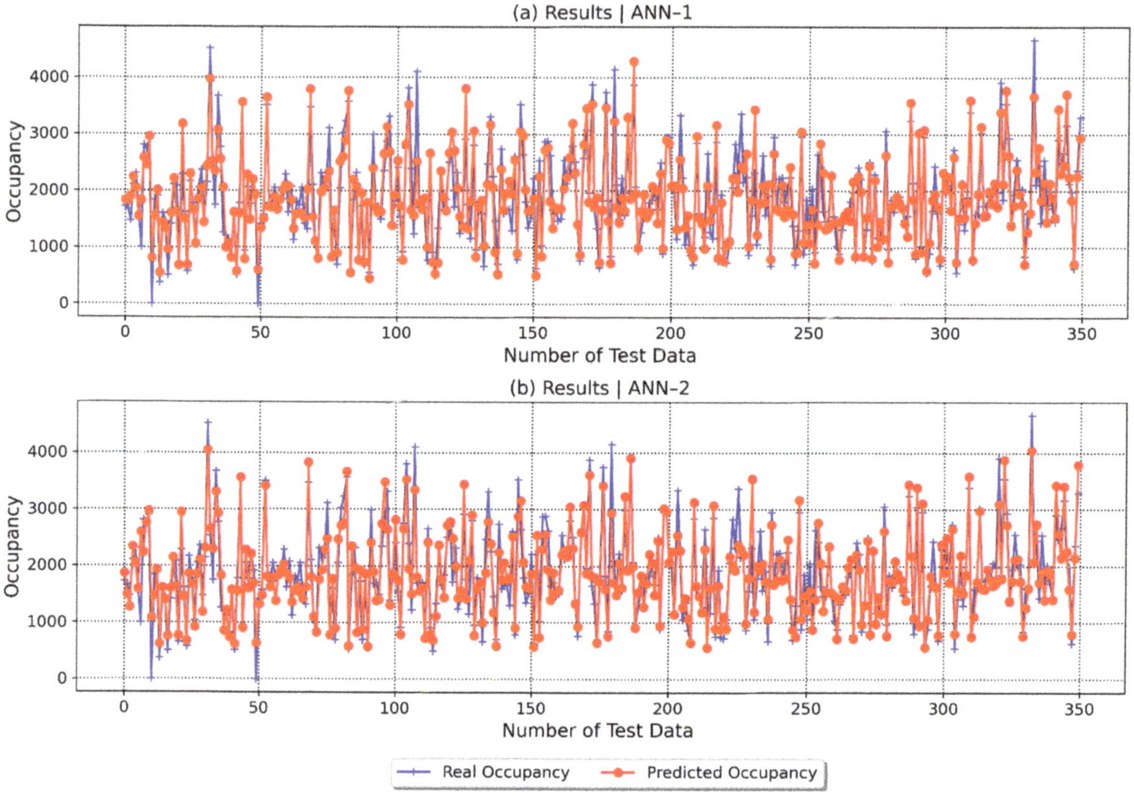

Figure 10. Actual and ANN-predicted occupancy results (ANN-1 and ANN-2).

Table 5. Actual and ANN-predicted occupancy results (hourly).

Time	Sunday, 18 August 2019		Monday, 19 August 2019		Thursday, 29 August 2019		Friday, 30 August 2019	
	Real	Pred.	Real	Pred.	Real	Pred.	Real	Pred.
10:00 a.m.	661	963	463	881	642	697	1269	769
11:00 a.m.	1346	1480	906	1322	1240	1345	1406	1602
12:00 p.m.	1448	1594	1418	2139	2194	2290	1738	1980
01:00 p.m.	2547	2412	1690	1751	1981	1657	2562	2445
02:00 p.m.	2921	3373	1643	1680	1489	1491	2601	3452
03:00 p.m.	3353	3384	1379	1648	1732	1557	2990	2870
04:00 p.m.	3181	3156	1547	1749	1722	1817	2518	2598
05:00 p.m.	2455	2833	1494	1787	1622	1565	2428	2498
06:00 p.m.	2339	2351	1907	2134	2034	1793	2701	2411
07:00 p.m.	2126	1892	1806	1833	2362	1821	2262	1930
08:00 p.m.	1644	1491	1496	1519	2102	1518	1805	1305
09:00 p.m.	777	704	727	635	887	903	818	553

From Table 5 and Figure 11, it is clear that the prediction values show a harmonious performance against time parameters. Moreover, although 30 August was a Friday, this analysis managed to approximate actual values with an accuracy of about 87% as an important measure of the success of predictions.

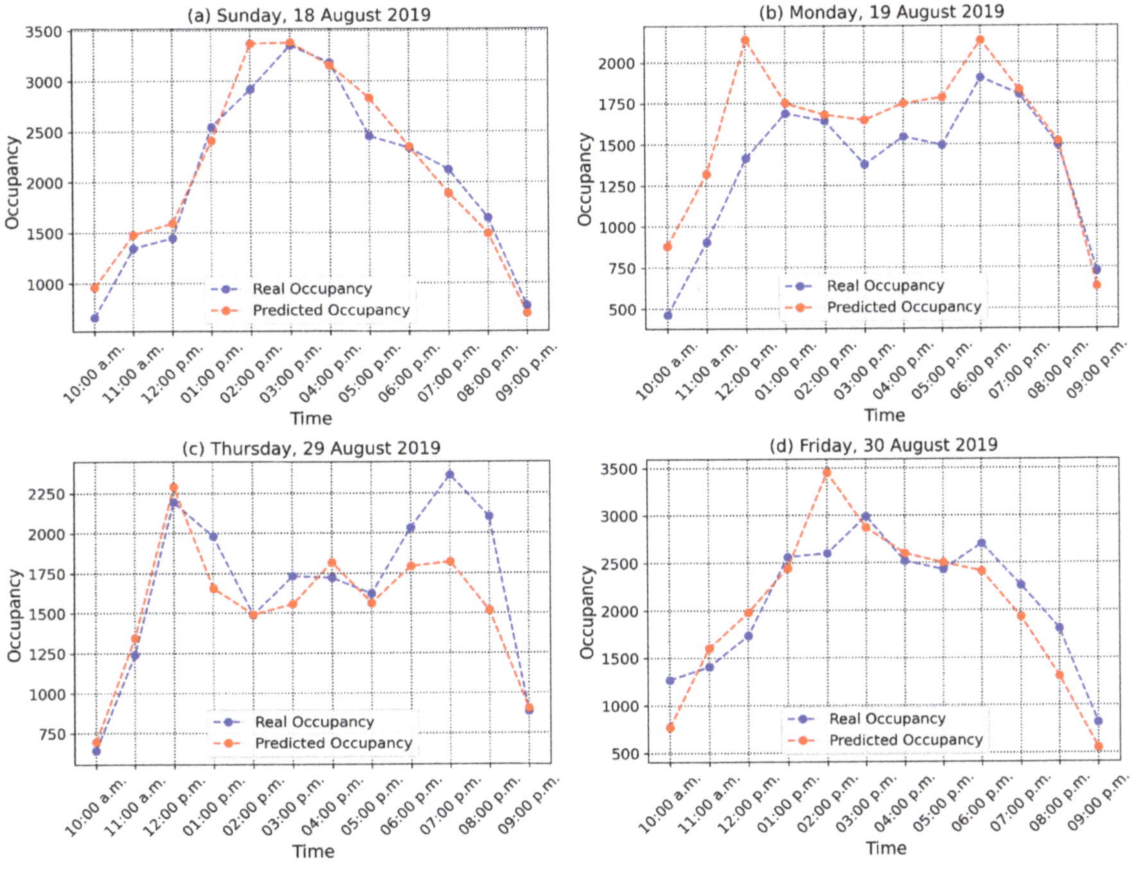

Figure 11. Actual and ANN-predicted occupancy results (hourly).

6.2. Energy Analysis Results

The energy analyses aim to demonstrate the effectiveness of the proposed HVAC control algorithm and compare it with traditional control systems. For this purpose, indoor temperature results and daily energy consumption values for the four scenarios were obtained from the IDA-ICE software. Two days are selected for detailed energy analysis and comparison of indoor temperatures according to energy simulation scenarios, Monday, 29 August 2019, and Saturday, 7 June 2019; these are illustrated in Figures 11 and 12, respectively.

Since the S1 scenario represents the full-powered HVAC at all times, the indoor temperature remains constant with small fluctuations at 24 °C for both 29 August and 7 June (Figures 12a and 13a, respectively), as expected. When the indoor temperature results of the S2 scenario, which represents the sensor-based traditional control approach, are examined for 29 August (Figure 12b), the temperature is found to vary between 23 and 25 °C across wide intervals. This is because basic thermostats allow the temperature to fluctuate a few degrees from the fixed temperature to reduce the frequency with which the cooling device is turned on and off. Consequently, it is seen that the HVAC control mechanism fails to respond to the rapid increase in outdoor temperature and occupancy numbers between 10 and 11 o'clock, and when the maximum occupancy number is reached, the indoor temperature values stay outside the comfort limits. Additionally, although the fluctuations for 7 June (Figure 13b) show a similar pattern, low outdoor temperatures cause the indoor temperatures to return to comfort limit values more quickly.

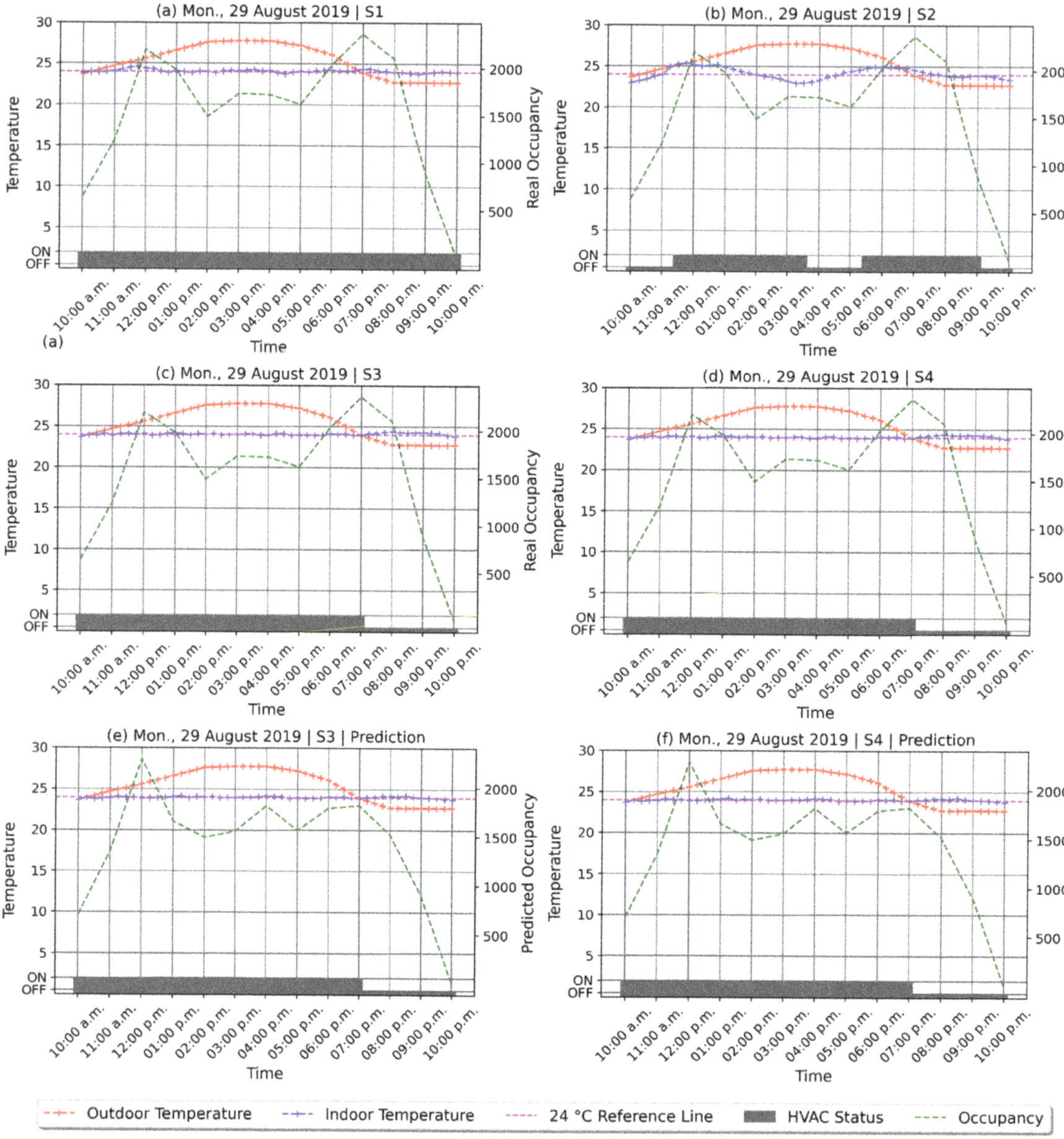

Figure 12. Comparison of indoor temperatures of scenarios for Monday, 29 August 2019.

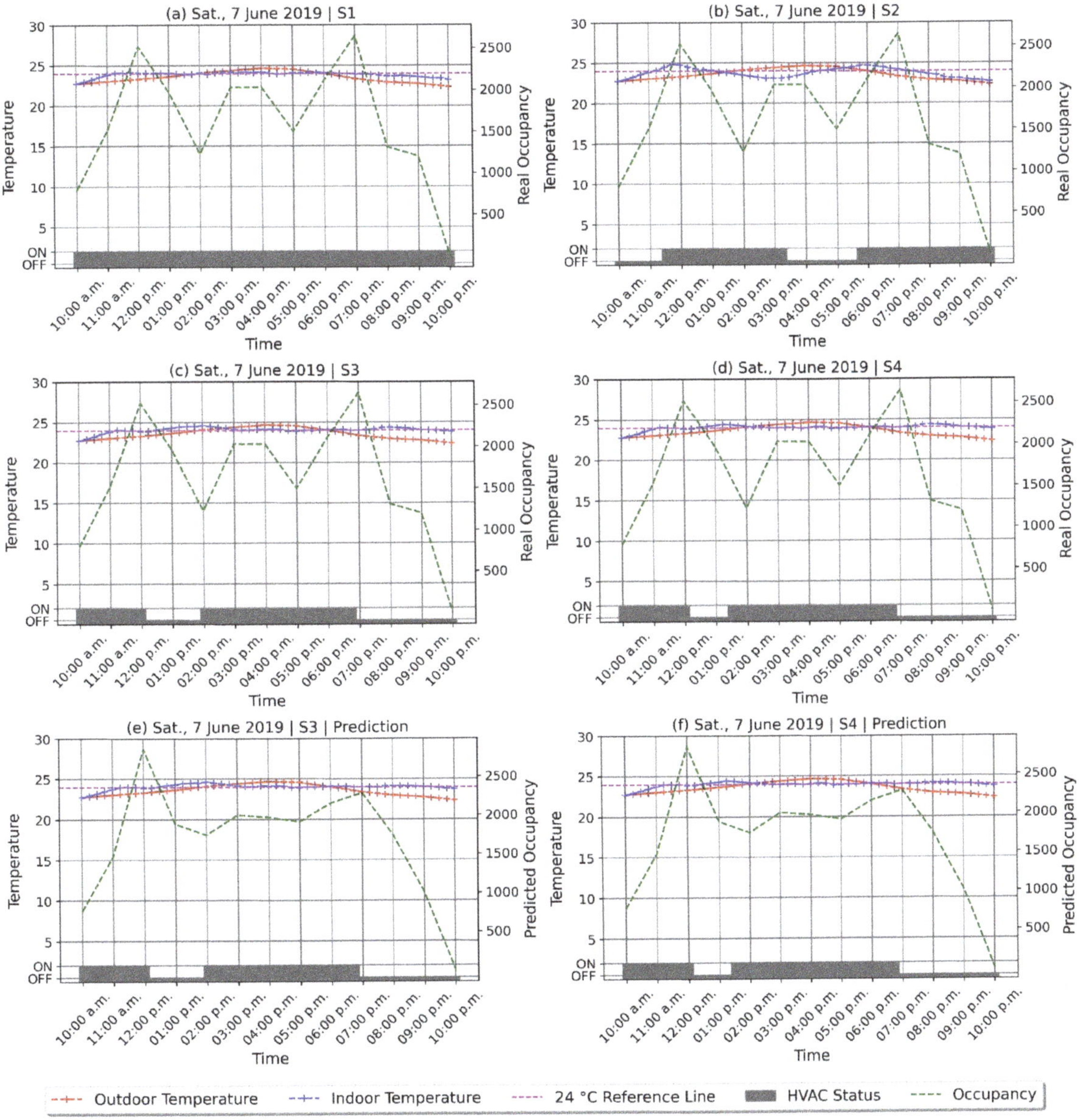

Figure 13. Comparison of indoor temperatures of scenarios for Saturday, 7 June 2019.

The S3 scenario represents the energy simulation according to our new control approach without pre-cooling, which is Algorithm 1, with the HVAC control adjusting according to the occupancy rate. Although there was a decrease in occupancy between 01:00 p.m. and 02:00 p.m. and between 04:00 p.m. and 05:00 p.m. on 29 August, the cooling status remained on as the outdoor temperature was higher than the setpoint (Figure 12c). At 07:00 p.m., as occupancy started to decrease and the air temperature dropped, the cooling went off, and the indoor temperature increased due to the occupancy. As a result of the dramatic decrease in the number of people, this increase ended before exceeding the comfort level. Furthermore, since the HVAC became operational in response to the rise in occupancy, the indoor air temperature remained mostly at the comfort level.

The scenario S4 for 29 August (Figure 12d) produces the same result as does S3 because conditions that would activate the pre-cooling did not arise on this day. Similarly, there is no difference in the application of the algorithm in the simulation with the estimated occupancy values on 29 August (Figure 12e,f) because the increase and decrease trends are captured correctly by ANN. Depending on the difference in occupancy values between real and predicted, small changes are observed in temperature changes and fluctuations. For instance, indoor temperatures do not rise as high for the simulations with predicted values as for the simulation with real values after the cooling is off because the predicted values are smaller than others for those time intervals.

In the S3 scenario for 7 June (Figure 13c), the cooling is switched off between 11:00 a.m. and 01:00 p.m. because the outdoor temperature was below the setpoint with a decrease in the number of people between these hours. An indoor temperature increase is observed to occur naturally in this period, but the low outdoor temperature prevents this increase from reaching significant levels.

Likewise, with the increase in the number of people, the cooling becomes active again from 01:00 p.m. Similar to the scenario for 29 August, cooling is deactivated by the algorithm in the hours close to the shopping mall closing time. While the actual occupancy numbers increase, estimated occupancy values decrease between 03:00 p.m. and 04:00 p.m. However, the S3 scenario simulation with estimated occupancy (Figure 13e) follows the same cooling status as the simulation with actual occupancy (Figure 13c) since air temperature is above the setpoint between these hours. This situation is critical to minimize the failures due to inaccurate estimation by ANN.

The main difference between S3 (Figure 11c,e) and S4 (Figure 11d,f) is that cooling status is active at 12:30 p.m. for S4. The reason for this is that the pre-cooling algorithm is activated under suitable conditions at S4. While there is a decrease in both real and predicted occupancy numbers between 12:00 p.m. and 01:00 p.m., there is an increase of more than 250 people between 01:00 p.m. and 02:00 p.m.. The algorithm starts the cooling 30 min before this increase in occupancy to prevent comfort disturbances caused by the rapid increase. As a result of pre-cooling, the indoor temperature falls to the setpoint level at the beginning of the occupancy increase, contrary to the S3 scenarios. Similar to the simulations performed for 29 August, the actual and predicted occupancy numbers lead to slight differences in the simulations.

When the daily energy consumption results are examined for 29 August (Figure 14), scenario S1 has the greatest consumption with 4090.61 kWh, as expected. Scenario 2 provides an energy saving of approximately 30% compared to S1; with a consumption value of 2279.26 kWh; however, scenarios S3 and S4 consume 22% less energy than S2. When the daily energy consumption results are analyzed for 7 June (Figure 14), the energy consumption trends generally show a similar pattern to that of 29 August. The S2 scenario uses almost 30% less energy than the S1, while S3 and S4 provide an energy saving of approximately 10% over S2. The savings presented by the HVAC control scenario are lower in June than August because delays resulting from the sensor-based approaches at low air temperatures affect the energy efficiency less.

There are also some minor naturally based differences between the simulations performed according to actual and estimated occupancy numbers because the simulation tool adjusts the HVAC power depending on the occupancy. Regarding the ANN values, it is natural to obtain a lower energy consumption because of simulations with predicted values for 29 August, since the average of the predicted occupancy number, 1485.27, is lower than the real occupancy average, 1613.57. Similarly, energy consumption values of simulations with predicted occupancy are greater than simulations with real occupancy because the average of the predicted occupancy, 1755.32, is greater than the real occupancy average, 1679.64.

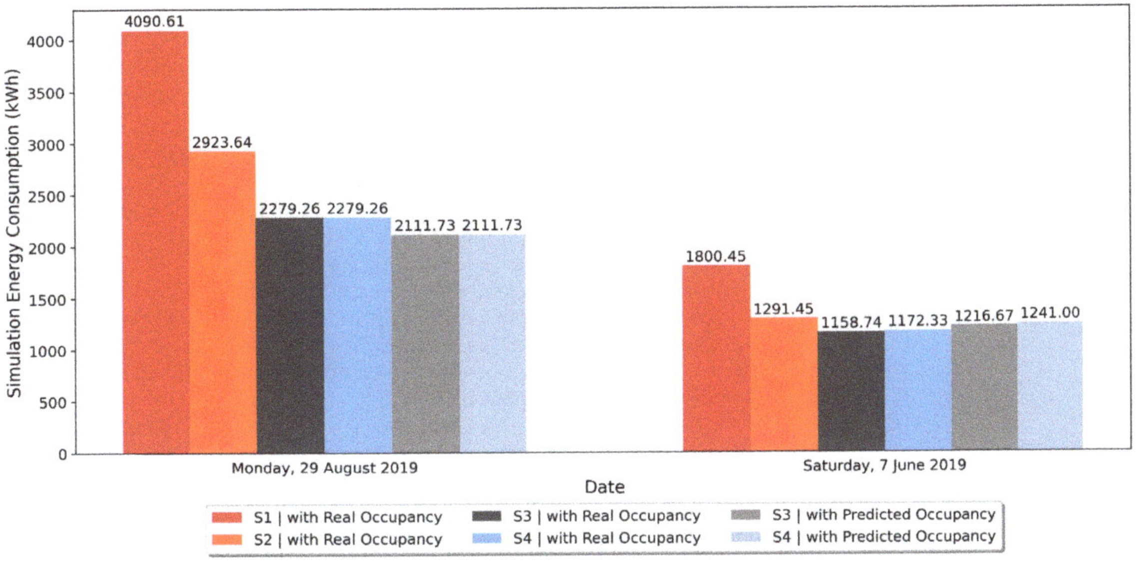

Figure 14. Comparison of energy consumption values of scenarios.

7. Conclusions

This paper presented an analysis of different HVAC control approaches according to their level of development using energy simulations in which ANN features as the focus of the study due to the need for the sensor-free control mechanism. The ANN analysis was performed using real occupancy and weather information collected for each day and hour, with the energy simulations performed for four scenarios using IDA-ICE software.

The ANN results showed that the prediction of occupancy numbers according to time intervals could be calculated with almost 87% accuracy. This accuracy rate was achieved with a limited dataset, and estimation precision should be expected to increase with stronger datasets developed over time. Further, the ANN prediction responded to different parameters, such as special days. This allows the proposed HVAC control algorithm to be used year-round, without exceptions.

According to Wong and Li (2010), "total energy use" is the top selection criterion, followed by "system reliability and stability", "operating and maintenance cost", and "control of indoor humidity and temperature". Since our control strategy is based on data, not real-time detection tools, while it reduces energy consumption, it also very positively affects reliability and operating cost. Different scenarios varying according to level of development were used to measure the effectiveness of our new HVAC control mechanism. A detailed examination of energy simulation results has revealed that the scenarios representing our AI-based occupant-centric control approach (S3 and S4) save a minimum of 10% energy consumption as compared to the traditional sensor-based approach (S2) and a minimum of 35% on those with full-powered HVAC at all times (S1). In the months when the outside temperature is high, these rates reach approximately 20% and 40%, respectively, because traditional approaches allow the indoor temperature to fluctuate excessively, causing an increase in the power consumed for cooling.

Another significant result is that there were only very slight differences in indoor temperature and energy consumption results between simulations performed with predicted and real occupancy numbers. This shows that using estimated values in the HVAC control algorithm does not significantly change the energy consumption or comfort level. Manifestly, the transformation of control approaches proposed has great potential for energy savings.

A few limitations should be noted. First, the proposed control algorithms (Algorithms 1 and 2) were not designed with any complexity; the study design was selected with relatively simple algorithms to show the savings to be made in a simple way. In cases where occupancy tends to decrease slightly for long periods and the outdoor temperature is low, for example, the cooling may remain off for a long time, a situation that was not represented here. In such cases, the occupancy not being very low could cause the interior temperature to rise (i.e., even though the air temperature is low). To avoid such a situation, the algorithm can easily be made more complex with the addition of further parameters, such as occupancy limit and cooling-off time limit.

Second, even though day-ahead weather forecasts mostly make perfect predictions for the following day, some days might fall outside the acceptable margin of error. Such a situation could cause a decrease in the comfort level or inefficiency in the energy consumption, albeit only for very limited periods (or very few days). However, and similarly not considered in this study, existing sensors might be used as an aid tool to measure the real situation and included in the algorithm (as stated) to prevent both these shortcomings.

As a major condition of the experimental design and thus a third limitation, only the cooling function of the HVAC was investigated. Regarding further research, therefore, a control algorithm can also be developed for heating. Then, the method for HVAC control introduced in this study may be applied to the shopping mall by real experimental setup and the results observed in reality. Furthermore (as indicated), more complex control algorithms can be developed according to the specific occupancy pattern of the building studied.

Finally, this study differs from others in considering prediction occupancy numbers with ANN as the main focus in order that significant energy savings can be achieved with a simple control algorithm. For this reason, the study can be a pioneer in terms of a new HVAC system with low installation cost and high energy efficiency. This research can play a major role in guiding the AI-based occupant-centric control tool for sustainable development, which can be used as a standalone control mechanism as it improves.

Author Contributions: Conceptualization, A.Y. and O.B.T.; methodology, A.Y., K.S.Ś., Y.A., M.K. and B.K.; software, A.Y. and K.S.Ś.; validation, M.K. and B.K.; formal analysis, Y.A.; investigation, A.Y.; resources, B.K., Y.E.A. and O.B.T.; data curation, A.Y.; writing—original draft preparation, A.Y.; writing—review and editing, Y.A.; visualization, A.Y.; supervision, O.B.T.; project administration, O.B.T. All authors have read and agreed to the published version of the manuscript.

Funding: This research received no external funding.

Institutional Review Board Statement: Not applicable.

Informed Consent Statement: Not applicable.

Data Availability Statement: Some or all data, models, or codes that support the findings of this study are available from the corresponding author upon reasonable request.

Conflicts of Interest: The authors declare no conflict of interest.

References

1. IEA (International Energy Agency). Energy Efficiency: Buildings. 2019. Available online: https://www.iea.org/topics/energyefficiency/buildings/# (accessed on 11 October 2022).
2. Yang, L.; Yan, H.; Lam, J.C. Thermal Comfort and Building Energy Consumption Implications—A Review. *Appl. Energy* **2014**, *115*, 164–173. [CrossRef]
3. Alcalá, R.; Benítez, J.M.; Casillas, J.; Cordón, O.; Pérez, R. Fuzzy Control of HVAC Systems Optimized by Genetic Algorithms. *Appl. Intell.* **2003**, *18*, 155–177. [CrossRef]
4. Mirinejad, H.; Sadati, S.H.; Ghasemian, M.; Torab, H. Control Techniques in Heating, Ventilating and Air Conditioning Systems. *J. Comput. Sci.* **2008**, *4*, 777–783. [CrossRef]
5. Gholamzadehmir, M.; del Pero, C.; Buffa, S.; Fedrizzi, R.; Aste, N. Adaptive-Predictive Control Strategy for HVAC Systems in Smart Buildings—A Review. *Sustain. Cities Soc.* **2020**, *63*, 102480. [CrossRef]
6. Mizumoto, M. Realization of PID Controls by Fuzzy Control Methods. *Fuzzy Sets Syst.* **1995**, *70*, 171–182. [CrossRef]
7. Soyguder, S.; Karakose, M.; Alli, H. Design and Simulation of Self-Tuning PID-Type Fuzzy Adaptive Control for an Expert HVAC System. *Expert Syst. Appl.* **2009**, *36 Pt 1*, 4566–4573. [CrossRef]

8. Chiou, Y.C.; Lan, L.W. Genetic Fuzzy Logic Controller: An Iterative Evolution Algorithm with New Encoding Method. *Fuzzy Sets Syst.* **2005**, *152*, 617–635. [CrossRef]
9. Mirinejad, H.; Welch, K.C.; Spicer, L. A Review of Intelligent Control Techniques in HVAC Systems. In Proceedings of the 2012 IEEE Energytech, Cleveland, OH, USA, 29–31 May 2012. [CrossRef]
10. Egilegor, B.; Uribe, J.P.; Arregi, G.; Pradilla, E.; Susperregi, L. A Fuzzy Control Adapted by a Neural Network to Maintain a Dwelling within Thermal Comfort. *Proc. Build. Simul.* **1997**, *97*, 87–94.
11. Kruse, R.; Klawonn, F.; Nauck, D. Learning from Fuzzy Rules. *Inform. Forsch. Und Entwickl.* **1997**, *12*, 2–6. [CrossRef]
12. Wu, Y.; Zhang, B.; Lu, J.; Du, K.-L. Fuzzy Logic and Neuro-Fuzzy Systems: A Systematic Introduction. *Int. J. Artif. Intell. Expert Syst.* **2011**, *2*, 47–80.
13. Malki, H.A.; Li, H.; Chen, G. New Design and Stability Analysis of Fuzzy Proportional-Derivative Control Systems. *IEEE Trans. Fuzzy Syst.* **1994**, *2*, 245–254. [CrossRef]
14. Ying, H. Practical Design of Non-linear Fuzzy Controllers with Stability Analysis for Regulating Processes with Unknown Mathematical Models. *Automatica* **1994**, *30*, 1185–1195. [CrossRef]
15. Wu, Z.Q.; Mizumoto, M. PID Type Fuzzy Controller and Parameters Adaptive Method. *Fuzzy Sets Syst.* **1996**, *78*, 23–35. [CrossRef]
16. Patel, A.V.; Mohan, B.M. Analytical Structures and Analysis of the Simplest Fuzzy PI Controllers. *Automatica* **2002**, *38*, 981–993. [CrossRef]
17. Li, H.X.; Zhang, L.; Cai, K.Y.; Chen, G. An Improved Robust Fuzzy-PID Controller with Optimal Fuzzy Reasoning. *IEEE Trans. Syst. Man Cybern. Part B: Cybern.* **2005**, *35*, 1283–1294. [CrossRef]
18. Ole Fanger, P. Thermal Comfort: Analysis and Applications in Environmental Engineering. *Appl. Ergon.* **1972**, 3, 181. [CrossRef]
19. Liang, J.; Du, R. Design of Intelligent Comfort Control System with Human Learning and Minimum Power Control Strategies. *Energy Convers. Manag.* **2008**, *49*, 517–528. [CrossRef]
20. Gacto, M.J.; Alcalá, R.; Herrera, F. Evolutionary Multi-Objective Algorithm to Effectively Improve the Performance of the Classic Tuning of Fuzzy Logic Controllers for a Heating, Ventilating and Air Conditioning System. In Proceedings of the IEEE SSCI 2011: Symposium Series on Computational Intelligence—GEFS 2011: 2011 IEEE 5th International Workshop on Genetic and Evolutionary Fuzzy Systems, Paris, France, 11–15 April 2011; pp. 73–80. [CrossRef]
21. Nowak, M.; Urbaniak, A. Utilization of Intelligent Control Algorithms for Thermal Comfort Optimization and Energy Saving. In Proceedings of the 2011 12th International Carpathian Control Conference, ICCC, Velke Karlovice, Czech Republic, 25–28 May 2011; pp. 270–274. [CrossRef]
22. Wei, T.; Wang, Y.; Zhu, Q. Deep Reinforcement Learning for Building HVAC Control. In Proceedings of the 54th Annual Design Automation Conference, Austin, TX, USA, 18–22 June 2017; Volume 12828. [CrossRef]
23. Du, Y.; Zandi, H.; Kotevska, O.; Kurte, K.; Munk, J.; Amasyali, K.; Mckee, E.; Li, F. Intelligent Multi-Zone Residential HVAC Control Strategy Based on Deep Reinforcement Learning. *Appl. Energy* **2021**, *281*, 116117. [CrossRef]
24. Pasgianos, G.D.; Arvanitis, K.G.; Polycarpou, P.; Sigrimis, N. A Non-linear Feedback Technique for Greenhouse Environmental Control. *Comput. Electron. Agric.* **2003**, *40*, 153–177. [CrossRef]
25. Moradi, H.; Saffar-Avval, M.; Bakhtiari-Nejad, F. Non-linear Multivariable Control and Performance Analysis of an Air-Handling Unit. *Energy Build.* **2011**, *43*, 805–813. [CrossRef]
26. Al-Assadi, S.A.K.; Patel, R.V.; Zaheer-Uddin, M.; Verma, M.S.; Breitinger, J. Robust Decentralized Control of HVAC Systems Using H ∞-Performance Measures. *J. Frankl. Inst.* **2004**, *341*, 543–567. [CrossRef]
27. Anderson, M.; Buehner, M.; Young, P.; Hittle, D.; Anderson, C.; Tu, J.; Hodgson, D. MIMO Robust Control for HVAC Systems. *IEEE Trans. Control Syst. Technol.* **2008**, *16*, 475–483. [CrossRef]
28. Dong, B. Non-Linear Optimal Controller Design for Building HVAC Systems. In Proceedings of the IEEE International Conference on Control Applications, Yokohama, Japan, 8–10 September 2010; pp. 210–215. [CrossRef]
29. Mossolly, M.; Ghali, K.; Ghaddar, N. Optimal Control Strategy for a Multi-Zone Air Conditioning System Using a Genetic Algorithm. *Energy* **2009**, *34*, 58–66. [CrossRef]
30. Yan, Y.; Zhou, J.; Lin, Y.; Yang, W.; Wang, P.; Zhang, G. Adaptive Optimal Control Model for Building Cooling and Heating Sources. *Energy Build.* **2008**, *40*, 1394–1401. [CrossRef]
31. Serale, G.; Fiorentini, M.; Capozzoli, A.; Bernardini, D.; Bemporad, A. Model Predictive Control (MPC) for Enhancing Building and HVAC System Energy Efficiency: Problem Formulation, Applications and Opportunities. *Energies* **2018**, *11*, 631. [CrossRef]
32. Kusiak, A.; Tang, F.; Xu, G. Multi-Objective Optimization of HVAC System with an Evolutionary Computation Algorithm. *Energy* **2011**, *36*, 2440–2449. [CrossRef]
33. Kusiak, A.; Xu, G.; Zhang, Z. Minimization of Energy Consumption in HVAC Systems with Data-Driven Models and an Interior-Point Method. *Energy Convers. Manag.* **2014**, *85*, 146–153. [CrossRef]
34. Wei, X.; Kusiak, A.; Li, M.; Tang, F.; Zeng, Y. Multi-Objective Optimization of the HVAC (Heating, Ventilation, and Air Conditioning) System Performance. *Energy* **2015**, *83*, 294–306. [CrossRef]
35. Biyik, E.; Brooks, J.D.; Sehgal, H.; Shah, J.; Gency, S. Cloud-Based Model Predictive Building Thermostatic Controls of Commercial Buildings: Algorithm and Implementation. In Proceedings of the American Control Conference, Chicago, IL, USA, 1–3 July 2015; pp. 1683–1688. [CrossRef]
36. Kelman, A.; Ma, Y.; Borrelli, F. Analysis of Local Optima in Predictive Control for Energy Efficient Buildings. *J. Build. Perform. Simul.* **2013**, *6*, 236–255. [CrossRef]

37. Huang, H.; Chen, L.; Hu, E. A New Model Predictive Control Scheme for Energy and Cost Savings in Commercial Buildings: An Airport Terminal Building Case Study. *Build. Environ.* **2015**, *89*, 203–216. [CrossRef]
38. Garnier, A.; Eynard, J.; Caussanel, M.; Grieu, S. Predictive Control of Multizone Heating, Ventilation and Air-Conditioning Systems in Non-Residential Buildings. *Appl. Soft Comput. J.* **2015**, *37*, 847–862. [CrossRef]
39. Barzin, R.; Chen, J.J.J.; Young, B.R.; Farid, M.M. Application of Weather Forecast in Conjunction with Price-Based Method for PCM Solar Passive Buildings—An Experimental Study. *Appl. Energy* **2016**, *163*, 9–18. [CrossRef]
40. Alibabaei, N.; Fung, A.S.; Raahemifar, K. Development of Matlab-TRNSYS Co-Simulator for Applying Predictive Strategy Planning Models on Residential House HVAC System. *Energy Build.* **2016**, *128*, 81–98. [CrossRef]
41. Afram, A.; Janabi-Sharifi, F. Theory and Applications of HVAC Control Systems—A Review of Model Predictive Control (MPC). *Build. Environ.* **2014**, *72*, 343–355. [CrossRef]
42. Afram, A.; Janabi-Sharifi, F.; Fung, A.S.; Raahemifar, K. Artificial Neural Network (ANN) Based Model Predictive Control (MPC) and Optimization of HVAC Systems: A State-of-the-Art Review and Case Study of a Residential HVAC System. *Energy Build.* **2017**, *141*, 96–113. [CrossRef]
43. Trčka, M.; Hensen, J.L.M. Overview of HVAC System Simulation. *Autom. Constr.* **2010**, *19*, 93–99. [CrossRef]
44. Afroz, Z.; Shafiullah, G.M.; Urmee, T.; Higgins, G. Modeling Techniques Used in Building HVAC Control Systems: A Review. *Renew. Sustain. Energy Rev.* **2018**, *83*, 64–84. [CrossRef]
45. Huang, H.; Chen, L.; Hu, E. A Neural Network-Based Multi-Zone Modelling Approach for Predictive Control System Design in Commercial Buildings. *Energy Build.* **2015**, *97*, 86–97. [CrossRef]
46. Javed, A.; Larijani, H.; Ahmadinia, A.; Emmanuel, R.; Mannion, M.; Gibson, D. Design and Implementation of a Cloud Enabled Random Neural Network-Based Decentralized Smart Controller with Intelligent Sensor Nodes for HVAC. *IEEE Internet Things J.* **2017**, *4*, 393–403. [CrossRef]
47. Sala-Cardoso, E.; Delgado-Prieto, M.; Kampouropoulos, K.; Romeral, L. Activity-Aware HVAC Power Demand Forecasting. *Energy Build.* **2018**, *170*, 15–24. [CrossRef]
48. Yang, S.; Wan, M.P.; Chen, W.; Ng, B.F.; Zhai, D. An Adaptive Robust Model Predictive Control for Indoor Climate Optimization and Uncertainties Handling in Buildings. *Build. Environ.* **2019**, *163*, 106326. [CrossRef]
49. Zhou, B.; Chikkala, J.; Schmitt, R. A Load-Adaptive and Predictive Control of Energy-Efficient Building Automation in Production Environment. *Procedia CIRP* **2019**, *79*, 245–250. [CrossRef]
50. Erickson, V.L.; Lin, Y.; Kamthe, A.; Brahme, R.; Surana, A.; Cerpa, A.E.; Sohn, M.D.; Narayanan, S. Energy Efficient Building Environment Control Strategies Using Real-Time Occupancy Measurements. In Proceedings of the BUILDSYS 2009—1st ACM Workshop on Embedded Sensing Systems for Energy-Efficiency in Buildings, Berkeley CA, USA, 3 November 2009; pp. 19–24. [CrossRef]
51. Erickson, V.L.; Cerpa, A.E. Occupancy Based Demand Response HVAC Control Strategy. In Proceedings of the BuildSys'10—2nd ACM Workshop on Embedded Sensing Systems for Energy-Efficiency in Buildings, Zurich, Switzerland, 2 November 2010; pp. 7–12. [CrossRef]
52. Oldewurtel, F.; Sturzenegger, D.; Morari, M. Importance of Occupancy Information for Building Climate Control. *Appl. Energy* **2013**, *101*, 521–532. [CrossRef]
53. Li, N.; Calis, G.; Becerik-Gerber, B. Measuring and Monitoring Occupancy with an RFID Based System for Demand-Driven HVAC Operations. *Autom. Constr.* **2012**, *24*, 89–99. [CrossRef]
54. Yang, Z.; Ghahramani, A.; Becerik-Gerber, B. Building Occupancy Diversity and HVAC (Heating, Ventilation, and Air Conditioning) System Energy Efficiency. *Energy* **2016**, *109*, 641–649. [CrossRef]
55. Capozzoli, A.; Piscitelli, M.S.; Gorrino, A.; Ballarini, I.; Corrado, V. Data Analytics for Occupancy Pattern Learning to Reduce the Energy Consumption of HVAC Systems in Office Buildings. *Sustain. Cities Soc.* **2017**, *35*, 191–208. [CrossRef]
56. Aftab, M.; Chen, C.; Chau, C.K.; Rahwan, T. Automatic HVAC Control with Real-Time Occupancy Recognition and Simulation-Guided Model Predictive Control in Low-Cost Embedded System. *Energy Build.* **2017**, *154*, 141–156. [CrossRef]
57. Shi, J.; Yu, N.; Yao, W. Energy Efficient Building HVAC Control Algorithm with Real-Time Occupancy Prediction. *Energy Procedia* **2017**, *111*, 267–276. [CrossRef]
58. Peng, Y.; Rysanek, A.; Nagy, Z.; Schlüter, A. Using Machine Learning Techniques for Occupancy-Prediction-Based Cooling Control in Office Buildings. *Appl. Energy* **2018**, *211*, 1343–1358. [CrossRef]
59. Nikdel, L.; Janoyan, K.; Bird, S.D.; Powers, S.E. Multiple Perspectives of the Value of Occupancy-Based HVAC Control Systems. *Build. Environ.* **2018**, *129*, 15–25. [CrossRef]
60. Ahmadi-Karvigh, S.; Becerik-Gerber, B.; Soibelman, L. Intelligent Adaptive Automation: A Framework for an Activity-Driven and User-Centered Building Automation. *Energy Build.* **2019**, *188–189*, 184–199. [CrossRef]
61. Pang, Z.; Chen, Y.; Zhang, J.; O'Neill, Z.; Cheng, H.; Dong, B. Nationwide HVAC Energy-Saving Potential Quantification for Office Buildings with Occupant-Centric Controls in Various Climates. *Appl. Energy* **2020**, *279*, 115727. [CrossRef]
62. Azuatalam, D.; Lee, W.-L.; de Nijs, F.; Liebman, A. Reinforcement Learning for Whole-Building HVAC Control and Demand Response. *Energy AI* **2020**, *2*, 100020. [CrossRef]
63. Deng, Z.; Chen, Q. Development and Validation of a Smart HVAC Control System for Multi-Occupant Offices by Using Occupants' Physiological Signals from Wristband. *Energy Build.* **2020**, *214*, 109872. [CrossRef]

64. Jung, W.; Jazizadeh, F. Human-in-the-Loop HVAC Operations: A Quantitative Review on Occupancy, Comfort, and Energy-Efficiency Dimensions. *Appl. Energy* **2019**, *239*, 1471–1508. [CrossRef]
65. Jazaeri, J.; Gordon, R.L.; Alpcan, T. Influence of Building Envelopes, Climates, and Occupancy Patterns on Residential HVAC Demand. *J. Build. Eng.* **2019**, *22*, 33–47. [CrossRef]
66. Ryan, E.M.; Sanquist, T.F. Validation of Building Energy Modeling Tools under Idealized and Realistic Conditions. *Energy Build.* **2012**, *47*, 375–382. [CrossRef]
67. Crawley, D.B.; Hand, J.W.; Kummert, M.; Griffith, B.T. Contrasting the Capabilities of Building Energy Performance Simulation Programs. *Build. Environ.* **2008**, *43*, 661–673. [CrossRef]
68. Bring, A.; Sahlin, P.; Vuolle, M. Models for Building Indoor Climate and Energy Simulation, A Report of Task 22 Building Energy Analysis Tools. Report of IEA SHC Task. 1999. Available online: https://www.equa.se/dncenter/T22Brep.pdf (accessed on 10 October 2022).
69. Achermann, M.; Zweifel, G. RADTEST—Radiant Heating and Cooling Test Cases. 2003. Available online: http://www.equaonline.com/iceuser/validation/old_stuff/RADTEST_final.pdf (accessed on 10 October 2022).
70. ISO (International Organization for Standardization). ISO 15099:2003. Thermal Performance of Windows, Doors and Shading Devices—Detailed Calculations. 2003. Available online: https://www.iso.org/standard/26425.html (accessed on 10 October 2022).
71. Karlsson, F.; Rohdin, P.; Persson, M.L. Measured and Predicted Energy Demand of a Low Energy Building: Important Aspects When Using Building Energy Simulation. *Build. Serv. Eng. Res. Technol.* **2007**, *28*, 223–235. [CrossRef]
72. Loutzenhiser, P.G.; Manz, H.; Moosberger, S.; Maxwell, G.M. An Empirical Validation of Window Solar Gain Models and the Associated Interactions. *Int. J. Therm. Sci.* **2009**, *48*, 85–95. [CrossRef]
73. Hilliaho, K.; Lahdensivu, J.; Vinha, J. Glazed Space Thermal Simulation with IDA-ICE 4.61 Software—Suitability Analysis with Case Study. *Energy Build.* **2015**, *89*, 132–141. [CrossRef]
74. Salvalai, G. Implementation and Validation of Simplified Heat Pump Model in IDA-ICE Energy Simulation Environment. *Energy Build.* **2012**, *49*, 132–141. [CrossRef]
75. Mazzeo, D.; Matera, N.; Cornaro, C.; Oliveti, G.; Romagnoni, P.; de Santoli, L. EnergyPlus, IDA ICE and TRNSYS Predictive Simulation Accuracy for Building Thermal Behaviour Evaluation by Using an Experimental Campaign in Solar Test Boxes with and without a PCM Module. *Energy Build.* **2020**, *212*, 109812. [CrossRef]
76. Milić, V.; Ekelöw, K.; Moshfegh, B. On the Performance of LCC Optimization Software OPERA-MILP by Comparison with Building Energy Simulation Software IDA ICE. In *Build. Environ.*; 2018; Volume 128, pp. 305–319. [CrossRef]
77. *Design Science Research in Information Systems*; Vaishnavi, V.; Kuechler, W.; Petter, S. (Eds.) Association for Information Systems: Atlanta, GA, USA, 2019.
78. Wieringa, R.J. *Design Science Methodology: For Information Systems and Software Engineering*; Springer: Berlin/Heidelberg, Germany, 2014. [CrossRef]
79. Markus, M.L.; Majchrzak, A.; Gasser, L. A design theory for systems that support emergent knowledge processes. 2002, MIS Quarterly 26, 179–212. *MIS Q.* **2002**, *26*, 179–212.
80. Peffers, K.; Tuunanen, T.; Gengler, C.; Rossi, M.; Hui, W.; Wirtanen, V.; Bragge, J. The design science research process: A model for producing and presenting information systems research. In Proceedings of the First International Conference on Design Science Research in Information Systems and Technology DESRIST, Claremont, CA, USA, 24–25 February 2006.
81. The Association of Real Estate and Real Estate Investment Companies of Turkey (GYODER). 2019. Available online: https://www.gyoder.org.tr/yayinlar/sektorel-yayinlar (accessed on 1 May 2021).
82. STATISTA 2021 (Number of Shopping Centers in Europe 2017, by Country). Available online: https://www.statista.com/statistics/912126/shopping-center-numbers-by-country-europe/ (accessed on 10 October 2022).
83. Wijayasekara, D.; Manic, M.; Sabharwall, P.; Utgikar, V. Optimal Artificial Neural Network Architecture Selection for Performance Prediction of Compact Heat Exchanger with the EBaLM-OTR Technique. *Nucl. Eng. Des.* **2011**, *241*, 2549–2557. [CrossRef]
84. Hagan, M.T.; Demuth, H.B.; Beale, M.H.; de Jesus, O. Neural Network Design. In *Neural Networks in a Softcomputing Framework*, 2nd ed.; Springer Science & Business Media: Berlin/Heidelberg, Germany, 2014.
85. Cartwright, H. (Ed.) *Artificial Neural Networks*; Springer: New York, NY, USA, 2021.

energies

MDPI

Article

Semantic Segmentation Algorithm-Based Calculation of Cloud Shadow Trajectory and Cloud Speed

Shitao Wang, Mingjian Sun and Yi Shen *

Department of Control Science and Engineering, Harbin Institute of Technology, Harbin 150001, China
* Correspondence: shen@hit.edu.cn; Tel.: +86-451-86413411-8602; Fax: +86-451-86418378

Abstract: Cloud covering is an important factor affecting solar radiation and causes fluctuations in solar energy production. Therefore, real-time recognition and the prediction of cloud covering and the adjustment of the angle of photovoltaic panels to improve power generation are important research areas in the field of photovoltaic power generation. In this study, several methods, namely, the principle of depth camera measurement distance, semantic segmentation algorithm, and long- and short-term memory (LSTM) network were combined for cloud observation. The semantic segmentation algorithm was applied to identify and extract the cloud contour lines, determine the feature points, and calculate the cloud heights and geographic locations of the cloud shadows. The LSTM algorithm was used to predict the trajectory and speed of the cloud movement, achieve accurate and real-time detection, and track the clouds and the sun. Based on the results of these methods, the shadow area of the cloud on the ground was calculated. The recursive neural LSTM network was also used to predict the track and moving speed of the clouds according to the cloud centroid data of the cloud images at different times. The findings of this study can provide insights to establish a low-cost intelligent monitoring predicting system for cloud covering and power generation.

Keywords: solar energy; semantic segmentation algorithm; cloud moving prediction; cloud shadow; cloud speed

Citation: Wang, S.; Sun, M.; Shen, Y. Semantic Segmentation Algorithm-Based Calculation of Cloud Shadow Trajectory and Cloud Speed. *Energies* **2022**, *15*, 8925. https://doi.org/10.3390/en15238925

Academic Editors: Luis Hernández-Callejo, Sergio Nesmachnow and Sara Gallardo Saavedra

Received: 24 October 2022
Accepted: 23 November 2022
Published: 25 November 2022

1. Introduction

Solar energy is a widely distributed and sustainable source of energy worldwide. Photovoltaic power generation technology can directly convert light energy into electrical energy through the photovoltaic effect, and it has the advantages of no pollution, safe use, and convenient maintenance. With continuous technical improvement and cost reduction, photovoltaic power generation has increased rapidly. In 2005, the global cumulative installed photovoltaic capacity exceeded 5 GW. According to "Snapshot of Global PV Markets 2020" [1] issued by the International Energy Agency, by the end of 2019, the global installed capacity exceeded 600 GW, and the average annual growth rate was 41%. In the past three years (2019–2022), the annual installed capacity has exceeded 100 GW. Figure 1 shows the global installed photovoltaic capacity over the past 10 years (2011–2019).

Large-scale photovoltaic projects require real-time monitoring of power quality and operating information while maintaining optimal scheduling. Therefore, it is essential to ensure the accurate forecasting of generation capacity, especially short-term and real-time forecasting [2]. Therefore, dynamically adjusting the solar panel according to weather type, cloud occlusion, and the radiation angle of sunlight to maximize the power generated by photovoltaic modules has always been an important research topic in the field of photovoltaic power generation [3–5]. Changes in photovoltaic power generation are almost proportional to the changes in radiation intensity, which are directly affected by cloud occlusion. Different weather types and cloud cover lead to considerable changes in the power generated by photovoltaic systems and power grid fluctuations [6–9].

However, in the field of photovoltaic power generation, it has always been challenging to accurately predict the weather type and cloud movement [10,11].

Cloud cover is usually analyzed based on the shape, size, distribution and evolution, and height of the clouds. The cloud shape changes easily; therefore, clouds should be monitored continuously and in real-time. Traditional artificial observation is based on subjective judgment and observation experience, so it cannot accurately predict the cloud shade. With research advances, observation technology has developed significantly. Contemporary observation technologies for different space environments include satellite- or space-based equipment for atmospheric observations and ground-based equipment for near-Earth observations. In large-scale ground-based photovoltaic power stations, power prediction is mainly based on short-term and real-time monitoring of the weather in the plant area. Considering the area of the power station, the speed and economic cost of ground equipment required for cloud monitoring as well as the monitoring method should be selected for cloud observation. Table 1 lists the various cloud height observation methods.

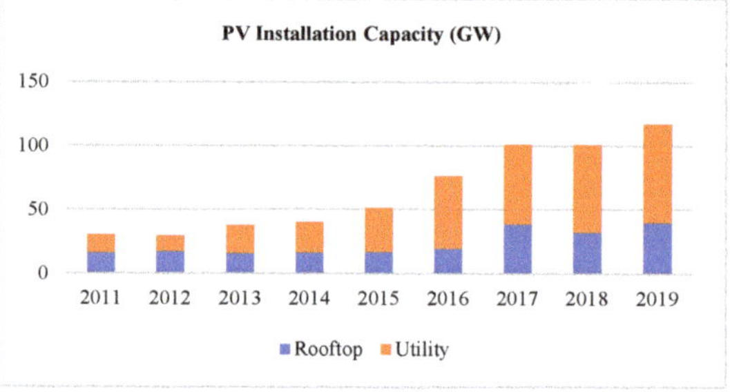

Figure 1. Global photovoltaic installed capacity during 2011–2019.

Table 1. Comparison of different cloud height observation methods.

	Manual	Radiosonde	Cloud Meter	LiDAR	Weather Radar
Observation range	Whole sky	Single point	Single point	Single point	Single point
Monitoring range	Maximum visibility	8–10 km	10–12 km	1–12 km	15 km
Frequency	0.5–6 h	6–12 h	Continuous	Continuous	Continuous
Accuracy	20–30%	100–200 m	2%	2%	60 m
Sub attribute	Cloud amount and type	Cloud top height	-	Cloud microphysical properties	Cloud top height and cloud microphysical properties
Difficulty	Low	Medium	Medium	High	High
Automation	No	No	Yes	Yes	Yes

Among the methods mentioned in Table 1, cloud meter and radar measurement are widely used; however, these methods have the disadvantages of high cost and inconsistent measurement results; moreover, it is challenging to obtain the edge profiles of clouds and predict the cloud shading range. Therefore, in this study, we developed and investigated a new low-cost prediction method that combined sky images and machine learning methods to obtain an accurate cloud height, extract the edge contours of clouds, measure the shade range (i.e., cloud cover), and predict and analyze the moving direction and speed of the clouds.

In this study, the weather type was detected and identified in real-time by using artificial intelligence algorithms and deep learning networks. In sky images, the existence of cloud shielding, range of shielding, and moving speed and track of clouds are determined to obtain insights for guiding the angle of photovoltaic panels and increasing power generation, providing a basis for the real-time prediction of photovoltaic power generation.

2. Predicting Cloud Shadow Moving Trajectory and Speed

2.1. Method for Cloud Monitoring

For cloud monitoring, multiple cloud cameras are distributed across a photovoltaic power station field to obtain aerial images that contain high-resolution spatiotemporal information about solar radiation. Various software can be used to process the information in the control room. Thereafter, the predicted value of solar energy is obtained. Several researchers have conducted related studies. In 2013, Tao et al. used a pair of CCD (charge coupled device) digital cameras to set a baseline length of 60 m to form a cloud base-height measurement system with binocular imaging [12–14]. The Harris corner detector was used to extract the corner features of the images, and then the relative disparity was obtained according to the matching feature points. The principle of photogrammetry was used to calculate the height of the cloud base. In 2013, Zhang et al. used industrial cameras and image processing technologies for cloud monitoring [15,16]. Cloud height was calculated based on the dual-camera measuring distance principle. The same feature points were obtained using CSIFT (color scale-invariant feature transform) and SIFT (scale-invariant feature transform) methods for object matching to detect the cloud speed. In 2015, Peng et al. used support vector machine classifiers to identify cloud clusters from multiple TSI images and evaluated the essential height and movement of each cloud cluster [17–21]. In 2018, the German DLR Solar Energy Research Institute developed the WobaS system [22–26]. This system comprises 2–4 cloud cameras that are used to capture sky images. These images are evaluated and the cloud speed and future distribution are calculated, enabling the successful prediction of solar radiation values in the next 15 min.

Typical cloud detection and measurement methods based on dual imaging systems use similar hardware; these methods often utilize two or more cameras (especially fisheye cameras with large viewing angles and TSI devices) and apply a similar triangle principle to calculate the distance between the cloud and the cameras (depth camera principle). These methods can thus achieve high-resolution images at low equipment cost. However, the software used in these methods are often different from those used in conventional machine learning algorithms or early deep learning algorithms [27–31] However, the recognition accuracy and feature matching of clouds and sun were insufficient. The results showed significant calculation errors in cloud parameters such as cloud height, cloud area, cloud shadow, and cloud speed.

Deep learning algorithms are mainly divided into three categories:

A convolutional neural network (CNN) is commonly used for image data analysis and processing such as image classification, target detection, and semantic segmentation (i.e., Mask R-CNN and YOLACT). A recurrent neural network (RNN) such as a long- and short-term memory (LSTM) network is often used for text analysis or natural language processing. A generative adricative network (GAN) is typically used for data generation or unsupervised learning applications such as generating similar original data; 3D-GAN is used to generate high-quality 3D objects [32–35]. In 2018, He et al. proposed the Mask R-CNN method based on faster R-CNN, which is an extension of Mask [36–38].

In this study, first, the edge contours of clouds were obtained, and then, the feature points on the edges were obtained using the PSPNet semantic segmentation algorithm [39] based on the images obtained from the CMOS imaging system. Furthermore, the LSTM algorithm was used to obtain the cloud parameters such as cloud height and moving track and speed of cloud shadow, which were then combined with geographic information to predict the cloud shadow occlusion on the ground.

2.2. Cloud Edge Contour Extraction and Feature Point Recognition

This study requires distinguishing the cloud and non-cloud parts of a picture, that is, to classify each pixel to form the boundary of a cloud. The sky and clouds have different colors; therefore, we considered using color features to classify each pixel to form a boundary. The texture features of the sky and clouds are also different; thus, texture features can also be used for classification. Texture features can be extracted using the gray-gradient co-occurrence matrix (GGCM).

Several types of neural networks can realize semantic segmentation; herein, the PSPNet network semantic segmentation algorithm (Figure 2) [11] was used for classification. In the PSPNet network, the netscope space pyramid pool structure was adopted, as shown in Figure 3.

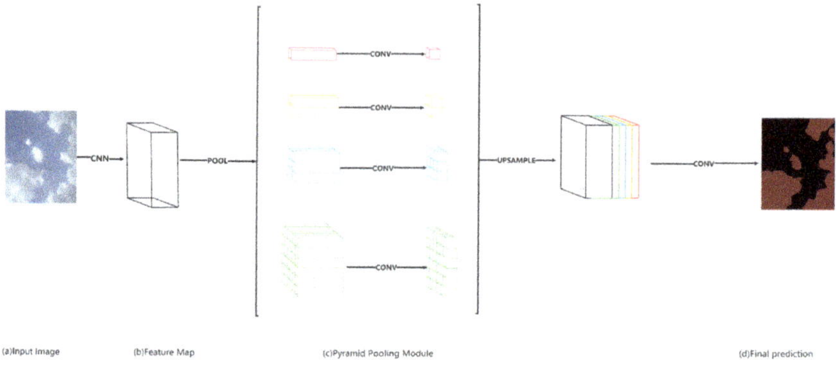

Figure 2. PSPNet network.

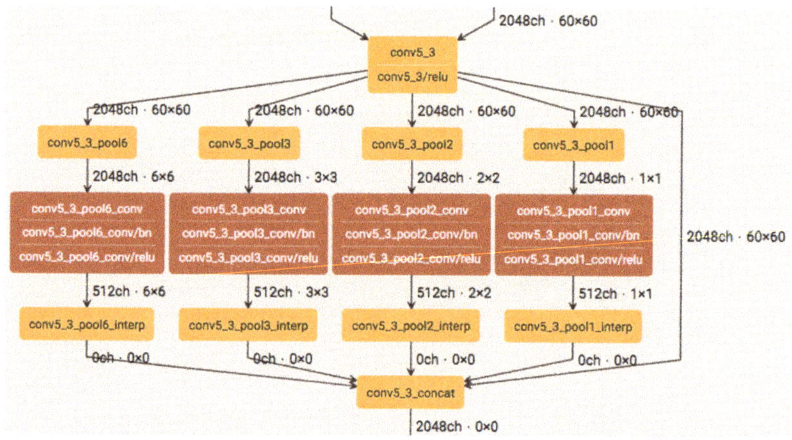

Figure 3. Netscope space pyramid pool structure.

PSPNet is a modification of the basic RESNET architecture, and it uses hole convolution. First, it pools the features and then processes them at the same resolution in the whole encoder network (one-fourth of the original image input) until a spatial pooling module is obtained. Auxiliary loss is considered in the middle layer of RESNET to optimize the overall learning, and the global context model is aggregated in the spatial pyramid pooling layer at the top of the modified RESNET encoder.

In this study, 3800 cloud pictures were annotated using lableme software and given as input to the network model for training. The training process is as follows:

Step 1: The weight and deviation of neurons in each layer is initialized.

Step 2: Forward propagation: the image is converted into a matrix input in RGB format, the linear combination value is obtained through the weight and deviation of neurons in each layer, and then the activation function is applied to the linear combination value.

Step 3: The loss function is used to calculate the error between the output value of forward propagation and the annotated images, and the weight and deviation of neurons in each layer are optimized using the back propagation algorithm according to the error.

Step 4: Steps 2 and 3 are repeated iteratively to reduce the error to a specified value, and the weight and deviation of each layer of neurons are saved to obtain a well-trained model.

The cloud images are given as input into the model, and the training results are shown in Figure 4.

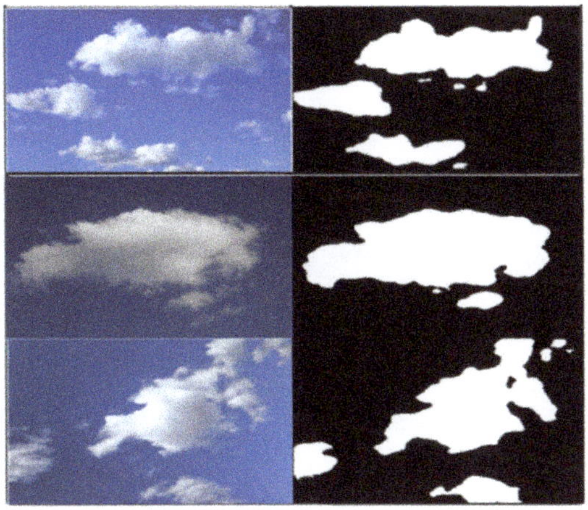

Figure 4. Example results of PSPNet.

2.3. Cloud Movement Trajectory and Velocity Recognition Based on LSTM Network

Cloud moving can be considered as a time series prediction problem. Contemporary deep learning methods mainly use RNNs. In this study, a LSTM network model was used, which comprises an input layer, a hidden layer, and an output layer. The internal structure of the hidden layer is shown in Figure 5.

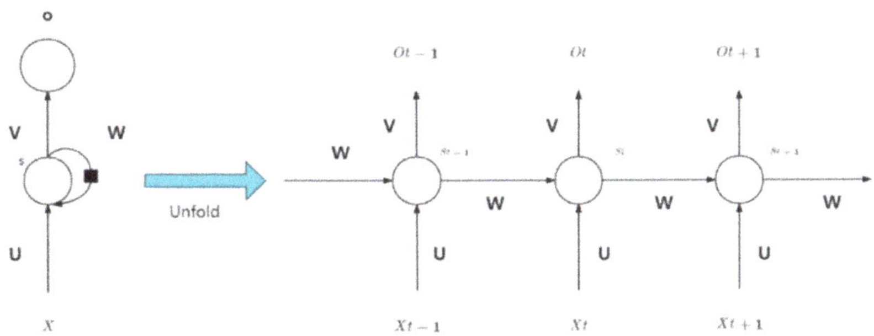

Figure 5. Internal structure of the hidden layer.

In Figure 5, $t-1$, t, and $t+1$ are continuous time series, X is the input sample, S_t is the memory of the sample at time t, and $S_t = f(W \times S_{t-1} + U \times X_t)$, where W represents the weight of the last time, U represents the weight of the input sample at the moment, and V represents the weight of the output.

For general initialization, the start time is considered as $t = 1$, input is $S_0 = 0$, and W, U, and V are initialized randomly; then, Equation (1) is used for prediction.

$$\begin{aligned} h_1 &= Ux_1 + Ws_0 \\ s_1 &= f(h_1) \\ o_1 &= g(Vs_1) \end{aligned} \tag{1}$$

where f and g are activation functions.

As time progresses, the state s_1 is considered the memory state at start time t_1, and these parameters then participate in the next predicting activity, as shown in Equation (2).

$$\begin{aligned} h_2 &= Ux_2 + Ws_1 \\ s_2 &= f(h_2) \\ o_2 &= g(Vs_2) \end{aligned} \tag{2}$$

Finally, the final output value is obtained using Equations (2) and (3).

$$\begin{aligned} h_t &= Ux_t + Ws_{t-1} \\ s_t &= f(h_t) \\ o_t &= g(Vs_t) \end{aligned} \tag{3}$$

LSTM updates the weight parameters W, U, and V using the loss function. For each time sequence, LSTM produces an error value e_t. The total error value E is calculated using Equation (4).

$$\begin{aligned} E &= \sum\nolimits_t e_t \\ \nabla U &= \frac{\partial E}{\partial U} = \sum\nolimits_t \frac{\partial e_t}{\partial U} \\ \nabla V &= \frac{\partial E}{\partial V} = \sum\nolimits_t \frac{\partial e_t}{\partial V} \\ \nabla W &= \frac{\partial E}{\partial W} = \sum\nolimits_t \frac{\partial e_t}{\partial W} E = \sum\nolimits_t e_t \end{aligned} \tag{4}$$

2.4. Calculating the Cloud Height and Shadow

2.4.1. Method of Calculating the Cloud Height and Shadow

Two cameras with the same internal parameters were placed in parallel so that their optical axes were parallel to each other and the cameras faced vertically upward. Another pair of coordinate axes were collinear. The two imaging planes were coplanar. The optical centers of the two cameras were at a fixed distance of d. Figure 6 shows a schematic of binocular stereo vision.

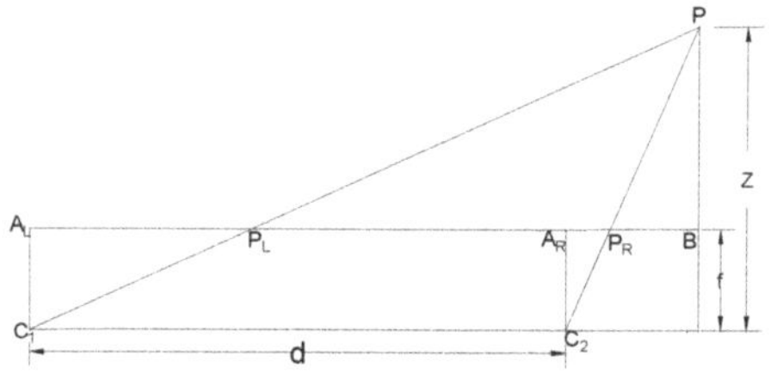

Figure 6. Schematic of binocular stereo vision.

In the above camera arrangement method, we assumed that the coordinate system of camera C_1 was $O_1X_1Y_1Z_1$, the coordinate system of camera C_2 was $O_2X_2Y_2Z_2$, the focal length of the two cameras was f, and the distance of the camera was d. The coordinates of any space point p photographed by two cameras at the same time are expressed as (x_1, y_1, z_1) in the C_1 coordinate system and (x_2, y_2, z_2) in the C_2 coordinate system. The image coordinates of space point p in camera C_1 are (u_1, v_1). The coordinate of the image point in camera C_2 is (u_2, v_2). Therefore, the ratio of p to camera C_1's X wheelbase from x_1 and u_1 is equal to the ratio of the Y wheelbase from y_1 and u_1 to camera C_1, and this ratio is equal to the ratio of the camera's focal length f and p. to camera C_1's Z wheelbase from z_1. The same is true for p and camera C_2. From this, the 3D space depth camera uses Equations (5) and (6).

$$\begin{cases} \frac{f}{z_1} = \frac{u_1}{x_1} = \frac{v_1}{y_1} \\ \frac{f}{z_2} = \frac{u_2}{x_2} = \frac{v_2}{y_2} \end{cases} \tag{5}$$

$$\begin{cases} X = x_1 = x_2 + d \\ Y = y_1 = y_2 \\ Z = z_1 = z_2 \end{cases} \tag{6}$$

These two equations are combined as follows.

$$x_1 - x_2 = d$$
$$x_1 = \frac{z_1}{f}u_1 = \frac{z}{f}u_1 \tag{7}$$
$$x_2 = \frac{z_2}{f}u_2 = \frac{z}{f}u_2$$

The binocular 3D vision method is used to reconstruct the 3D space points using Equations (8)–(12).

$$d = \frac{z}{f}(u_1 - u_2) \tag{8}$$

$$X = x_1 = \frac{z}{f}u_1 = \frac{u_1}{u_1 - u_2}d \tag{9}$$

$$Y = y_1 = \frac{z}{f}v_1 = \frac{v_1}{u_1 - u_2}d \tag{10}$$

$$Z = \frac{f}{u_1 - u_2}d \tag{11}$$

$$S_{act} = \left(\frac{Z}{f}\right)^2 S_{img} \tag{12}$$

2.4.2. Calculation Method of Solar Irradiation Angle

To calculate the illumination angle of sunlight, first, the altitude and azimuth of the Sun are calculated. The relationship between the altitude angle and the latitude angle and the time angle is obtained from the geometric relationship of the Sun and the Earth using Equation (13):

$$\sin\varphi = \sin\varnothing \sin\delta + \cos\varnothing \cos\delta \cos\omega \tag{13}$$

where $\varnothing$ is the local latitude; δ is the declination angle; and ω is the hour angle.

The solar declination angle (δ) is the angle between the Sun and the Earth center line and the equatorial plane. As the Earth moves around the Sun, the declination angle changes accordingly. The declination angle is representative of the season and fluctuates between $-23°26'$ and $+23°26'$, and it repeats the cycle in years. The approximate declination angle is calculated using Equation (14):

$$\delta = 23.45 \times \sin\left(360 \times \frac{284 + n}{365}\right) \tag{14}$$

where n represents the date serial number (based on 1 year), and it is in the range of 1–365. For a leap year, the value of n will be 1–366, and the denominator 365 will be changed to 366.

Azimuth is represented by γ, and it can be considered the approximate angle between the shadow and the meridian of a straight line erected on the ground under the Sun, that is, the angle between the shadow cast by the light falling on the ground and the local meridian. γ is set to 0 in the due north of the target, continues to expand clockwise, and changes in the range 0–360°. The relevant measurement work was carried out in a clockwise direction, the starting destination of the solar azimuth was set in the north of the reference object, the ending destination was considered the incident direction of sunlight, and the required angle was measured in a clockwise direction.

The relationship between the azimuth, altitude angle, declination angle, dimension, and time angle is expressed using the following equations.

$$\sin\gamma = \frac{\cos\delta \sin\omega}{\cos\varphi} \quad (15)$$

$$\cos\gamma = \frac{\sin\alpha \sin\varnothing - \sin\delta}{\cos\varphi \cos\varnothing} \quad (16)$$

The solar time angle ω in Equations (13) and (15) can be obtained using the following equations.

$$\omega = 15(ST - 12) \quad (17)$$

$$ST = LT + Z \quad (18)$$

where ST is true solar time, LT is the local time, and Z is the time zone; the 24 h format is used to calculate time.

The projection area of a cloud on the ground is predicted by calculating the cloud height, the edge contour of cloud, and the illumination angle of sunlight.

3. Results and Discussion

3.1. Verification Experiment and Results of Object Shadow Casting

Real-time measurement of clouds and cloud shadows is challenging; therefore, we used fixed objects such as a flagpole to replace clouds for the experiments. A local coordinate system was established with the flagpole as the origin. First, the relative position of the flagpole and the camera was estimated, and then the relative position of the flagpole and the shadow was determined. The estimated results were compared with the actual measurement to verify the effectiveness of the cloud shadow position calculation in the local coordinate system. There was only a rotation and translation transformation relationship between the local coordinate system and the world coordinate system; therefore, the effectiveness in the local coordinate system is equal to that in the world coordinate system. The experimental method is presented in Figure 7.

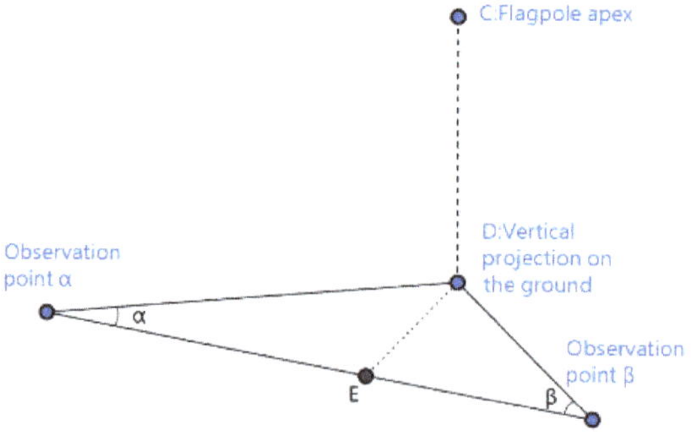

Figure 7. Relative position model of the flagpole and camera.

The experimental steps are as follows:

Step 1. Two adjustable level platforms are set up under the flagpole, the level ruler is placed on the level platform, and adjusted to the level of the water platform.

Step 2. The cameras are placed on a horizontal platform in parallel, and they capture images in a vertically upward position.

Step 3. The distance between the two water platforms is measured and recorded.

Step 4. The length and azimuth of the shadow is measured and recorded.

Step 5. The azimuth and distance of the flagpole relative to the two cameras are measured and recorded.

Step 6. The length and azimuth of the shadow and the azimuth and distance of the flagpole relative to the two cameras are calculated.

Step 7. The local coordinate system is built with the flagpole as the origin, and the calculation results and measurement results are expressed in the local coordinate system for comparison.

The relative position between the flagpole and the camera can be determined by the distance between the flagpole and the camera and the angle between the flagpole and the camera and the two cameras. First, the angle between the flagpole and the camera line and the two camera lines is calculated, that is, the angles α and β, respectively (Figure 8). α and β are calculated using the images captured using cameras A and B. We carried out camera correction; therefore, the connecting line between the two observation points can be considered the transverse dividing line passing through the center point in the photos taken by cameras A and B. O is considered the center point for taking photos, and P is the imaging point of the flagpole vertex in the photo; it was assumed that the pixel coordinates of points O and P were (x_0, y_0) and (x, y), respectively. Then,

$$\sin\alpha = \frac{|y - y_0|}{\sqrt{(x - x_0)^2 + (y - y_0)^2}} \tag{19}$$

$$\alpha = \arcsin(\sin(\alpha)) \tag{20}$$

The α obtained using the above equation is in agreement with that obtained in Figure 7. β was obtained in a similar manner.

Using the obtained values of α and β, the distance between the flagpole and the camera was calculated, as shown in Figure 7, $DE \perp AB$, $DE = AB/(\cot\alpha + \cot\beta)$. Then, $AD = DE/\tan\alpha$, where AD is the horizontal distance between observation point A and flagpole vertex C. Similarly, the distance BD between the flagpole and observation point B could be calculated. Thus, we determined the relative positions of the flagpole and the camera.

Because the Sun is sufficiently far from the Earth, the sunlight reaching the Earth can be considered parallel light. Therefore, for the same object, the length of its shadow is determined by the solar altitude angle. A larger solar altitude angle implies a shorter shadow, smaller solar altitude angle, and longer shadow. As shown in Figure 8, when the cloud height h and the solar altitude angle α are known, then, the shadow length $d = H/\tan\alpha$.

The direction in which the shadow extends is opposite to the direction of the Sun; thus, the direction of the shadow can be calculated by the Sun azimuth, as shown in Figure 9; α is the azimuth of the Sun with 0° due north, and β is the shadow azimuth with 0° due north. $B = \alpha - 180°$.

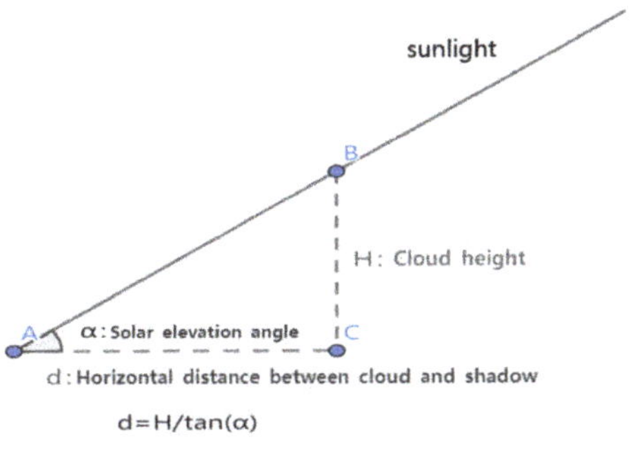

Figure 8. Demonstration of shadow length.

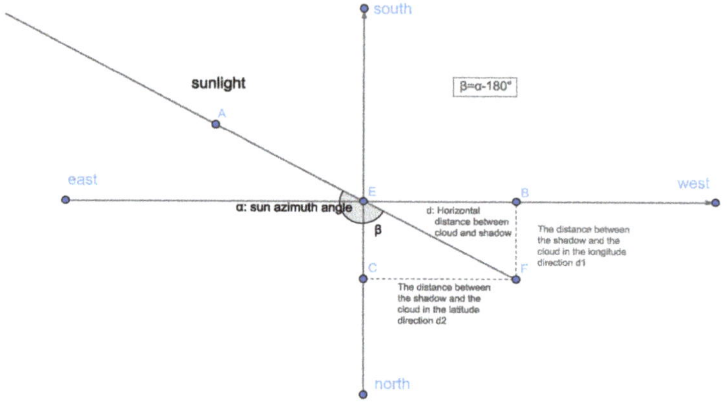

Figure 9. Demonstration of the shadow orientation.

The experimental contents and steps are as follows.

Two platforms were placed under the flagpole and adjusted to horizontal using a level ruler. The cameras were placed on a horizontal platform, and the lens was placed facing vertically upward to capture the top of the flagpole. The distance between two horizontal platforms as well as the length and extension direction of the flagpole in the ground shadow were measured and recorded. Any distortion of the captured picture was corrected, and then the height h of the flagpole was calculated. The solar altitude angle at that time was calculated according to the longitude and latitude of the shooting location and the shooting time α and azimuth β. Then, the shadow length and extension direction were calculated. The local coordinate system was built with the flagpole as the origin, and the calculation results and measurement results were expressed in the local coordinate system for comparison.

The pictures taken by the left and right cameras are presented in Figure 10.

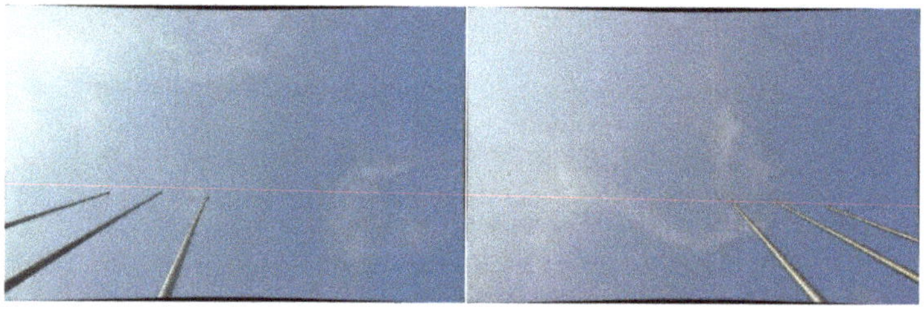

Figure 10. Pictures taken by the (**left**) and (**right**) cameras.

The pixel coordinates of the center point of the picture were (2144,1424); in the pictures taken by the left and right cameras, the pixels at the top of the middle flagpole were (2913,1849) and (1465,1797), respectively. Camera distance (baseline) was 7.35 m, the measured shadow length was 31.2 m, and the measured shadow orientation was 94° (0° due north). The angle between the connecting line of flagpole and the left camera and the connecting line of the two cameras was 28°, and the distance from the flagpole to the left camera was 3.9 m. The angle between the connecting line of the flagpole and the right camera and the connecting line of the two cameras was 30°, and the distance from the flagpole to the right camera was 3.8 m.

According to the principle of calculating the relative position between the flagpole and the cameras, the angle between the connecting line of the flagpole and the left camera and the connecting line of the two cameras was 28.93°, with an error rate of 3.32%. The distance from the left camera to the flagpole was 3.66 m, with an error rate of 6.15%. The angle of the connecting line between the flagpole and the right camera and the connecting line between the two cameras was 28.78°, with an error rate of 4.07%. The distance from the right camera to the flagpole was 3.69 m, with an error rate of 2.89%.

The calculated flagpole height was 16.3039 m. The longitude and latitude of the flagpole were 122.082920° E and 37.530085° N. The shadow was measured at 16:40:00 on 23 July 2021. The calculated solar altitude angle and azimuth angle were 27.2826° and 275.2011°, respectively.

According to the shadow length $d = H/\tan\alpha$, the calculated shadow length was 31.6117 m, and the error rate was ~1.32%.

Because the shadow azimuth equal to the sun azimuth minute 180° and the shadow azimuth was 95.2012°; therefore, three groups of experiments were carried out, and the results are shown in Table 2.

Table 2. Experimental results.

Group	P1	P2	BL	LS	SA	CLS	CSA	ER
1	(2633,1825)	(1465,1797)	5.73	31.2	94°	30.49	95.2012°	2.89%
2	(2733,1833)	(1465,1797)	6.5	31.2	94°	30.59	95.2012°	1.94%
3	(2913,1849)	(1465,1797)	7.35	31.2	94°	31.61	95.2012°	1.32%

P1 and P2—pixel coordinates of the top of the flag pole in the images taken by the left and right cameras, respectively; BL—length of the baseline, that is, distance between the two cameras (m); LS—measured shadow length of the flagpole (m); SA—azimuth angle of the flagpole shadow; CLS—flagpole shadow length predicted using the proposed method; CSA—flagpole shadow azimuth predicted using the proposed method; ER—error ratio of the predicted shadow length.

After many experiments, the experimental data of the solar altitude and solar azimuth were compared with the reference data, and the average errors were 0.0568° and 0.0629°. Therefore, we believe that the experimental method for calculating the solar altitude and solar azimuth is reliable.

3.2. *Verification Experiment and Results of Cloud Shadow Moving Track and Speed*

For continuously moving clouds, the LSTM network is used to predict the moving direction and speed of clouds from a set of continuous cloud images. Considering the influence of the change of solar orientation on the cloud shadow position, the cloud shadow cannot be predicted directly. The proposed method first predicts the cloud position and then calculates the cloud shadow position combined with the cloud height and solar orientation information. Because the cloud height changes, it is necessary to train the data pertaining to a change in cloud height. Using multiple groups of continuous cloud images, the cloud centroid and cloud height were obtained as the training set for training the LSTM network. The LSTM network brings an additional operation into the network through exquisite gate control, solving the problem of gradient disappearance.

After training the neural network, some data are used for prediction. After using the cloud centroid and cloud height calculated from continuous pictures as the input, the network predicts the cloud position and cloud height for a certain time in the future; then, the shadow position combined with the sun orientation information is calculated. This predicted position of the cloud is the position of the cloud centroid. The changes in cloud shape are irregular; therefore, the shape of the last input picture is used as the approximate shape of the prediction result.

Next, the prediction results were verified. For example, when the predicted position of the cloud was for 5 min later, then the cloud was photographed 5 min later to calculate the actual position and compare it with the predicted position.

According to the taken photos, it was observed that the cloud contour changed with time; thus, the contour information cannot adequately characterize a cloud (the cloud contours in the pictures taken at different times are different). Therefore, this study used the centroid of the cloud contour to identify the location characteristics of the cloud.

To calculate the center of mass, n contour mass points are set on the x–O–y coordinate plane as the mass points of $m_1, m_2, \ldots \ldots m_n$, respectively. Their coordinates can be (x_1, y_2), $(x_1, y_2), \ldots \ldots (x_n, y_n)$, respectively; then, these n particles comprise a particle system.

Furthermore, $\overline{x} = \frac{M_y}{M} = \frac{\sum_{i=1}^{n} m_i x_i}{\sum_{i=1}^{n} m_i}$, $\overline{y} = \frac{M_x}{M} = \frac{\sum_{i=1}^{n} m_i y_i}{\sum_{i=1}^{n} m_i}$, where M is the total mass of the contour point, and M_x and M_y are the static moments of the particle on the *x*- and *y*-axes, respectively; then point $(\overline{x}, \overline{y})$ is the required center of mass.

The centroid of the cloud was obtained by calling the API for computing the centroid in Opencv library function in PYTHON and giving the coordinate data of the cloud contour as the input to the proposed model.

The centroid data of the cloud was first obtained using the centroid acquisition method to recognize the cloud edge contour of the cloud image through the UNET network, and then the centroid position of the cloud was calculated according to the edge contour of the cloud, as shown in Figure 11.

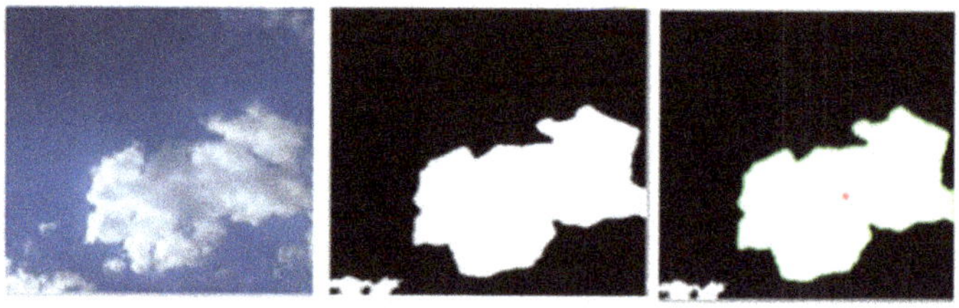

Figure 11. Method of calculating the centroid position.

From the images captured at 10 s intervals, we manually selected 11 images with whole clouds. Continuous photos were obtained by using the centroid acquisition method in succession; a data record contains 11 triples of information including cloud centroid longitude, cloud centroid dimension, and time, as shown in Figure 12.

```
id,Centroid_x,Centroid_y
0,2950,2958
0,2776,2816
0,2634,2757
0,2431,3619
0,2199,2474
0,2015,2368
0,1734,2264
0,1425,2151
0,1191,2014
0,1031,2020
0,635,1864
1,2715,2024
1,2533,1932
1,2328,1946
1,2056,1846,307605
1,1786,1747,438437
1,1631,1633,591926
1,1591,1421,720898
1,1408,1286,746991
1,1233,1166,587396
1,1037,1018,216669
1,878,965,13571
```

Figure 12. Cloud position data.

Figure 13 shows 1140 records of such data; the first 1040 pieces were used as the training dataset and the remaining 100 pieces were used as the test dataset. The data were divided into two parts. The centroid longitude and latitude from time 1 to time 10 were used as the training input set, and the centroid longitude and latitude from time 2 to time 11 were used as the verification result set. Similarly, the centroid longitude and latitude from time 1 to time 10 of each data in the test dataset were used as the input, and the output of the model was the centroid longitude and latitude prediction of the next time corresponding to each time.

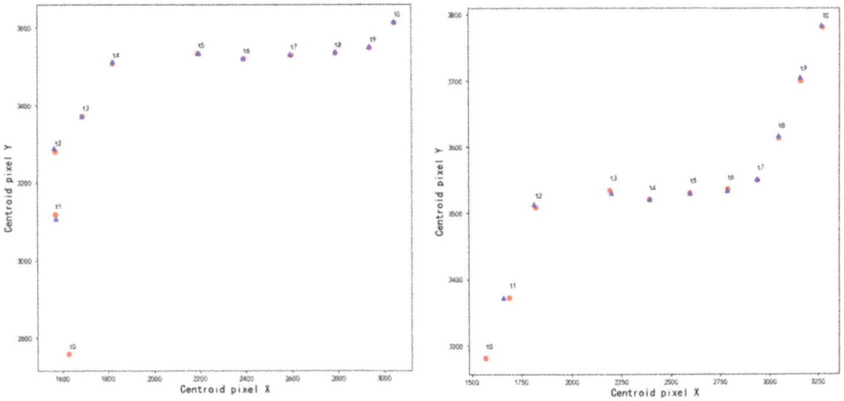

Figure 13. Comparison of the prediction of different cloud centroids at two times and corresponding real longitude and latitude.

The locus diagram of centroid points is drawn to demonstrate the real and predicted centroid longitude and latitude of the cloud, wherein the red dot is the real centroid longitude and latitude, the blue dot is the predicted centroid longitude and latitude, t0 is time 1, t1 is time 2, and so on. Figure 13 shows the comparison between the predicted results and the real longitude and latitude by tracking and predicting the centroids of different clouds at two different times. The red and blue dots in the figure are the true centroid longitude and latitude and the predicted centroid longitude and latitude, respectively.

After randomly selecting 100 centroid points, these were normalized and the mean square error was calculated, as shown in Figure 14. The blue points in the figure are the root mean square error of the predicted value. The root mean square error is the square root of the ratio of the sum of squares of prediction errors to the number of prediction times. A smaller value of the root mean square error implies a more accurate prediction.

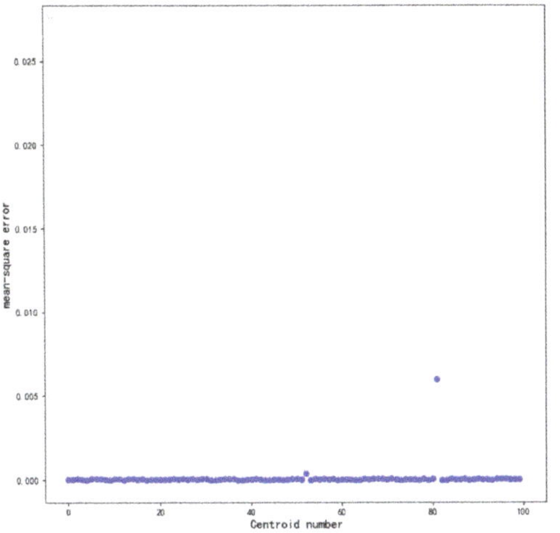

Figure 14. Mean square error of 100 centroids randomly selected; the blue points in the figure are the root mean square error of the predicted value.

Multiple groups of data were verified; then, the CNN-LSTM network model was used to predict the moving trajectory of the cloud more accurately.

4. Conclusions

This study presents a new low-cost and easy-to-implement method for predicting the influence of cloud on solar radiation.

This method can accurately predict the trajectory of a cloud and be used at solar power stations to effectively predict the location of cloud shadows in tens of minutes, thus enabling an adjustment of the solar panels to a suitable angle in advance. Compared with other implementations, it can save product costs and increase the rate of generation of solar panel energy. This research is conducive to the progress of related works of solar radiation and energy generation.

The proposed method also had some limitations. The sky camera's shooting field of vision is limited; thus, the calculation and prediction of clouds in the sky are affected to some extent and can only be predicted within a limited range. Considering this limitation, during the actual implementation, the cloud prediction range can be increased by deploying sky cameras at multiple points around the area of the photovoltaic power station.

Author Contributions: Conceptualization, S.W.; Methodology, M.S. and Y.S.; Validation, S.W. and M.S.; Formal analysis, S.W.; Investigation, S.W. and M.S.; Supervision, Y.S.; Writing—original draft preparation, S.W.; Writing—review and editing, Y.S. and M.S. All authors have read and agreed to the published version of the manuscript.

Funding: This research received no external funding.

Conflicts of Interest: The authors declare no conflict of interest.

References

1. Detollenaere, A.; Wetter, J.V.; Masson, G. Snapshot of Global PV Markets 2020. *IEA Photovolt. Power Syst. Programme* **2020**, *4*, 10–11. Available online: https://iea-pvps.org/wp-content/uploads/2020/04/IEA_PVPS_Snapshot_2020.pdf (accessed on 1 September 2022).
2. Behera, M.K.; Majumder, I.; Nayak, N. Solar photovoltaic power forecasting using optimized modified extreme learning machine technique. *Eng. Sci. Technol. Int. J.* **2018**, *21*, 428–438. [CrossRef]
3. Antonanzas, J.; Osorio, N.; Escobar, R.; Urraca, R.; Martinez-de-Pison, F.J.; Antonanzas-Torres, F. Review of photovoltaic power forecasting. *Sol. Energy* **2016**, *136*, 78–111. [CrossRef]
4. Perez, R.; Kivalov, S.; Schlemmer, J.; Hemker, K., Jr.; Renné, D.; Hoff, T.E. Validation of short and medium term operational solar radiation forecasts in the US. *Sol. Energy* **2010**, *84*, 2161–2172. [CrossRef]
5. Sfetsos, A.; Coonick, A.H. Univariate and multivariate forecasting of hourly solar radiation with artificial intelligence techniques. *Sol. Energy* **2000**, *68*, 169–178. [CrossRef]
6. Barbieri, F.; Rajakaruna, S.; Ghosh, A. Very short-term photovoltaic power forecasting with cloud modeling: A review. *Renew. Sustain. Energy Rev.* **2017**, *75*, 242–263. [CrossRef]
7. Hu, K.; Cao, S.; Wang, L.; Li, W.; Lv, M. A new ultra-short-term photovoltaic power prediction model based on ground-based cloud images. *J. Clean. Prod.* **2018**, *200*, 731–745. [CrossRef]
8. Bacher, P.; Madsen, H.; Nielsen, H.A. Online short-term solar power forecasting. *Sol. Energy* **2009**, *83*, 1772–1783. [CrossRef]
9. Ghimire, S.; Deo, R.C.; Raj, N.; Mi, J. Deep solar radiation forecasting with convolutional neural network and long short-term memory network algorithms. *Appl. Energy* **2019**, *253*, 113541. [CrossRef]
10. Si, Z.; Yang, M.; Yu, Y.; Ding, T. Photovoltaic power forecast based on satellite images considering effects of solar position. *Appl. Energy* **2021**, *302*, 117514. [CrossRef]
11. Lin, F.; Zhang, Y.; Wang, J. Recent advances in intra-hour solar forecasting: A review of ground-based sky image methods. *Int. J. Forecast.* **2021**. [CrossRef]
12. Wang, G.C.; Urquhart, B.; Kleissl, J. Cloud base height estimates from sky imagery and a network of pyranometers. *Sol. Energy* **2019**, *184*, 594–609. [CrossRef]
13. Hutchison, K.; Wong, E.; Ou, S.C. Cloud base heights retrieved during night-time conditions with MODIS data. *Int. J. Remote Sens.* **2006**, *27*, 2847–2862. [CrossRef]
14. Theocharides, S.; Makrides, G.; Livera, A.; Theristis, M.; Kaimakis, P.; Georghiou, G. Dayahead photovoltaic power production forecasting methodology based on machine learning and statistical post-processing. *Appl. Energy* **2020**, *268*, 115023. [CrossRef]
15. Li, Z.; Li, J.; Menzel, W.P.; Schmit, T.J.; Ackerman, S.A. Comparison between current and future environmental satellite imagers on cloud classification using MODIS. *Remote Sens. Environ.* **2007**, *108*, 311–326. [CrossRef]

16. Zhang, X.; Liu, K.; Wang, X.; Yu, C.; Zhang, T. Moving Shadow Removal Algorithm Based on HSV Color Space. *TELKOMNIKA Indones. J. Electr. Eng.* **2014**, *12*, 2769–2775. [CrossRef]
17. Peng, Z.; Yu, D.; Huang, D.; Heiser, J.; Yoo, S.; Kalb, P. 3D cloud detection and tracking system for solar forecast using multiple sky imagers. *Sol. Energy* **2015**, *118*, 496–519. [CrossRef]
18. Urquhart, B.; Kurtz, B.; Dahlin, E.; Ghonima, M.; Shields, J.E.; Kleissl, J. Development of a sky imaging system for short-term solar power forecasting. *Atmos. Meas. Tech.* **2015**, *8*, 875–890. Available online: https://amt.copernicus.org/articles/8/875/2015/amt-8-875-2015.pdf (accessed on 15 September 2022). [CrossRef]
19. Escrig, H.; Batlles, F.; Alonso, J.; Baena, F.; Bosch, J.; Salbidegoitia, I.; Burgaleta, J. Cloud detection, classification and motion estimation using geostationary satellite imagery for cloud cover forecast. *Energy* **2013**, *55*, 853–859. [CrossRef]
20. Héas, P.; Mémin, É. Three-dimensional motion estimation of atmospheric layers from image sequences. *IEEE Trans. Geosci. Remote Sens.* **2008**, *46*, 2385–2396. Available online: https://www.researchgate.net/publication/3205956_Three-Dimensional_Motion_Estimation_of_Atmospheric_Layers_From_Image_Sequences (accessed on 15 September 2022). [CrossRef]
21. Chu, Y.; Pedro, H.T.; Coimbra, C.F. Hybrid intra-hour DNI forecasts with sky image processing enhanced by stochastic learning. *Sol. Energy* **2013**, *98*, 592–603. [CrossRef]
22. Caldas, M.; Alonso-Suárez, R. Very short-term solar irradiance forecast using all-sky imaging and real-time irradiance measurements. *Renew. Energy* **2019**, *143*, 1643–1658. [CrossRef]
23. Wang, F.; Xuan, Z.; Zhen, Z.; Li, Y.; Li, K.; Zhao, L.; Shafie-khah, M.; Catalão, J.P. A minutely solar irradiance forecasting method based on real-time sky image-irradiance mapping model. *Energy Convers. Manag.* **2020**, *220*, 113075. [CrossRef]
24. El Alani, O.; Abraim, M.; Ghennioui, H.; Ghennioui, A.; Ikenbi, I.; Dahr, F.-E. Short term solar irradiance forecasting using sky images based on a hybrid CNN-MLP model. *Energy Rep.* **2021**, *7*, 888–900. [CrossRef]
25. Du, J.; Min, Q.; Zhang, P.; Guo, J.; Yang, J.; Yin, B. Short-Term Solar Irradiance Forecasts Using Sky Images and Radiative Transfer Model. *Energies* **2018**, *11*, 1107. [CrossRef]
26. Kong, W.; Jia, Y.; Dong, Z.Y.; Meng, K.; Chai, S. Hybrid approaches based on deep whole-sky-image learning to photovoltaic generation forecasting. *Appl. Energy* **2020**, *280*, 115875. [CrossRef]
27. Voyant, C.; Notton, G.; Kalogirou, S.; Nivet, M.L.; Paoli, C.; Motte, F.; Fouilloy, A. Machine learning methods for solar radiation forecasting: A review. *Renew. Energy* **2017**, *105*, 569–582. [CrossRef]
28. Gala, Y.; Fernández, Á.; Díaz, J.; Dorronsoro, J.R. Hybrid machine learning forecasting of solar radiation values. *Neurocomputing* **2016**, *176*, 48–59. [CrossRef]
29. Fernández, Á.; Gala, Y.; Dorronsoro, J.R. Machine learning prediction of large area photovoltaic energy production. In *Data Analytics for Renewable Energy Integration*; Springer: Cham, Switzerland, 2014; pp. 38–53. Available online: http://link.springer.com/10.1007/978-3-319-13290-7_3 (accessed on 15 September 2022).
30. Mellit, A.; Massi Pavan, A.; Lughi, V. Deep learning neural networks for short-term photovoltaic power forecasting. *Renew. Energy* **2021**, *172*, 276–288. [CrossRef]
31. Huang, G.B.; Zhu, Q.Y.; Siew, C.K. Extreme learning machine: Theory and applications. *Neurocomputing* **2006**, *70*, 489–501. [CrossRef]
32. Suh, S.; Kim, J.; Lukowicz, P.; Lee, Y.O. Two-stage generative adversarial networks for binarization of color document images. *Pattern Recognit.* **2022**, *130*, 108810. [CrossRef]
33. Terrén-Serrano, G.; Martínez-Ramón, M. Multi-layer wind velocity field visualization in infrared images of clouds for solar irradiance forecasting. *Appl. Energy* **2021**, *288*, 116656. [CrossRef]
34. Paoli, C.; Voyant, C.; Muselli, M.; Nivet, M. Forecasting of preprocessed daily solar radiation time series using neural networks. *Sol. Energy* **2010**, *84*, 2146–2160. [CrossRef]
35. Cao, J.; Lin, X. Study of hourly and daily solar irradiation forecast using diagonal recurrent wavelet neural networks. *Energy Convers. Manag.* **2008**, *49*, 1396–1406. [CrossRef]
36. Zhao, X.; Wei, H.; Wang, H.; Zhu, T.; Zhang, K. 3D-CNN-based feature extraction of ground-based cloud images for direct normal irradiance prediction. *Sol. Energy* **2019**, *181*, 510–518. [CrossRef]
37. Wu, K.; Xu, Z.; Lyu, X.; Ren, P. Cloud detection with boundary nets. *ISPRS J. Photogramm. Remote Sens.* **2022**, *186*, 218–231. [CrossRef]
38. Wang, Q.; Zhou, C.; Zhuge, X.; Liu, C.; Weng, F.; Wang, M. Retrieval of cloud properties from thermal infrared radiometry using convolutional neural network. *Remote Sens. Environ.* **2022**, *278*, 113079. [CrossRef]
39. Fang, H.; Lafarge, F. Pyramid scene parsing network in 3D: Improving semantic segmentation of point clouds with multi-scale contextual information. *ISPRS J. Photogramm. Remote Sens.* **2019**, *154*, 246–258. [CrossRef]

sustainability

MDPI

Article

Maritime Autonomous Surface Ships: Problems and Challenges Facing the Regulatory Process

Mohamad Issa [1,2,*], Adrian Ilinca [2], Hussein Ibrahim [3] and Patrick Rizk [2]

1 Institut Maritime du Québec à Rimouski, Rimouski, QC G5L 4B4, Canada
2 Département de Génie, Université du Québec à Rimouski, Rimouski, QC G5L 3A1, Canada
3 Centre de Recherche d'Innovation en Intelligence Énergétique, Sept-Îles, QC G4R 5B7, Canada
* Correspondence: missa@imq.qc.ca

Abstract: Technological innovation constantly transforms and redefines the human element's position inside complex socio-technical systems. Autonomous operations are in various phases of development and practical deployment across several transport domains, with marine operations still in their infancy. This article discusses current trends in developing autonomous vessels and some of the most recent initiatives worldwide. It also investigates the individual and combined effects of maritime autonomous surface ships (MASS) on regulations, technology, and sectors in reaction to the new marine paradigm change. Other essential topics, such as safety, security, jobs, training, and legal and ethical difficulties, are also considered to develop a solution for efficient, dependable, safe, and sustainable shipping in the near future. Finally, it is advised that holistic approaches to building the technology and regulatory framework be used and that communication and cooperation among various stakeholders based on mutual understanding are essential for the MASS to arrive in the maritime industry successfully.

Keywords: autonomous shipping; MASS; IMO; maritime law; Maritime Safety Committee; advanced sensor module; shore control center; cyber security threats

Citation: Issa, M.; Ilinca, A.; Ibrahim, H.; Rizk, P. Maritime Autonomous Surface Ships: Problems and Challenges Facing the Regulatory Process. *Sustainability* **2022**, *14*, 15630. https://doi.org/10.3390/su142315630

Academic Editors: Luis Hernández-Callejo, Sergio Nesmachnow and Sara Gallardo Saavedra

Received: 11 October 2022
Accepted: 22 November 2022
Published: 24 November 2022

1. Introduction

With fast-increasing technology, a new paradigm shift is occurring, considering alternative marine fuels that promise safer, greener, and more efficient ships than ever before in response to stringent international legislative requirements. The first change occurred during the First Industrial Revolution in the 1800s when mechanical power was introduced, and vessels began to be driven by steam-powered coal engines. The Second Industrial Revolution began in the early 1900s when the advent of diesel engines improved the efficiency and reliability of ships by using oil as a new fuel. The internet–digital revolution, representing the Third Industrial Revolution, introduced computerized ship control in the 1970s. With the introduction of gas as a fuel, such as liquefied natural gas (LNG) [1–5], we are taking a step closer to the new paradigm linked with cyber-physical systems and autonomy as part of "Shipping 4.0 [6–8]".

Porathe et al. [9] present four reasons why autonomous shipping is seen as a feasible choice: (1) the efforts to reduce transportation costs; (2) the need for a better onboard working environment for crews and the prevention of future seafarer shortages; (3) the need to reduce emissions on a worldwide scale; and (4) the desire to improve shipping safety. According to a 2010 report submitted to the International Maritime Organization (IMO) by the Baltic and International Maritime Council (BIMCO) and the International Shipping Federation (ISF), the shipping industry is expected to face tightening labor markets, with recurrent shortages of ship officers [10], due to hazardous working conditions and extended periods away from land. Under the fiercely competitive economy of scale, the shipping industry has seen downward pressure on freight rates and excess capacity. Reduced ship pollution and emissions and improved ship safety are more important than ever with the emergence of low- or zero-carbon alternative fuels [11].

Under these conditions, the launch of maritime autonomous surface ships (MASS) will be a watershed moment that will either disrupt or precipitate a paradigm shift in the shipping sector and maritime transport system. Therefore, communication and coordination among stakeholders, particularly those involved in the maritime and port industries, would be required for the safe, effective, and efficient adoption and operation of MASS. As a result, critical concerns related to autonomous shipping and their impact on policy, technology, and industry should be investigated together with their interaction for a successful introduction and smooth settlement of MASS and associated infrastructures in the marine industry.

On the regulatory side, the IMO agreed to conduct a regulatory scoping exercise (RSE) to assess the safe, secure, and environmentally sound operation of MASS [12]. However, the RSE would be a complicated issue because it would touch a few areas, including safety, security, contacts with ports, pilotage in the event of an incident, and the marine environment. In addition, international maritime conventions, such as the International Convention for the Safety of Life at Sea (SOLAS), the International Regulations for the Prevention of Collisions at Sea (COLREG), and the Standards for Training and Certification of Watchkeepers (STCW), apply to MASS [13]. Therefore, IMO Member States will be asked to review the scope of their domestic laws considering the RSE.

Technological development will improve ships' control capabilities, communication, and interfaces using the newest information and communications technology (ICT) systems. As a result, they will soon be operated by remote land-based or offshore services [14]. Unmanned watercraft have already been deployed for military, aeronautical, and research purposes. Deep-sea exploration also uses submersible unmanned vehicles, such as autonomous underwater vehicles (AUV) and remotely operated vehicles (ROV), which are still being developed. However, regarding safety, efficiency, and environmental protection, the technology that replaces manning must outperform the personnel [15].

On the industrial side, autonomous vehicles are already being developed in various means of transportation, such as airplanes, trains, and automobiles. Therefore, MASS is expected to significantly impact shipbuilding, equipment, and devices, as well as shipping and port infrastructures in the maritime industry. Furthermore, autonomy, automation, unmanned operation, big data, enterprise-grade connectivity, and analytics will steadily grow in the maritime industry [16]. As a result, good communication and coordination with essential stakeholders, particularly the shipping, shipbuilding, and port industries, are required to implement MASS properly.

To the best of the authors' knowledge, several review studies have discussed briefly or deeply the regulatory challenges concerning MASS. The authors' discussions and interviews with maritime experts such as naval officers, senior marine engineers, and naval architects inspired this essay. In this paper, the authors have chosen to focus on all the effects that MASS may have on the maritime industry at the human level (such as training and education), legislative level (definition of transparent laws and regulations), and technological level (such as security navigation). The paper presents some reflections on the obstacles and issues that need to be clarified soon. It does not deal with data based on experiments, calculations, or quantified scenarios. The primary motivation of the authors is to present the magnitude of these challenges and the work that remains to be done to achieve safe autonomous surface ship navigation worldwide.

The structure of the present paper is as follows. We first introduced the latest projects on global trends for building autonomous vessels. Second, the impact of MASS on regulations, technology, and industries has been explored, as well as their relationships to uncover both previous and future efforts to prepare for the new maritime paradigm change. Finally, other essential problems, e.g., safety, security, jobs, training, ethics, liability, and insurance, were explored to obtain greater insight regarding future shipping that is efficient, reliable, safe, and sustainable.

2. Global Autonomous Vessel Developments

The shipping industry has recently faced changes due to the Fourth Industrial Revolution. One such transition is the AI (artificial intelligence), robots, IoT (internet of things), and autonomous vehicles paradigm shift in technological progress [17]. Big data and the achievements of the Third Industrial Revolution have been integrated with AI and IoT technology to enable smart shipping. Autonomous ships, e-Navigation, and smart ports are further examples of marine transportation advancements.

Many companies, including Rolls Royce, DNV, the Norwegian University of Science and Technology (NTNU), and Norway's Kongsberg, have announced ambitious intentions to create all-electric and autonomous container ships by 2020, as shown in Table 1. Other groups worldwide are working on similar, if not competing, concepts and systems to enable unmanned operations and infrastructure initiatives, such as autonomous ports and high-speed communications.

Table 1. Maritime full-electric and autonomous vessels projects since 2012.

Project Name and Period	Developers	Characteristics
The e5 Project—Asahi tanker (2019–2022)	Asahi Tanker Co., Ltd., Exeno Yamamizu Corporation, Mitsui O.S.K. Lines Ltd., and Mitsubishi Corporation	Tonnage: 499 Propelled by electric motors driving two azimuth pods (2×300 kW) and two tunnel thrusters (2×68 kW) Batteries rated 3.5 MWh Maximum speed: 10 knots
Yara Birkeland (2017–2021)	Yara, Kongsberg, NTNU, and DNV	120TEU containership (length:80 m) Fully autonomous Propelled by electric motors driving two azimuth pods and two tunnel thrusters Batteries rated 7–9 MWh Maximum speed: 13 knots
AAWA (2015–2018)	Rolls Royce, DNV, Aalto University, University of Turku, Inmarsat, etc.	Standard development of safety, policy, economy, and collision avoidance module
Revolt (2014–2018)	DNV and NTNU	100TEU containership (length: 60 m) Autonomous Fully battery-powered (300 kWh) Maximum speed: 6 knots
MUNIN (2012–2015)	consists of eight partners with both scientific and industrial backgrounds located in Germany, Norway, Sweden, Iceland and Ireland	Development of a concept for an autonomous dry bulk carrier (length: 200 m) Development of sensor, navigation, and communication systems

In 2012, the European Commission-funded project Maritime Unmanned Navigation through Intelligence in Networks (MUNIN) began looking into unmanned ships' feasibility in various areas, including technical maturity, economic benefits, social impact, and safety during deep-sea voyages [18,19]. Following the MUNIN project, DNV and NTNU launched the Revolt as a specific research project to build an autonomous, zero-emission, and short-sea vessel to help manage traffic congestion in urban regions on the EU's road network [20,21].

The Advanced Autonomous Waterborne Applications Initiative (AAWA), founded by Rolls-Royce in 2015, is another notable initiative related to autonomous vessels. This project brought together a diverse group of stakeholders, including universities, ship designers, equipment manufacturers, and classification societies, to examine the economic, social, legal, regulatory, and technological barriers that must be overcome for autonomous ships to become a reality. Its goal is to provide preliminary designs for the future generation of innovative ship solutions, complete with technical specifications [22].

The Yara Birkeland is one of the most recent autonomous ship initiatives. Yara and Kongsberg built the world's first totally electric container feeder vessel. Reducing up to 40 thousand truck travel in densely populated urban areas is estimated to significantly cut NO_X and CO_2 emissions while enhancing road safety and alleviating traffic congestion [23,24].

Last but not least, the e5 project is a Japanese consortium dedicated to developing renewable energy-powered commercial ships. The name "e5" refers to the partnership's five "focus points": electrification, environment, evolution, efficiency, and economics. The e5 Tanker claims to be the world's first entirely electric oil tanker, with a 3.5 MWh battery that can "operate non-stop for 10 h on a half-capacity battery", according to the company [25]. In addition, the ship will have a high level of automation [26] and will be charged using wind and solar energy to cut emissions further [27].

3. Problems and Challenges Facing the Regulatory Process

The fact that all technical shipping rules relating to the safety of navigation, environmental protection, and training/watchkeeping standards were designed with the idea that humans would do some functions must be reviewed in the context of autonomous vessels. A few instances are sufficient to demonstrate the flaws in the current regulatory structure if applied to the MASS operations without modification. Chapter V, regulation 24 of the International Convention for the Safety of Life at Sea (SOLAS) 1974 requires that manual control of the ship's steering be established promptly in dangerous navigational situations, or an autonomous ship without a crew will be unable to comply with this law [28]. Regulations that need human judgment are a more complex matter. It is unclear how this law would apply to vessels designed to make navigational decisions using algorithms based on data collected from their sensors. Rule 2 of the International Regulations for Preventing Collisions at Sea (COLREGs) 1972, for example, states that nothing shall exonerate any vessel, or the owner, master, or crew thereof, from the consequences of any neglect or any precaution which may be required by ordinary practice of seamen. Those developing the new technology often remind us that deep learning based programs are flexible and react to and from the new patterns which are programmed to identify, meaning that a program could learn situational awareness and the subjective aspects of COLREGs. Even so, this poses a significant challenge to those who seek to regulate the matter.

Additionally, there are severe risks in today's fully automated ships, such as sensor defects and software errors. For example, aviation incidents involving the Boeing 737 MAX in 2018 and 2019 are examples where the airplane's angle sensors gave the altitude control system inaccurate information. As a result, the airplane crashed because it was challenging to bypass the mechanism manually. As a result, under the current blame system, harm brought on by improper algorithms may be categorized as a product defect (and hence a technical failure) and negligence (based on the root issue).

Although there is no definitive answer at this time regarding how, if at all, regulations like the COLREGs will be modified for MASS application, it is a crucial topic of discussion in the maritime sector. The implementation of the COLREGs with MASS is facing numerous obstacles, based on the information at the time this article was written. For most of the rules, participants preferred the original COLREGs. However, some rules were preferred with modest modifications. Most of these findings are consistent with the regulatory scoping exercise conclusions from the IMO. Adding or refining meanings for terminology, e.g., "master and crew", "the common practice of sailor", "crew ashore", and "lookout" were among the most popular revisions.

Additionally, an all-around colored MASS-identifying light was selected to add different traffic separation schemes that are required for MASS. Since almost 75% of participants preferred more than one amendment over the original regulation, it was clear that participants were amenable to some adjustment. Additionally, those who have had more practice using the COLREGs demonstrated a modest propensity toward selecting the revised rules compared to participants who had had less practice. To better train seamen for the future as the maritime sector adopts autonomy, it is crucial that MASS and its impact on the COLREGs and other IMO instruments are further investigated immediately.

3.1. Impact on Regulation

Despite the rapid advancement of science and technology in the marine industry, autonomous vessels must unquestionably adhere to international standards in order to operate securely between nations and even seabed areas outside of national authority. Although some parts of manned vessel regulation, such as some clauses of the International Safety Management (ISM) Code, may be compatible with unmanned vessels, there is a need for unique international rules to consider the characteristics of unmanned vessels. A request for RSE was recently submitted to the Maritime Safety Committee (MSC) and was incorporated into the MSC work plan at MSC 98 [29] to ensure MASS safety, security, and environmental soundness. The RSE for MASS aims to determine the degree of autonomy that may affect existing regulatory frameworks to address MASS operations. The degrees of autonomy at MSC 100 [30] were divided into four phases to help with the RSE process (see Figure 1). One should emphasize that MASS can operate in multiple levels of autonomy during a single voyage.

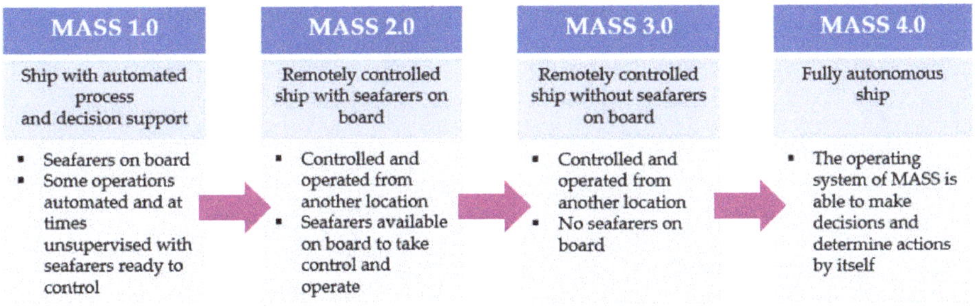

Figure 1. MASS's level of automation, according to IMO.

All conventions seem obsolete, and new regulatory standards will be needed. It is recommended that all IMO committees and subcommittees work together using the goal-based approach. The MSC recently authorized a revision of generic principles for producing IMO goal-based standards (GBS) to set safety goals and functional requirements while considering the whole MASS lifetime [31]. Risk assessment and software quality assurance (SQA) will be necessary, in addition to the GBS, for MASS's safety in both the real and virtual worlds.

Autonomous shipping is a new technology requiring an international regulatory or harmonization between existing regulatory rules for all states' territorial waters. The matter is further complicated as rules and regulations are embodied in several international agreements over the last century or so, in some cases after years of negotiations conducted by the international community. Until an international consensus on regulating this new technology is reached, it is doubtful that autonomous ships will operate in international waters beyond any state's territorial waters [32].

3.2. Impact on Technology

Demonstrating that autonomous systems are at least as safe as piloted ship systems and providing the ship shore control center (SSCC) with enough situation awareness represent one of the most challenging issues in building the technology for MASS. The ship systems should be remotely monitored and managed by the operators of the SCC to obtain essential information through satellite at short intervals in case of emergencies such as rescue attempts or evasive maneuvers. If the autonomous system fails, the SSCC should include a smart alarm system and the capacity to transition to manual control mode. Figure 2 depicts the MASS and SSCC systems, their essential equipment and operations, and the relationship formed by satellite data.

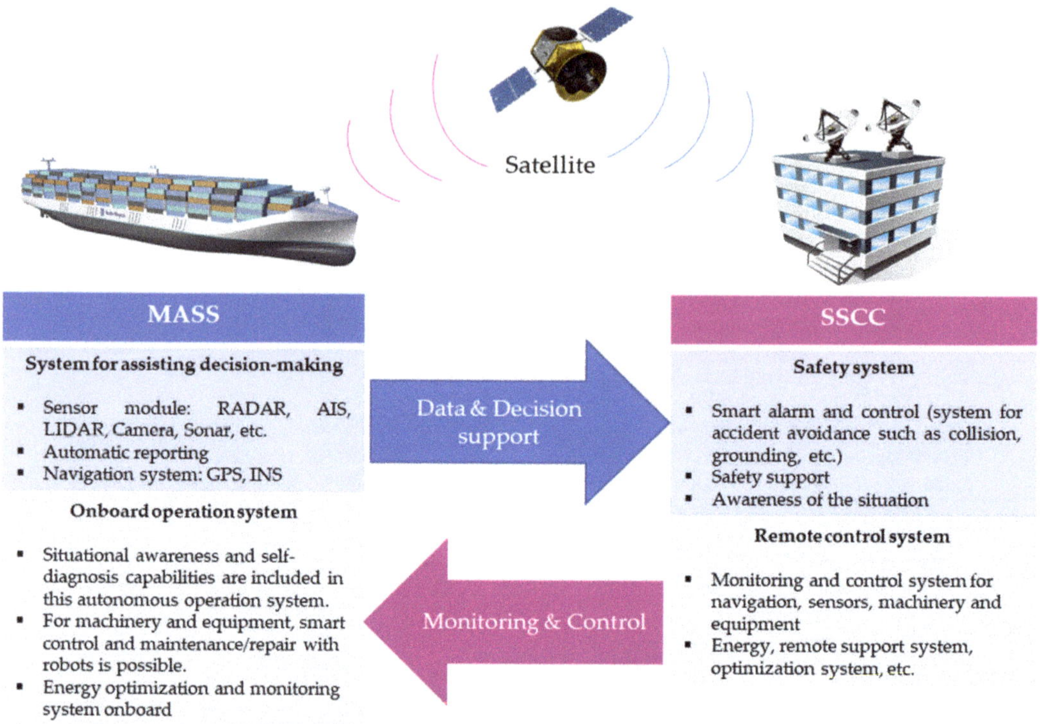

Figure 2. The relationship between MASS and the ship-shore-control-center (SSCC).

The sensors' dependability must be ensured through design approval, remote and on-premises testing, and monthly inspections, particularly for sensors that support monitoring and decisions from SCC. Sensor failures pose a significant risk to the system's safety. Therefore, the most significant safety sensors should consider redundancy, diagnostics, prognosis, and homogeneous and heterogeneous redundancy. It's worth noting that heterogeneous redundancy is more dependable than others because it can eliminate sensor-type dependency [33]. A more extensive elicitation of experts could also be advantageous to overcome various concerns connected to threats affecting autonomous ships' safe and efficient operations due to a lack of failure data and easy access to the data.

The Relevance of Cyber Risks Management for Shipping Operations

Based on their complexity, transportation systems may have the following four levels of cyber systems:

The first is the perceptual layer, which uses components, such as wireless sensors and GPS, to connect the cyber and physical worlds. The second type is network systems, which it's used to convey data (e.g., satellite networks and the internet mobile communication network). The third tier is the support layer, which includes cloud computing and intelligent computing, and the fourth layer is the application layer, which connects people and the physical world to cyber systems (e.g., intelligent transportation and environmental monitoring), see Figure 3. All these four layers are present in the context of the modern vessel. Such integration is achieved utilizing Ethernet Industrial Protocols that collect and process data via wireless and fiber optic sensors, cameras, radars, satellite communications, and cloud computing.

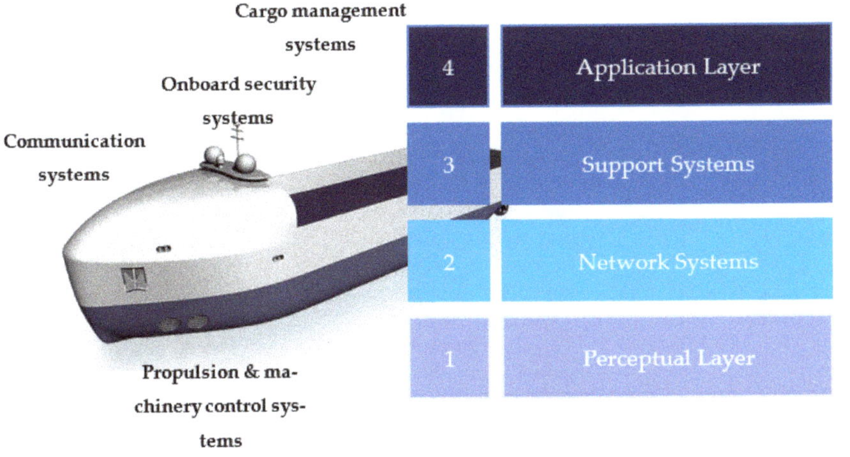

Figure 3. Illustration of the main systems integrated into modern vessels and the four levels of cyber systems.

While the integration of technology promises to make sea transportation safer, more ecologically friendly, and entertaining while lowering costs, it also raises the risk of disrupted vessel operations. The rising use of information technology (IT) systems during marine transportation eliminates the need for the perpetrator to bypass physical security measures, as happened in the 9/11 attacks. Taking control of a vessel or disrupting its operation can now potentially be achieved electronically by remotely interfering with any other of these four layers. Interference can be achieved in a variety of ways, the most prominent of which are as follows:

- Injecting malicious software, such as malware, viruses, trojans, and worms, into a vessel's IT-controlled power management system or navigational system, which could corrupt chart data stored in an electronic chart display and information system (ECDIS); a failure occurring during ship software maintenance [34]. Such instances may force the vessel to remain in port until the malicious software is removed, fresh ECDIS computers are installed, and a classification surveyor is present [35].
- Spoofing or jamming a vessel's positioning systems, such as the Global Positioning System (GPS) or Global Navigation Satellite System (GNSS), or the tracking system of containers.
- Infecting the ship's primary server with ransomware, encrypting sensitive files and apps relating to customs, passengers, and scheduling, and allowing only ransom payment to unlock them [36].
- By-pass the firewalls between the vessel's public and safety-critical network and gain access to operationally critical data and processes.

The IMO was alarmed by two events in particular: the first occurred in 2017, when at least 20 vessels in the Black Sea appeared in the automatic identification system (AIS) 20 miles inland, close to a Russian airport [37], and the second occurred between 2011 and 2013 when a criminal gang infiltrated the container tracking system at the Port of Antwerp located in Flanders (Belgium) and stole containers in which illicit substances were hidden, unbeknownst to their owners [38].

The IMO, alerted by such instances, emphasized the necessity for enterprise-wide cyber risk management by all industry stakeholders, including public authorities and commercial companies. Given that interconnectivity is the fundamental pillar of digitalized and autonomous operations, it's understandable that those recommendations would be addressed to such a large audience. However, perhaps the most critical recommendation of the IMO is that a cyber risk management program is included in safety management systems.

3.3. Impact on Industry

The shipping industry has relied on the knowledge and experience of ship crews for hundreds of years. With unmanned vessels, autonomous technology is designed to revolutionize the marine sector. Small autonomous boats have already entered operation, while larger vessel technology is still developing. It is time for the marine industry to embrace autonomy and comprehend how it will influence the industry's future and how to utilize it best. MASS will affect ship design, shipbuilding, and port infrastructure, including services and interfaces. On-shore shipping ports will be transformed by automation, from port infrastructure and cargo handling to land-based logistics and transportation. One of the logistics industry's goals is to provide fast service, which allows shippers and customers to adjust dispatches and receive deliveries from this self-contained logistics transport chain on the fly [39].

Communication and cooperation among MASS stakeholders based on mutual understanding will be critical to the MASS's successful introduction to the marine industry. Figure 4 depicts the main stakeholders and their relationships. Stakeholders in the maritime sector would include seafarers onboard and ashore, insurance companies, cargo and bunkering corporations, research institutions, universities, and training centers. Furthermore, autonomous vessels will transform existing industries by introducing system integration and control, system management and maintenance, SSCC operation and management, fleet management, cybersecurity, big data analysis, smart sensors, and communication. Furthermore, to make autonomous ships effective and dependable, development, alteration, and interpretation of maritime rules and regulations, as well as communication and cooperation among stakeholders, are essential for the MASS to be successful.

Figure 4. The main stakeholders for ships.

3.4. Impact on Jobs and Training

While the marine business is rapidly expanding, finding suitably skilled sailors is a constant challenge. Lloyds Register [40], in particular, forecasted severe shortages of skilled officers and crews by 2025. Furthermore, the introduction of MASS has generated concerns about the seafarers and positions to be replaced by AI and autonomous systems. However, this change will trigger new business and jobs for highly qualified crews and operators, particularly those with knowledge of technology, IT systems, engineering, and public relations and regulations [41].

Crewmembers' training needs to focus on different skills and competencies, from seafaring skills and automation and communication engineering knowledge, where the engineering support team ensures the communication between the shore team and the automated ship in an efficient, bidirectional way [42]. In addition, watchkeeping personnel and companies have an essential role in ensuring safe faring. For practicing challenging safety situations, a well-designed simulator is used, but the only problem with that simulator is the inability to create real-time challenging safety situations, which require creativity and deep knowledge of seagoing accidents. Thus, ship operators require a combination of nautical and technological expertise, such as voyage planning, digital and port approaches for communication duties, mooring, unmooring, ship monitoring, and docking [40–42].

The use of automation could mitigate the predicted worker shortage. Many maritime jobs will be transferred to land-based SSCC due to remote and autonomous operations, allowing the industry to recruit new people who find a marine career onshore more appealing. It is also expected that autonomous ships will improve seafarers' quality of life. The difficulties of staying on board for extended periods and the risks of marine mishaps will be reduced if ships are controlled from the shore.

MASS on-shore operators receive relevant training and education under the International Convention on Standards of Training, Certification, and Watchkeeping for Seafarers (STCW). However, in light of the declining number of seafarers, it may also be essential to explore developing new STCW Convention qualifying criteria or new knowledge, understanding, and proficiencies. Thus, while applying a reliable maritime education and training (MET), qualified trainers must be considered along with their ability to teach and assess their trainees. An effective training methodology must hold a cognitive, psychomotor, and affective learning approach with clear objectives corresponding to the domain and level of the required competencies. Moreover, the trainers must be creative and engage the trainees in the learning process by promoting a leadership spirit in an appropriate way, i.e., seeing, thinking, and applying what is learned. Finally, continuous educational research and training must be provided to face future challenges in shipping while applying MASS.

3.5. Issue of Laws and Ethics

The industry has embraced advanced and new technologies to boost productivity, cut costs, and increase safety. As there is a mutual influence of regulations and technologies, effective and timely regulatory procedures are essential for the industry to profit from the benefits of the technology entirely. Traditionally, liability has been given to human individuals or organizations that are considered legal entities, such as shipping companies. An algorithm is not regarded as a moral or legal agent, and assigning blame for wrongdoing is impossible. This issue was thoroughly analyzed in the automotive industry. The testing of classic examples of moral problems is part of the argument about the safety of self-driving cars [41]. The ISM Code (SOLAS Chapter IX) requirements to establish a legal organization responsible for the safe operation of ships and pollution prevention, for example, will continue to apply to the MASS [42].

The development and use of autonomous ships will raise a wide range of ethical challenges. Human communication has dominated ship operations in the past, but the implementation of MASS includes man–machine and machine–machine communication. This implementation's risk or change assessment should include analysis and protocols of cases in which machine communication fails or is denied. The definition of legal liability boundaries, particularly the establishment of reasonable criteria and scopes of responsibility between shipowner and manufacturer, is required, as well as an appropriate security structure for insurance coverage.

As an example, consider the following inquiry on the ethical issue. It was thought that a MASS would take the most cost-effective path. However, a manned passenger ship capsized near the MASS, and communication systems between the MASS and the manned ship were unavailable or misdirected, leaving the crews and passengers on the capsized ship with little choice but to wait for assistance. Unfortunately, the MASS may be unable to distinguish the passenger ship in dangerous circumstances. Who is liable for failing to recognize the ship and perform rescue duties?

4. Discussion

IMO convention's standards are structured into several categories. Many of those standards will need revision as some may be obsolete.

One is detailed control requirements. The existing SOLAS method assumes a physical navigation bridge with an officer of the watch stationed on it, from which the vessel may be controlled immediately. This is the basis for several distinct criteria. For example, there are standards for steering gear, propulsion controls, propeller pitch controls, and watertight compartment controls to be supplied on the bridge, as well as voice communications provisions. Another example of this regulation is the need for pilot transfers to be supervised by a certified person with bridge communication. These will need to be updated primarily to allow for shore-based control.

Second, the precise criteria for electronic communications systems presume that there is a crew on board who is in regular contact with the shore and other vessels. Both radar and onboard radio people, a VHF unit on the bridge, constant radio watch, and other requirements are among them. Communications systems include the facilities for sending distress calls by at least two separate and independent means, equipment capable of receiving shore-to-ship distress alerts and transmitting and receiving ship-to-ship distress alerts, and search and rescue coordinating communications. It also requires on-scene communications, maritime safety information, general radio communications to and from shore-based radio systems, and bridge-to-bridge communications. There are further needs for the master to convey any navigational dangers he encounters, in addition to the hardware requirements [43–45]. Considering all these obligations, most of these communications must be preserved. No one wants the requirement for radio watch on distress frequencies, the ability to send distress calls, or inter-ship communication to go away in the case of autonomous shipping. Even the ability to receive maritime safety notices on board may be helpful, if only because they will need to be relayed if the shore-based controller is outside the transmitting station's range. However, the standards will need to be changed to relate to radio signals being relayed to and from the shore-based controller via the vessel rather than to someone on the vessel.

Third, clarification is needed as to numerous references in the IMO conventions to the master. The Comite Maritime International (CMI) has produced a spreadsheet in its submission to the IMO identifying provisions in the IMO regulations that will need clarification or amendment to deal with unmanned vessels. It also identifies numerous provisions with the comment interpretation of the master.

Fourth, the International Maritime Organization must adopt new regulations to deal with autonomous vessels that do not have a crew on board. Finally, training and certification standards for remote onshore controllers will need to be added to the STCW. These should only be found in countries that have signed the MARPOL convention.

SOLAS must also be addressed in terms of what it does not include. For example, the features required of the communications and remote-control devices used to manage the vessel while at sea are entirely dependent on autonomous shipping. Therefore, it will have to deal with issues such as the following in-depth:

- The reliability of propulsion and other machinery, such as steering gear, will have to be controlled for long periods, possibly weeks, from a distance with limited possibilities of interim maintenance.

- The reliability of ancillary mechanisms used to start, stop, control, and reverse the propulsion machinery. This will be especially crucial while maneuvering, avoiding collisions, and stopping the vessel when necessary. A failure in any of these areas might be catastrophic.
- The method of notifying shore controllers of risks to the vessel or cargo, such as fire, smoke, or water ingress.
- Similarly, pollution detection and avoidance devices. For example, there would be sensitive devices to detect leaks of bunker oil, as well as fail-safe technology for self-sealing tanks and transferring oil from a burst tank to another storage location.
- The communication arrays and other electronics to be carried, and their capabilities.
- The methods for transmitting signals to the vessel, their effectiveness, and any backup or spare capacity that may be necessary.

5. Conclusions

Regarding safety, security, and environmental protection conventions and regulations for autonomous surface ships, there are new and distinct concerns to be addressed. As a result, before MASS is introduced into commercial shipping, more holistic, worldwide, and unified approaches for new regulatory frameworks to the MASS must assure the prevention of marine accidents and environmental protection. It is also crucial to comprehend the MASS's impact on legislation, technology, and industries and the interactions among relevant players. While some preliminary studies have been completed, various projects are underway or planned worldwide to develop pilot ships, competing concepts and systems to support unmanned operations, and infrastructure initiatives such as autonomous ports and high bandwidth communications. The MASS should be monitored and managed remotely by the SSCC's operators, with a smart alarm system receiving critical information through satellite. The MASS and SSCC systems and sensors must be designed and built, and their synergetic effects must be carefully examined. Onboard equipment and devices will need to be interconnected to efficiently gather, manage, and analyze data from the MASS. They will be heavily modularized to avoid failures and have a high degree of redundancy and endurance. The MASS will affect ship design, shipbuilding, and port infrastructure, including services and interfaces.

Communication and cooperation among numerous stakeholders based on mutual understanding would be critical for a successful introduction of the MASS to the maritime industries, including shipping, shipbuilding, equipment production, and classification societies. MASS can modify pirate, terrorist, and criminal behavior patterns. By establishing new inspection procedures, technical and institutional considerations should be made to increase security. While the number of seafarers is expected to decline, developing qualification criteria for MASS onshore operators and providing relevant training and education will be critical. Regarding legal and ethical concerns, the time it takes for technology to mature vs. the time it takes to implement relevant legislation and procedures may negatively impact the timely adoption of innovations. A quantitative analysis of the influence of the MASS on technologies and industries, including economic consequences, will be addressed as part of future work.

Author Contributions: Conceptualization, M.I. and A.I.; methodology M.I.; validation: A.I.; investigation, M.I. and P.R.; resources, P.R. and H.I.; writing—original draft preparation, M.I.; writing—review and editing, M.I., P.R. and H.I.; visualization, M.I. and H.I.; supervision, A.I.; project administration, A.I. All authors have read and agreed to the published version of the manuscript.

Funding: Authors have received no external funding.

Institutional Review Board Statement: Not applicable.

Informed Consent Statement: Not applicable.

Data Availability Statement: Not applicable.

Conflicts of Interest: The authors declare no conflict of interest.

References

1. Issa, M.; Ibrahim, H.; Ilinca, A.; Hayyani, M. A Review and Economic Analysis of Different Emission Reduction Techniques for Marine Diesel Engines. *Open J. Mar. Sci.* **2019**, *9*, 148–171. [CrossRef]
2. Sideri, O.; Papoutsidakis, M.; Lilas, T.; Nikitakos, N.; Papachristos, D. Green shipping onboard: Acceptance, diffusion & adoption of LNG and electricity as alternative fuels in Greece. *J. Shipp. Trade* **2021**, *6*, 1–29.
3. Felício, J.A.; Rodrigues, R.; Caldeirinha, V. Green Shipping Effect on Sustainable Economy and Environmental Performance. *Sustainability* **2021**, *13*, 4256. [CrossRef]
4. Issa, M.; Ilinca, A. *Petrodiesel and Biodiesel Fuels for Marine Applications*; Petrodiesel Fuels; CRC Press: Boca Raton, FL, USA, 2021; Volume 3, pp. 1015–1033.
5. Issa, M. Optimisation Opérationnelle, Écologique, et Énergétique des Groupes Électrogènes Diesel. Ph.D. Thesis, Université du Québec à Rimouski, Québec, QC, Canada, 20 January 2020.
6. Burmeister, H.C.; Bruhn, W.; Rødseth, Ø.J.; Porathe, T. Autonomous unmanned merchant vessel and its contribution towards the e-Navigation implementation: The MUNIN perspective. *Int. J. E-Navig. Marit. Econ.* **2014**, *1*, 1–13. [CrossRef]
7. Rødseth, Ø.J. From concept to reality: Unmanned merchant ship research in Norway. In Proceedings of the Underwater Technology (UT), Busan, Republic of Korea, 21–24 February 2017.
8. Hogg, T.; Ghosh, S. Autonomous merchant vessels: Examination of factors that impact the effective implementation of unmanned ships. *Aust. J. Marit. Ocean Aff.* **2016**, *8*, 206–222. [CrossRef]
9. Porathe, T.; Prison, J.; Man, Y. Situation awareness in remote control centres for unmanned ships. In Proceedings of the Human Factors in Ship Design & Operation, London, UK, 26–27 February 2014; p. 93.
10. Lang, D. Update on the Worldwide Supplyand Demand for Seafarers. In *Report from the Baltic and International Maritime Council (BIMCO) and the International Shipping Federation (ISF)*; International Chamber of Shipping: London, UK, 2010.
11. Kim, M.; Hizir, O.; Turan, O.; Day, S.; Incecik, A. Estimation of added resistance and ship speed loss in a seaway. *Ocean Eng.* **2017**, *141*, 465–476. [CrossRef]
12. International Maritime Organization (IMO). Report of the Maritime Safety Committee on Its Ninety-Eighth Session. MSC 98/23. 2017. Available online: http://www.imla.co/sites/default/files/msc_98-23_-_report_of_the_maritime_safety_committee_on_itsninety-eighth_session_secretariat.pdf (accessed on 3 November 2022).
13. IMO MSC. Regulatory Scoping Exercise for the use of Maritime Autonomous Surface Ships (MASS). MSC 99th Session, MSC, 99. 2018. Available online: https://comitemaritime.org/wp-content/uploads/2018/05/CMI-IWG-Submission-to-MSC-99.pdf (accessed on 1 November 2022).
14. DNV, G.L.; Vartdal, B.J.; Skjong, R.; St Clair, A.L. Remote-Controlled and Autonomous Ships in the Maritime Industry. 2018. Available online: https://www.dnv.com/maritime/publications/remote-controlled-autonomous-ships-paper-download.html (accessed on 11 October 2022).
15. Komianos, A. The autonomous shipping era. operational, regulatory, and quality challenges. *Trans. Nav. Int. J. Mar. Navig. Saf. Sea Transp.* **2018**, *12*, 335–348. [CrossRef]
16. Shahbakhsh, M.; Emad, G.R.; Cahoon, S. Industrial revolutions and transition of the maritime industry: The case of Seafarer's role in autonomous shipping. *Asian J. Shipp. Logist.* **2022**, *38*, 10–18. [CrossRef]
17. MUNIN. D9.3: Quantitative Assessment. FP7 GA-No. 314286. 2015. Available online: http://www.unmanned-ship.org/munin/wp-content/uploads/2015/10/MUNIN-D9-3-Quantitative-assessment-CML-final.pdf (accessed on 22 October 2022).
18. Karetnikov, V.; Ol'Khovik, E.; Ivanova, A.; Butsanets, A. Technology Level and Development Trends of Autonomous Shipping Means. In *Energy Management of Municipal Transportation Facilities and Transport*; Springer: Cham, Switzerland, 2019; pp. 421–432.
19. Ziajka-Poznańska, E.; Montewka, J. Costs and Benefits of Autonomous Shipping—A Literature Review. *Appl. Sci.* **2021**, *11*, 4553. [CrossRef]
20. Ait Allal, A.; Mansouri, K.; Youssfi, M.; Qbadou, M. Toward a study of environmental and social impact of autonomous ship. In *Euro-Mediterranean Conference for Environmental Integration*; Springer: Cham, Switzerland, 2017; pp. 1709–1711.
21. Ahvenjärvi, S. The human element and autonomous ships. *Trans. Nav. Int. J. Mar. Navig. Saf. Sea Transp.* **2016**, *10*. [CrossRef]
22. Zolich, A.; Palma, D.; Kansanen, K.; Fjørtoft, K.; Sousa, J.; Johansson, K.H.; Jiang, Y.; Dong, H.; Johansen, T.A. Survey on communication and networks for autonomous marine systems. *J. Intell. Robot. Syst.* **2019**, *95*, 789–813. [CrossRef]
23. Allal, A.A.; Mansouri, K.; Youssfi, M.; Qbadou, M. Toward energy saving and environmental protection by implementation of autonomous ship. In Proceedings of the 2018 19th IEEE Mediterranean Electrotechnical Conference (MELECON), Marrakech, Morocco, 2–7 May 2018; pp. 177–180.
24. Levander, O. Autonomous ships on the high seas. *IEEE Spectr.* **2017**, *54*, 26–31. [CrossRef]
25. Yuan, J.; Nian, V. *A Preliminary Evaluation of Marinized Offshore Charging Stations for Future Electric Ships*; ADBI Working Paper 1199; Asian Development Bank Institute: Tokyo, Japan, 2020. Available online: https://www.adb.org/publications/preliminary-evaluation-marinizedoffshore-charging-stations-future-electric-ships (accessed on 22 October 2022).
26. Verma, J.; Kumar, D. Recent developments in energy storage systems for marine environment. *Mater. Adv.* **2021**. [CrossRef]
27. Schröder-Hinrichs, J.U.; Hollnagel, E.; Baldauf, M. From Titanic to Costa Concordia—A century of lessons not learned. *WMU J. Marit. Aff.* **2012**, *11*, 151–167. [CrossRef]
28. International Maritime Organization. Draft Report of the Maritime Safety Committee on Its Ninety-Eighth Session (continued) [For items 1, 2, 3 (part), 4 (part), 5 (part), 6 (part), 7, 8, 9, 10, 11, 12 (part) and 15 (part) see MSC 98/WP. 1]. *Regulation* **2017**, *2*, 1–6.

29. IMO. Regulatory Scoping Exercise for the Use of Maritime Autonomous Surface Ships (MASS). MSC 100/WP.8. 2018. Available online: https://maiif.org/wp-content/uploads/2019/06/MSC-100_20-Annex-20-1.pdf (accessed on 23 October 2022).
30. IMO. Generic Guidelines for Developing IMO Goal-Based Standards. MSC.1/Circ.1394/Rev.2. 2019. Available online: https://imorules.com/GUID-CEEB55CC-E9D9-4CFB-B515-6B767644BA62.html (accessed on 11 November 2022).
31. IMO. 2019. Interim Guidelines for MASS Trials. MSC 101/5/5. Available online: https://www.register-iri.com/wp-content/uploads/MSC.1-Circ.1604.pdf (accessed on 11 November 2022).
32. Soyer, B.; Tettenborn, A. (Eds.) *New Technologies, Artificial Intelligence and Shipping Law in the 21st Century*; Taylor & Francis: Abingdon, UK, 2019.
33. Shenoi, R.A.; Bowker, J.A.; Dzielendziak, A.S.; Lidtke, A.K.; Zhu, G.; Cheng, F.; Westgarth, R. Global Marine Technology Trends 2030. 2015. Available online: https://www.lr.org/en/insights/global-marine-trends-2030/global-marine-technology-trends-2030/ (accessed on 11 November 2022).
34. Lloyds Register (LR); QinetiQ; University of Southampton. Global Marine Technology Trends 2030: Autonomous Systems. 2017. Available online: https://cdn.southampton.ac.uk/assets/imported/transforms/content-block/UsefulDownloads_Download/F9AFACCCB8B444559D4212E140D886AF/68481%20Global%20Marine%20Technology%20Trends%20Autonomous%20Systems_FINAL_SINGLE_PAGE.pdf (accessed on 12 October 2022).
35. Alcaide, J.I.; Llave, R.G. Critical infrastructures cybersecurity and the maritime sector. *Transp. Res. Procedia* **2020**, *45*, 547–554. [CrossRef]
36. Daum, O. Cyber security in the maritime sector. *J. Mar. L. Com.* **2019**, *50*, 1.
37. Furumoto, K.; Kolehmainen, A.; Silverajan, B.; Takahashi, T.; Inoue, D.; Nakao, K. Toward automated smart ships: Designing effective cyber risk management. In Proceedings of the 2020 International Conferences on Internet of Things (iThings) and IEEE Green Computing and Communications (GreenCom) and IEEE Cyber, Physical and Social Computing (CPSCom) and IEEE Smart Data (SmartData) and IEEE Congress on Cybermatics (Cybermatics), Rhodes, Greece, 2–6 November 2013; pp. 100–105.
38. To Move Drugs, Traffickers Are Hacking Shipping Containers. *Motherboard Tech by Vice*. Available online: https://www.vice.com/en/article/bmjgk8/how-traffickers-hack-shipping-containers-to-move-drugs (accessed on 1 March 2020).
39. Mallam, S.C.; Nazir, S.; Sharma, A. The human element in future Maritime Operations–perceived impact of autonomous shipping. *Ergonomics* **2020**, *63*, 334–345. [CrossRef]
40. Lloyds Register. *Cyber-Enabled Ships: Deploying Information and Communications Technology in Shipping–Lloyds Register's Approach to Assurance*; Lloyds Register: London, UK, 2016.
41. Zhang, D.; Han, Z.; Zhang, K.; Zhang, J.; Zhang, M.; Zhang, F. Use of Hybrid Causal Logic Method for Preliminary Hazard Analysis of Maritime Autonomous Surface Ships. *J. Mar. Sci. Eng.* **2022**, *10*, 725. [CrossRef]
42. Pundars, B. Autonomous Shipping in Changing the Structures-Future Implications on Maritime Education and Training. Master's Thesis, Novia University of Applied Sciences, Vaasa, Finland, 2020.
43. Moore, W.H.; Roberts, K.H. Safety management for the maritime industry: The international safety management code. In *International Oil Spill Conference*; American Petroleum Institute: Washington, DC, USA, 1995; Volume 1995, No. 1; pp. 305–310.
44. Wróbel, K.; Gil, M.; Krata, P.; Olszewski, K.; Montewka, J. On the use of leading safety indicators in maritime and their feasibility for Maritime Autonomous Surface Ships. In Proceedings of the Institution of Mechanical Engineers, Part O: Journal of Risk and Reliability, Online, 2 July 2021.
45. Ramos, M.A.; Utne, I.B.; Mosleh, A. Collision avoidance on maritime autonomous surface ships: Operators' tasks and human failure events. *Saf. Sci.* **2019**, *116*, 33–44. [CrossRef]

sustainability

MDPI

Article

Solar Irradiance Probabilistic Forecasting Using Machine Learning, Metaheuristic Models and Numerical Weather Predictions

Vateanui Sansine [1,2,*], Pascal Ortega [1], Daniel Hissel [2] and Marania Hopuare [1]

1 GEPASUD, Université de Polynésie Française, Campus d'Outumaoro, 98718 Puna'auia, Tahiti, French Polynesia
2 FEMTO-ST/FCLAB, Université de Franche-Comté, CNRS, Rue Thierry Meg, CEDEX, F-90010 Belfort, France
* Correspondence: vateanui.sansine@doctorant.upf.pf; Tel.: +689-40-80-38-76

Abstract: Solar-power-generation forecasting tools are essential for microgrid stability, operation, and planning. The prediction of solar irradiance (SI) usually relies on the time series of SI and other meteorological data. In this study, the considered microgrid was a combined cold- and power-generation system, located in Tahiti. Point forecasts were obtained using a particle swarm optimization (PSO) algorithm combined with three stand-alone models: XGboost (PSO-XGboost), the long short-term memory neural network (PSO-LSTM), and the gradient boosting regression algorithm (PSO-GBRT). The implemented daily SI forecasts relied on an hourly time-step. The input data were composed of outputs from the numerical forecasting model AROME (Météo France) combined with historical meteorological data. Our three hybrid models were compared with other stand-alone models, namely, artificial neural network (ANN), convolutional neural network (CNN), random forest (RF), LSTM, GBRT, and XGboost. The probabilistic forecasts were obtained by mapping the quantiles of the hourly residuals, which enabled the computation of 38%, 68%, 95%, and 99% prediction intervals (PIs). The experimental results showed that PSO-LSTM had the best accuracy for day-ahead solar irradiance forecasting compared with the other benchmark models, through overall deterministic and probabilistic metrics.

Keywords: solar irradiance; forecasting; numerical weather predictions; machine learning; deep learning; metaheuristic models; optimization

Citation: Sansine, V.; Ortega, P.; Hissel, D.; Hopuare, M. Solar Irradiance Probabilistic Forecasting Using Machine Learning, Metaheuristic Models and Numerical Weather Predictions. *Sustainability* **2022**, *14*, 15260. https://doi.org/10.3390/su142215260

Academic Editors: Luis Hernández-Callejo, Sergio Nesmachnow and Sara Gallardo Saavedra

Received: 9 October 2022
Accepted: 31 October 2022
Published: 17 November 2022

1. Introduction

Global electricity demand is expected to rise by 2.4% in 2022, despite economic weaknesses and high prices [1]. This rise, driven by the growth of the world population, the industrialization of developing countries, and the worldwide process of urbanization [2], uses fossil fuels as the main power source. This has proven to be detrimental for the environment and the climate. Therefore, renewable energies have gained a lot of attention, especially photovoltaics (PVs), due to their accessibility, low cost, lifetime, and environmental benefits. Solar PV installations are growing faster than any other renewable energy. Indeed, PVs are forecast to account for 60% of the increase in global renewable capacity in 2022 [3]. In this context, PVs provide many environmental and economic benefits. However, uncontrollable factors such as the weather, seasonality, and climate lead to intermittent, random, and volatile PV power generation. These significant constraints still hinder the large-scale integration of PVs into the power grid and interfere with the reliability and stability of existing grid-connected power systems [4]. Thus, a reliable forecast of PV power outputs is essential to ensure the stability, reliability, and cost-effectiveness of the system [5]. Those forecasts are usually implemented through prediction of the global horizontal irradiance (GHI). There are three main groups of solar irradiance forecasting model [6]:

- Statistical models, which are based on historical data and their ability to extract information/patterns from the data to forecast time series.
- Physical models, which are based on sky images, satellite images, or numerical weather predictions (NWPs) to infer the dynamics of solar radiation through the atmosphere.
- Hybrid models, which exploit statistical and physical models to obtain forecasts with higher precision.

Machine learning algorithms, classified under statistical models, have become very popular for studies related to PV power-output forecasting, and play an important role in contemporary solar-irradiance forecasting for conventional grid management and for smaller and independent microgrids.

Ogliari et al. [7] compared two deterministic models with hybrid methods, a combination of an artificial neural network (ANN) and a clear sky radiation model, for PV power output forecasting. The models were trained on one year of measured data in a PV plant located in Milan, Italy. The results show that the hybrid method is the most precise for PV output forecasting, demonstrating advantages by combining physical models with machine learning algorithms.

Crisosto et al. [8] used a feedforward neural network (FFNN) with Levenberg–Marquardt backpropagation (LM–BP) to make predictions for one hour ahead with one-minute resolution in the city of Hanover, Germany. The model was trained on a four-year dataset including all-sky images, used for cloud cover computation, and measured global irradiance. For hourly average predictions, the FFNN-LM-BP showed the best results with an RMSE (Wh/m^2) = 65, and R^2 = 0.98, compared with the persistence model with an RMSE = 91 and R^2 = 0.91.

Yu et al. [9] used a long short-term memory (LSTM) model to predict GHI in three cities in the USA, namely, New York, Atlanta, and Hawaii. The time horizons of the model were one hour ahead and one day ahead. The model's performance was compared with other models such as the autoregressive integrated moving average (ARIMA), convolutional neural network (CNN), FFNN, and recurrent neural network (RNN). For hourly predictions, the LSTM model was more precise in all three states, with R^2 exceeding 0.9 on cloudy and partially cloudy days, whereas R^2 for the RNN was only 0.70 and 0.79 in Atlanta and Hawaii. For daily forecasting, LSTM outperformed the other models except in clear-sky days for New York, whereas for Hawaii and Atlanta, LSTM was better in every case.

However, it is difficult to improve the forecast from only one machine learning model, which sometimes suffers from instability originating from poor parameter choice, or from a reduced number of input variables. Ensemble learning is a popular development trend in artificial intelligence (AI) algorithms [10]. It combines independent models with stronger learners, which can achieve better stability and prediction effects compared with individual models [11].

Huang et al. [12] used gradient boosting regression (GBRT), extreme gradient lifting (XGboost), Gaussian process regression (GPR), and random forest (RF) models to carry out GHI predictions. Those ensemble models performed better than other stand-alone models such as decision tree (DT), backpropagation neural network (BPNN), and support vector machine regression (SVR). It is concluded that the stacking models—including GBRT, XGboost, GPR, and RF—are the best models to predict solar radiation.

Li et al. [13] used XGboost to implement point forecasts for solar irradiance and kernel density estimation (KDE) to generate probabilistic forecasts from the above prediction results. This method enabled the computation of confidence levels and demonstrated better results than other benchmark algorithms such as SVR and random forest.

To improve the efficiency of machine learning (ML) models, an increasing number of studies have used metaheuristic models in order to optimize the parameters of the considered GHI forecasting model.

Jia et al. [14] utilized particle swarm optimization (PSO) coupled with a Gaussian exponential model (GEM) to predict daily and monthly solar radiation (Rs). The hybrid PSO-GEM model showed the best results for Rs prediction.

Duan et al. [15] used NWP, together with the kernel-based nonlinear extension of Arps decline (KNEA) to predict solar irradiance. The KNEA algorithm is optimized by a metaheuristic algorithm called the Bat algorithm (BA). The proposed method for GHI forecasting is called the BA-KNEA. Duan et al. also implemented other hybrid models such as PSO-XGboost, BA-XGboost, and PSO-KNEA. The results showed that BA-KNEA is better at performing solar radiation forecasts.

In summary, ensemble learning models are an emerging trend in ML, proving to be appropriate tools for regression, and therefore, GHI forecasting. They have shown good results compared with deep learning models for day-ahead GHI point forecasts [12]. Moreover, ensemble methods can be further improved with metaheuristic models for parameter optimization, as well as the prevention of potential numerical instability from which various ML models suffer. However, one of the drawbacks of point forecasts is that they contain limited information about the volatility and randomness of solar irradiance. Point forecasts cannot satisfy the needs of a power system's optimized operation [13]. For this reason, considerable attention has been drawn to probabilistic forecasting, which enables the computation of prediction intervals to provide to grid dispatchers in order to facilitate grid operation.

This study focused on the implementation of daily probabilistic forecasts with hybrid models such as PSO-XGboost, PSO-LSTM, PSO-GBRT, and quantile mapping for the computation of prediction intervals. The hybrid models were compared with other reference models, namely, ANN, CNN, LSTM, RF, and GBRT.

The novelty of this work lies in the residual modeling implemented with an innovative hybrid model (PSO-LSTM), enabling us to compute prediction intervals with different confidence levels, and thus obtain probabilistic forecasts. To the best of our knowledge, no day-ahead probabilistic GHI predictions have been implemented with this method. Secondly, we demonstrate that using a deep learning approach combined with metaheuristic models can achieve higher accuracy than ensemble models, or their optimized versions.

In order to produce those forecasts, historical data measured on-site coupled with NWP were used in the training of GHI forecasting models. These forecasting tools are intended to control a combined cold- and power-generation system, comprising several energy production and storage sub-systems, the whole being powered by solar energy. This prototype is called RECIF (the French abbreviation for a microgrid for electricity and cold cogeneration), and has been developed within the framework of a project funded by the French National Agency for Research (ANR) and is being implemented at the University of French Polynesia (UPF).

The rest of the paper is organized as follows: the historical data and the implemented data processes are presented in Section 2, followed in Section 3 by a theoretical background of machine learning and metaheuristic models. The results, analysis, suggestions for future research, and perspectives are presented in Section 4. The conclusions and the principal results are presented in Section 5.

2. Materials and Methods

2.1. Input Variables

This study utilized historical data measured from the weather station set-up in the University of French Polynesia. Two years of measurements are at our disposal, from 2019 to 2020. Those measurements are crucial in the design and implementation of a reliable forecasting system based on machine learning algorithms. The meteorological variables are measured with a time step of 1 min. The GHI is measured with a BF5 pyranometer supplied by Delta Devices, which uses an array of photodiodes with a unique computer-generated shading pattern to measure the diffuse horizontal irradiance (DHI) and GHI [16]. This enables the computation of the direct normal irradiance (DNI) for a given solar zenith angle. The set of inputs chosen from the weather station, for the GHI day-ahead forecasting models, was as follows:

- Ambient temperature T (°C);
- Dew point temperature (°C);
- Relative humidity H (%);
- Atmospheric pressure (hPa);
- Wind velocity *WV* (m/s) and the wind direction WD (°);
- Amount of rain (mm);
- Solar irradiance variables such as GHI, DHI, and DNI;
- The clear-sky model, GHI_{CLS}, as the theoretical value of GHI in clear-sky conditions.

An overview of the data is presented in Table 1. The processing of the historical data is detailed in Section 2.2.

Table 1. Descriptive statistics including the mean, standard deviation (std), minimum/maximum values, and the quantiles for each meteorological variable.

	GHI	DHI	DNI	Temperature	Rel Humidity	Pressure	Dew Point	Wind Speed	Wind Direction	Rain	CLS
Count	898,460	898,460	898,460	898,460	898,460	898,460	898,460	898,460	898,460	898,460	898,460
Mean	219.1	80.0	139.2	26.1	76.8	1005.5	21.6	1.8	133.1	0.0	297.9
Std	320.5	125.2	265.1	2.3	9.2	2.4	1.8	1.2	93.2	0.0	381.0
Min	0.2	0.2	0.0	18.4	35.0	996.0	12.1	0.0	0.0	0.0	0.0
25%	0.2	0.2	0.0	24.3	70.0	1004.0	20.5	1.1	68.8	0.0	0.0
50%	1.9	1.6	0.0	25.9	77.6	1006.0	21.8	1.7	110.3	0.0	0.0
75%	370.1	110.0	120.6	27.8	83.7	1007.0	22.9	2.3	214.2	0.0	649.0
Max	1253	814.1	1156.4	33.5	99.2	1013.0	26.8	13.4	360.0	2.8	1145.8

In addition to these in situ measurements, numerical weather predictions (NWPs) were used to train our day-ahead forecasting models. The numerical weather prediction model AROME was implemented by Météo-France with a resolution of $0.025 \times 0.025°$ (2.5×2.5 km) in French Polynesia. These predictions have a maximum time horizon of 42 h and are updated every 12 h in French Polynesia. In Figure 1, each node (or grid point) of the AROME model for the north-eastern part of Tahiti is depicted, numbered from 1 to 34. Two years of NWP outputs are available, spanning from January 2019 to December 2020 with an hourly time-step. The GHI values predicted by AROME are only available from 9 am to 4 pm.

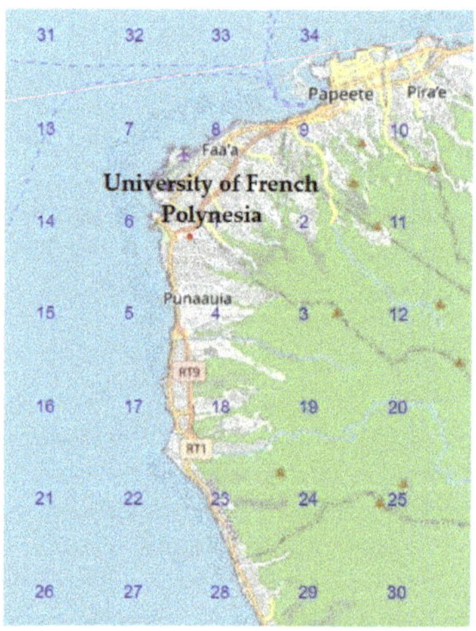

Figure 1. Points from the AROME grid (red dot represents the University).

2.2. Data Processing

This section explains the steps involved in processing the historical data and the AROME output. A vital step in data processing is to remove anomalous data that are caused by technical glitches from the sensors, such as negative and Not a Number (or NaN) values, or outliers. After the removal of NaN values, the outliers are detected through the interquartile range method (IQR). A sliding mean is applied to the 1 min time-step meteorological data, in order to obtain hourly values, and making correlation with the AROME output possible. The mean value at time t is computed from the 60 previous measurements.

In order to quantify the errors between the in situ measurements and the AROME model, and then determine which points of the AROME grid to use for the training of the machine learning algorithms, the following metrics were used: the mean square error (MSE), the root-mean-square error (RMSE), and the determination coefficient (R^2).

$$\text{MSE} = \frac{1}{N}\sum_{i=0}^{N}\left(y_{\text{measured},i} - y_{\text{predicted},i}\right)^2, \tag{1}$$

$$\text{RMSE} = \sqrt{\frac{1}{N}\sum_{i=0}^{N}\left(y_{\text{measured},i} - y_{\text{predicted},i}\right)^2}, \tag{2}$$

$$R^2 = 1 - \frac{\sum_{i=0}^{N}\left(y_{\text{measured},i} - y_{\text{predicted},i}\right)^2}{\sum_{i=0}^{N}\left(y_{\text{measured},i} - \overline{y}_{\text{measured}}\right)^2}, \tag{3}$$

where N is the number of observations, $y_{\text{measured},i}$ is the measurements, $\overline{y}_{\text{measured}}$ is the mean value of the measurements, and $y_{\text{predicted},i}$ is the predicted values. The results are presented in Figure 2.

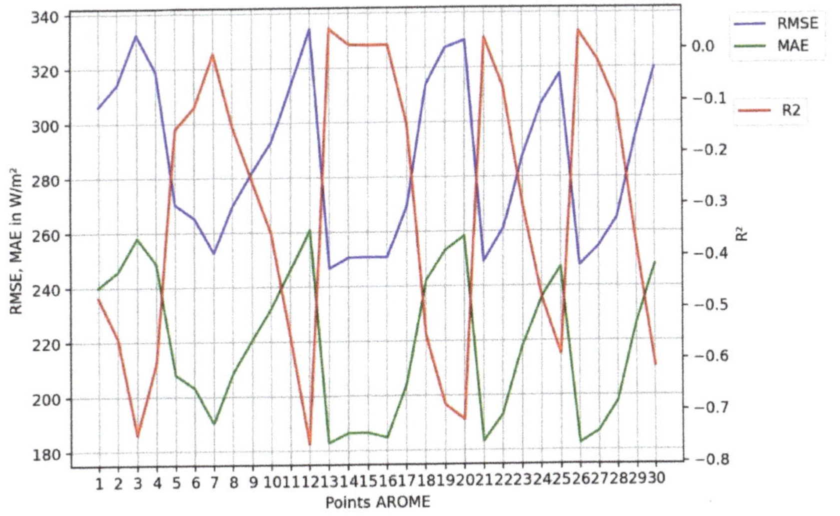

Figure 2. Errors between the measured GHI at the UPF and the AROME predictions for each grid point.

Points 31 to 34 were not used because they contained a great number of outliers in the first semester. The selected points were, arbitrarily, the points with some of the lowest correlation ($R^2 < -0.45$), i.e., n°3, 12, 20, 25, and points that exhibited positive correlation with the measured data, i.e., n°7, 13, 14, 15, 16, 17, 21, and 26.

The missing data were not replaced (through linear interpolation for example), but the consecutiveness of the dates of the data was ensured in the construction of the input data (or input vector); thus, no missing values were processed in the machine learning algorithms.

The night hours were removed from the measured data; consequently, the GHI forecasts implemented in this study were only performed for the hours between 6 am and 8 pm. After correlating the measurements and the AROME output, the merged data were normalized according to Equation (4):

$$X_{normalized} = \frac{X - X_{min}}{X_{max} - X_{min}}, \tag{4}$$

The data were then split into 70% as training data, 20% as validation data, and 10% as testing data.

3. Theoretical Background

3.1. Long Short-Term Memory (LSTM)

In recent years, LSTM has been widely applied to implement GHI forecasting [17,18]. One of the main advantages, compared with a classical RNN, is that LSTM models can deal with long-term dependencies found in the data without having problems such as vanishing gradients [19] using forget gates.

As shown in Figure 3, a typical LSTM network consists of one cell and three gates (an input gate, forget gate, and output gate). The input gate adjusts the amount of new data stored in the unit. The output gate determines which information to obtain from the cell, while the forget gate determines which information can be discarded [15]. Each gate uses either tanh or sigmoid as activation functions.

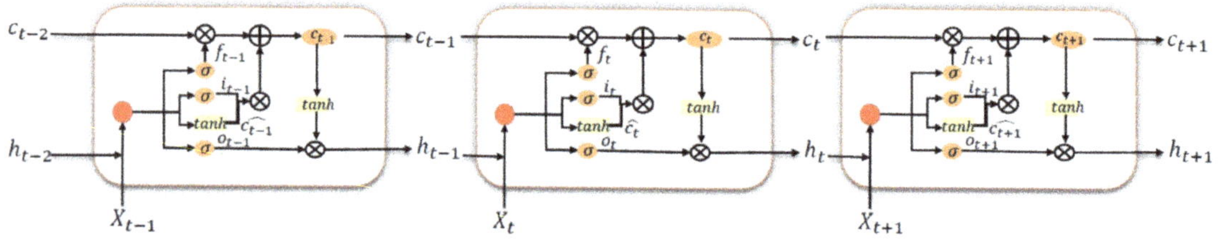

Figure 3. Basic structure of an LSTM model [15].

The input gate can be calculated with Equation (5) [15]:

$$gate(f_i) = \sigma_s(w_i x_t + u_i h_{t-1} + b_i), \tag{5}$$

where σ_s is the sigmoid activation function, h_{t-1} is the cell output at the previous time-step, W_i and U_i are weight factors, and b_i is the bias.

The forget gate can be computed with Equation (6) [15]:

$$gate(f_t) = \sigma_s(w_t x_t + u_t h_{t-1} + b_t), \tag{6}$$

where W_t and U_t are weight factors and b_t is the bias.

The output is finally computed with Equation (7) [15]:

$$gate(f_o) = \sigma_s(w_o x_t + u_o h_{t-1} + b_o), \tag{7}$$

where W_o and U_o are weight factors and b_o is the bias.

In this study, an LSTM model and an optimized LSTM (PSO-LSTM) model were used to implement daily GHI forecasting. They were compared with other models for probabilistic predictions. The parameters used for optimization are listed in Section 3.4.

The implemented LSTM model was composed of two LSTM models for day-ahead forecasting. One model was to process historical data; the second model was used to process AROME outputs. The outputs of the two LSTM models were concatenated, before being processed by a classical ANN.

3.2. Particle Swarm Optimization (PSO)

Particle swarm optimization was first proposed by Kennedy and Eberhart [20]. This algorithm simulates the predatory actions of a swarm of animals to find the best solution. A massless swarm of particles is created, with only two parameters: their position and speed. Each particle searches for the optimal solution separately in the search space and records it as the current individual extremum. The position of the extremum is shared with other particles in the whole swarm. If one individual extreme value is the best out of all other extremes, it is recorded as the global optimal solution. The global optimal solution is updated every time a particle finds a better extremum.

All the particles in the swarm adjust their velocity and position according to the current extremum already seen by the individual and the current global optimal solution shared by the whole swarm. The formulas for updating the position and speed of the PSO algorithm are shown in Equations (8) and (9) [20]:

$$\mathrm{X}_{\mathrm{i,t}} = \mathrm{X}_{\mathrm{i,t-1}} + \mathrm{V}_{\mathrm{i,t}}, \quad (8)$$

$$\mathrm{V}_{\mathrm{i,t}} = I_W \times \mathrm{V}_{\mathrm{i,t-1}} \times c_1 \times \theta_1 \times (pbest_i - \mathrm{X}_{\mathrm{i,t-1}}) + c_2 \times \theta_2 \times (gbest_i - \mathrm{X}_{\mathrm{i,t-1}}), \quad (9)$$

where $\mathrm{X}_{\mathrm{i,t}}$ is the position of the i-th particle during the t-th iteration, and $\mathrm{V}_{\mathrm{i,t}}$ is the speed of the i-th particle during the t-th iteration. c_1 and c_2 are called the cognitive (personal) and social (global) coefficients, respectively. The coefficients control the exploitation of the individual extremum found by each particle and the levels of exploration made by the swarm in the entire search space. θ_1 and θ_2 are random data, in the range [0, 1]. $pbest_i$ is the best location of the i-th particles among all iterations. $pbest_i$ is the best global location of all particles. I_W is random data initialized in the range [0, 1]

3.3. XGboost

XGboost is a machine learning algorithm realized by gradient lifting technology, and is the first parallel gradient enhanced tree (GBDT) algorithm. XGboost is based on classification and regression tree (CART) theory [21]. It provides parallel tree boosting and is one of the leading machine learning algorithms for regression, classification, and ranking problems. The XGboost model is built by adding trees iteratively. The predicted values of the i-th sample in the t-th iteration can be expressed as follows [21]:

$$\hat{y}_{i,t} = \hat{y}_{i,t-1} + f_t(X_i), \quad (10)$$

where $f_t(X_i)$ represents the addition needed to improve the model. The tree is added iteratively to minimize the objective function, which can be expressed as [21]:

$$obj^{(t)} = \sum_{i=1}^{n} L(y_i, \hat{y}_{i,t-1} + f_t(X_i)) + \Omega(f_t), \quad (11)$$

where $obj^{(t)}$ is the loss function [21].

$$\Omega(f_t) = \gamma T + \frac{1}{2}\lambda \sum_{j=1}^{T} w_j^2, \quad (12)$$

γ and λ are parameters that represent the model complexity. T is the number of leaves, and w_j is a weight parameter.

3.4. Hybrid Models

In this study, a hybrid model, PSO-XGboost, was implemented in order to obtain point forecasts of the GHI. The PSO algorithm is used to choose the best parameters for the XGboost algorithm. Seven important parameters for the XGboost model were chosen, as listed in Table 2. Those parameters were also used by Yu et al. [10] in order to estimate daily reference evapotranspiration values. The parameter "number of trees" has been added, because it is also an important parameter for XGboost.

Table 2. Parameters used in the optimization of XGboost.

Parameters.	Range	Meaning
Learning rate	[0.0001, 0.1]	Step size at each iteration while moving toward the minimization of a loss function.
Number of trees	[1, 500]	Number of trees in XGboost.
Maximum depth	[1, 500]	Maximum depth of a tree. The higher this value, the more likely the model is to overfit.
Subsample	[0.2, 1]	Subsample ratio of the training instances.
Colsample_by_tree	[0.2, 1]	Subsample ratio of columns when constructing each tree.
Min_child_weight	[0, 1]	Minimum instance weight needed in a child.
Gamma	[0.0001, 0.01]	Minimum loss reduction required to make further partitions on a leaf node of the tree.

The ML models were used, in this case, to solve a regression problem; therefore, we set R^2 to be the main metric of the PSO algorithm. R^2 is a positive-oriented metric; thus, the practical objective function used here was $1-R^2$. Indeed, the more the precision of the results increases, the closer R^2 is to 1, which also represents a minimum in the objective function $1-R^2$. Twenty particles are used for the PSO algorithm in order to limit computation time, and to explore the entire research space. The flow chart of the hybrid model is presented in Figure 4.

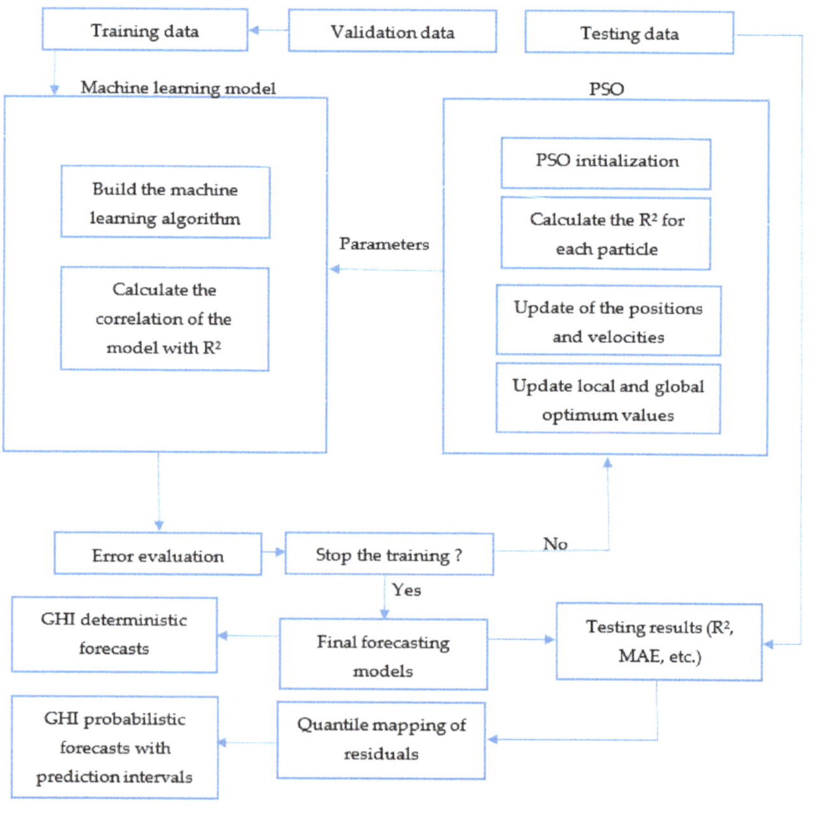

Figure 4. Flow chart of the hybrid models.

A PSO-Gradient boosting model was also implemented, but with fewer parameters than the PSO-XGboost. Only the maximum depth, the learning rate, number of trees, and subsample were used in this instance. The other parameters seen for PSO-XGboost were not available for the gradient boosting algorithm. As stated above, a hybrid PSO-LSTM model was also implemented for daily GHI forecasting. The parameters chosen for the optimization are presented in Table 3.

Table 3. Parameters of the LSTM model for particle swarm optimization.

Parameters	Range	Meaning
LSTM cells in the first LSTM model	[1, 1000]	Number of cells in the first LSTM model.
LSTM cells in the second LSTM model	[1, 1000]	Number of cells in the second LSTM model.
Dropout rate in the first LSTM model	[0.0001, 0.5]	Rate for dropout regularization in the first LSTM model.
Dropout rate in the second LSTM model	[0.0001, 0.5]	Rate for dropout regularization in the second LSTM model.
Dense in the first layer of the ANN	[1, 1000]	Number of neurons in the first layer of the ANN.
Dense in the second layer of the ANN	[1, 1000]	Number of neurons in the second layer of the ANN.
Dropout rate in the ANN	[0.0001, 0.5]	Rate for dropout regularization in the ANN.
Learning rate	$[10 \times 10^{-10}, 10 \times 10^{-2}]$	Step size at each iteration while moving toward a minimum of a loss function.
Epochs	[1, 100]	Number of times the algorithm is trained on the training data.
Validation split	[0.1, 0.5]	Split between training and validation data.

3.5. Residual Modeling

Probabilistic forecasting was implemented in this study through residual modeling. For each individual hour, the residuals were computed and assumed to have either a Gaussian or a Laplacian distribution. This method was inspired by He et al. [22]. The quantiles of the residuals were computed and taken as prediction intervals (PIs). To compute the different quantiles for all the considered distributions, we first needed to consider their cumulative distribution function (cdf) $\mathrm{F}_{\mathrm{Residus}}(\mathrm{x})$ in Equation (13).

$$\forall x \in \mathbb{R},\ F_{Residus}(x) = \mathbb{P}(Residus \leq x), \tag{13}$$

The inverse of the cdf is called the percent point function or quantile function $Q(q)$, and is provided in Equation (14):

$$\forall q \in [0,1],\ Q(q) = F^{-1}_{Residus}(x) = inf\{x \in \mathbb{R}, F_{Residus}(x) \geq\ q\ \}, \tag{14}$$

where $Q(0.25)$, $Q(0.5)$, and $Q(0.75)$ are the first quantile, the median, and the third quantile, respectively. The specific quantile function corresponded to a specific distribution (Gaussian or Laplacian). The PIs were calculated at different confidence levels or CLs. In this study, the 38%, 68%, 95%, and 99% PIs were derived from this inverse cdf for the Gaussian distribution in Equation (15). For the Laplacian distribution, the PIs could be derived using Equation (16), defined in [22]:

$$P_{cl+\frac{1-cl}{2}} = \sigma_t Q(cl), \tag{15}$$

$$P_{cl+\frac{1-cl}{2}} = -\sigma_t ln(2(1-cl)), \tag{16}$$

where σ_t is the standard deviation of the distribution (Laplacian or Gaussian). Given the symmetry of those distributions, the upper bounds, U_t, and lower bounds, L_t, were derived using Equations (17)–(19) [22]:

$$U_t = P_{cl+\frac{1-cl}{2}}, \tag{17}$$

$$L_t = P_{\frac{1-cl}{2}}, \tag{18}$$

$$L_t = U_t, \tag{19}$$

3.6. Metrics for Probabilistic Forecasting

The quality of probabilistic forecasts was quantified using three different metrics, namely, the prediction interval coverage percentage (PICP), the prediction interval normalized average width (PINAW), and the coverage width-based criterion (CWC), as defined in [13].

The PICP, detailed in Equations (20) and (21), indicates how many real values lie within the bounds of the prediction interval:

$$\text{PICP} = \frac{1}{N}\sum_{i=1}^{N} \delta_i, \tag{20}$$

$$\delta_i = \begin{cases} 1 \ if \ y_i \in [L_i, U_i] \\ 0 \ if \ y_i \notin [L_i, U_i] \end{cases}, \tag{21}$$

The PINAW, shown in Equation (22), quantitatively measures the width of the different PIs:

$$\text{PINAW} = \frac{1}{NR}\sum_{i=1}^{N}(U_i - L_i) \tag{22}$$

where R is a normalizing factor. The PINAW represents the quantitative width of the PIs; thus, a lower value of PINAW represents better performance for the prediction intervals.

The CWC, shown in Equations (23) and (24), combines the PICP and PINAW to optimally balance the probability and coverage.

$$\text{CWC} = \text{PINAW}\left(1 + \gamma(\text{PICP})e^{-\rho(PICP-\mu)}\right), \tag{23}$$

$$\gamma(PICP) = \begin{cases} 0 \ if \ PICP \geq \mu \\ 1 \ if \ PICP < \mu \end{cases}, \tag{24}$$

where μ is the preassigned PICP which is to be satisfied, and ρ is a penalizing term. When the preassigned PICP is not satisfied, the CWC increases exponentially. The CWC is a negatively oriented metric, meaning the lower the value, the better.

4. Results

4.1. Preliminary Results

Firstly, before implementing any hybrid model, it is necessary to quantify whether the AROME predictions are effective in increasing the accuracy of our forecasting models. Secondly, a study was also performed to determine how many days should be input into the models, so that we have optimal precision in daily forecasts. These two preliminary results are shown in Table 4 and were only performed for the XGboost model and for lagged terms, from 1 day prior to 5 days prior. The employed metrics were MAE, RMSE, and R^2.

Table 4. Results for lagged days and NWP with the XGboost model.

XGboost Model	Number of Days Prior	MAE (W/m^2)	RMSE (W/m^2)	R^2
Without AROME	1 day	135.19	206.02	0.73
	2 days	136.34	212.47	0.67
	5 days	115.22	182.97	0.74
With AROME	1 day	121.21	194.50	0.76
	2 days	126.29	198.2	0.72
	5 days	**110.24**	**176.47**	**0.76**

The results show that the use of AROME does increase the prediction accuracy for daily GHI forecasts. Indeed, for the same number of lagged days, the results with AROME are always better than without AROME, in terms of MAE, R^2, and RMSE.

With the AROME data, the best values in terms of MAE and RMSE are 110.24 W/m^2 and 176.47 W/m^2 and $R^2 = 0.76$, respectively, for 5 days prior. For this reason, 5 days prior was taken as a standard way to implement our GHI forecasting tools.

However, it would be interesting to carry out the same study with more lagged days at inputs of the machine learning algorithms. In order to carry out such studies, more historical data and AROME outputs are needed.

The results show a decrease in accuracy for 2 days prior. One possible explanation for this decrease is that the default parameters of the XGboost algorithm might be not ideal for daily GHI predictions for 2 days prior.

4.2. Hybrid Models Results

Tables 5 and 6 present the parameters found by PSO for the XGboost and LSTM algorithms, respectively. Once the optimal parameters are found, the optimized models are tested on the testing data.

Table 5. Optimal parameters for XGboost.

Parameters	Value
Learning rate	0.1
Number of trees	400
Maximum depth	400
Subsample	0.71
Colsample_by_tree	0.99
Min_child_weight	0.96
Gamma	0.01

Table 6. Optimal parameters for LSTM.

Parameters	Value
LSTM cells in the first LSTM model	208
LSTM cells in the second LSTM model	5
Dropout rate in the first LSTM model	0.4
Dropout rate in the second LSTM model	0.15
Density in the first layer of the ANN	712
Density in the second layer of the ANN	786
Dropout rate in the ANN	0.72
Learning rate	0.09
Epochs	50
Validation split	0.5

The results for all the models used for daily GHI predictions are summarized in Table 7 with deterministic metrics, and in Table 8 with probabilistic metrics.

Table 7. Deterministic metrics for all implemented models used for daily GHI predictions.

Models	MAE (W/m^2)	RMSE (W/m^2)	R^2
ANN	120.50	179.40	0.75
CNN	125.66	188.77	0.73
LSTM	115.69	184.74	0.74
Random Forest	106.19	166.42	0.79
Gradient Boosting	111.20	174.54	0.77
XGboost	110.24	176.47	0.76
PSO-Gradient Boosting	105.06	167.24	0.79
PSO-LSTM	**99.37**	154.84	**0.82**
PSO-XGboost	105.02	**153.69**	**0.82**

Table 8. Probabilistic metrics for the implemented models.

Models	Predicted Intervals	Gaussian Distribution			Laplacian Distribution		
		PICP (%)	PINAW (%)	CWC (%)	PICP (%)	PINAW (%)	CWC (%)
ANN	38%	39.81	9.50	9.50	30.95	6.78	13.57
	68%	71.03	18.79	18.79	64.48	16.45	32.90
	95%	94.05	32.77	65.54	94.83	35.30	70.60
	99%	99.60	42.62	42.62	99.80	44.57	44.57
CNN	38%	42.06	10.03	10.03	39.88	7.62	7.62
	68%	71.63	19.37	19.37	67.4	17.30	34.60
	95%	94.64	33.89	67.79	94.24	37.06	74.12
	99%	98.81	45.35	90.71	98.21	46.50	93.00
LSTM	38%	47.81	9.48	9.48	39.88	6.93	6.93
	68%	75.40	18.00	18.00	70.04	15.67	15.67
	95%	93.45	31.37	62.74	94.24	33.85	67.69
	99%	98.21	41.97	83.94	98.02	44.03	88.06
Random Forest	38%	48.61	9.68	9.68	43.85	7.06	7.06
	68%	77.58	18.78	18.78	69.44	15.96	15.96
	95%	94.25	32.35	64.70	93.85	33.47	66.95
	99%	98.61	42.28	84.57	98.02	43.41	86.82
Gradient boosting	38%	48.81	9.99	9.99	43.06	7.28	7.28
	68%	74.80	19.0	19.0	70.63	16.62	16.62
	95%	93.85	32.55	65.11	93.45	34.68	69.37
	99%	98.61	42.31	84.62	97.42	43.97	87.95
XGboost	38%	53.17	9.96	9.96	42.86	7.18	7.18
	68%	75.79	18.94	18.94	69.84	16.3	16.30
	95%	92.46	32.13	64.27	93.85	33.98	67.96
	99%	98.61	41.84	83.69	98.41	43.38	86.76

Table 8. *Cont.*

Models	Predicted Intervals	Gaussian Distribution			Laplacian Distribution		
		PICP (%)	PINAW (%)	CWC (%)	PICP (%)	PINAW (%)	CWC (%)
PSO-LSTM	38%	43.06	8.6	8.6	39.28	6.3	6.3
	68%	73.61	16.86	16.86	69.05	14.68	14.68
	95%	94.63	29.57	59.13	94.44	31.74	63.48
	99%	99.00	40.1	40.1	98.61	42.44	84.87
PSO-Gradient Boosting	38%	52.38	9.63	9.63	43.84	6.93	6.92
	68%	75.20	18.41	18.41	72.42	15.81	15.81
	95%	93.25	31.50	63.01	93.05	33.10	66.20
	99%	98.61	41.40	82.81	97.81	43.19	86.38
PSO-XGboost	38%	45.63	8.84	8.84	39.48	6.6	6.6
	68%	74.01	17.13	17.13	69.84	15.31	15.31
	95%	94.44	30.39	60.78	94.84	33.15	66.30
	99%	98.81	40.61	81.22	98.21	43.23	86.47

For the deterministic metrics, we first note that the use of PSO increases the accuracy of standalone models such as LSTM, GBDT, and XGboost. Indeed, there were decreases in MAE and RMSE and an increase in R^2 when considering standalone models with their optimized versions.

The deterministic metrics also show that the hybrid PSO-XGboost method is the best for implementing daily forecasting, in terms of RMSE = 153.69 W/m^2 and R^2 = 0.82. However, the PSO-LSTM model is also strong, but in terms of MAE = 99.37 W/m^2, as well as R^2 = 0.82. Neither of the two models has any significant advantage over the other.

In order to choose the best model, a Taylor diagram was drawn (Figure 5) for all implemented machine learning models. It can be seen that PSO-LSTM was slightly better than all the other models for deterministic predictions, because it was closer to the observation than the other models. The standard deviation was also the same for the observation and PSO-LSTM (red dotted line), meaning that it appropriately represented the variability in solar irradiance.

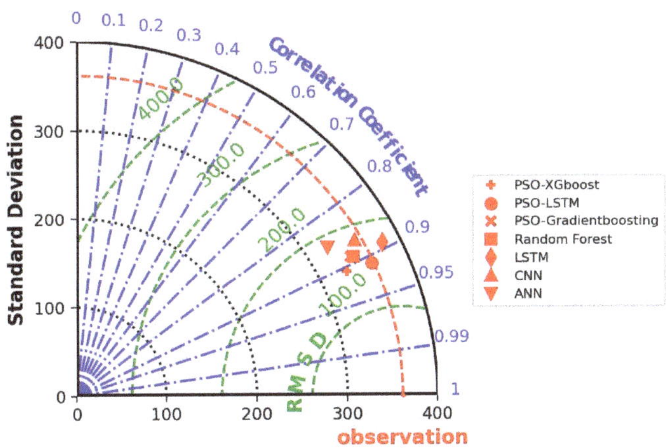

Figure 5. Taylor diagram for deterministic comparison between models.

For all models, we can see that the RMSE is greater than the MAE, which is the manifestation of high variance in individual errors. Indeed, because the RMSE is a quadratic scoring rule, it tends to assign high weight to large errors, whereas the MAE gives the same weight to all errors, independently of their magnitude. This variance has been studied thanks to residual modelling and the generation of prediction intervals.

For the probabilistic forecasts the PICP, PINAW, and CWC values were computed for all forecasting models. Highlighted in black in Table 8 are the best CWC values, for 38%, 68%, 95%, and 99% PIs. PSO-LSTM is the best algorithm for all prediction intervals. For 38%, 68%, 95%, and 99% PIs, the CWC values are 6.3, 14.68, 59.13, and 40.1, respectively. Notably, for 38% and 68%, the best fit was the Laplacian distribution, whereas for the 95% and 99% PIs, the best fit was the Gaussian distribution. The proposed methods were implemented in Python 3.7, with the machine learning package Tensorflow 2.2.0. Duan et al. [15] also used PSO-XGboost for predicting solar radiation in four different locations in China. After training with four different datasets, the four R^2 values for 1-day-ahead forecasting were 0.816, 0.84, 0.787, and 0.755. Those values are not far from our own PSO-XGboost algorithm, with an $R^2 = 0.82$. In our case, the PSO-LSTM model was even better than the PSO-XGboost, demonstrating that deep learning models can still outperform ensemble learning models for day-ahead forecasting and, to the best of our knowledge, no PSO-LSTM has ever been used with quantile mapping to obtain day-ahead GHI probabilistic forecasting. The accuracy of point forecasts depends, however, on the global structure of the LSTM, meaning that a simpler structure from an LSTM model might not have the same results.

Figure 6 shows the PSO-LSTM predictions with the corresponding prediction intervals. We can see that the GHI measurements do stay within the prediction intervals; however, we can see that the prediction intervals are quite large. For this problem, it would be interesting to use another method for computing the confidence levels (CLs), which are smaller than the prediction intervals computed in this paper. Li et al. [13] used kernel density estimation (KDE) for confidence level computation, which gave PINAW values of 15.45, 17.03, and 19.55 for 80%, 85%, and 90% CIs, respectively. This is considerably smaller than the PIs in this article, which are approximately equal to 30 for 95% PIs.

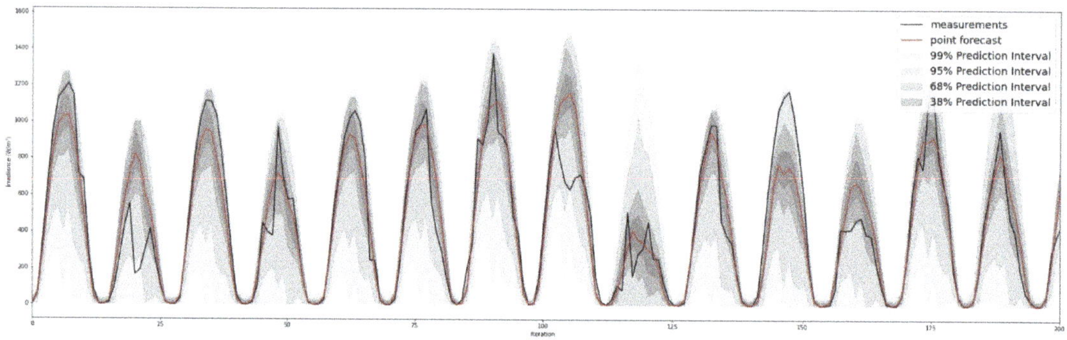

Figure 6. PSO-LSTM probabilistic forecasting.

4.3. *Perspectives and Future Research*

For the PSO algorithm, the higher the number of particles, the better the exploration of the entire search space; however, the computation time increases accordingly. In order to reduce computation time, we limited ourselves to 20 particles in the swarm. According to Eberhart et al. [23], population sizes ranging from 20 to 50 are optimal in terms of minimalizing the number of evaluations (population size multiplied by the number of iterations) needed to obtain a sufficient solution. Nevertheless, it would be interesting to see the result for GHI day-ahead predictions with the number of particles in a range from 20 to 500 particles for maximum exploration ability of the search space.

As presented in Section 4.1, the maximum precision was obtained for five days of measurements at the input of the models. It is assumed that the more information (lagged days), the better the precision of the forecasting models. For this reason, it would be interesting to carry out a study with more lagged days fed into the models. However, a constraint arises when more lagged days are used as input vectors. Indeed, with the available data, using more lagged days would greatly reduce the number of training samples. To retain a sufficient number of days for the training of our forecasting models, while simultaneously increasing the number of lagged days, more meteorological data and AROME outputs are needed.

Testing another meta-heuristic model also seems a promising way to improve GHI forecasts. Duan et al. [15] used the Bat algorithm for parameter optimization. Other bio-inspired optimization processes could be implemented, such as the grey wolf optimizer (GWO), whale optimization algorithm (WOA), or salp swarm algorithm (SSA). Duan et al. also showed that the KNEA algorithm is appropriate for providing accurate point forecasts. Therefore, a hybrid model with the metaheuristic models listed above, coupled with the KNEA algorithm, seems to be a very good way to implement daily GHI forecasts. As mentioned in the last section, combining the computation of confidence levels with the KDE method represents a very efficient way of obtaining better probabilistic forecasting from the aforementioned hybrid models.

5. Conclusions

The accurate forecasting of solar irradiance is paramount for photovoltaic power generation. In this study, the solar irradiance forecasts from the operational weather prediction model (AROME), implemented by Météo-France, were compared with in situ measurements for error quantification. In order to drastically improve the forecasting accuracy on-site, to control an isolated solar-powered microgrid called RECIF, implemented in Tahiti, ML algorithms were coupled with a metaheuristic particle swarm optimization (PSO) model for parameter optimization. The novelty of this paper resides in the implementation of probabilistic forecasting by combining an innovative hybrid model (PSO-LSTM) with quantile mapping. Mapping of the residuals allowed us to generate 38%, 68%, 95%, and 99% prediction intervals (PIs) with two different distributions, for probabilistic forecasting. Nine machine learning models were used for comparison purposes, namely, artificial neural network (ANN), convolutional neural network (CNN), long short-term memory (LSTM), random forest (RF), gradient boosting (GBRT), XGboost, PSO-LSTM, PSO-GBRT, and PSO-XGboost. PSO-LSTM was superior to all other models with MAE = 99.37 W/m^2, RMSE = 154.84 W/m^2, and R^2 = 0.82, coupled with a Taylor diagram. The PSO-LSTM model was also the best for all probabilistic metrics, exhibiting a Laplacian distribution for 38% and 68% prediction intervals, with CWC values equal to 6.33 and 14.68, respectively. Furthermore, the PSO-LSTM model showed the best results, exhibiting a Gaussian distribution for 95% and 99% prediction intervals, with CWC values equal to 59.13 and 40.1, respectively. This demonstrates that deep learning models coupled with metaheuristic models can outperform the ensemble learning method for day-ahead GHI forecasting.

Author Contributions: Conceptualization, V.S.; methodology, V.S., D.H. and P.O.; software, V.S. and P.O.; validation, P.O. and D.H.; formal analysis, V.S., P.O. and D.H.; investigation, V.S.; resources, P.O. and M.H.; data curation, V.S. and M.H.; writing—original draft preparation, V.S.; writing—review and editing, V.S., P.O. and D.H.; supervision, P.O. and D.H.; project administration, P.O. and D.H.; funding acquisition, P.O. and D.H. All authors have read and agreed to the published version of the manuscript.

Funding: This study was supported by the EIPHI Graduate School (contract ANR-17-EURE-0002) and the Region Bourgogne Franche-Comté. We thank the National Agency of Research (ANR-18-CE05-0043) for purchasing the equipment needed for this investigation. We also thank the FEMTO-ST laboratory and the University of French Polynesia, for funding this research.

Data Availability Statement: Data are provided within this article.

Conflicts of Interest: The authors declare no conflict of interest.

References

1. IEA. *Electricity Market Report—July 2022*; IEA: Paris, France, 2022. Available online: https://www.iea.org/reports/electricity-market-report-july-2022 (accessed on 2 September 2022).
2. IEA. *World Energy Outlook 2017—Excutive Summary*; Technical Report; IEA: Paris, France, 2017. Available online: https://www.iea.org/publications/freepublications/publication/WEO_2017_Executive_Summary_English_version.pdf (accessed on 3 September 2022).
3. IEA. *Renewable Energy Market Update—May 2022*; IEA: Paris, France, 2022.
4. Ramsami, P.; Oree, V. A hybrid method for forecasting the energy output of photovoltaic systems. *Energy Convers. Manag.* **2015**, *95*, 406–413. [CrossRef]
5. Ehara, T. Overcoming PV Grid Issues in the Urban Areas. 2009. Available online: https://iea-pvps.org/wp-content/uploads/2020/01/rep10_06.pdf (accessed on 2 January 2022).
6. Akhter, M.N.; Mekhilef, S.; Mokhlis, H.; Shah, N.M. Review on forecasting of photovoltaic power generation based on machine learning and metaheuristic techniques. *IET Renew. Power Gener.* **2019**, *13*, 1009–1023. [CrossRef]
7. Ogliari, E.; Dolara, A.; Manzolini, G.; Leva, S. Physical and hybrid methods comparison for the day ahead PV output power forecast. *Renew. Energy* **2017**, *113*, 11–21. [CrossRef]
8. Crisosto, C.; Hofmann, M.; Mubarak, R.; Seckmeyer, G. One-Hour Prediction of the Global Solar Irradiance from All-Sky Images Using Artificial Neural Networks. *Energies* **2018**, *11*, 2906. [CrossRef]
9. Yu, Y.; Cao, J.; Zhu, J. An LSTM short-term solar irradiance forecasting under complicated weather conditions. *IEEE Access* **2019**, *7*, 145651–145666. [CrossRef]
10. Yu, J.; Zheng, W.; Xu, L.; Zhangzhong, L.; Zhang, G.; Shan, F. A PSO-XGBoost Model for Estimating Daily Reference Evapotranspiration in the Solar Greenhouse. *Intell. Autom. Soft Comput.* **2020**, *26*, 989–1003. [CrossRef]
11. Zhang, C.; Ma, Y. (Eds.) *Ensemble Machine Learning*; Springer: Boston, MA, USA, 2012. [CrossRef]
12. Huang, L.; Kang, J.; Wan, M.; Fang, L.; Zhang, C.; Zeng, Z. Solar Radiation Prediction Using Different Machine Learning Algorithms and Implications for Extreme Climate Events. *Front. Earth Sci.* **2021**, *9*, 596860. [CrossRef]
13. Li, X.; Ma, L.; Chen, P.; Xu, H.; Xing, Q.; Yan, J.; Lu, S.; Fan, H.; Yang, L.; Cheng, Y. Probabilistic solar irradiance forecasting based on XGBoost. *Energy Rep.* **2022**, *8*, 1087–1095. [CrossRef]
14. Jia, Y.; Wang, H.; Li, P.; Su, Y.; Wang, F.; Huo, S. Particle swarm optimization algorithm with Gaussian exponential model to predict daily and monthly global solar radiation in Northeast China. *Environ. Sci. Pollut. Res.* **2022**. [CrossRef]
15. Duan, G.; Wu, L.; Liu, F.; Wang, Y.; Wu, S. Improvement in Solar-Radiation Forecasting Based on Evolutionary KNEA Method and Numerical Weather Prediction. *Sustainability* **2022**, *14*, 6824. [CrossRef]
16. Badosa, J.; Wood, J.; Blanc, P.; Long, C.N.; Vuilleumier, L.; Demengel, D.; Haeffelin, M. Solar irradiances measured using SPN1 radiometers: Uncertainties and clues for development. *Atmos. Meas. Tech.* **2014**, *7*, 4267–4283. [CrossRef]
17. Ghimire, S.; Deo, R.C.; Raj, N.; Mi, J. Deep solar radiation forecasting with convolutional neural network and long short-term memory network algorithms. *Appl. Energy* **2019**, *253*, 113541. [CrossRef]
18. Mutavhatsindi, T.; Sigauke, C.; Mbuvha, R. Forecasting Hourly Global Horizontal Solar Irradiance in South Africa Using Machine Learning Models. *IEEE Access* **2020**, *8*, 198872–198885. [CrossRef]
19. Hochreiter, S.; Bengio, Y.; Frasconi, P.; Schmidhuber, J. Gradient Flow in Recurrent Nets: The Difficulty of Learning LongTerm Dependencies. In *A Field Guide to Dynamical Recurrent Networks*; IEEE: Manhattan, NY, USA, 2001; pp. 237–243. [CrossRef]
20. Kennedy, J.; Eberhart, R. Particle swarm optimization. In Proceedings of the ICNN'95—International Conference on Neural Networks, Perth, WA, Australia, 27 November–1 December 1995; Volume 4, pp. 1942–1948. [CrossRef]
21. Chen, T.; Guestrin, C. XGBoost: A scalable tree boosting system. In Proceedings of the 22nd ACM SIGKDD International Conference on Knowledge Discovery and Data Mining, San Francisco, CA, USA, 13–17 August 2016; pp. 785–794. [CrossRef]
22. He, H.; Lu, N.; Jie, Y.; Chen, B.; Jiao, R. Probabilistic solar irradiance forecasting via a deep learning-based hybrid approach. *IEEJ Trans. Electr. Electron. Eng.* **2020**, *15*, 1604–1612. [CrossRef]
23. Eberhart; Shi, Y. Particle swarm optimization: Developments, applications and resources. In Proceedings of the 2001 Congress on Evolutionary Computation (IEEE Cat. No.01TH8546), Seoul, Republic of Korea, 27–30 May 2001; Volume 1, pp. 81–86. [CrossRef]

energies

MDPI

Article

System Frequency Control Method Driven by Deep Reinforcement Learning and Customer Satisfaction for Thermostatically Controlled Load

Rusi Chen [1], Haiguang Liu [1], Chengquan Liu [2,*], Guangzheng Yu [2], Xuan Yang [3] and Yue Zhou [3]

1 State Grid Hubei Electric Power Research Institute, Wuhan 430077, China
2 Department of Electrical Engineering, Shanghai University of Electric Power, Shanghai 201306, China
3 State Grid Hubei Electric Power Co., Ltd., Wuhan 430072, China
* Correspondence: y20103028@mail.shiep.edu.cn

Abstract: The intermittence and fluctuation of renewable energy aggravate the power fluctuation of the power grid and pose a severe challenge to the frequency stability of the power system. Thermostatically controlled loads can participate in the frequency regulation of the power grid due to their flexibility. Aiming to solve the problem of the traditional control methods, which have limited adjustment ability, and to have a positive influence on customers, a deep reinforcement learning control strategy based on the framework of soft actor–critic is proposed, considering customer satisfaction. Firstly, the energy storage index and the discomfort index of different users are defined. Secondly, the fuzzy comprehensive evaluation method is applied to evaluate customer satisfaction. Then, the multi-agent models of thermostatically controlled loads are established based on the soft actor–critic algorithm. The models are trained by using the local information of thermostatically controlled loads, and the comprehensive evaluation index fed back by users and the frequency deviation. After training, each agent can realize the cooperative response of thermostatically controlled loads to the system frequency only by relying on the local information. The simulation results show that the proposed strategy can not only reduce the frequency fluctuation, but also improve customer satisfaction.

Keywords: thermostatically controlled load; frequency regulation; customer satisfaction; soft actor–critic; energy storage index; discomfort index

Citation: Chen, R.; Liu, H.; Liu, C.; Yu, G.; Yang, X.; Zhou, Y. System Frequency Control Method Driven by Deep Reinforcement Learning and Customer Satisfaction for Thermostatically Controlled Load. *Energies* **2022**, *15*, 7866. https://doi.org/10.3390/en15217866

Academic Editors: Luis Hernández-Callejo, Sergio Nesmachnow and Sara Gallardo Saavedra

Received: 18 September 2022
Accepted: 18 October 2022
Published: 24 October 2022

1. Introduction

With the increasing proportion of renewable energy in the power grid, the characteristics of intermittence and fluctuation will bring considerable challenges to the active power balance and frequency stability of the power grid [1]. The traditional power system maintains the balance of the system by adjusting the output of the generating side units. The regulation method is relatively simple and will generate additional economic and environmental costs [2]. In addition, with the increase in power load and the extensive access to renewable energy, the regulation capacity of the power generation side gradually decreases [3]. The power system with renewable energy as the main body can utilize advanced information technology to integrate and dispatch demand-side resources to provide a variety of auxiliary services [4,5]. Therefore, reasonable control of demand-side resources can supplement the traditional system frequency regulation, and thus enhance the stability of the power system [6].

In the demand-side resources, the thermostatically controlled load (TCL) is a kind of electric equipment controlled by a thermostat, which can realize electric heating conversion and adjustable temperature, including in heat pumps, electric storage water heaters (ESWHs), refrigerators, and heating, ventilation and air conditioning (HVAC) systems [7]. TCL can be used to provide frequency regulation services, and is mainly based on the following three points. Firstly, it is widely distributed in residential, commercial and industrial buildings, with adjustable potential. Secondly, it has sufficient thermal storage capacity and can be regarded as distributed energy storage equipment. Thirdly, the control method is flexible and can respond to the power demand of the system in time [8]. Therefore, in order to fully excavate the frequency regulation potential of flexible resources on the demand side and maintain the grid frequency within a certain offset range, it is necessary to conduct in-depth research on the control strategy of large-scale TCLs on the demand side.

In the current research, there are mainly three control methods for TCLs to participate in ancillary services: centralized control, decentralized control and hybrid control [9–11]. In centralized control, the control center sends control signals to all controlled loads, but it needs to build a large number of communication channels, leading to high control costs. Hu et al. [9] established a hierarchical centralized load tracking control framework that coordinates demand-side heterogeneous TCL aggregators and uses a state–space model for modeling. The decentralized control decentralizes the judgment mechanism of load control to the local control terminal, and pre-sets the procedures or thresholds at the local control terminal. When the demand-side device detects important parameter changes, the load acts according to the pre-set strategy. Because the judgment of decentralized control is performed at the local port, the demand for communication is low and the response speed is fast. However, the control effect is largely influenced by the user behavior and the error of the detection device. Delavari and Kamwa [10] applied a multi-objective optimization approach for optimizing each load setting to reduce the amount of load response required and trigger the load based on the frequency response index of decentralized control. The hybrid control combines the features of centralized and decentralized control, and establishes a control framework of "centralized parameter setting–decentralized decision making", and coordinates large-scale users and grid control centers through load aggregators (LAs). Song et al. [11] built a two-stage control model based on hybrid control to participate in energy market trading. Based on hybrid control, Wang et al. [12] used TCLs to mitigate PV and load variations in microgrid communities. The above methods require a communication network between the control center and all aggregates, increasing the cost and difficulty of demand-side load control.

In the research on the participation of TCLs in auxiliary service, Ref. [13] built a dynamic model and verified the performance of a variable-frequency heat pump in providing frequency modulation services by using direct load control. This paper mainly studied the dynamic response performance of a single air-conditioning system, but focused less on the coordinated control of large-scale air-conditioning loads. Ref. [14] established a virtual energy storage model of variable-frequency air conditioning, shielded part of the model information through a hierarchical control framework and simplified the downlink control by using a unified broadcast signal. However, in this paper, the adjustable capacity of the air conditioning cluster will be sacrificed in order to simplify the downlink control. There are two main control modes of TCL, namely direct switch and temperature setting [15]. Ref. [16] realized frequency adjustment based on load direct switch. The advantage of this method is that the tracking accuracy of the system is high and the influence on the user's comfort is low within the range of load regulation ability. The disadvantage is that when the indoor temperature of the load is concentrated near the temperature boundary, the equipment will be frequently switched on and off, which will not only fail to complete the adjustment task, but also reduce the service life of the equipment [17]. Temperature setting can avoid the above disadvantages, but its limitation is that the tracking effect of power depends on the designed controller including a minimum variance controller [18], a sliding mode controller [19] and an internal model controller [20]. In addition, its limitations are also shown in the large range of temperature changes, which will have an impact on the comfort of users [21]. Pallonetto et al. [22] established a residential building energy management system (EMS) based on a combination of optimization techniques and machine learning models. This EMS reduces energy consumption while maintaining thermal comfort. Therefore, it is important to consider the influence of consumer satisfaction in the process of load response control to dispatch the motivation of user participation in demand response.

The above-mentioned documents all adopt a single control method to schedule and control TCL, but they often fail to meet the application requirements. This is not only because of their inherent defects, but also because of the different requirements of users. Direct switching is suitable for loads with high temperature requirements. At this time, the user expects the start and stop frequency of the equipment to be kept at a low level to extend the service life of the equipment. The load with a low temperature requirement is suitable for temperature setting. At this time, the user expects the temperature change to be maintained within a certain range to improve the comfort. Therefore, it is of practical significance to combine the two control modes. Ref. [23] proposed a hybrid control strategy based on a parallel structure. This strategy can improve the tracking accuracy of the system and reduce the number of switches in the equipment. However, the temperature changes widely, which will reduce the user's comfort. In recent years, the deep reinforcement learning algorithm has provided a new solution to the frequency control problem of power system.

With its strong search and learning ability, the deep reinforcement learning algorithm has the potential of online optimization of decision making in the face of complex nonlinear frequency control problems. The Q-learning algorithm of deep reinforcement learning is used to realize the cooperative control of distributed generation units, thus eliminating the frequency deviation of the system in [24,25]. Ref. [26] combined electric water heater buffer models with domain randomization to reduce the initialization time of Q-learning in demand response control. Ruelens et al. [27] applied batch reinforcement learning to coordinate the power consumption of users with thermostatically controlled loads. However, the Q-learning algorithm can only discretely select control actions from low-dimensional action domains, so it cannot deal with problems with continuous variables [28]. Ref. [29] proposed a deep reinforcement learning algorithm that acts on the continuous action domain, thus realizing the adaptive control of load frequency. An energy management scheme for AC control based on the deep deterministic policy gradient (DDPG) algorithm is proposed in [30]. However, this algorithm is only suitable for the optimal control of a single generator set, and is not suitable for the control of large-scale thermostatically controlled load. In [31], a distributed soft actor–critic (DSAC)-based data-driven frequency control method is proposed; the DSAC model estimates the distribution of value function over returns instead of only estimating the mean. The DSAC method based on entropy regularization has a faster learning speed compared to the traditional expectation-based reinforcement learning methods.

In view of the above problems, we take the thermostatically controlled load as the frequency control object, and based on the deep reinforcement learning algorithm, propose a frequency stability control method with the participation of thermostatically controlled load, considering customer satisfaction. Firstly, considering the operation characteristics of thermostatically controlled load with different control types, the energy storage index and the discomfort index are established, and the fuzzy comprehensive discrimination method is used to evaluate the customer satisfaction. Then, in order to realize the frequency cooperative control of large-scale thermostatically controlled load, a multi-agent control model is established based on the soft actor–critic algorithm to realize the continuous action control of thermostatically controlled load. Through the multi-agent reinforcement learning model considering customer satisfaction, the frequency response control of each thermostatically controlled load cluster can be coordinated online. The main contributions of this paper are as follows:

1. The influence of consumer satisfaction on the frequency response of TCLs is considered, and indicators reflecting satisfaction for TCLs with different control modes are established.
2. For the control problem of frequency response control for large-scale TCLs, this paper proposes a deep reinforcement learning control method based on the SAC algorithm.

This method can rely only on local load and frequency data to achieve real-time cooperative control of large-scale TCLs, which reduces the communication pressure in the scheduling process.

The remainder of this paper is organized as follows. In Section 2, the TCL dynamic model and control methods are formulated. Section 3 is concerned with the comprehensive control index of TCLs considering customer satisfaction. In Section 4, the frequency response control of TCLs based on SAC deep reinforcement learning is modeled. Case studies are provided in Section 5. Finally, conclusions are summarized in Section 6.

2. TCL Dynamic Model and Control Methods

2.1. TCL Dynamic Model

The first-order ordinary differential equation model considering indoor environment, outdoor environment and building characteristics has high accuracy and simple calculation, and is widely used in practice [32,33]. State variables T_i and virtual variables s_i are introduced into the model. The operating characteristics of the i-th TCL in the cooling mode can be expressed as:

$$\frac{dT_i(k)}{dk} = \frac{1}{C_i R_i}(T_\infty(k) - T_i(k) - s_i(k)R_iP_i) \tag{1}$$

Among them, the change rule of $s_i(k)$ is as follows:

$$s_i(k+\Delta k) = \begin{cases} 0 & s_i(k) = 1 \text{ and } T_i(k) \leq T_i^{\min} \\ 1 & s_i(k) = 0 \text{ and } T_i(k) \geq T_i^{\max} \\ s_i(k) & \text{others} \end{cases} \tag{2}$$

$$\begin{cases} T_i^{\min} = T_i^{set} - \frac{\delta}{2} \\ T_i^{\max} = T_i^{set} + \frac{\delta}{2} \end{cases} \tag{3}$$

where $T_\infty(k)$ and $T_i(k)$ are outdoor temperature and indoor temperature, respectively; C_i, P_i and P_i are the equivalent heat capacity, equivalent heat resistance and energy transfer rate of the i-th TCL, respectively; $s_i(k)$ indicates the load switch state, on state $s_i(k) = 1$ and off state $s_i(k) = 0$. $T_i^{\max}$ and $T_i^{\min}$ are the upper and lower limits of the temperature during load operation, respectively. T_i^{set} is the temperature setting value, δ is the temperature dead zone interval and is a constant. k and Δk are operation time and control period, respectively. By solving Equation (1):

$$T_i(k) = T_\infty(k) - s_i(k)R_iP_i - (T_\infty(k) - s_i(k)R_iP_i - T_i(0))e^{-\frac{k}{C_iR_i}} \tag{4}$$

where $T_i(0)$ represents the initial indoor temperature. For a load cluster composed of N TCLs, the aggregate power consumption is the sum of the rated power of all loads.

$$P_{total}(k) = \sum_{i=1}^{N} P_i^n s_i(k) \tag{5}$$

$$P_i^n = \frac{P_i}{\eta_i} \tag{6}$$

where P_i^n is the rated power of the i-th TCL and η_i is the energy conversion efficiency coefficient of the i-th TCL.

Figure 1 shows the frequency response model of the power system with TCL load participating in frequency modulation, where T_{Ga} and T_{Gb} are the time constants of the governor and the turbine, respectively. The governor and the turbine are the instantaneous characteristic compensation links, which are expressed by the lead lag transfer function between the time constants T_1 and T_2. T_R is the TCL response delay time constant, T_c is the communication delay time constant and R_{eq} is the unit adjustment rate. ΔP_G, ΔP_L, H and D represent the total output power, the disturbance power, inertia time constant and load damping coefficient of the system, respectively. Δf is the frequency offset.

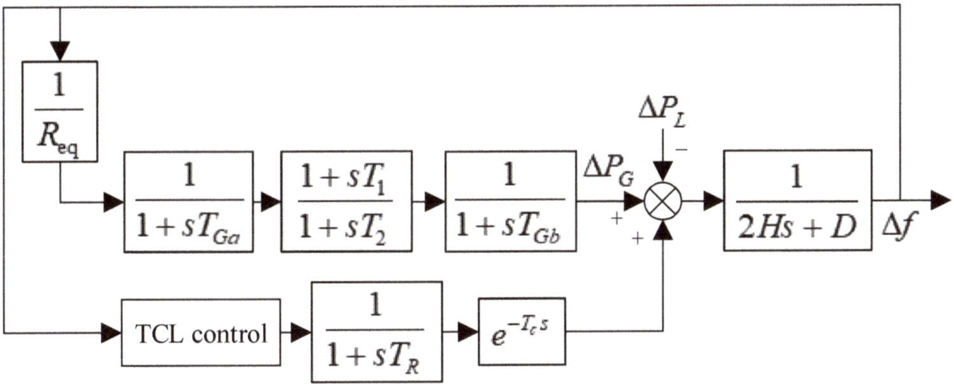

Figure 1. Frequency modulation model of power system with TCL participation.

2.2. TCL Control Methods

The control methods of TCL are mainly divided into direct switch control and temperature setting control. The TCL operation characteristics of the direct switch are shown in Figure 2a. The temperature setting value of the load remains unchanged, and the dispatching command directly acts on the equipment switch during the operation time $k = k_0$. The advantage of this method is that it can accurately track the power within the adjustable temperature range and has little impact on the user's comfort. However, when the indoor temperature approaches the temperature boundary, it will cause frequent switching, thus reducing the service life of the equipment. The TCL operating characteristics of the temperature setting are shown in Figure 2b. The dispatching command increases the temperature setting value at time $k = k_0$. Since the temperature dead zone of the load is unchanged, its operating range will change, thus indirectly changing the switching state of the equipment. Since the indoor temperature of the load is always uniformly distributed in the temperature dead zone, the indoor temperature will not approach the temperature boundary, but it has a notable impact on the user's comfort, and its tracking effect depends on the designed controller.

In practical application, the reasonable distribution of power regulation can make load clusters with different control modes cooperate with each other, which can not only realize accurate tracking of power, but also avoid their limitations and meet the different needs of users. Many factors need to be considered when allocating the power regulation amount. On the one hand, it is necessary to meet the requirements of the power system for frequency regulation and ensure the tracking accuracy within a certain range. On the other hand, for the load with direct switch, the number of switches should be reduced as much as possible to prolong the service life of the equipment. For the load with temperature setting, the temperature change should be reduced as much as possible to improve the user's comfort. There is often a highly nonlinear relationship between these factors and power distribution, and the frequency regulation requires high real-time performance. Therefore, the conventional optimization method cannot obtain the optimal power allocation.

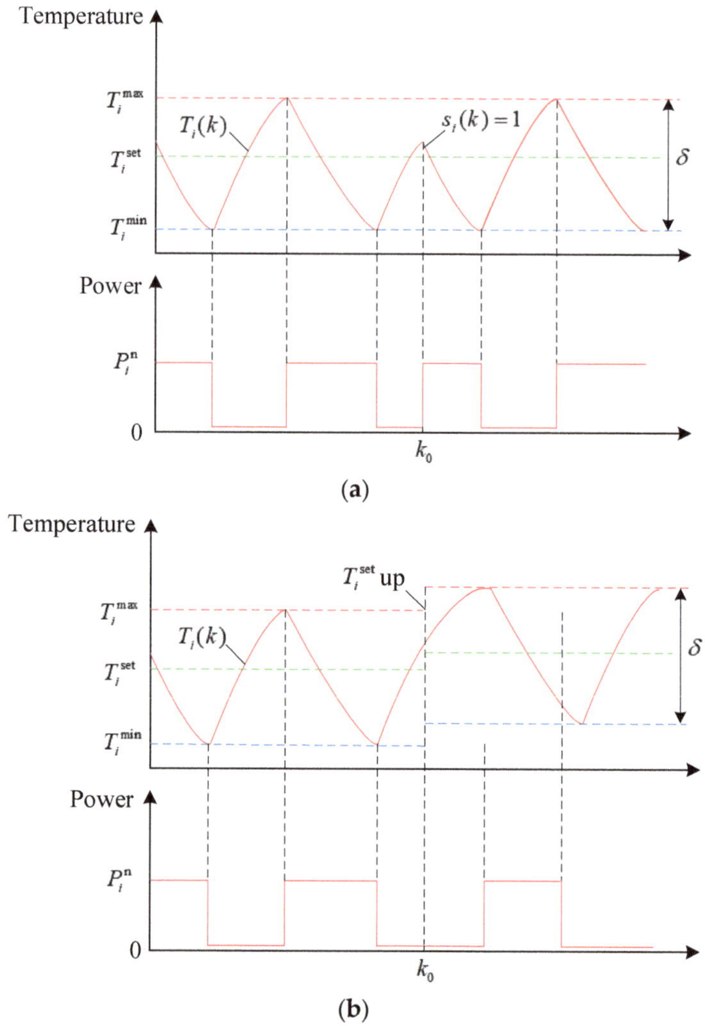

Figure 2. Operation characteristics of TCL in different control modes. (**a**) Direct switch control; (**b**) temperature setting control.

3. Comprehensive Control of TCL Considering Customer Satisfaction

3.1. Calculation of TCL Load Regulation Index

The temperature setting value of the direct switch load is almost constant, and the influence on the user's comfort is negligible. Its control mode directly acts on the equipment switch. When the indoor temperature is close to the temperature boundary, the load regulation capacity will decrease, and the start and stop frequency of the equipment will be higher and the service life will be shortened. When the indoor temperature is close to the temperature setting value, the load regulation capacity will rise. At this time, the start and stop frequency of the equipment is low and the service life is extended. In order to characterize the regulation ability of TCL and provide a basis for evaluating the service life of equipment, we use the definition of battery state of charge for reference, and define the energy storage index C_s for TCL cluster with a direct switch under refrigeration mode:

$$C_s = \frac{2}{\delta}\left| \frac{\sum_{i=1}^{N} T_{\max} - T_i}{N} - \frac{\delta}{2} \right| \quad T_i \in (T_i^{\min}, T_i^{\max}) \tag{7}$$

According to the definition of C_s, the closer C_s is to 0, the closer the indoor temperature is to the temperature set value. At this time, the temperature distribution of TCL is relatively uniform, its adjustable potential is large and the switch switching is not frequent. The closer C_s is to 1, the closer the indoor temperature is to the upper and lower temperature limits. At this time, the temperature distribution of TCL is more concentrated, its adjustable potential is small and the switch switching is more frequent. Therefore, when the agent outputs the control command, it should make C_s close to 0 as much as possible, so as to reduce the start and stop frequency of the equipment.

The tracking effect of the load depends on the designed controller, and because the set value changes, the temperature changes in a large range, which will reduce the comfort level of the user. To characterize the user's comfort, the discomfort index C_u is defined as follows for TCL cluster with temperature setting:

$$C_u = \left| T_i^{set} - T_i^{set}(0) \right| \tag{8}$$

where $T_i^{set}(0)$ indicates the initial temperature setting value. According to the definition of C_u, the more the temperature setting value deviates from the initial temperature setting value, the higher the user's discomfort level. Therefore, when the agent outputs the control command, it should make C_u as close to 0 as possible to reduce the user's discomfort.

3.2. Customer Satisfaction Assessment Method Based on Regulation Index

According to the above analysis, users' needs are different under different control methods. In order to comprehensively evaluate customer satisfaction, the fuzzy comprehensive evaluation (FCE) method is adopted for evaluation. The specific operations are as follows.

(1) Construct the factor set of customer satisfaction $U = \{C_s, C_u\}$.
(2) Build a customer satisfaction evaluation set V = {satisfied, more satisfied, general, less satisfied, dissatisfied}.
(3) Determine the weight of each factor. Since the factor set in this paper is composed of two factors, C_s and C_u, which have the same importance to users, the weight is $A = [a_1, a_2] = [0.5, 0.5]$.
(4) The fuzzy evaluation matrix is established. Firstly, the degree to which each factor belongs to each comment is evaluated. Since most things follow a normal distribution, the membership function is selected as a Gaussian function.

$$r_{sp}(y_s) = e^{-\left(\frac{y_s - u_{sp}}{\sigma_{sp}}\right)^2} \tag{9}$$

where y_s is the input of the s-th factor, C_s and C_u. u_{sp} and σ_{sp} are the mean and standard deviation of the s-th factor and the p-th comment, respectively. Then, the fuzzy evaluation matrix R is:

$$R = \begin{bmatrix} r_{11} & r_{12} & r_{13} & r_{14} & r_{15} \\ r_{21} & r_{22} & r_{23} & r_{24} & r_{25} \end{bmatrix} \tag{10}$$

(5) Fuzzy comprehensive evaluation is carried out. The fuzzy evaluation set is:

$$B = A \circ R = [b_1 \ b_2 \ b_3 \ b_4 \ b_5] \tag{11}$$

where $\circ$ represents the operation of the fuzzy matrix. Since the weighted average fuzzy synthesis operator has an obvious weight effect and strong comprehensive degree, and can make full use of the information of R, the element b_p is:

$$b_p = \min\left(1, \sum_{s=1}^{2} a_s r_{sp}\right) \tag{12}$$

(6) Evaluate customer satisfaction. In order to make the level continuous and quantitative, the level rank corresponding to the matrix B element is set as 1, 2, 3, 4 and 5, and the customer satisfaction m is defined as:

$$m = \frac{b_1 + 2b_2 + 3b_3 + 4b_4 + 5b_5}{b_1 + b_2 + b_3 + b_4 + b_5} \tag{13}$$

According to the definition of m, the smaller m is, the higher the user's satisfaction.

In practical application, the agent should not only consider customer satisfaction, but also meet the requirements of the power system for frequency regulation, that is, to ensure the tracking accuracy within a certain range. In order to evaluate the tracking performance of the system, the root mean square error index E_{RMS} is defined as:

$$E_{\mathrm{RMS}} = \sqrt{\frac{\sum\limits_{\Delta k=1}^{N_s} (e(\Delta k))^2}{N_s (P_{\mathrm{target}}^{\max} - P_{\mathrm{target}}^{\min})^2}} \times 100\% \tag{14}$$

where N_s is the number of control cycle Δk, $e(\Delta k)$ is an error signal in a control period, and $P_{\mathrm{target}}^{\max}$ and $P_{\mathrm{target}}^{\min}$ are the minimum value and the maximum value of the tracking power signal, respectively.

According to the definition of E_{RMS}, the smaller E_{RMS} is, the higher the tracking accuracy of the system. In order to comprehensively evaluate the regulation effect and provide the basis for the optimization of the power distribution signal, the comprehensive evaluation index J is defined as:

$$J = (1 - \lambda) E_{\mathrm{RMS}} + \lambda m \tag{15}$$

where λ is the proportion of satisfaction.

In fact, priority should be given to ensuring the frequency stability of the power grid: when the tracking accuracy is within a certain range, customer satisfaction can be considered; otherwise, it will not be considered. The relationship between λ and E_{RMS} is as follows:

$$\lambda = \begin{cases} G_1 & 0 < E_{\mathrm{RMS}} \le F_1 \\ G_2 & F_1 < E_{\mathrm{RMS}} \le F_2 \\ G_3 & F_2 < E_{\mathrm{RMS}} \le F_3 \\ 0 & E_{\mathrm{RMS}} > F_3 \end{cases} \tag{16}$$

where F_1, F_2, F_3, G_1, G_2 and G_3 are constants, which can be set according to actual operation conditions.

4. Frequency Response Control of TCL Based on SAC Deep Reinforcement Learning

4.1. A Deep Reinforcement Learning Model of Soft Actor–Critic

Reinforcement learning is adaptive learning in the way of trial and error of agents. The agent interacts with the environment continuously, and takes actions to change the environmental state by acquiring the environmental state. The agent obtains corresponding rewards or punishments as the update guidance of the model parameters, so as to obtain the maximum cumulative rewards in continuous learning. Through this perception action evaluation learning method, the agent continuously obtains knowledge in the interaction process, and constantly adjusts and improves its action strategy to adapt to the environment, and finally gives a better task execution strategy. The environment interaction mode is generally described by the Markov decision process (MDP) and is composed of five tuples (S, A, P, r, γ), namely state space S, action space A, state transition probability P, return function r and discount factor γ.

In this paper, the deep reinforcement learning based on the flexible actor evaluator framework is used to control the frequency response of TCL. The framework of the proposed control model is shown in Figure 3. In the iterative calculation at time t, the actor first generates the action a_t through the policy network according to the operating state s_t of the TCL cluster observed at this time. After that, the TCL cluster performs state transition according to the control strategy at this time, and reaches the state s_{t+1} at the next time. At the same time, the system environment calculates the reward $r(s_t, a_t)$ at time t and feeds it back to the agent, who records $(s_t, a_t, r(s_t, a_t), s_{t+1})$ in the experience pool. Then, the action strategy sampling of the actor and the system state are input to the critical at the same time, and the action value function $Q(s_t, a_t)$ is output to evaluate the strategy. This process is carried out circularly, and the actor and the judge update their neural network parameters through the gradient descent method, so as to realize the adaptive learning of the model. During the training process, the accumulated return of the agent in the response period will gradually increase and eventually become stable. By introducing the maximum entropy encouragement strategy, the SAC reinforcement learning algorithm can improve the robustness of the algorithm and speed up the training speed. It can make accurate and effective control decisions for large-scale temperature control loads in the complex power supply and demand environment.

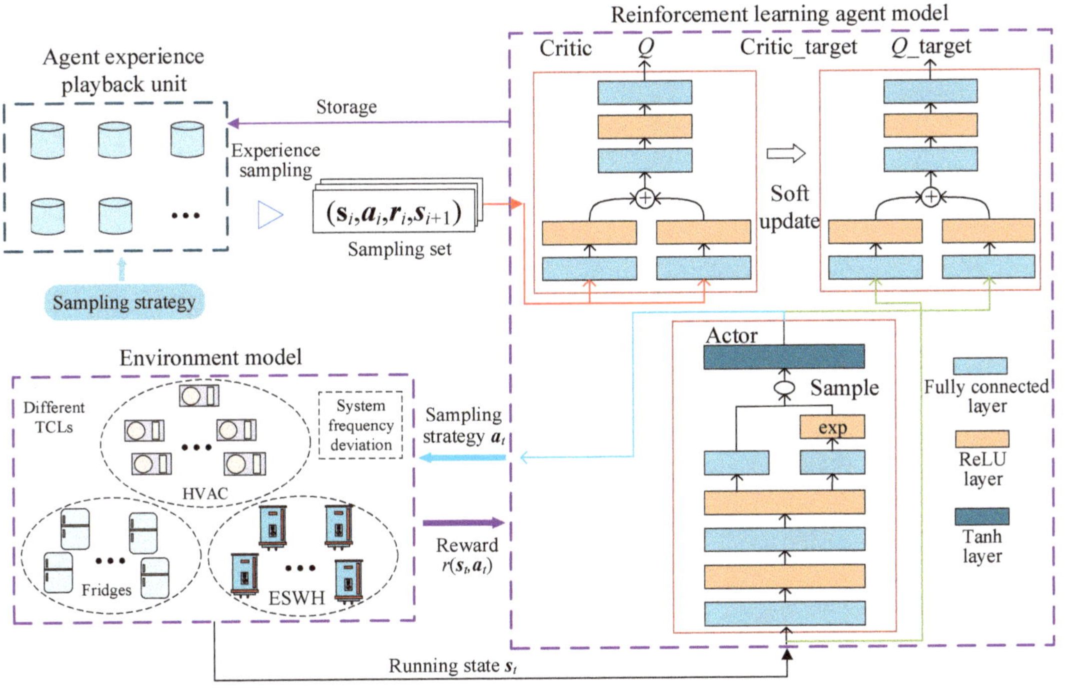

Figure 3. Frequency response control framework of large-scale TCLs based on deep reinforcement learning.

4.2. SAC Deep Reinforcement Learning Method

4.2.1. SAC Objective Function

The objective function of SAC requires the strategy to maximize the policy entropy while maximizing the cumulative return, so as to avoid greedy sampling in the learning process and falling into local optimization. According to this, the objective function $\pi^*_{\max}$ is constructed as shown in the equation.

$$\pi^*_{\max} = \arg\max_{\pi} \sum_{t=1}^{T} E_{(s_q,a_q)\sim p_\pi}(r(s_q,a_q) + \alpha H(\pi(\cdot|s_q))) \tag{17}$$

where $E(\cdot)$ is the expected function, π is a policy, s_q is the state space of the q-th agent, a_q is the action space of the TCL and $r(s_q,a_q)$ is the reward function of the q-th agent. The state action trajectory $(s_q,a_q) \sim p_\pi$ formed by the strategy π. α is the temperature term, which determines the influence of entropy on reward. $H(\pi(\cdot|s_q))$ is the entropy term of the strategy in the state, and its calculation method is shown in Equation (18).

$$H(\pi(a_q|s_q)) = -\int_{a_q} \pi(a_q|s_q)\log(\pi(a_q|s_q))da_q = E_{a_q\sim p_\pi}(-\log(\pi(a_q|s_q))) \tag{18}$$

4.2.2. SAC Iteration Strategy

The value function in the reinforcement learning process is shown in Equation (19), which is used for strategic value evaluation $Q(s_q,a_q)$. The bellman backup operator is shown in Equation (20) and is used for policy updating.

$$Q(s_q,a_q) = r(s_q,a_q) + \gamma E_{s_{t+1}\sim p}(Q(s_{q+1},a_{q+1})) \tag{19}$$

$$T^\pi Q(s_q,a_q) \triangleq r(s_q,a_q) + \gamma E_{s_{t+1}\sim p}(V(s_{q+1})) \tag{20}$$

where $E_{s_{t+1}\sim p}$ is the expected function of the state space, T^π is the bellman backup operator under the policy π and γ is the discount factor of the reward. $V(s_{q+1})$ is a new value function of the state, and its calculation method is shown in the following equation.

$$V(s_q) = E_{a_q\sim\pi}(Q(s_q,a_q) - \log\pi(a_q|s_{q+1})) \tag{21}$$

Meanwhile, there are

$$Q^{k+1} = T^\pi Q^k \tag{22}$$

where Q^k is the value function of the kth calculation.

Equation (23) can be realized by iterating Equations (20) and (22) continuously.

$$\lim_{k\to\infty} Q^k = \hat{Q} \tag{23}$$

where $\hat{Q}$ is the soft Q-value.

4.2.3. SAC Policy Update

The strategy updating method in the calculation process is shown in Equation (24).

$$\pi_{new} = \arg\min_{\pi\in\Pi} D_{KL}\left(\pi(\cdot|s_q)||\frac{\exp\left(\frac{1}{\alpha}Q^{\pi_{old}}(s_q,\cdot)\right)}{Z^{\pi_{old}}(s_q)}\right) \tag{24}$$

where KL is divergence, Π is a policy set and $Q^{\pi_{old}}(\cdot)$ is the value function under the old strategy π_{old}. $Z^{\pi_{old}}(s_q)$ is the partition function under the old strategy π, which is used to normalize the distribution.

4.2.4. Construction of SAC Algorithm

The SAC algorithm needs to construct neural networks, including Q-value network and strategy network. The Q-value network outputs a single value through several layers of neural networks, and the strategy network outputs a Gaussian distribution. In this process, the neural network will be updated. The Q-value network parameter has an update strategy as shown in Equation (25), and the policy network parameter has an update strategy as shown in Equation (26).

$$J_Q(\theta) = E_{(s_q,a_q,s_{q+1})\sim D}\left(\frac{1}{2}\left(Q(s_q,a_q) - \left(r(s_q,a_q) + \gamma V_{\overline{\theta}}(s_{q+1})\right)\right)^2\right) \tag{25}$$

$$J_\pi(\phi) = D_{KL}\left(\pi(\cdot|s_q)||\exp\left(\frac{1}{\alpha}Q_\theta(s_q,\cdot) - \log Z(s_q)\right)\right) \tag{26}$$

where θ is the network parameter of Q-value, ϕ is the policy network parameter and $V_{\overline{\theta}}$ and Q_θ are the new value function and the value function after substituting the Q-value network parameters. $Z(s_q)$ is the partition function of the state.

The temperature parameter is an important parameter to assist in maximizing entropy, which can maximize the exploration of action space. A reasonable temperature parameter setting is helpful to realize iterative testing of all feasible actions. Therefore, the update of the temperature parameter is as shown in Equation (27).

$$J(\alpha) = E_{a_q\sim\pi_q}(-\alpha\log\pi_q(a_q|\pi_q) - \alpha H_0) \tag{27}$$

where π_q is the control strategy of the q-th agent. H_0 is the entropy term.

The expressions (25)–(27) are all obtained by calculation. Throughout the whole process, the Q-value network parameters, strategy network parameters and temperature parameters are continuously updated through deep neural network learning, which can make the model converge continuously and solve the optimal strategy.

4.3. Design of SAC Deep Reinforcement Learning Model for TCL Frequency Response

In this paper, the SAC algorithm is used to solve the control strategy problem of large-scale temperature control load participating in system frequency regulation. The structural model of the proposed control method is shown in Figure 3. The agent in the figure is an agent based on a deep neural network. The environment of the controller is the frequency Δf deviation of the power system, the differentiation and integration of Δf, the baseline power signal P_{base} of the aggregated TCL, the aggregated power consumption P_{total} and the automatic generation control signal P_{AGC}. The automatic generation control signal is a series of positive and negative power signals, which represents the active power deviation between the supply and demand of the power system. The baseline power signal of aggregated TCL is the sum of the rated power $P^n_{i.set}$ of TCL cluster under a certain temperature setting value.

$$P_{base} = \sum_{i=1}^{N} P^n_{i.set} \tag{28}$$

$$P^n_{i.set} = \frac{T_\infty - T_i^{set}}{\eta_i R_i} \tag{29}$$

A tracking power signal P_{target} can be generated by superimposing the P_{AGC} and P_{base}. The tracking error signal e can be generated by subtracting the P_{target} and P_{total}, that is

$$P_{\text{target}} = P_{\text{AGC}} + P_{base} \tag{30}$$

$$e = P_{\text{target}} - P_{\text{total}} \tag{31}$$

The agent obtains the response power of TCL cluster after optimization according to the environmental information. Then, the direct switch control cluster and the temperature setting control cluster complete the adjustment task according to the obtained control signals. The method includes two stages: offline pre-learning and online application. In the offline pre-learning stage, the pre-learning process will iteratively update all parameters of the agent. During each self-learning iteration, the agent will conduct action exploration (i.e., generate different commands) to interact with the environment. After exploration, the parameters of the agent will be updated according to the system frequency deviation and the reward function of the TCL controller. With an appropriate reward function R and considering environmental constraints, the gradient of the actor (i.e., the gradient of the control target relative to the parameters of the agent) will be calculated and used to update all the parameters of the agent. In the online application stage, the intelligent personnel will calculate the operation value (i.e., generate command) according to the observation value and parameters obtained by themselves for each control cluster.

For the frequency response model of large-scale temperature control load, the negative value of the comprehensive evaluation index J is taken as the reward function of the agent, that is

$$R = J = (1 - \lambda)E_{\mathrm{RMS}} + \lambda m \tag{32}$$

By introducing the system frequency deviation and customer satisfaction into the reward function, the obtained control strategy can improve the tracking accuracy of the system, reduce the switching frequency of the equipment, reduce the temperature change and improve the customer satisfaction.

5. Result Analysis

5.1. Example Introduction and Scenario Setting

In order to verify the effectiveness of the method proposed in this paper for large-scale thermostatically controlled loads to participate in power grid frequency control, we take the distribution network with large-scale thermostatically controlled loads as an example, and set the disturbance power of the regional power grid as the net load power, that is, the difference between the original load power and the power generated by new energy, such as photovoltaic and wind power. The disturbance power in the simulation time is shown in Figure 4. As an important thermostatically controlled load, HVAC accounts for a large proportion and is easy to control and manage. Therefore, we selected 2000 HVAC units for simulation experiments. The load parameter settings are shown in Table 1. The initial indoor temperature of the load is evenly distributed in the temperature dead zone, and the temperature dead zone is set to 1.2 °C. P_{AGC} is the actual frequency regulation signal from the PJM power market in the United States, which changes every four seconds.

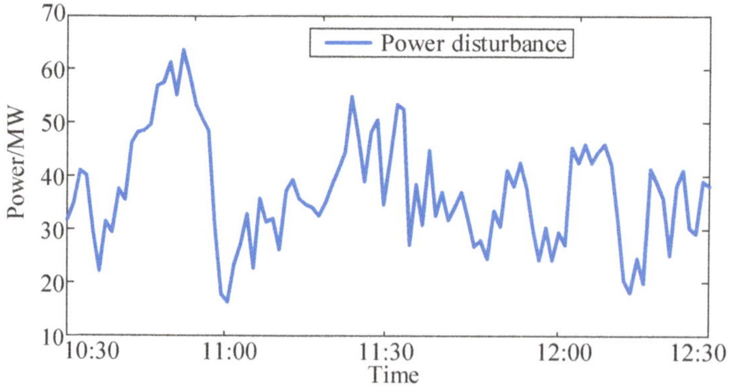

Figure 4. Variation of disturbance power within 2 h.

Table 1. Setting of load parameters.

Parameters	Meaning	Value
C_i	Equivalent heat capacity	N (2, 0.01)
R_i	Equivalent thermal resistance	N (2, 0.01)
P_i	Energy transfer rate	N (14, 0.01)
T_∞	Outdoor temperature	32
$T_i^{set}(0)$	Initial temperature setting	20
δ	Temperature dead zone	1.2
η_i	Energy conversion efficiency coefficient	2.5

Note: N (2, 0.01) represents the normal distribution with the mean value of 2 and the variance of 0.01, and the others are the same.

The parameter settings in the customer satisfaction evaluation and comprehensive evaluation indicators are shown in Table 2. When the value of ERMS is (0, 2%], the frequency regulation effect of the power system is better. At this time, customer satisfaction is mainly considered. When the value of E_{RMS} is (2, 3%], λ is 0.5, indicating that the root mean square error index and the customer satisfaction have the same impact on the system regulation. When the value of E_{RMS} is (3%, 5%], the frequency regulation effect of the power system is poor, and the root mean square error index is mainly considered at this time. When E_{RMS} is greater than 5%, λ is 0, indicating that the system regulation will no longer consider the customer satisfaction, and will focus on improving the tracking accuracy.

Table 2. Setting of customer satisfaction evaluation and comprehensive evaluation parameters.

Parameters	Value	Parameters	Value
u_{s1}	0	F_1	2%
u_{s2}	0.25	F_2	3%
u_{s3}	0.5	F_3	5%
u_{s4}	0.75	G_1	0.8
u_{s5}	1	G_2	0.5
σ_{sp}	0.2	G_3	0.3

5.2. Frequency Control Effect Analysis Considering Customer Satisfaction

Figure 5 compares the changes in customer satisfaction before and after customer satisfaction is included in the agent optimization process. As can be seen from Figure 5, on the one hand, the peak value of customer satisfaction increased from 3.5 before optimization to 4.7 after optimization. On the other hand, within the simulation time range, customer satisfaction after optimization is higher than that before optimization. Figure 6 compares the frequency control effect of the method proposed in this paper before and after considering customer satisfaction. As can be seen from Figure 6, since the customer satisfaction index is added to the control target, the algorithm's penalty weight for frequency deviation is relatively reduced, resulting in an increase in the maximum frequency deviation of the system compared with that before considering customer satisfaction. Considering customer satisfaction will limit the number of TCLs participating in frequency response, resulting in a poor system frequency control effect, but the impact is not significant. Therefore, the method proposed in this paper can better balance the frequency deviation control and customer satisfaction.

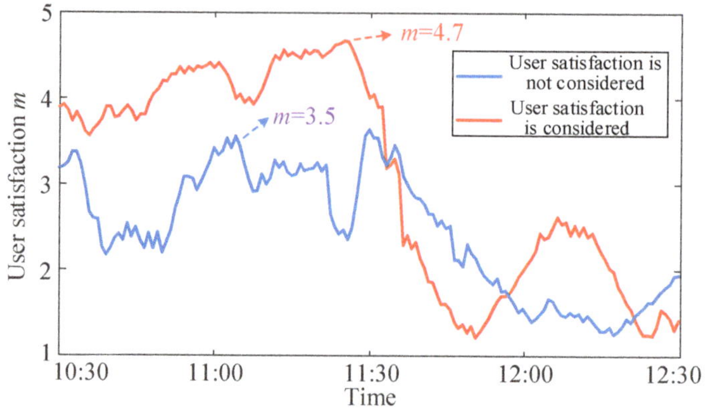

Figure 5. Comparison of customer satisfaction before and after optimization.

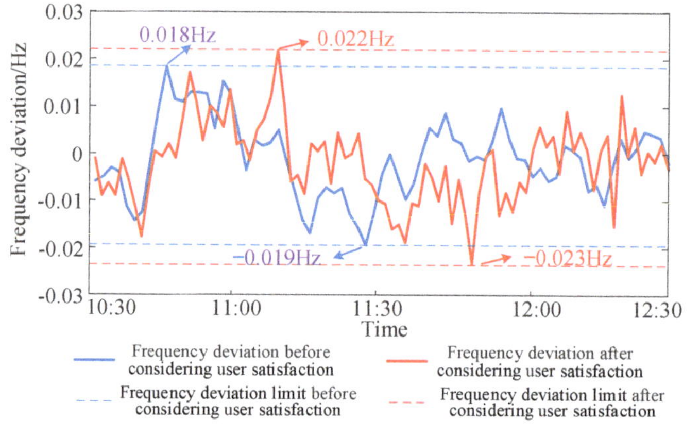

Figure 6. Frequency deviation before and after considering customer satisfaction.

5.3. Frequency Control EFFECT Analysis Based on SAC Deep Reinforcement Learning

In order to verify the effectiveness of the proposed SAC deep reinforcement learning algorithm in the collaborative control of large-scale thermostatically controlled load compared with the traditional PID method, we used the traditional PID controller, the PID controller optimized by particle swarm optimization (PSO) algorithm parameters, and the algorithm proposed in this paper to conduct simulation experiments on the thermostatically controlled load controlled by the direct switch control and the temperature setting control. The number of HVAC experiments is 2000. The comparison of the three control methods in the system frequency control effect is shown in Figure 7.

It can be seen from Figure 7 that under the regulation control of the conventional PID controller, there is a system frequency deviation of about 0.047 Hz in the period when the disturbance power fluctuates violently. Compared with the traditional PID controller, the frequency effect of the PID controller is improved after PSO algorithm parameter optimization, but the effect is still not ideal. The frequency deviation of the system fluctuates within the range of (−0.037 Hz, 0.038 Hz). The algorithm proposed in this paper can keep the system frequency deviation within (−0.02 Hz, 0.023 Hz), and can significantly improve the frequency stability of the power grid.

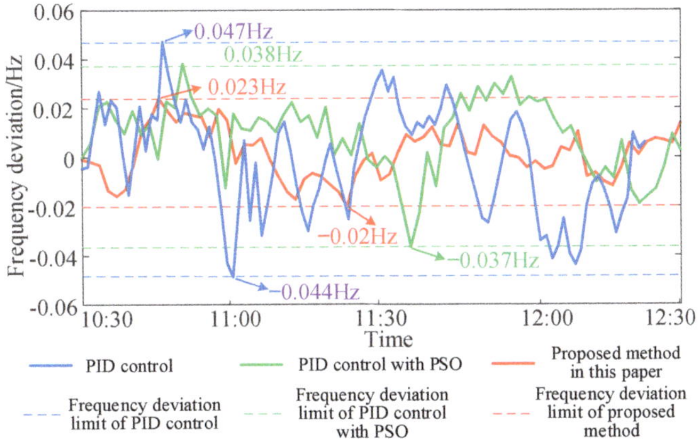

Figure 7. Comparison of frequency control effect between the proposed algorithm and PID controller.

5.4. Comparative Analysis of the Algorithm and Traditional Deep Reinforcement Learning

This section shows the advantages of the algorithm in this paper compared with the data-driven deep Q network (DQN) algorithm and the DDPG algorithm from the aspects of system frequency control effect and algorithm convergence speed. The proposed algorithm and DDPG are both deep reinforcement learning algorithms based on continuous action space. As shown in Figure 8, the frequency control effect of the two algorithms is significantly better than that of the DQN algorithm designed based on the discrete action space. After integrating the advantages of model driven and data driven, the algorithm proposed in this paper further improves the real-time frequency control effect in the continuous action domain compared with the fully trained DDPG algorithm.

Figure 9 compares the iterative convergence process of the cumulative reward value of the DQN algorithm, the DDPG algorithm and the algorithm proposed in this paper. Among them, after about 250 and 300 iteration cycles, the cumulative reward value of DQN algorithm and DDPG algorithm tends to be stable and will not continue to increase. It is worth noting that each iteration cycle in this paper is an empirical trajectory containing 200 iterations. That is, the two algorithms need 50,000 and 60,000 iterations, respectively, to converge. However, the proposed algorithm only needs about 150 iteration cycles (30,000 iterations) to complete the parameter training of the deep neural network, and the oscillation amplitude of the convergence curve of the algorithm is the smallest.

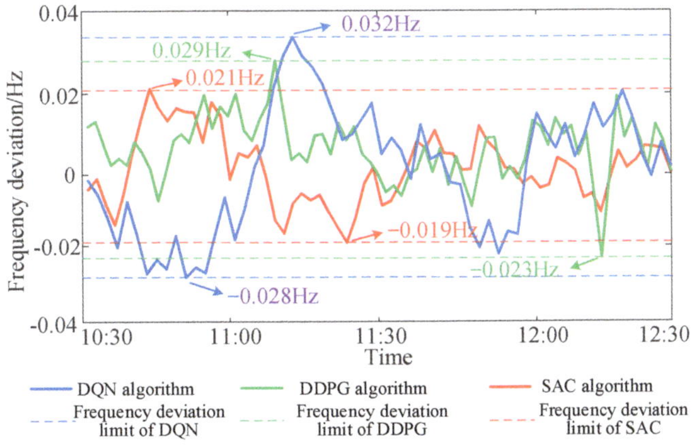

Figure 8. Comparison of frequency control effect between different algorithms.

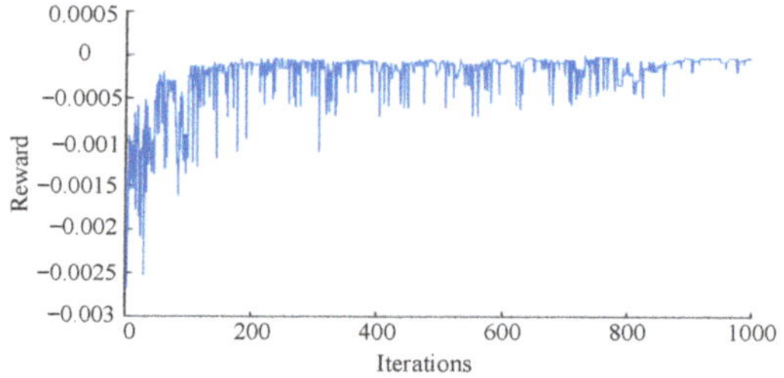

(**a**) Iterations of DQN

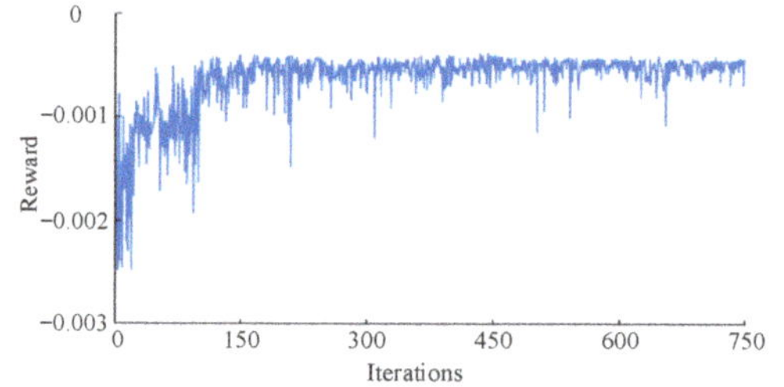

(**b**) Iterations of DDPG

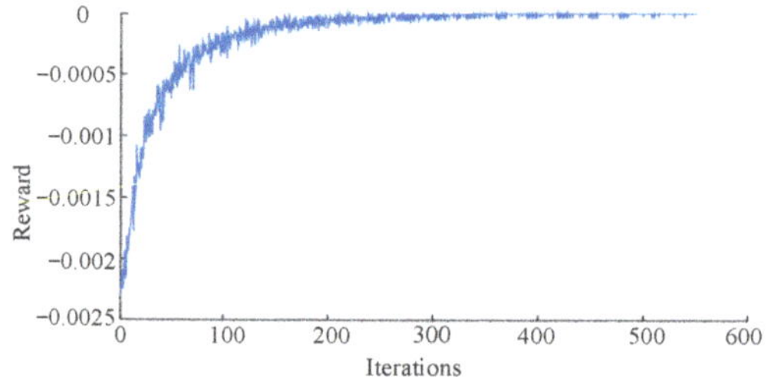

(**c**) Iterations of SAC

Figure 9. Iterative convergence process of cumulative reward value: (**a**) DQN algorithm; (**b**) DDPG algorithm; (**c**) SAC algorithm in this paper.

6. Conclusions

In this paper, considering the support of large-scale thermostatically controlled load on the demand side to the frequency of power grid, based on the deep reinforcement learning of soft actor–critic, a frequency cooperative control method of thermostatically controlled load considering customer satisfaction is proposed to solve the frequency control problem of a power system with large-scale thermostatically controlled load on the demand side participating in frequency regulation. In the example analysis, a distribution network is taken as the research object, and the performance of different algorithms is compared and verified based on time domain simulation. The simulation results show that compared with the existing deep reinforcement learning methods, this algorithm has obvious advantages in system frequency control, customer satisfaction and algorithm training time.

The algorithm proposed in this paper is mainly used to solve the frequency cooperative control problem under the participation of large-scale thermostatically controlled load in the distribution network, and does not consider the application of other flexible resources on the demand side in the system frequency modulation. The next step is to explore the application and practice of multi-agent deep reinforcement learning method in the frequency response of demand-side flexible resources based on the operation mechanism and mathematical modeling of demand-side flexible resources.

Author Contributions: Conceptualization, R.C. and G.Y.; Data curation, C.L. and X.Y.; Investigation, H.L. and Y.Z.; Methodology, R.C.; Software, H.L.; Supervision, G.Y. and Y.Z.; Validation, G.Y.; Visualization, X.Y.; Writing—original draft, R.C. and H.L.; Writing—review and editing, C.L., X.Y. and Y.Z. All authors have read and agreed to the published version of the manuscript.

Funding: This work was supported by the Science and Technology Project of State Grid Hubei Electric Power Research Institute: "Research on frequency dynamic prediction and active control strategy of high proportion new energy power system under mutational weather" (project number B31532225680).

Data Availability Statement: The data in this paper are from a real distribution network and involve a confidentiality agreement. The dataset in this paper is not publicly available.

Conflicts of Interest: The authors declare no conflict of interest.

Abbreviation

TCL	thermostatically controlled load
ESWH	heat pump, electric storage water heater
HVAC	heating, ventilation and air conditioning
LA	load aggregator
EMS	energy management system
DDPG	deep deterministic policy gradient
SAC	soft actor–critic
DQN	Deep Q network
DSAC	distributed soft actor–critic
FCE	fuzzy comprehensive evaluation
MDP	Markov decision process
PSO	particle swarm optimization
T_i	The indoor temperature of the *i*-th TCL
T_∞	The outdoor temperature
C_i	The equivalent heat capacity of the *i*-th TCL
R_i	The equivalent heat resistance of the *i*-th TCL
P_i	The energy transfer rate of the *i*-th TCL
$s_i(k)$	The load switch state of TCL
$T_i^{\max}$	The upper limits of the temperature during load operation
$T_i^{\min}$	The lower limits of the temperature during load operation
T_i^{set}	the temperature setting value
δ	the temperature dead zone interval
P_i^n	the rated power of the *i*-th TCL
η_i	the energy conversion efficiency coefficient
C_s	the energy storage index for TCL cluster with direct switch control
C_u	the discomfort index for TCL cluster with temperature setting control
r_{sp}	the membership function of the *s*-th factor and the *p*-th comment
y_s	the input of the *s*-th factor
u_{sp}	the mean of the *s*-th factor and the *p*-th comment
σ_{sp}	the standard deviation of the *s*-th factor and the *p*-th comment
b_p	the element of fuzzy evaluation set *B*
m	the customer satisfaction index
E_{RMS}	the root mean square error index of frequency
$e(\Delta k)$	the error signal in a control period
$P_{\mathrm{target}}^{\max}$	the maximum value of the tracking power signal
$P_{\mathrm{target}}^{\min}$	the minimum value of the tracking power signal
J	the comprehensive evaluation index
λ	the proportion of satisfaction
Δf	the frequency deviation of the power system
P_{base}	the baseline power signal of the aggregated TCLs
P_{total}	the aggregated power consumption of TCLs
P_{AGC}	the automatic generation control signal

References

1. Li, C.; Wu, Y.; Sun, Y.; Zhang, H.; Liu, Y.; Liu, Y.; Terzija, V. Continuous Under-Frequency Load Shedding Scheme for Power System Adaptive Frequency Control. *IEEE Trans. Power Syst.* **2020**, *35*, 950–961. [CrossRef]
2. Guan, M. Scheduled Power Control and Autonomous Energy Control of Grid-Connected Energy Storage System (ESS) with Virtual Synchronous Generator and Primary Frequency Regulation Capabilities. *IEEE Trans. Power Syst.* **2020**, *37*, 942–954. [CrossRef]
3. Hosseini, S.A.; Toulabi, M.; Dobakhshari, A.S.; Ashouri-Zadeh, A.; Ranjbar, A.M. Delay Compensation of Demand Response and Adaptive Disturbance Rejection Applied to Power System Frequency Control. *IEEE Trans. Power Syst.* **2020**, *35*, 2037–2046. [CrossRef]
4. Trovato, V.; Sanz, I.; Chaudhuri, B.; Strbac, G. Advanced control of thermostatic loads for rapid frequency response in Great Britain. *IEEE Trans. Power Syst.* **2017**, *32*, 2106–2117. [CrossRef]
5. Bao, Y.; Li, Y.; Hong, Y.; Wang, B. Design of a hybrid hierarchical demand response control scheme for the frequency control. *IET Gener. Transm. Distrib.* **2015**, *9*, 2303–2310. [CrossRef]
6. Mallada, E.; Zhao, C.; Low, S. Optimal load-side control for frequency regulation in smart grids. *IEEE Trans. Autom. Control* **2017**, *62*, 6294–6309. [CrossRef]
7. Yu, Y.; Quan, L.; Mi, Z.; Lu, J.; Chang, S.; Yuan, Y. Improved Model Predictive Control with Prescribed Performance for Aggregated Thermostatically Controlled Loads. *J. Mod. Power Syst. Clean Energy* **2022**, *10*, 430–439. [CrossRef]
8. Wang, J.; Shi, Y.; Zhou, Y. Intelligent demand response for industrial energy management considering thermostatically controlled loads and EVs. *IEEE Trans. Ind. Informatics* **2019**, *15*, 3432–3442. [CrossRef]
9. Hu, J.; Cao, J.; Chen, M.Z.Q.; Yu, J.; Yao, J.; Yang, S.; Yong, T. Load Following of Multiple Heterogeneous TCL Aggregators by Centralized Control. *IEEE Trans. Power Syst.* **2017**, *32*, 3157–3167. [CrossRef]
10. Delavari, A.; Kamwa, I. Improved Optimal Decentralized Load Modulation for Power System Primary Frequency Regulation. *IEEE Trans. Power Syst.* **2018**, *33*, 1013–1025. [CrossRef]
11. Song, M.; Sun, W.; Wang, Y.; Shahidehpour, M.; Li, Z.; Gao, C. Hierarchical Scheduling of Aggregated TCL Flexibility for Transactive Energy in Power Systems. *IEEE Trans. Smart Grid* **2020**, *11*, 2452–2463. [CrossRef]
12. Wang, Y.; Tang, Y.; Xu, Y.; Xu, Y. A Distributed Control Scheme of Thermostatically Controlled Loads for the Building-Microgrid Community. *IEEE Trans. Sustain. Energy* **2020**, *11*, 350–360. [CrossRef]
13. Kim, Y.J.; Fuentes, E.; Norford, L.K. Experimental Study of Grid Frequency Regulation Ancillary Service of a Variable Speed Heat Pump. *IEEE Trans. Power Syst.* **2016**, *31*, 3090–3099. [CrossRef]
14. Song, M.; Gao, C.; Yan, H.; Yang, J. Thermal Battery Modeling of Inverter Air Conditioning for Demand Response. *IEEE Trans. Smart Grid* **2018**, *9*, 5522–5534. [CrossRef]
15. Zhao, H.; Wu, Q.; Huang, S.; Zhang, H.; Liu, Y.; Xue, Y. Hierarchical Control of Thermostatically Controlled Loads for Primary Frequency Support. *IEEE Trans. Smart Grid* **2018**, *9*, 2986–2998. [CrossRef]
16. Luo, F.; Dong, Z.; Meng, K.; Wen, J.; Wang, H.; Zhao, J. An operational planning framework for large-scale thermostatically controlled load dispatch. *IEEE Trans. Ind. Inform.* **2017**, *13*, 217–227. [CrossRef]
17. Zhang, W.; Lian, J.; Chang, C.Y.; Kalsi, K. Aggregated Modeling and Control of Air Conditioning Loads for Demand Response. *IEEE Trans. Power Syst.* **2013**, *28*, 4655–4664. [CrossRef]
18. Zhang, R.; Chu, X.; Zhang, W.; Liu, Y. Active Participation of Air Conditioners in Power System Frequency Control Considering Users' Thermal Comfort. *Energies* **2015**, *8*, 10818–10841. [CrossRef]
19. Bashash, S.; Fathy, H.K. Modeling and Control of Aggregate Air Conditioning Loads for Robust Renewable Power Management. *IEEE Trans. Control Syst. Technol.* **2013**, *21*, 1318–1327. [CrossRef]
20. Zhao, H.; Zhao, J.; Shu, T.; Pan, Z. Hybrid-Model-Based Deep Reinforcement Learning for Heating, Ventilation, and Air-Conditioning Control. *Front. Energy Res.* **2021**, *8*, 610518. [CrossRef]
21. Chen, Z.; Shi, J.; Song, Z.; Yang, W.; Zhang, Z. Genetic Algorithm Based Temperature-Queuing Method for Aggregated IAC Load Control. *Energies* **2022**, *15*, 535. [CrossRef]
22. Pallonetto, F.; de Rosa, M.; Milano, F.; Finn, D.P. Demand response algorithms for smart-grid ready residential buildings using machine learning models. *Appl. Energy* **2019**, *239*, 1265–1282. [CrossRef]
23. Tindemans, S.H.; Strbac, G. Low-Complexity Decentralized Algorithm for Aggregate Load Control of Thermostatic Loads. *IEEE Trans. Ind. Appl.* **2021**, *57*, 987–998. [CrossRef]
24. Ma, K.; Yuan, C.; Liu, Z.; Yang, J.; Guan, X. Hybrid control of aggregated thermostatically controlled loads: Step rule, parameter optimization, parallel and cascade structures. *IET Gener. Transm. Distrib.* **2016**, *10*, 4149–4157. [CrossRef]
25. Xi, L.; Li, H.; Zhu, J.; Li, Y.; Wang, S. A Novel Automatic Generation Control Method Based on the Large-Scale Electric Vehicles and Wind Power Integration Into the Grid. *IEEE Trans. Neural Netw. Learn. Syst.* 2022. [CrossRef] [PubMed]
26. Peirelinck, T.; Hermans, C.; Spiessens, F.; Deconinck, G. Domain Randomization for Demand Response of an Electric Water Heater. *IEEE Trans. Smart Grid* **2020**, *12*, 1370–1379. [CrossRef]
27. Ruelens, F.; Claessens, B.J.; Vandael, S.; de Schutter, B.; Babuška, R.; Belmans, R. Residential demand response of thermostatically controlled loads using batch reinforcement learning. *IEEE Trans. Smart Grid* **2017**, *8*, 2149–2159. [CrossRef]

28. Liu, H.; Xu, F.; Fan, P.; Liu, L.; Wen, H.; Qiu, Y.; Ke, S.; Li, Y. Load Frequency Control Strategy of Island Microgrid with Flexible Resources Based on DQN. In Proceedings of the 2021 IEEE Sustainable Power and Energy Conference (iSPEC), Nanjing, China, 23–25 December 2021; pp. 632–637. [CrossRef]
29. Yan, Z.; Xu, Y. A multi-agent deep reinforcement learning method for cooperative load frequency control of a multi-area power system. *IEEE Trans. Power Syst.* **2020**, *35*, 4599–4608. [CrossRef]
30. Yu, L.; Xie, W.; Xie, D.; Zou, Y.; Zhang, D.; Sun, Z.; Zhang, L.; Zhang, Y.; Jiang, T. Deep Reinforcement Learning for Smart Home Energy Management. *IEEE Internet Things J.* **2019**, *7*, 2751–2762. [CrossRef]
31. Xie, J.; Sun, W. Distributional Deep Reinforcement Learning-Based Emergency Frequency Control. *IEEE Trans. Power Syst.* **2021**, *37*, 2720–2730. [CrossRef]
32. Yan, Z.; Xu, Y. Data-driven load frequency control for stochastic power systems: A deep reinforcement learning method with continuous action search. *IEEE Trans. Power Syst.* **2019**, *34*, 1653–1656. [CrossRef]
33. Acharya, S.; Moursi, M.S.E.; Al-Hinai, A. Coordinated Frequency Control Strategy for an Islanded Microgrid with Demand Side Management Capability. *IEEE Trans. Energy Convers.* **2018**, *33*, 639–651. [CrossRef]

sustainability

MDPI

Systematic Review

Identifying Key Components in Implementation of Internet of Energy (IoE) in Iran with a Combined Approach of Meta-Synthesis and Structural Analysis: A Systematic Review

Mir Hamid Taghavi [1], Peyman Akhavan [2,*], Rouhollah Ahmadi [3,*] and Ali Bonyadi Naeini [1]

[1] School of Management, Economics and Progress Engineering, Iran University of Science and Technology, Tehran 16846-13114, Iran
[2] Department of Industrial Engineering, Qom University of Technology, Qom 1519-37195, Iran
[3] School of Advanced Technologies, Iran University of Science and Technology, Tehran 16846-13114, Iran
* Correspondence: akhavan@qut.ac.ir (P.A.); ahmadi@iust.ac.ir (R.A.)

Abstract: The increasing consumption of energy and the numerous obstacles in the way of its extraction, including diminishing fossil fuels and the turn towards renewable energies, environmental changes, a tendency towards systems of information networks, rising costs of energy and advancement of technology have made the need for new technologies aimed at efficient management of energy more imminent. The Internet of Energy (IoE) technology has been recognized as a novel and efficient strategy that provides the necessary tools for optimal energy management. The present study was carried out with the purpose of identifying key components in implementation of IoE in Iran. This study is practical in its goal and descriptive-explorative in its methodology. First, the data were categorized using the qualitative method of meta-synthesis and using the Sandelowski and Barroso method. The statistical population of the study was the scholarly finding of 2010–2021 and 55 papers were sampled from the published works. The kappa coefficient was used to determine reliability and quality control. The kappa coefficient calculated with SPSS equals 0.87, which falls in the "excellent" category. Second, the frequency and importance of each component was determined using the Shannon entropy technique. The purpose of this method is to measure the weight or importance of each component based on frequency and to identify the key components. Third, the MICMAC structural analysis method was used to evaluate the influence/dependence of components by eight experts in the field of energy and determine strategic components. The purpose of this step is to compare the results with the results of the second step of the research. The results show that 82 indicators play a role in implementation of the concept of IoE; these indicators can be divided into ten axial categories of rules and regulations, individual and human factors, funding, technological infrastructure, cultural and social factors, security factors, technological factors, knowledge factors, learning style, and management factors. In the Shannon entropy method, technological infrastructure, management factors, and rules and regulations are the most significant, respectively. In MICMAC structural analysis, the components of managerial factors, technological infrastructure, and financing have the largest share in influence and dependence, respectively. Conclusion: The two components of management factors and technological infrastructure can be considered as key and strategic components in implementation of IoE in Iran.

Keywords: IoE; optimal energy management; sustainable development; meta-synthesis; MICMAC analysis

Citation: Taghavi, M.H.; Akhavan, P.; Ahmadi, R.; Bonyadi Naeini, A. Identifying Key Components in Implementation of Internet of Energy (IoE) in Iran with a Combined Approach of Meta-Synthesis and Structural Analysis: A Systematic Review. *Sustainability* **2022**, *14*, 13180. https://doi.org/10.3390/su142013180

Academic Editors: Luis Hernández-Callejo, Sergio Nesmachnow and Sara Gallardo Saavedra

Received: 4 May 2022
Accepted: 28 September 2022
Published: 14 October 2022

1. Introduction

In the 21st century, the growing demand for energy and the widespread use of fossil fuels and traditional energy sources have been challenged by factors like the energy crisis, environmental pollution, and global warming [1]. In 2011, 82% of energy was generated from fossil fuels [2]. On the one hand, global energy demand in 2018 increased by 2.3 percent compared to 2017, which is the largest increase since 2010. As a result, CO_2 emissions from the energy sector set a new record in 2018. According to figures released by the International Energy Agency (IEA), global energy demand will increase by more than two-thirds by 2035 [3]. Compared to the pre-industrial temperature levels, global warming has reached 1.5 degrees Celsius, and, if this trend continues, will exceed 2 degrees Celsius and have a negative impact on the planet and human life [2]. Undoubtedly, such an increase in energy demand will put an additional burden on the old energy infrastructure, which will lead to serious problems of grid congestion and reduced energy quality. The usual grid structure faces reliability problems due to the lack of real-time monitoring, automation techniques, error detection, transparency, and flexibility [3,4]. The demand for renewable energy sources, such as solar and wind energy, is increasing significantly as a solution for the problem of traditional energy sources. In addition to protecting the environment, such an approach will also meet future energy demands [5].

Although renewable energy sources have advantages, such as sustainable development and environmental conservation, they have disadvantages too. It is difficult to accurately predict the amount of energy generation from renewable energy sources and it mainly depends on environmental conditions [6,7]. Moreover, with the existing electricity infrastructure, energy from renewable sources cannot be fully efficient. China, for example, generates most of its energy in a green way but still faces an energy crisis because it cannot deliver the energy it needs to its large population. The gradual shift to decentralized renewable sources also shows that electricity generation depends on the seasons, and this unpredictable nature of electricity generation requires new demand-supply management techniques [8,9]. In addition, a key question for peer-to-peer energy trading is making it possible to share and connect the existing energy infrastructure with decentralized renewable energy sources. This matter requires systematic management and intelligent control, in addition to renewable energy sources, Distributed Generation (DG), flexibility, and transparency to achieve a smart, sustainable, and coordinated energy market. Distributed Generation (DG) of electricity provides several advantages, such as high efficiency and environmental protection, reduction of transmission and distribution losses, supporting the local power grid, and improving system stability. A better way to understand the potential benefits of DG is to take a system approach that considers generation and associated loads as a subsystem or "micro-grid" [10]. Micro-Grid (MG) ecosystems are increasingly being utilized to integrate smart grids with renewable energy sources such as wind power, photovoltaics, hydro turbines, biogas, etc. A micro-grid is a set of micro-resources such as micro-turbines, fuel cells, photovoltaic systems, storage systems, and wind turbines that provides distributed energy generation [2,11,12]. It can be connected to the utility grid (grid mode) or used independently and separated from the utility grid (island mode) [10]. The micro-grid also allows local energy exchange in the smart grid and reduces the waste due to energy transmission. In short, micro-grids are considered a solution to meet the challenges facing traditional power systems [2,11,12].

Smart grid technology is created in the context of micro-grids. The smart grid provides a platform for the production, distribution, storage, and transmission of energy and creates a reliable, transparent, flexible, and automated power system. A smart grid system with balanced generation and consumption of energy ensures energy sustainability [2,11,13]. Since decentralized renewable energy sources are widely used in micro-grids, achieving a stable power balance is difficult [4]. Therefore, there is a more imminent need to find a solution for the demand–supply balance, optimal management of energy, sustainable development, and all the problems mentioned before. IoE is of great importance as a future solution for the optimal management of energy production and consumption. IoE provides access to large amounts of decentralized energy sources by considering micro-grids as infrastructure in future energy systems. The purpose of this study is to identify the infrastructure components to implement the new concept of IoE in Iran.

The existing studies have presented concepts related to the IoE in a scattered manner. Therefore, a comprehensive classification of the factors affecting this concept helps to understand it correctly. On the other hand, rapid global and technological developments in all areas such as IoE are inevitable. That is why future study approaches, including scenario analysis, become more important. In order to use this approach, it is necessary to identify the key drivers of that area so that future scenarios can be designed based on them, as well as appropriate policies for each scenario. In addition to identifying and classifying the components affecting IoE, the present study ranks them and determines the strategic components that can be used as an input for future research.

The present study answers two basic questions: What are the fundamental components in implementation of the concept of IoE? What are the key drivers of the implementation of IoE concept? In response to these questions, in response to these questions, two approaches of meta-synthesis and MICMAC analysis were used. In the first approach, after going through the screening process of papers based on the Critical Appraisal Skills Program (CASP), the number of relevant papers for review was determined. Then, based on the review of research literature and library studies, a set of key parameters for the implementation of IoE were extracted in the form of main and sub-categories. At this stage, using the MAXQDA software, the research parameters were coded and the frequency of each of them was determined. In the next step, the validity of extracted parameters was measured based on the opinions of ten experts. Having expertise and related education in the area of IoE have been the two main factors in the selection of research experts. Finally, strategic components were determined based on the Shannon entropy method and MICMAC structural analysis.

2. Theoretical Foundations of Research

The term IoE was first coined in 2011 by renowned American researcher Jeremy Rifkin. In his book, *The Third Industrial Revolution*, he points to the role of IoE in reducing fossil fuel consumption, increasing the use of decentralized energy sources, and decreasing environmental pollutions [14].

IoE combines the two concepts of the smart grid and Internet of Things (IoT). IoT is a concept in which every object can be identified, accessed, and even remotely controlled through the Internet and via the Internet Protocol (IP). This concept, based on smart grids, has been developed and introduced to the scientific community as IoE [3,15]. IoE refers to a robust understanding of IoT, big data, artificial intelligence technologies, and computing capabilities in centralized and decentralized energy management systems with the aim of optimizing the efficiency of existing energy infrastructure. It also facilitates coordination between renewable energy sources, smart grids, micro-grids, electric vehicles, and control centers, with the primary goal of improving efficiency, flexibility, and energy support [4,16]. In other words, IoE provides a real-time interface between the smart grid and a large set of equipment, and by processing data and information, creates the capacity for optimal energy production and storage while balancing energy production and consumption in the smart grid [17]. IoE is a paradigm that transforms current grid systems from centralized and one-way energy production to sustainable, flexible, efficient, reliable, and highly secure energy girds [1]. Using IoE paradigm provides a complete set of benefits. First, with the balanced effect of IP-based networking, it is possible to coordinate interactions with a large number of ICT technologies. In addition, Machine-to-Machine (M2M) interactions decentralize the control process, which in turn destroys the central communication network. Finally, interactive communication is the key to success in the global free energy market [18].

The development of renewable energies, along with the growth of information and communication technology, are the two driving and key elements in the field of IoE. Therefore, IoE can be seen as an energy efficiency system that enables the distribution of clean energy through systems of information and communication technology and can be studied as a smart grid [7,19]. Smart energy control, energy security, demand-side management of energy, increasing the use of renewable energies and their integration, reducing energy loss, reducing blackouts due to reduced energy production, the possibility of real-time monitoring, reducing operating and maintenance costs, increasing energy efficiency, system flattening, resource management, and self-organization are the main benefits of IoE [2,4,5].

3. Empirical Foundations of Research

The concept of IoE has gained the attention of various sectors, including universities, industries, and government departments [4]. For example, in the ARTEMIS IoE project in Europe, 38 companies from 10 European countries are developing IoE technology by focusing on Electric Mobility Infrastructure and smart grids [20]. In the United States, the Center for Future Renewable Electric Energy Delivery and Management (FREEDM), established by the National Science Foundation (NSF), has created new energy distribution infrastructure with the ability to plug and play decentralized renewable energy sources. In their view, IoE is considered a tool for flexible and automatic distribution of electricity [21]. In 2015, the President of China introduced IoE as a green solution to the global electricity demands [22]. In the same year, China launched a project called the Global Energy Internet (GEI), which works by developing smart grids to connect decentralized renewable energy sources by exchanging their information over the Internet [8]. In 2016, as well, the Global Energy Interconnection Development and Cooperation Organization (GEIDCO) in China introduced IoE as a sustainable new source of energy [23].

Numerous studies have been conducted on the new concept of IoE and its application in optimal energy management. Miglani et al. (2020) [4] introduced IoE as an essential technology required in the energy sector to not only manage demand response and peer-to-peer energy trading, but also provide smart grid security. In this study, the use of blockchain technology in the context of IoE is considered an important tool in creating a decentralized structure, countering cyber-attacks, and maintaining smart grid security.

Hossein Motlagh et al. (2020) [2] examined the widespread uses of IoT technology in the energy sector (production, transmission, distribution, and consumption of energy). They also offer blockchain technology as a solution to the challenges of IoE, such as privacy and security.

Taghavi et al. (2021) [17], expressing the need for optimal energy management in the country due to the increased likelihood of facing an energy crisis in the near future, considered the new paradigm of IoE as a suitable solution and presented an IoE model for optimal energy management.

Sani et al. (2019) [23] considered the structure of the existing smart grids in the field of energy to be insufficient, and therefore proposed a cybersecurity structure for IoE. This structure introduces an identity-security mechanism called "I-ICAAAN" (Integrity, Confidentiality, Availability, Authorization, Authenticity, and Nonrepudiation), a secure communication protocol and a smart security system for energy management. Such a structure provides sufficient privacy and security for data and components of the network. It defines IoE as a software platform for controlling, monitoring, and managing the entire smart network through two-way interaction between all sources of energy production and consumption.

Nguyen et al. (2018) [24] propose a building energy management system (BEMS) based on IoE to manage issues such as large volumes of building energy data and energy overload problems in the future. Based on the studies, the most important key components for the implementation of IoE were identified as follows (Table 1).

Table 1. Key components in implementation of IoE.

Axial Category	Primary (Open) Code	Sources
Rules and regulations	Defining institutional rules and regulations related to the IoE, defining crimes and penalties related to their violation	[11,13,25–27]
	Monitoring the proper implementation and enforcement of these laws	[4,5,28]
	Protecting intellectual property rights in the production and dissemination of information	[8,9,12]
	creating protection laws	[29–33]
	Creating the appropriate legal and political environment	[34–36]
	Having laws and policies related to information security on confidential data	[11,12,21,23,33,36]
	Having a regulatory system and defining its role	[4,9,29,37]
	Creating encouraging laws and documents to raise awareness	[9,27,32]
	Developing a national vision of the IoE	[7–9,33,38,39]
	Setting standards and the appropriate frameworks	[9,20,35]
	Facilitating insurance laws related to IoE entrepreneurs	[30,32]
	Facilitating cooperation rules of domestic and foreign companies in the field of IoE	[7,13,21,23,38,40,41]
Individual and human factors	Acceptance of changes and comprehensive participation in implementation and application	[30,41,42]
	Recognizing the technological capability and the capacity to benefit from new technologies	[3,43,44]
	Comprehending the technology's usefulness	[36,45]
	Understanding the technology's ease of application	[36,45]
	Recognizing the social consequences (social role) of technology	[38,41,46]
	Training and benefiting from experts in the field of the new technology	[8,9,31,43]
	Training and raising users' awareness about security threats and vulnerabilities	[11,12,47]
Financing	Gathering the required funds to invest in transformative energy and digital technologies	[9,12,20,21,25,36,39,48]
	Gathering the required funds for research and training of human resources	[9,12,20,21,25,36,39]
	Gathering the required funds for monitoring and maintenance	[8,9,20,21,25,49]
	Having a network economy	[21,25,39,45,49]
Technological infrastructure	Existence of powerful microprocessors and Internet servers in the country	[8,23,30,32,38,43,50–53]
	Bandwidth	[4,8,29,32,36,38,42,49]
	Free access of the final consumers to the Internet	[5,8,17,23,33,54]
	Online monitoring of energy consumption	[3,5,17,24,33,54]
	Optimization of energy production infrastructure	[9,11,17,24,54]
	Optimization of energy conversion mechanisms	[12,17,37,51]
	Systematic processes for energy distribution	[5,9,17,25,38,48,55]
	Communication between the energy supply and consumption chains	[4,8,17,29,38,48]
	Connection of various tools to the Internet	[9,17,29,34,56]
	Access to modern hardware equipment	[5,12,17,23,25,38,40,42,52,54]
	Existence of appropriate software infrastructure	[5,12,17,23,25,38,40,42,52,54,57]
	Possibility of storing data related to energy production and consumption	[17,22,23,25,38,40,52]
	Sharing information on units covered by energy consumption	[11,17,25,42,58]
	Standardization of required technologies (Localization)	[2,4,25,27,30,34,46,58]
	Technological integration	[2,4,25,27,30,34,41,46,58]
	Technical support and system monitoring	[9,12,25,26,31,57]
	Network control in a wide range	[9,11–13,27,28,53]

Table 1. *Cont.*

Axial Category	Primary (Open) Code	Sources
Cultural and social factor	Awareness of the Internet culture and optimal energy management	[6,8,21,40]
	Exchanging information in society and raising public awareness	[29,37,56]
	Building trust in society and transparency in sharing data and information on energy consumption	[1,39,43,48]
	Identifying new sources of awareness	[4,29]
	Paying attention to indigenous and social cultures	[30,41,58]
	Organizational interactions for integration of data and information	[7,23,24,33,45]
Safety Factors	Existence of information security technologies to prevent cyber-attacks and hacker intrusions	[7,23,33,36,40,48]
	Data security, content protection, and prevention of forgery and misuse of information sources	[6,8,23,25,32,33]
	Access to information and data based on roles and responsibilities	[9,11,28,38]
	Prevention of identity forgery or improper authentication, protection of the network against the intrusion of unauthorized agents	[9,11,23,28]
	Continuous auditing and monitoring of security events	[23,25,45]
	Securing hardware and software equipment	[4,9,11,23,33]
Technological resources	Blockchain technology and decentralized governance	[2–4,6,43,51,57]
	Essential technologies for energy replacement	[23,25,27,32,34,41,58]
	Technology for storing energy (energy storage batteries)	[4,9,26,33,57]
	Smart technology of multi-energy (integration of various energy sectors)	[1,28,38,41]
	Multiple renewable energy sources alternatives	[9,25–27,32,34,41,58]
	Renewable energy production technologies	[9,22,25–27,32,34,41,58]
	Large-scale supply of renewable energies	[6,37,52]
	Smart technology of traffic and transportation networks	[5,9]
Knowledge resources	Knowledge and expertise in planning a wide range of Internet of Energy networks	[11,45]
	Domestic research and development	[45]
	Knowledge and expertise in the field of artificial intelligence	[45]
	Knowledge and expertise in information and communication technology	[45]
	International knowledge and expertise in sustainable development	[45]
	International knowledge and expertise in renewable energy	[4,9]
	International knowledge and expertise of IoT	[29,31]
	International energy management knowledge	[4]
	Maintenance knowledge and expertise	[4]
Learning style	Learning by doing	[5]
	Learning by Interacting	[5,9]
	Learning by using	[5,9]
	Organizational science and technology system	[20]
Management factors	Required determination and commitment for implementation of the Internet of Energy on a macro level	[6,11,12,15,17,21,22,25,33,35,42, 50]
	The existence of a flat organizational structure	[6,9,11,20,21,25,26,30]
	Management of uncertainty	[1,2,8,9,20,25,26,32,59]
	Demand response management of consumers	[9,20,24,25,35,49,53,60]
	Consideration of risks associated with the implementation of new technologies	[8,9,11,12,25,30,37,43,51–53]
	Application of a whole system approach	[1,4,11,21,24,27,28,30,37]
	Strategic infrastructure planning (Strategic investments)	[2,5,6,32,39,41,44,45,55,58]
	Planning for energy management of smart homes	[3,22,24,39,44,45,61]
	Planning for energy management of smart cities	[3,24,36,38,39,60,61]

4. Research Method

The present study is practical in its purpose and descriptive-explorative in its methodology. In the first step of the research, the Sandelowski and Barroso method (2007) [62] was used to identify the components of the implementation of the concept of IoE. Meta-synthesis is a qualitative method based on a systematic review of literature to gain in-depth knowledge of the phenomenon under study. With the expansion of research in various fields of science and the confrontation of the scientific community with an explosion of information, researchers have, in practice, come to the conclusion that it is mostly not possible to be aware, up-to-date, and a master in all aspects of a field. Therefore, synthesis methods that offer the essence of research on a particular subject in a systematic and scientific way to researchers have become increasingly popular (Tables S1 and S2). Meta-synthesis evaluates other research; hence, it is called an evaluation of evaluations. Meta-synthesis is not merely an integrated review of the literature, but an analysis of the findings of these studies [63].

In the second step of the research, a futuristic approach called structural analysis was used. The potential of this method in using qualitative data along with quantitative data has made it one of the most widely used methods in research about the future. In this step, the matrix of analysis of the interaction of variables is completed by forming a panel consisting of eight experts in the field of energy. Then, in the framework of MIKMAK forecasting software, the influence and dependence (direct and indirect) of each variable on others are measured and strategic or key driving variables are obtained. MIKMAK software is one of the best software designed to implement structural analysis. The output of the software, in the form of tables and graphs, can help in understanding the system relationships and how they will work in future [64].

5. Research Findings

5.1. Meta-Synthesis

In this step, papers and studies conducted from 2010 to 2021 in the field of IoE were studied and analyzed. The Web of Science, Science Direct, Google Scholar, Springer, Emerald, ResearchGate, and Scopus databases were used to collect and categorize papers based on content, using two keywords of "Internet of Energy" and "Energy Internet" in the title; a total of 417 studies were found. Then, the process of reviewing papers, including the title, abstract, content, and research methodology began, the purpose of which was to exclude studies that were not relevant to the research questions. The review process is summarized in Figure 1.

The next step was to evaluate the methodological quality of the research, which aimed to eliminate studies in which the researcher did not trust the findings. The most commonly used tool for assessing the quality of primary studies in qualitative research is the Critical Appraisal Skills Program (CASP), which helps identify the accuracy, validity, and importance of qualitative studies by asking ten questions. These questions focus on the following: 1. Research objectives, 2. Methodological logic, 3. Research design, 4. Sampling method, 5. Data collection, 6. Reflectivity, 7. Ethical considerations, 8. Accuracy of data analysis, 9. Clarity of results and findings, and 10. Value of research [63].

In using this tool, studies were assigned a score of 1 to 5 on the above criteria after being studied. Based on the 50-point scale of CASP, the researcher proposed a scoring system according to Table 2 and categorized the studies based on their methodological quality. Studies that scored below the "good" category (score 31) were excluded from the project [65].

Table 2. Scoring system of the Critical Appraisal Skills Program (CASP).

Very Weak	Weak	Medium	Good	Very Good
0–11	11–20	21–30	31–40	41–50

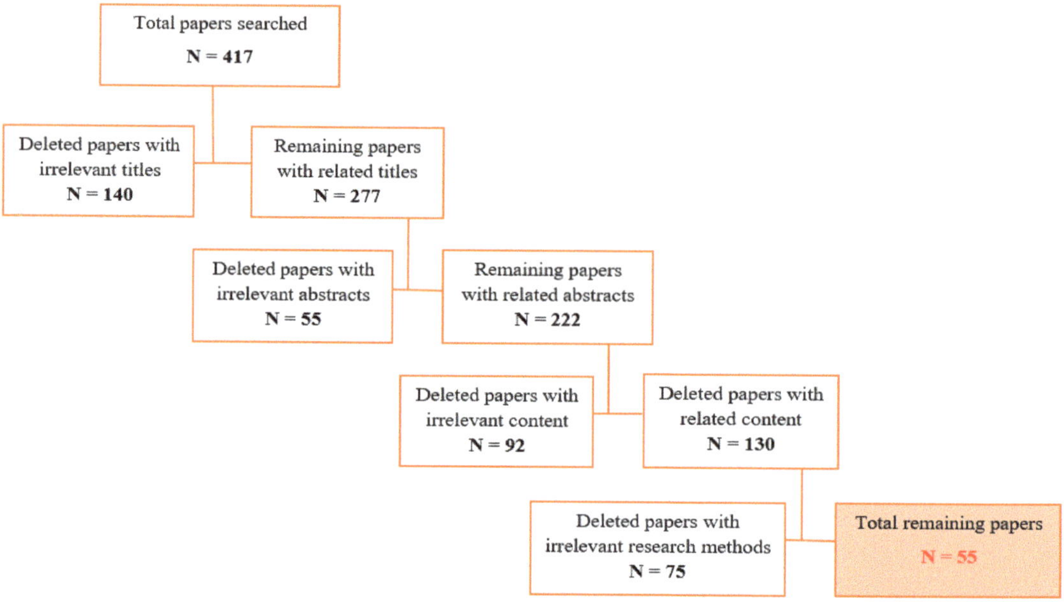

Figure 1. Screening process of papers.

In this study, the 55 studies that had survived the first round of filtering based on title, abstract, content, and research methodology were then evaluated using the CASP system (Table 3). After assigning scores to each study based on the given criteria and eliminating studies with a score of less than 31, 55 studies were finally accepted to enter the evaluation process, of which 11 studies were assigned to the "very good" category and 44 studies to the "good" category. Therefore, after a few rounds of filtering, 362 papers were eliminated from the initial 417 and 55 found their way to analysis (Figure S1, Table 4). After evaluating the papers, the data were categorized as primary codes (open code) with reference to the source and frequency. Each of the codes was then classified according to their meaning in terms of similar concepts, which helped identify the main components of the research Table 1.

Table 3. The result of the Critical Appraisal Skills Program (CASP).

Paper Code	Research Purposes	Methodological Logic	Research Design	Sampling Method	Data Collection	Reflectivity	Ethical Considerations	Accuracy of Data Analysis	Clarity of Results and Findings	Value of Research	Total Points
C01	4	4	3	4	4	4	5	3	3	4	38
C02	3	4	4	3	3	2	5	3	2	2	31
C03	3	3	4	2	4	3	5	4	5	4	37
C04	3	4	5	4	3	4	5	4	4	4	40
C05	5	4	4	3	4	4	5	3	4	3	39
C06	5	4	5	4	4	5	5	4	4	4	44
C07	2	3	2	4	4	3	5	3	3	3	32
C08	3	3	4	3	3	3	5	3	2	3	32

Table 3. *Cont.*

Paper Code	Research Purposes	Methodological Logic	Research Design	Sampling Method	Data Collection	Reflectivity	Ethical Considerations	Accuracy of Data Analysis	Clarity of Results and Findings	Value of Research	Total Points
C09	3	4	3	3	2	3	5	4	3	2	32
C10	4	4	3	4	3	3	5	4	3	4	37
C11	2	4	3	2	3	3	5	4	2	3	31
C12	2	3	2	4	3	4	5	3	4	3	33
C13	2	3	4	4	3	2	5	4	3	2	32
C14	3	3	2	2	2	3	5	3	2	2	27
C15	3	4	4	3	4	4	5	4	4	3	38
C16	3	4	4	4	4	4	5	3	4	4	39
C17	4	3	3	3	3	4	5	4	4	4	37
C18	4	4	4	4	4	4	5	4	4	4	41
C19	4	4	4	4	3	4	5	4	4	4	40
C20	2	3	2	3	4	3	5	2	3	2	29
C21	3	4	3	4	4	3	5	4	4	3	37
C22	2	3	4	4	3	4	5	3	4	3	35
C23	5	5	4	5	4	4	5	5	4	4	45
C24	4	3	2	3	2	3	5	3	3	2	30
C25	4	4	4	3	4	4	5	4	3	4	39
C26	2	3	3	4	4	3	5	4	3	3	34
C27	3	3	4	2	3	3	5	2	3	2	30
C28	5	4	3	4	4	4	5	4	4	4	41
C29	5	4	4	4	4	4	5	4	4	4	42
C30	3	3	4	3	4	4	5	4	3	3	36
C31	5	5	4	4	4	4	5	4	4	4	43
C32	4	4	4	5	4	4	5	4	4	4	42
C33	4	4	3	4	4	3	5	4	4	4	39
C34	3	4	4	3	4	3	5	4	3	3	36
C35	4	4	4	4	3	4	5	4	3	3	38
C36	3	3	3	4	4	4	5	3	3	3	35
C37	4	4	3	3	3	4	5	4	3	4	37
C38	5	4	4	4	3	4	5	4	4	3	40
C39	4	4	4	4	4	4	5	4	4	4	41
C40	4	4	4	4	3	4	5	4	4	4	40
C41	2	3	2	3	4	3	5	2	3	2	29
C42	3	4	3	4	4	3	5	4	4	3	37
C43	2	3	4	4	3	4	5	3	4	3	35
C44	5	5	4	5	4	4	5	5	4	4	45
C45	4	3	2	3	2	3	5	3	3	2	30

Table 3. *Cont.*

Paper Code	Research Purposes	Methodological Logic	Research Design	Sampling Method	Data Collection	Reflectivity	Ethical Considerations	Accuracy of Data Analysis	Clarity of Results and Findings	Value of Research	Total Points
C46	4	4	4	3	4	4	5	4	3	4	39
C47	2	3	3	4	4	3	5	4	3	3	34
C48	3	3	4	2	3	3	5	2	3	2	30
C49	5	4	3	4	4	4	5	4	4	4	41
C50	4	4	3	4	4	4	5	3	3	4	38
C51	3	4	4	3	3	2	5	3	2	2	31
C52	3	3	4	2	4	3	5	4	5	4	37
C53	3	4	5	4	3	4	5	4	4	4	40
C54	5	4	4	3	4	4	5	3	4	3	39
C55	5	4	5	4	4	5	5	4	4	4	44

Table 4. List of papers evaluated using the Critical Assessment Skills Program (CASP).

Paper Code	Title
C01	Delivering future-proof energy infrastructure
C02	Internet of Energy (IoE) and High-Renewables Electricity System Market Design
C03	Distributed network security framework of energy Internet based on Internet of Things
C04	Optimal energy management strategies for Energy Internet via deep reinforcement learning approach
C05	Design and optimization of integrated energy management network system based on Internet of Things technology
C06	Blockchain for Internet of Energy management: Review, solutions, and challenges
C07	Does Internet development improve green total factor energy efficiency? Evidence from China
C08	Using the Internet of Things in smart energy systems and networks
C09	Research on the Medium and Long Term Development Framework of Smart Grid under the Background of Energy Internet
C10	An Internet of Energy framework with distributed energy resources, prosumers and small-scale virtual power plants: An overview
C11	Energy Internet—A New Driving Force for Sustainable Urban Development
C12	Energy aware smart city management system using data analytics and Internet of Things
C13	Energy management solutions in the Internet of Things applications: Technical analysis and new research directions
C14	Entropy theory of distributed energy for Internet of Things
C15	Centralized, decentralized, and distributed control for Energy Internet
C16	Application and assessment of Internet of Things toward the sustainability of energy systems: Challenges and issues
C17	Energy Internet in China
C18	Energy Internet—Towards Smart Grid 2.0
C19	An Internet of Things based energy efficiency monitoring and management system for machining workshop

Table 4. *Cont.*

Paper Code	Title
C20	Energy management based on Internet of Things: practices and framework for adoption in production management
C21	Energy Internet blockchain technology
C22	Energy Management Strategies for RES-enabled Smart-grids empowered by an Internet of Things (IOT) Architecture
C23	The Internet of Energy: Smart Sensor Networks and Big Data Management for Smart Grid
C24	Internet of Things Role in Renewable Energy Resources
C25	Optimal sharing energy of a complex of houses through energy trading in the Internet of Energy
C26	Does the Internet development affect energy and carbon emission performance?
C27	Digitalization and energy: How does Internet development affect China's energy consumption?
C28	Dynamic assessment of Energy Internet's emission reduction effect—a case study of Yanqing, Beijing
C29	An overview of "Energy + Internet" in China
C30	Energy Internet: The business perspective
C31	Modeling of the Internet of Energy (IoE) for Optimal Energy Management with an Interpretive Structural Modeling (ISM) Approach
C32	Internet of Things (IOT) and the Energy Sector
C33	The Internet of Energy: A Web-Enabled Smart Grid System
C34	A Review of Internet of Energy Based Building Energy Management Systems: Issues and Recommendations
C35	Energy Management in Smart Cities Based on Internet of Things: Peak Demand Reduction and Energy Savings
C36	Towards an Internet of Energy
C37	Discussion on Energy Internet and Its Key Technology
C38	An integrated approach for multi-objective optimization and MCDM of Energy Internet under uncertainty
C39	A comprehensive review of Energy Internet: basic concept, operation and planning methods, and research prospects
C40	Energy Harvesting for the Internet-of-Things: Measurements and Probability Models
C41	Cyber security framework for Internet of Things-based Energy Internet
C42	The Energy and Emergy of the Internet
C43	Optimal Charging Control of Energy Storage and Electric Vehicle of an Individual in the Internet of Energy with Energy Trading
C44	Information and resource management systems for Internet of Things: Energy management, communication protocols, and future applications
C45	Research on operation and management muti-node model of mega city Energy Internet
C46	Energy Internet forums as acceleration phase transition intermediaries
C47	Energy-Efficient Device Architecture and Technologies for the Internet of Everything
C48	Internet of Things for Modern Energy Systems: State-of-the-Art, Challenges, and Open Issues
C49	An Overview of Internet of Energy (IoE) Based Building Energy Management System
C50	Integration of electric vehicles and management in the Internet of Energy
C51	Green Energy Management of the Energy Internet Based on Service Composition Quality
C52	IoT Technologies for Augmented Human: a Survey
C53	The Development of the Energy Internet of Things in Energy Infrastructure
C54	Energy Internet and We-Energy
C55	Architecture of the Internet of Energy Network: An Application to Smart Grid Communications

5.1.1. Analytical Quality Control

In qualitative research, the concept of trustworthiness is used instead of the concepts of reliability and validity. In this regard, to control the extracted concepts, the coding of the two researchers was compared. To evaluate the degree of agreement between two coders (by two people or using two tools or at two different times) and, therefore, to evaluate internal reliability, the Kappa interclass correlation was used in SPSS. The kappa index value is calculated to be 0.87, which is in the range of excellent agreement (0.81–1) [66].

5.1.2. Shannon Entropy

The steps for data analysis based on the Shannon entropy method are as follows:

- First, the frequency of each of the identified indicators should be determined based on content analysis:
- The desired frequency matrix should be normalized. For this purpose, the linear normalization method is used (Equation (1)):

$$n_{ij} = \frac{x_{ij}}{\sum x_{ij}} \tag{1}$$

- The entropy value of each indicator (E_j) is calculated based on Equation (3):

$$k = \frac{1}{Ln(a)};\ \mathrm{a} = \text{Number of indicators} \tag{2}$$

$$E_j = -k\sum[n_{ij}LN(n_{ij})] \tag{3}$$

- The significance coefficient of each indicator must be calculated. Whatever W_j has a higher value is more significant (Equation (4)):

$$W_j = \frac{E_j}{\sum E_j} \tag{4}$$

To calculate the weight of each of the components, the total weight of its codes was calculated, and the ranking took place based on the weights obtained in Table 5.

Table 5. Determining significance and emphasis of the studied research on the effective factors.

Axial Category	Primary Code	Abundance	$\sum P_{ij} \times knP_{ij}$	Uncertainty (Ej)	Significance Factor (Wj)	Rank of Indicators	Component Rank
Rules and regulations	Defining institutional rules and regulations related to the IoE, defining crimes and penalties related to their violation	5	−0.0536	0.0122	0.0126	8	3
	Monitoring the proper implementation and enforcement of these laws	3	−0.0359	0.0081	0.0085	10	
	Protecting intellectual property rights in the production and dissemination of information	3	−0.0359	0.0081	0.0085	10	
	Creating protection laws	5	−0.0536	0.0122	0.0126	8	
	Creating the appropriate legal and political environment	4	−0.0451	0.0102	0.0106	9	
	Having laws and policies related to information security on confidential data	6	−0.0617	0.0140	0.0145	7	

Table 5. *Cont.*

Axial Category	Primary Code	Abundance	$\sum P_{ij} \times knP_{ij}$	Uncertainty (Ej)	Significance Factor (Wj)	Rank of Indicators	Component Rank
	Having a regulatory system and defining its role	4	−0.0451	0.0102	0.0106	9	
	Creating encouraging laws and documents to raise awareness	3	−0.0359	0.0081	0.0085	10	
	Developing a national vision of the IoE	6	−0.0617	0.0140	0.0145	7	
	Setting standards and the appropriate frameworks	3	−0.0359	0.0081	0.0085	10	
	Facilitating insurance laws related to IoE entrepreneurs	2	−0.0259	0.0059	0.0061	11	
	Facilitating cooperation rules of domestic and foreign companies in the field of IoE	7	−0.0694	0.0157	0.0163	6	
Individual and human factors	Acceptance of changes and comprehensive participation in implementation and application	3	−0.0359	0.0081	0.0085	10	7
	Recognizing the technological capability and the capacity to benefit from new technologies	3	−0.0359	0.0081	0.0085	10	
	Comprehending the technology's usefulness	2	−0.0259	0.0059	0.0061	11	
	Understanding the technology's ease of application	2	−0.0259	0.0059	0.0061	11	
	Recognizing the social consequences (social role) of technology	3	−0.0359	0.0081	0.0085	10	
	Training and benefiting from experts in the field of the new technology	4	−0.0451	0.0102	0.0106	9	
	Training and raising users' awareness about security threats and vulnerabilities	3	−0.0359	0.0081	0.0085	10	
Financing	Gathering the required funds to invest in transformative energy and digital technologies	8	−0.0767	0.0174	0.0181	5	6
	Gathering the required funds for research and training of human resources	7	−0.0694	0.0157	0.0163	6	
	Gathering the required funds for monitoring and maintenance	6	−0.0617	0.0140	0.0145	7	
	Having a network economy	5	−0.0536	0.0122	0.0126	8	
Technological infrastructure	Existence of powerful microprocessors and Internet servers in the country	10	−0.0904	0.0205	0.0213	3	1
	Bandwidth	8	−0.0767	0.0174	0.0181	5	
	Free access of the final consumers to the Internet	6	−0.0617	0.0140	0.0145	7	
	Online monitoring of energy consumption	6	−0.0617	0.0140	0.0145	7	
	Optimization of energy production infrastructure	5	−0.0536	0.0122	0.0126	8	
	Optimization of energy conversion mechanisms	4	−0.0451	0.0102	0.0106	9	
	Systematic processes for energy distribution	7	−0.0694	0.0157	0.0163	6	
	Communication between the energy supply and consumption chains	6	−0.0617	0.0140	0.0145	7	
	Connection of various tools to the Internet	7	−0.0694	0.0157	0.0163	6	

Table 5. *Cont.*

Axial Category	Primary Code	Abundance	$\sum P_{ij} \times knP_{ij}$	Uncertainty (Ej)	Significance Factor (Wj)	Rank of Indicators	Component Rank
	Access to modern hardware equipment	10	−0.0904	0.0205	0.0213	3	
	Existence of appropriate software infrastructure	11	−0.0969	0.0220	0.0228	2	
	Possibility of storing data related to energy production and consumption	7	−0.0694	0.0157	0.0163	6	
	Sharing information on units covered by energy consumption	5	−0.0536	0.0122	0.0126	8	
	Standardization of required technologies (Localization)	8	−0.0767	0.0174	0.0181	5	
	Technological integration	9	−0.0837	0.0190	0.0197	4	
	Technical support and system monitoring	6	−0.0617	0.0140	0.0145	7	
	Network control in a wide range	7	−0.0694	0.0157	0.0163	6	
Cultural and social factors	Awareness of the Internet culture and optimal energy management	4	−0.0451	0.0102	0.0106	9	7
	Exchanging information in society and raising public awareness	3	−0.0359	0.0081	0.0085	10	
	Building trust in society and transparency in sharing data and information on energy consumption	4	−0.0451	0.0102	0.0106	9	
	Identifying new sources of awareness	2	−0.0259	0.0059	0.0061	11	
	Paying attention to indigenous and social cultures	3	−0.0359	0.0081	0.0085	10	
	Organizational interactions for integration of data and information	5	−0.0536	0.0122	0.0126	8	
Safety factors	Existence of information security technologies to prevent cyber-attacks and hacker intrusions	6	−0.0617	0.0140	0.0145	7	5
	Data security, content protection, and prevention of forgery and misuse of information sources	6	−0.0617	0.0140	0.0145	7	
	Access to information and data based on roles and responsibilities	4	−0.0451	0.0102	0.0106	9	
	Prevention of identity forgery or improper authentication, protection of the network against the intrusion of unauthorized agents	4	−0.0451	0.0102	0.0106	9	
	Continuous auditing and monitoring of security events	3	−0.0359	0.0081	0.0085	10	
	Securing hardware and software equipment	5	−0.0536	0.0122	0.0126	8	
Technological resources	Blockchain technology and decentralized governance	7	−0.0694	0.0157	0.0163	6	4
	Essential technologies for energy replacement	6	−0.0617	0.0140	0.0145	7	
	Technology for storing energy (energy storage batteries)	5	−0.0536	0.0122	0.0126	8	
	Smart technology of multi-energy (integration of various energy sectors)	4	−0.0451	0.0102	0.0106	9	
	Multiple renewable energy sources alternatives	8	−0.0767	0.0174	0.0181	5	

Table 5. *Cont.*

Axial Category	Primary Code	Abundance	$\sum P_{ij} \times knP_{ij}$	Uncertainty (Ej)	Significance Factor (Wj)	Rank of Indicators	Component Rank
	Renewable energy production technologies	9	−0.0837	0.0190	0.0197	4	
	Large-scale supply of renewable energies	3	−0.0359	0.0081	0.0085	10	
	Smart technology of traffic and transportation networks	2	−0.0259	0.0059	0.0061	11	
Knowledge resources	Knowledge and expertise in planning a wide range of Internet of Energy networks	2	−0.0259	0.0059	0.0061	11	8
	Domestic research and development	1	−0.0146	0.0033	0.0034	12	
	Knowledge and expertise in the field of artificial intelligence	1	−0.0146	0.0033	0.0034	12	
	Knowledge and expertise in information and communication technology	1	−0.0146	0.0033	0.0034	12	
	International knowledge and expertise in sustainable development	1	−0.0146	0.0033	0.0034	12	
	International knowledge and expertise in renewable energy	2	−0.0259	0.0059	0.0061	11	
	International knowledge and expertise of IoT	2	−0.0259	0.0059	0.0061	11	
	International energy management knowledge	1	−0.0146	0.0033	0.0034	12	
	Maintenance knowledge and expertise	1	−0.0146	0.0033	0.0034	12	
Learning style	Learning by doing	1	−0.0146	0.0033	0.0034	12	9
	Learning by Interacting	2	−0.0259	0.0059	0.0061	11	
	Learning by using	2	−0.0259	0.0059	0.0061	11	
	Organizational science and technology system	1	−0.0146	0.0033	0.0034	12	
Management factors	Required determination and commitment for implementation of the Internet of Energy on a macro level	12	−0.1032	0.0234	0.0243	1	2
	The existence of a flat organizational structure	8	−0.0767	0.0174	0.0181	8	
	Management of uncertainty	9	−0.0837	0.0190	0.0197	5	
	Demand response management of consumers	8	−0.0767	0.0174	0.0181	5	
	Consideration of risks associated with the implementation of new technologies	11	−0.0969	0.0220	0.0228	2	
	Application of a whole system approach	9	−0.0837	0.0190	0.0197	4	
	Strategic infrastructure planning (Strategic investments)	10	−0.0904	0.0205	0.0213	3	
	Planning for energy management of smart homes	7	−0.0694	0.0157	0.0163	6	
	Planning for energy management of smart cities	7	−0.0694	0.0157	0.0163	6	

5.2. *Structural Analysis Using MICMAK Software*

In this step, the ten components extracted in the previous step are placed in a 10 by 10 matrix and evaluated, based on the opinion of experts, by being assigned numbers between 0–3 in accordance with Table 6. The final availability matrix after expert scores is shown in Figure 2. Based on the findings obtained from Table 7, it can be said that the matrix filling index is 88%, which indicates the high degree of connectivity and influence of the identified variables with and on each other.

Table 6. Evaluation table of relationships between variables.

Effectless	**Low Effect**	**Medium Effect**	**High Effect**
Zero	One	Two	Three

.	1 : Ru	2 : IHF	3 : FI	4 : TI	5 : CSF	6 : SF	7 : TR	8 : KR	9 : LS	10 : MF
1 : Ru	0	1	3	2	2	3	1	2	1	3
2 : IHF	1	0	2	1	1	2	1	2	3	2
3 : FI	1	0	0	3	1	2	2	3	1	2
4 : TI	1	1	2	0	2	3	2	2	2	2
5 : CSF	1	2	1	2	0	1	1	1	1	2
6 : SF	1	1	1	3	1	0	2	1	1	2
7 : TR	1	1	1	1	2	2	0	2	1	2
8 : KR	1	2	1	2	1	1	2	0	2	1
9 : LS	0	1	1	2	2	1	1	1	0	2
10 : MF	2	2	3	2	2	2	2	2	2	0

Figure 2. Final availability matrix of research variables.

Table 7. Preliminary analysis of interaction matrix.

Indicators	**Matrix Size**	**Number of Iterations**	**Number of Zeros**	**Number of Ones**	**Number of Twos**	**Number of Threes**	**Total**	**Fillrate**
Values	10	2	12	39	40	9	88	88%

5.2.1. Determining the Degree of Direct Influence and Dependence of Components

Based on the matrix of direct effects, the sum of rows and columns of the matrix indicates the degree of influence and dependence of the components, respectively. As can be seen in Table 8, the component of management factors has the greatest influence on other factors, and the components of laws and regulations and technological infrastructure come in second and third places. Based on the software results on the level of dependence, the component of technological infrastructure has been dependent on other components the most; management factors and security factors come in second and third in terms of dependence.

Table 8. The degree of direct influence and dependence of components.

Rank	**Components**	**Influence**	**Components**	**Dependence**
1	Management Factors	19	Technological Infrastructure	18
2	Rules	18	Management Factors	18
3	Technological Infrastructure	17	Security Factors	17
4	Individual and Human Factors	15	Knowledge Resources	16
5	Financing	15	Financing	15
6	Security Factors	13	Cultural and Social Factors	14
7	Technological Resources	13	Technological Resources	14
8	Knowledge Resources	13	Learning Style	14
9	Cultural and Social Factors	12	Individual and Human Factors	11
10	Learning Style	11	Rules	9
	Totals	146	Totals	146

5.2.2. Location of Components in the Zones of the Influence and Dependence Map

Variables are divided into four types based on their location in one of the four areas of the influence-dependence map Figure 3:

(1) Zone (1) components (linkage or strategic variables): These components have two common characteristics of high degrees of influence and dependence, and any small change in these components will cause fundamental changes in the system. Based on the output of MICMAK software, components of technological infrastructure and management factors are located in this area.
(2) Zone (2) components (influential variables): Zone 2 components influence the system more than they are dependent on it. The rules and regulations component is located in this area.
(3) Zone (3) components (Independent variables): The components of this zone have, on average, lower degrees of influence and dependence. A change in these variables does not cause a serious change in the system. Individual and human factors fall within this zone.
(4) Zone (4) components (dependent variables): The components of this zone have little influence on the system and are themselves subject to changes in other variables. The components in this zone include financing, technological resources, knowledge resources, security factors, cultural and social factors, and learning style.

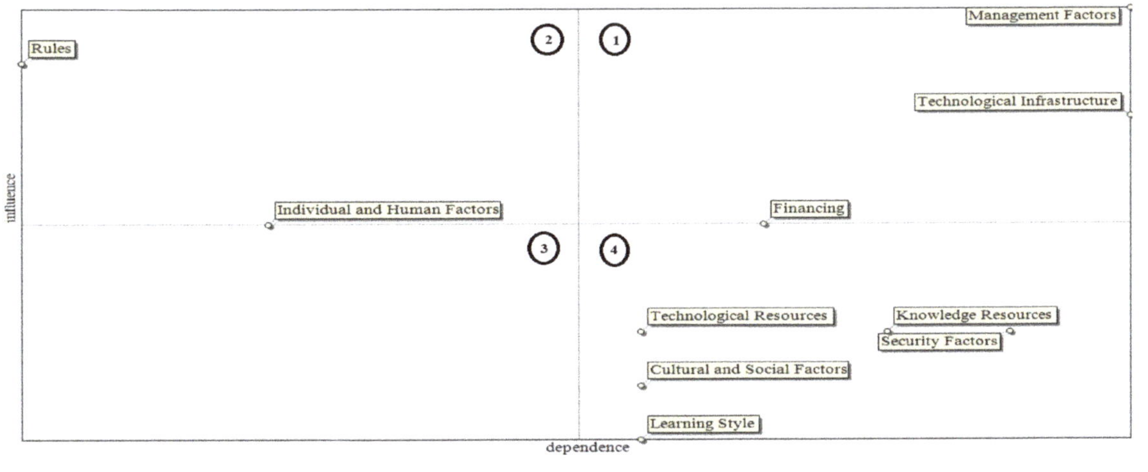

Figure 3. Diagram of system stability/instability.

5.2.3. Analyzing the Graph of Influence

The graph of influence shows the relationships between the components and how they influence each other. This graph is shown in the form of red and blue lines, the end of which is shown by an arrow and indicates the direction of the component's influence. Red lines indicate strong influence of factors on each other and blue lines, with differences in thickness, showing moderate to weak relationships (Figure 4).

The status of relationships in the graph of influence indicates that the variables of management factors, laws and regulations, and technological infrastructure have been the source of the most severe influences and increased their role in the system. Management factors, technological infrastructure, and security factors are also strongly influenced by other components of the system. Table 9 shows the share of each component in influence and dependence and Figure 5 shows the movement of each component.

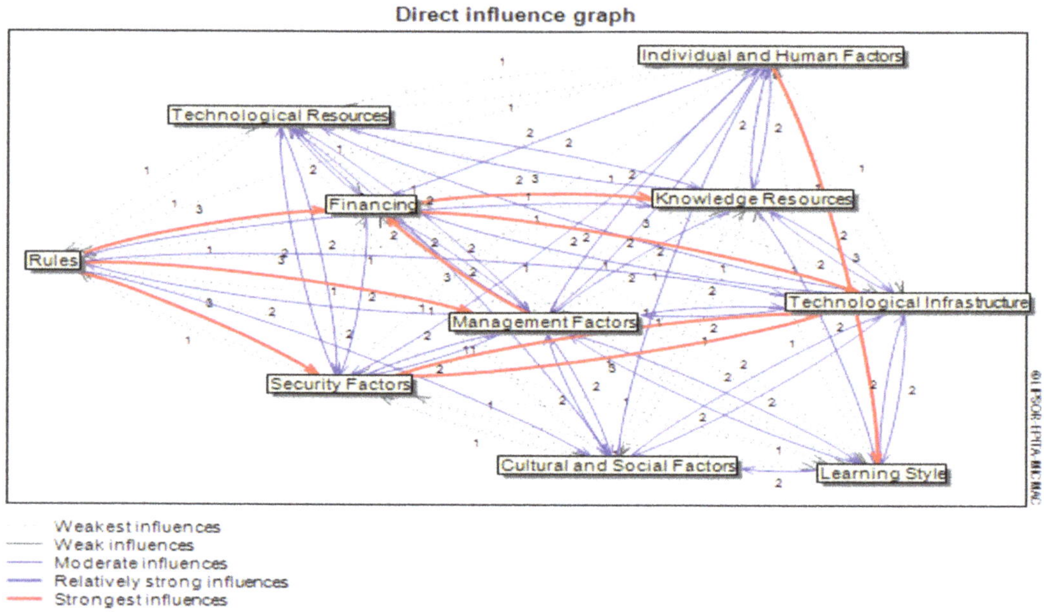

Figure 4. The influence cycle graph.

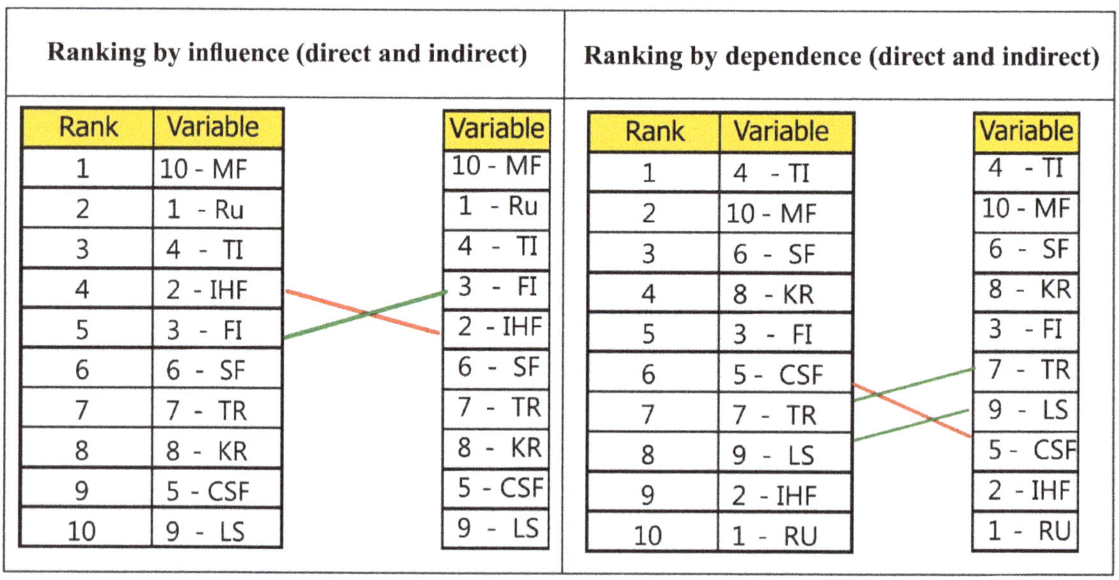

Ranking by influence (direct and indirect)

Rank	Variable	Variable
1	10 - MF	10 - MF
2	1 - Ru	1 - Ru
3	4 - TI	4 - TI
4	2 - IHF	3 - FI
5	3 - FI	2 - IHF
6	6 - SF	6 - SF
7	7 - TR	7 - TR
8	8 - KR	8 - KR
9	5 - CSF	5 - CSF
10	9 - LS	9 - LS

Ranking by dependence (direct and indirect)

Rank	Variable	Variable
1	4 - TI	4 - TI
2	10 - MF	10 - MF
3	6 - SF	6 - SF
4	8 - KR	8 - KR
5	3 - FI	3 - FI
6	5 - CSF	7 - TR
7	7 - TR	9 - LS
8	9 - LS	5 - CSF
9	2 - IHF	2 - IHF
10	1 - RU	1 - RU

Figure 5. Movement of components in direct and indirect influence and dependence.

Table 9. Arrangement of components with the largest contribution to direct influence and dependence.

Rank	Label	Direct Influence	Label	Direct Dependence	Label	Indirect Influence	Label	Indirect Dependence
1	MF	1301	TI	1232	MF	1268	TI	1224
2	Ru	1232	MF	1232	Ru	1239	MF	1181
3	TI	1164	SF	1164	TI	1128	SF	1139
4	IHF	1027	KR	1095	FI	1031	KR	1082
5	FI	1027	FI	1027	IHF	1004	FI	1005
6	SF	890	CSF	958	SF	921	TR	1000
7	TR	890	TR	958	TR	894	LS	965
8	KR	890	LS	958	KR	883	CSF	964
9	CSF	821	IHF	753	CSF	857	IHF	782
10	LS	753	Ru	616	LS	769	Ru	653

6. Discussion and Conclusions

The Internet of Energy (IoE) technology as a novel solution has changed the methods of production, transmission, and consumption of energy and has affected human life. IoE plays an essential role as an efficient tool to increase energy efficiency, recover the economy of energy and sustainable development. In order to answer the research questions, two approaches of meta-synthesis and MICMAC analysis were used. First, after the screening process of papers based on Critical Appraisal Skills Program (CASP), relevant papers were identified and carefully reviewed. Then, the research parameters were coded using MAXQDA software to determine their frequency and classification. The kappa coefficient is a statistic in qualitative research that shows the robustness of the methodology by measuring the agreement of experts on the extracted codes. In this research, the Kappa coefficient value is 0.87, which is in the excellent range and indicates the reliability of the method. There is also a consensus among experts in the field of IoE about the research parameters. In the next step, the importance of each component was determined using the Shannon entropy and MICMAC structural analysis methods. In the Shannon entropy method, based on the frequency of components and calculating the significance coefficient for each of them, the components can be ranked. In the MICMAK structural analysis method, the influence and dependence levels of the components were obtained, which resulted in determining the strategic components that have the largest share in influence and dependence. In other words, the accuracy of the results can be ensured by comparing the results obtained from the Shannon entropy and MICMAK structural analysis methods. The results show that 82 indicators under the umbrella of ten axial components are involved in the implementation of IoE: rules and regulations, individual and human factors, financing, technological infrastructure, cultural and social factors, security factors, technological resources, knowledge resources, learning style, and managerial factors. In the Shannon entropy method, technological infrastructure (1), management factors (2), rules and regulations (3), technological resources (4), security factors (5), financing (6), cultural and social factors and individual and human factors (7), knowledge resources (8), and learning style (9) are the most significant, respectively. In MICMAK structural analysis, the components of management factors (1), technological infrastructure (2), security factors and financing (3), knowledge resources (4), rules and regulations and technological resources (5), cultural and social factors and individual and human factors (6), and learning style (7) have the largest share in influence and dependence, respectively. Conclusion: The two components of management factors and technological infrastructure are the most important in both methods and can be considered as key and strategic components, which is consistent with the findings of researchers, such as Taghavi et al., 2021 [17]; Miglani et al., 2020 [4]; Hua et al., 2019 [5]; Qiu et al., 2019 [19]; Sun, 2019 [45]; Lombardi et al., 2018 [47]; and

Town et al., 2018 [38]. On the other hand, individual and human factors and cultural and social factors together are of equal importance, which is in accordance with the findings of Umer et al., 2019 [46]; Pirmagomedov and Koucheryavy, 2019 [36]; and Mahapatra, 2018 [41]. In both methods, the learning style has less priority.

One of the important points in qualitative research is that the basis of such research is the opinions of experts. Undoubtedly, the emergence of new studies in the area of IoE introduce new parameters that keep the way open for future research.

Today, Scenario-Based Strategic Planning (SBSP) is one of the most important and key tools in the field of future studies that has attracted the attention of many researchers. SBSP outlines a more realistic future for individuals and helps them make future decisions. The use of this tool requires the identification of key drivers in the subject under study. The output of this research can be a good criterion for future works of researchers. Therefore, it is suggested that researchers use the results of this study on the subject of future studies regarding the IoE. Blockchain technology is another emerging technology that is influential in various fields such as energy. In their study, Azizi et al. (2021) [67] mentioned the use of Internet of Things (IoT) and blockchain in the smart supply chain. In addition, as another suggestion to researchers, studying the application of blockchain technology in the field of IoE is another interesting topic that can pave the way for future research.

Supplementary Materials: The following supporting information can be downloaded at: https://www.mdpi.com/article/10.3390/su142013180/s1, Figure S1: PRISMA 2020 flow diagram for systematic reviews which included searches of databases and registers; Table S1: PRISMA 2020 Main Checklist; Table S2: PRISMA 2020 Abstract Checklist [68].

Author Contributions: Conceptualization, P.A., R.A. and M.H.T.; methodology, A.B.N. and M.H.T.; software, M.H.T.; validation, P.A., R.A. and A.B.N.; formal analysis, P.A. and M.H.T.; investigation, R.A. and A.B.N.; resources, P.A., R.A., A.B.N. and M.H.T.; data curation, M.H.T.; writing—original draft preparation, M.H.T.; writing—review and editing, P.A., R.A. and A.B.N.; visualization, A.B.N.; supervision, P.A. and R.A.; project administration, P.A., R.A. and A.B.N. All authors have read and agreed to the published version of the manuscript.

Funding: This research received no external funding.

Institutional Review Board Statement: Not applicable.

Informed Consent Statement: Not applicable.

Data Availability Statement: Not applicable.

Conflicts of Interest: The authors declare no conflict of interest.

References

1. Hong, Z.; Feng, Y.; Li, Z.; Wang, Y.; Zheng, H.; Li, Z.; Tan, J. An integrated approach for multi-objective optimisation and MCDM of energy internet under uncertainty. *Future Gener. Comput. Syst.* **2019**, *97*, 90–104. [CrossRef]
2. Hossein Motlagh, N.; Mohammadrezaei, M.; Hunt, J.; Zakeri, B. Internet of Things (IoT) and the energy sector. *Energies* **2020**, *13*, 494. [CrossRef]
3. Hannan, M.A.; Faisal, M.; Ker, P.J.; Mun, L.H.; Parvin, K.; Mahlia, T.M.I.; Blaabjerg, F. A review of internet of energy based building energy management systems: Issues and recommendations. *IEEE Access* **2018**, *6*, 38997–39014. [CrossRef]
4. Miglani, A.; Kumar, N.; Chamola, V.; Zeadally, S. Blockchain for Internet of Energy management: Review, solutions, and challenges. *Comput. Commun.* **2020**, *151*, 395–418. [CrossRef]
5. Hua, H.; Qin, Y.; Hao, C.; Cao, J. Optimal energy management strategies for energy Internet via deep reinforcement learning approach. *Appl. Energy* **2019**, *239*, 598–609. [CrossRef]
6. Pourbabak, H.; Chen, T.; Su, W. Centralized, decentralized, and distributed control for energy internet. In *The Energy Internet*; Woodhead Publishing: Cambridge, UK, 2018; pp. 3–19.
7. Qi, J.; Wu, D. Green energy management of the energy Internet based on service composition quality. *IEEE Access* **2018**, *6*, 15723–15732. [CrossRef]
8. Ren, S.; Hao, Y.; Xu, L.; Wu, H.; Ba, N. Digitalization and energy: How does internet development affect China's energy consumption? *Energy Econ.* **2021**, *98*, 105220. [CrossRef]
9. Wu, H.; Hao, Y.; Ren, S.; Yang, X.; Xie, G. Does internet development improve green total factor energy efficiency? Evidence from China. *Energy Policy* **2021**, *153*, 112247. [CrossRef]

10. Yoldaş, Y.; Önen, A.; Muyeen, S.M.; Vasilakos, A.V.; Alan, I. Enhancing smart grid with micro-grids: Challenges and opportunities. *Renew. Sustain. Energy Rev.* **2017**, *72*, 205–214. [CrossRef]
11. Ahmad, T.; Zhang, D. Using the internet of things in smart energy systems and networks. *Sustain. Cities Soc.* **2021**, *68*, 102783. [CrossRef]
12. Khatua, P.K.; Ramachandaramurthy, V.K.; Kasinathan, P.; Yong, J.Y.; Pasupuleti, J.; Rajagopalan, A. Application and assessment of internet of things toward the sustainability of energy systems: Challenges and issues. *Sustain. Cities Soc.* **2020**, *53*, 101957. [CrossRef]
13. Wang, D.; Zhong, D.; Souri, A. Energy management solutions in the Internet of Things applications: Technical analysis and new research directions. *Cogn. Syst. Res.* **2021**, *67*, 33–49. [CrossRef]
14. Rifkin, J. *The Third Industrial Revolution: How Lateral Power Is Transforming Energy, the Economy, and the World*; Palgrave MacMillan: New York, NY, USA, 2011.
15. Bui, N.; Castellani, A.P.; Casari, P.; Zorzi, M. The internet of energy: A web-enabled smart grid system. *IEEE Netw.* **2012**, *26*, 39–45. [CrossRef]
16. Huang, A.Q.; Crow, M.L.; Heydt, G.T.; Zheng, J.P.; Dale, S.J. The future renewable electric energy delivery and management (FREEDM) system: The energy internet. *Proc. IEEE* **2010**, *99*, 133–148. [CrossRef]
17. Taghavi, M.H.; Akhavan, P.; Ahmadi, R.; Bonyadi Naeini, A. Modeling of the Internet of Energy (IoE) for Optimal Energy Management with an Interpretive Structural Modeling (ISM) Approach, Iranian Journal of Information. *Process. Manag.* **2021**, *36*, 1049–1080.
18. Poyner, I.K.; Sherratt, R.S. Privacy and security of consumer IoT devices for the pervasive monitoring of vulnerable people. In Proceedings of the Living in the Internet of Things: Cybersecurity of the IoT, London, UK, 28–29 March 2018; pp. 1–5.
19. Qiu, C.; Cui, S.; Yao, H.; Xu, F.; Yu, F.R.; Zhao, C. A novel QoS-enabled load scheduling algorithm based on reinforcement learning in software-defined energy internet. *Future Gener. Comput. Syst.* **2019**, *92*, 43–51. [CrossRef]
20. Kafle, Y.R.; Mahmud, K.; Morsalin, S.; Town, G.E. Towards an internet of energy. In Proceedings of the 2016 IEEE International Conference on Power System Technology (POWERCON), Wollongong, NSW, Australia, 28 September–1 October 2016; pp. 1–6.
21. Feng, C.; Liao, X. An overview of "Energy + Internet" in China. *J. Clean. Prod.* **2020**, *258*, 120630. [CrossRef]
22. Yang, S.X.; Zhu, C.X.; Qiao, L.; Chi, Y.Y. Dynamic assessment of Energy Internet's emission reduction effect—a case study of Yanqing, Beijing. *J. Clean. Prod.* **2020**, *272*, 122663. [CrossRef]
23. Sani, A.S.; Yuan, D.; Jin, J.; Gao, L.; Yu, S.; Dong, Z.Y. Cyber security framework for Internet of Things-based Energy Internet. *Future Gener. Comput. Syst.* **2019**, *93*, 849–859. [CrossRef]
24. Nguyen, V.T.; Vu, T.L.; Le, N.T.; Jang, Y.M. An Overview of Internet of Energy (IoE) Based Building Energy Management System. In Proceedings of the 2018 International Conference on Information and Communication Technology Convergence (ICTC), Jeju, Korea, 17–19 October 2018; pp. 852–855.
25. Strbac, G.; Konstantelos, I.; Pollitt, M.; Green, R. Delivering future-proof energy infrastructure. In *Report for National Infrastructure Commission*; University of Cambridge & Imperial College London: London, UK, 2016.
26. Strielkowski, W.; Streimikiene, D.; Fomina, A.; Semenova, E. Internet of energy (IoE) and high-renewables electricity system market design. *Energies* **2019**, *12*, 4790. [CrossRef]
27. Zhihong, J.; Jian, H.; Wenzhou, L.; Zhe, C.; Ning, L.; Siyuan, W.; Chang, L. Energy internet—A new driving force for sustainable urban development. *Energy Procedia* **2018**, *152*, 1206–1211. [CrossRef]
28. Xiao, Z.; Wang, J.; Yang, X.; Xiao, Y.; Liu, H.; Wu, Y.; Chen, J. Research on the medium and long-term development framework of smart grid under the background of energy Internet. *Procedia Comput. Sci.* **2020**, *175*, 769–773. [CrossRef]
29. Chen, X.; Li, C.; Tang, Y.; Xiao, Q. An Internet of Things based energy efficiency monitoring and management system for machining workshop. *J. Clean. Prod.* **2018**, *199*, 957–968. [CrossRef]
30. Liu, G.; Qu, L.; Zeng, R.; Gao, F. Energy internet in china. In *The Energy Internet*; Woodhead Publishing: Cambridge, UK, 2019; pp. 265–282.
31. Shrouf, F.; Miragliotta, G. Energy management based on Internet of Things: Practices and framework for adoption in production management. *J. Clean. Prod.* **2015**, *100*, 235–246. [CrossRef]
32. Wang, Z.L. Entropy theory of distributed energy for internet of things. *Nano Energy* **2019**, *58*, 669–672. [CrossRef]
33. Zhang, J. Distributed network security framework of energy internet based on internet of things. *Sustain. Energy Technol. Assess.* **2021**, *44*, 101051. [CrossRef]
34. Al-Ali, A.R. Internet of things role in the renewable energy resources. *Energy Procedia* **2016**, *100*, 34–38. [CrossRef]
35. Cao, Y. Energy Internet blockchain technology. In *The Energy Internet*; Woodhead Publishing: Cambridge, UK, 2019; pp. 45–64.
36. Pirmagomedov, R.; Koucheryavy, Y. IoT technologies for Augmented Human: A survey. *Internet Things* **2019**, *14*, 100120. [CrossRef]
37. Mahmud, K.; Khan, B.; Ravishankar, J.; Ahmadi, A.; Siano, P. An internet of energy framework with distributed energy resources, prosumers and small-scale virtual power plants: An overview. *Renew. Sustain. Energy Rev.* **2020**, *127*, 109840. [CrossRef]
38. Mahmud, K.; Town, G.E.; Morsalin, S.; Hossain, M.J. Integration of electric vehicles and management in the internet of energy. *Renew. Sustain. Energy Rev.* **2018**, *82*, 4179–4203. [CrossRef]
39. Muhanji, S.O.; Flint, A.E.; Farid, A.M. *eIoT: The Development of the Energy Internet of Things in Energy Infrastructure*; Springer Nature: Berlin/Heidelberg, Germany, 2019; p. 160. [CrossRef]

40. Lin, C.C.; Deng, D.J.; Kuo, C.C.; Liang, Y.L. Optimal charging control of energy storage and electric vehicle of an individual in the internet of energy with energy trading. *IEEE Trans. Ind. Inform.* **2017**, *14*, 2570–2578. [CrossRef]
41. Mahapatra, C. Energy-Efficient Device Architecture and Technologies for the Internet of Everything. Ph.D. Thesis, University of British Columbia, University Endowment Lands, Vancouver, BC, Canada, 2018.
42. Rana, M. Architecture of the internet of energy network: An application to smart grid communications. *IEEE Access* **2017**, *5*, 4704–4710. [CrossRef]
43. Chen, Z.; Liu, Q.; Li, Y.; Liu, S. Discussion on energy internet and its key technology. *J. Power Energy Eng.* **2017**, *5*, 1–9. [CrossRef]
44. Lin, B.; Zhou, Y. Does the Internet development affect energy and carbon emission performance? *Sustain. Prod. Consum.* **2021**, *28*, 1–10. [CrossRef]
45. Sun, Q. *Energy Internet and We-Energy*; Springer: Berlin/Heidelberg, Germany, 2019.
46. Umer, T.; Rehmani, M.H.; Kamal, A.E.; Mihaylova, L. Information and resource management systems for Internet of Things: Energy management, communication protocols and future applications. *Future Gener. Comput. Syst.* **2018**, *92*, 1021–1027. [CrossRef]
47. Lombardi, F.; Aniello, L.; De Angelis, S.; Margheri, A.; Sassone, V. A blockchain-based infrastructure for reliable and cost-effective IoT-aided smart grids. In Proceedings of the Living in the Internet of Things: Cybersecurity of the IoT, London, UK, 28–29 March 2018. [CrossRef]
48. Cao, Y.; Li, Q.; Tan, Y.; Li, Y.; Chen, Y.; Shao, X.; Zou, Y. A comprehensive review of Energy Internet: Basic concept, operation and planning methods, and research prospects. *J. Mod. Power Syst. Clean Energy* **2018**, *6*, 399–411. [CrossRef]
49. Zhou, K.; Yang, S.; Shao, Z. Energy internet: The business perspective. *Appl. Energy* **2016**, *178*, 212–222. [CrossRef]
50. Babar, M.; Khattak, A.S.; Jan, M.A.; Tariq, M.U. Energy aware smart city management system using data analytics and Internet of Things. *Sustain. Energy Technol. Assess.* **2021**, *44*, 100992. [CrossRef]
51. Lin, C.C.; Wu, Y.F.; Liu, W.Y. Optimal sharing energy of a complex of houses through energy trading in the Internet of energy. *Energy* **2021**, *220*, 119613. [CrossRef]
52. Mahapatra, C.; Moharana, A.K.; Leung, V. Energy management in smart cities based on internet of things: Peak demand reduction and energy savings. *Sensors* **2017**, *17*, 2812. [CrossRef]
53. Raghavan, B.; Ma, J. The energy and emergy of the internet. In Proceedings of the 10th ACM Workshop on Hot Topics in Networks, Cambridge, MA, USA, 14–15 November 2011; pp. 1–6.
54. Hyysalo, S.; Juntunen, J.K.; Martiskainen, M. Energy Internet forums as acceleration phase transition intermediaries. *Res. Policy* **2018**, *47*, 872–885. [CrossRef]
55. Cao, J.; Yang, M. Energy internet-towards smart grid 2.0. In Proceedings of the 2013 Fourth International Conference on Networking and Distributed Computing, Los Angeles, CA, USA, 21–24 December 2013; pp. 105–110.
56. Jaradat, M.; Jarrah, M.; Bousselham, A.; Jararweh, Y.; Al-Ayyoub, M. The internet of energy: Smart sensor networks and big data management for smart grid. *Procedia Comput. Sci.* **2015**, *56*, 592–597. [CrossRef]
57. Ren, G.; Yu, M.; Yin, D.; Huang, S.; Xu, H.; Yuan, M. Design and optimization of integrated energy management network system based on internet of things technology. *Sustain. Comput. Inform. Syst.* **2021**, *30*, 100502. [CrossRef]
58. Ziogou, C.; Voutetakis, S.; Papadopoulou, S. Energy Management Strategies for RES-enabled Smart-grids empowered by an Internet of Things (IoT) Architecture. In *Computer Aided Chemical Engineering*; Elsevier: Amsterdam, The Netherlands, 2017; Volume 40, pp. 2473–2478.
59. Smart, G.; Atkinson, J.; Mitchell, J.; Rodrigues, M.; Andreopoulos, Y. Energy harvesting for the Internet-of-Things: Measurements and probability models. In Proceedings of the 23rd International Conference on Telecommunications (ICT), Thessaloniki, Greece, 16–18 May 2016; pp. 1–6.
60. Shakerighadi, B.; Anvari-Moghaddam, A.; Vasquez, J.C.; Guerrero, J.M. Internet of things for modern energy systems: State-of-the-art, challenges, and open issues. *Energies* **2018**, *11*, 1252. [CrossRef]
61. Yao, C.; Zhang, Z.; Yu, J.; Feng, W.; Zhou, C.; Han, Y. Research on operation and management multi-node model of mega city energy Internet. *Glob. Energy Interconnect.* **2018**, *1*, 130–136.
62. Sandelowski, M.; Barroso, J. *Handbook for Synthesizing Qualitative Research*; Springer Publishing Company: Berlin/Heidelberg, Germany, 2006.
63. De Ridder, D.T.; Lensvelt-Mulders, G. Taking stock of self-control: A meta-analysis of how trait self-control relates to a wide range of behaviors. In *Self-Regulation and Self-Control*; Routledge: England, UK, 2018; pp. 221–274.
64. Omran, A.; Khorish, M.; Saleh, M. Structural analysis with knowledge-based MICMAC approach. *Int. J. Comput. Appl.* **2014**, *86*, 39–43. [CrossRef]
65. Chenail, R.J. Learning to appraise the quality of qualitative research articles: A contextualized learning object for constructing knowledge. *Qual. Rep.* **2011**, *16*, 236–248. [CrossRef]
66. Honorton, C.; Ferrari, D.B.; Hansen, G. Meta-analysis of Forced-Choice Precognition Experiments (1935–1987). In *The Star Gate Archives: Reports of the United States Government Sponsored Psi Program, 1972–1995. Volume 2: Remote Viewing*; McFarland: McFarland, CA, USA, 2018; Volume 291, pp. 1985–1995.

67. Azizi, N.; Malekzadeh, H.; Akhavan, P.; Haass, O.; Saremi, S.; Mirjalili, S. IoT-Blockchain: Harnessing the Power of Internet of Thing and Blockchain for Smart Supply Chain. *Sensors* **2021**, *21*, 6048. [CrossRef]
68. Page, M.J.; McKenzie, J.E.; Bossuyt, P.M.; Boutron, I.; Hoffmann, T.C.; Mulrow, C.D.; Shamseer, L.; Tetzlaff, J.M.; Akl, E.A.; Brennan, S.E. The PRISMA 2020 Statement: An Updated Guideline for Reporting Systematic Reviews. MetaArXiv. 14 September 2020. Available online: www.prisma-statement.org (accessed on 3 May 2022).

sustainability

MDPI

Article

Neuro-Cybernetic System for Forecasting Electricity Consumption in the Bulgarian National Power System

Kostadin Yotov, Emil Hadzhikolev, Stanka Hadzhikoleva * and Stoyan Cheresharov

Faculty of Mathematics and Informatics, University of Plovdiv Paisii Hilendarski, 236 Bulgaria Blvd., 4027 Plovdiv, Bulgaria
* Correspondence: stankah@uni-plovdiv.bg

Abstract: Making forecasts for the development of a given process over time, which depends on many factors, is in some cases a difficult task. The choice of appropriate methods—mathematical, statistical, or artificial intelligence methods—is also not obvious, given their great variety. This paper presented a model of a forecasting system by comparing the errors in the use of time series on the one hand, and artificial neural networks on the other. The model aims at multifactor predictions based on forecast data on significant factors, which were obtained by automated testing of different methods and selection of the methods with the highest accuracy. Successful experiments were conducted to forecast energy consumption in Bulgaria, including for household consumption; industry consumption, the public sector and services; and total final energy consumption.

Keywords: electricity consumption; forecast energy consumption; forecasting system

Citation: Yotov, K.; Hadzhikolev, E.; Hadzhikoleva, S.; Cheresharov, S. Neuro-Cybernetic System for Forecasting Electricity Consumption in the Bulgarian National Power System. *Sustainability* **2022**, *14*, 11074. https://doi.org/10.3390/su141711074

Academic Editors: Luis Hernández-Callejo, Sergio Nesmachnow and Sara Gallardo Saavedra

Received: 25 July 2022
Accepted: 31 August 2022
Published: 5 September 2022

1. Introduction

The forecasting of the future is extremely important for the effective management of a process or system. Forecasting is about predicting the future as accurately as possible, given all of the information available, including historical data and knowledge of any future events that might impact the forecasts [1]. From a scientific point of view, forecasting is a scientifically based assumption about the future state and development of processes, events, indicators, etc. [2]. Considering the possibility of the existence of many different forecasts for the development of a given process in the future, forecasting can be defined as a reasonable assumption of possible options for development in a given area and the probability that they will be realized.

The synergy between mathematics and computer science has led to the development of a wide variety of algorithms, approaches, methods, and tools for forecasting. Widely used, with application in various fields are mathematical and statistical methods including regression and clustering [1,3], time series [4,5], polynomial approximations [6], fuzzy collaborative methods [7], as well as many methods for artificial intelligence predicting, such as machine learning [8,9], etc. On the one hand, this diversity provides an opportunity to choose a specific approach to solving a given task, but, on the other hand, it makes it difficult to find the most effective solution.

In the process of our work on multifactor and multi-step forecasting of energy consumption in the Republic of Bulgaria, we came to the need to forecast many socio-economic factors through which to make the final forecast. The functions, by which the individual factors change, as well as the energy consumption, can have a variety of linear and nonlinear forms, where the appropriate forecasting methods for each of them may be different. Determining the most accurate forecast values for the factors would have a positive effect on the accuracy of forecasting the target value, which in our case is energy consumption.

The automation of the process of choosing the most effective method for any individual factor or target value contributes to the acceleration of the process and the improvement of the forecast accuracy. Finding the most effective forecasting method requires experimenting with several different approaches that are combined and compared. Making more effort to solve the prognostic task could save future effort, time, and money. Fast, but essentially inaccurate forecasts often lead to unreasonable investments or unrealized opportunities.

2. Predicting Electricity Consumption—State of the Art

Whether the forecast is long-term, medium-term, or short-term, predicting electricity consumption has a key role to play in investment planning, the introduction of new capacities or the decommissioning of unnecessary ones, and the assessment of the behavior of the entire economic system. Effective modeling of electricity consumption is becoming a vital task aimed at avoiding costly mistakes in unreasonable investments, shutting down important facilities, improperly scheduled repairs or short-sightedness in exports or imports. Therefore, we should not be surprised that in the literature on the subject there are many proposed options for dealing with forecasting problems.

Official studies focused on the development of the energy sector in Bulgaria, factors influencing electricity consumption and approaches for forecasting the consumption and factors have been conducted by teams from the Bulgarian Academy of Sciences (BAS) and Risk Management Lab. Traditional methods such as correlation and regression analyses have been used in both studies.

In the research, conducted in BAS, a basic and in-depth analysis has been made of the impact of individual factors influencing electricity consumption. Taken into consideration are the country's gross domestic product, gross value added by economy sectors, population size, number of employees, income, prices, changes in temperature until 2040 according to the Bulgarian National Institute of Meteorology and Hydrology, etc. Forecasts have been made based on three different scenarios according to different expectation for changes in factors [10].

The team of Risk Management Lab creates mathematical and statistical models for forecasting the electricity balance (including as an element of itself and forecasting electricity consumption). The study examines specific factors in terms of electricity consumptions by households and the industry [11].

Due to the great social and economic importance of forecasting electricity consumption, many scientists have proposed different types of forecasting models to solve the problem in the last few decades. The methods for forecasting electricity consumption can be defined in several categories:

- Statistical models for analysis—correlation methods and regression models—BAS strategy [10], Risk Management Lab [11], Mohamed and Bodger [12];
- Time series—Lee, Gaik and Yee [13], Sun, Zhang et al. [14], Chou and Tran [15];
- Granger gray forecasting systems—Ding, Hipel and Dang [16], Lee and Tong [17], Huang, Wang, et al. [18];
- Neural networks—Chung [19], Chernykh, Chechushkov and Panikovskaya [20], Khosravani, Castilla, et al. [21], Yoo and Myriam [22], Hu and Yi-Chung [19], Jahn [23];
- Advanced machine learning methods—Bouktif, Fiaz, Ouni and Serhani [24], Alamaniotis, Bargiotas and Tsoukalas [25], Alamaniotis [26], Chou, Truong [27], Kong, Dong, Jia, Hill, Xu and Zhang [28], Moradzadeh, Moayyed, Zakeri, Mohammadi-Ivatloo and Aguiar [29];
- Bagging and boosting methods—Khwaja, Anpalagan, Naeem and Venkatesh [30], Khwaja, Zhang, Anpalagan and Venkatesh [31], Cao, Wan, Zhang, Li and Song [32];
- Hybrid methods—Karatasou, Santamouris and Geros [33], Cervená and Schneider [34], Farsi, Amayri, Bouguila and Eicker [35], Chung, Gu and Yoo [36], Sajjad, Khan, Ullah, Hussain et al. [37], Alamo, Medina, Ruano et al. [38] etc.

Each of the considered approaches has its advantages and disadvantages in different specific tasks and situations. This shows that the creation of an automated forecasting system in which multiple forecasting models can be implemented and compared could bring many benefits in forecasting not only electricity consumption but also other quantities.

3. Materials and Methods

3.1. Forecasting Process

In the general case, the solution of forecasting tasks is performed in a similar way and the process can be reduced to several stages—data collection and processing, a study of solution methods, solution, and analysis of the results [1]. The algorithm is iterative and individual stages can be performed and overlap repeatedly over time.

Data collection is often associated with preliminary research in the subject area, which provides additional information related to the set task. At this stage, an idea of the possible factors influencing the predicted values is formed. In the best case, the data on the selected factors are provided by the stakeholder or can be collected from one or more sources. In other cases, new technical equipment and software applications must be built in order to collect them.

The merging and synchronization of data by time, location, seasons, or other criteria is an integral activity when using multiple data sources providing data for various factors.

Data analysis and processing include activities to check and clear incorrect input data; conversion of data into structures suitable for modeling; and graphical representation of the data through which trends, periodicity, seasonality, etc., can be detected. In some cases, data behavior may be affected by various methodological and technological differences in data collection, as well as social, societal, climatic, or other changes. This implies the use of various methods for analysis and subsequent forecasting of the formed segments.

In the stage of research of methods for solving the problem of forecasting, various existing methods and algorithms are considered or specific ones are created. Both standard statistical and artificial intelligence methods and models can be used as forecasting methods. The number of methods and their use may depend on the requirements of the specific task, on factors related to the environment, performers, etc. An important part of this stage is the definition of indicators for evaluating the effectiveness of forecasting methods. The evaluation of efficiency may include various parameters depending on the volume of data, technical capabilities of computer systems, cost, speed, etc.

In the last stages, one or several of the perspective models for forecasting are selected and applied for the solution of the problem. After evaluating the results, individual steps can be repeated many times in order to achieve better results.

This whole process requires a lot of time and resources. This raises the question—is it possible to fully or partially automate the forecasting process, including the conducting of experiments with various mathematical and artificial intelligence methods, comparing errors, and choosing the best solution? This is the purpose of our work that is presented in this article.

3.2. Multifactor Forecasting System through Automated Selection of the Best Methods

The successful solution of the set task for forecasting energy consumption in the national power system is directly related to the correct assessment of the factors influencing energy consumption. These are macroeconomic and demographic indicators, social parameters, weather conditions, and others. This work investigated several factors influencing electricity consumption: gross domestic product; energy intensity; population size; population income; price of electricity; expected temperatures for the respective period; energy efficiency; and electricity consumption in a preceding period.

The automation of various stages and activities of the forecasting process requires careful planning. Such a system must support the following basic capabilities:

- mechanisms for integrating different forecasting methods into the system;
- potential to work with a different number of factors;
- automated search and selection of effective forecasting methods when solving a specific task.

In our proposed model for an automated search of effective forecasting methods, two main approaches are used:

- Complex forecasting, which is multivariate time series forecasting. It is applicable to multifactor forecasting, in which preliminary forecasting of individual factors is performed;
- Simple forecasting, which is univariate time series forecasting.

The long-term goal of our research is to develop a neurocybernetic system for energy consumption forecasting that supports various mathematical and artificial intelligence methods. In the first version, we used a neural network in its basic version. Further work on the system involves adding other forecasting methods.

3.3. Mathematical Model of the Forecasting Task

The main approach to work in the forecasting system through artificial neural networks can be formally presented as follows:

Let us assume we have the following event:

$$S = S\left(\vec{X_1},\ \vec{X_2},\ \ldots \vec{X_k}\right),$$

whose outcome is determined by k influencing factors $\vec{X_1},\ \vec{X_2},\ \ldots \vec{X_k}$. Let a sample of values be provided for each of the factors $\vec{X_i}$

$$\vec{X_i}:\ \left(x_{i,1},\ x_{i,2}, \ldots x_{i,j_i}\right),\ i = 1, 2, \ldots k,\ \ i, j_i \in N.$$

The main steps performed by the system when using complex forecasting are the following:

(1) Automatically searches for and constructs a neural network NN with input vectors $\vec{X_i}$, approximating S with minimal error E_{NN};

(2) Based on the samples $\left(x_{i,1},\ x_{i,2}, \ldots x_{i,j_i}\right)$ forecasts the future behavior of the factors $\vec{X_i}$ using p in a number of different factors $\{m_t\}_{t=1\ldots p}$;

(3) For each of the factors $\vec{X_i}$ the efficiency of the used forecasting methods is compared separately;

(4) For the final forecast value of $\vec{X_i}$, the forecast of the respective most effective method is selected;

(5) Through the forecasted values of the factors $\vec{X_i}$ and the method M, the outcome of the event is predicted.

Various criteria can be used to evaluate the effectiveness of forecasting methods, such as decision error, cost, speed, computer resources used, etc. In our study, we chose to call the method M_1 more effective than M_2 if it has a lesser prediction error, i.e.,

$$E_{M_1} \leq E_{M_2}. \quad (1)$$

Condition (1) allows us to introduce a formula for the efficiency of a forecasting method as the inversely proportional values of the error that occurs when forecasting with it:

$$efficiency = \frac{1}{error}. \quad (2)$$

Absolute error (AE) or root mean square error, obtained when forecasting on an n-tuple data array, can be used. For single point forecasts created on $(n-1)$ the sample element, the absolute forecasting error for the last element is:

$$AE = |Real_n - Forecast(n)|, \quad (3)$$

where

- $Real_n$ is the actual value of the measured value,
- $Forecast(n)$ is the predicted value proposed by the method used.

For multi-step forecasts for the last k elements of the data array, one of the most popular error metrics were used—the mean absolute error (MAE), mean square error (MSE), and symmetric mean absolute percentage error (SMAPE) [39]. They are calculated by the formulas:

$$\text{MAE} = \frac{1}{k}\sum_{i=n-k+1}^{n}|real_i - forecast(i)|, \quad (4)$$

$$\text{MSE} = \frac{1}{k}\sum_{i=n-k+1}^{n}[real_i - forecast(i)\,]^2. \quad (5)$$

$$\text{SMAPE} = \frac{100\%}{n}\sum_{i=1}^{n}\frac{|Forecast - Real|}{|Forecast| + |Real|} \quad (6)$$

3.4. System Architecture

The modules in a forecasting software system automate the main activities (Figure 1). The user of the system accesses the individual modules through a common interface provided by the Manager module. In addition to the connection with the user, it provides management and control over the other modules.

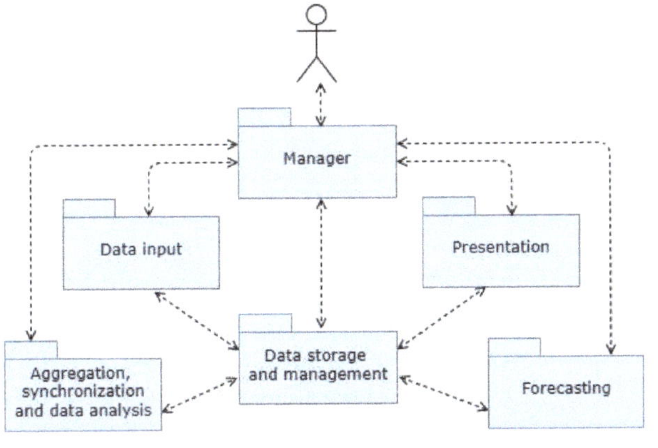

Figure 1. Main modules and the connections between them in an automated forecasting system.

The data entry module assists the user in entering data and their initial classification.

Merging and synchronization tools are useful in cases where data from different sources are used. At the user's choice, through functions or parameters, the data are converted into a format suitable for making forecasts. The data analysis is supported by graphical tools integrated in the module and standard statistical methods, providing the user with opportunities for additional classification and clearing of incorrect data, as well as preparation of various data models.

The forecasting module presents an opportunity to choose one or more forecasting methods, as well as an opportunity to evaluate the most effective method. An important feature that the system must support is easy integration of new forecasting and evaluation methods in the module.

The presentation module contains graphical tools for visualization of the results obtained from the most effective method, as well as from all other methods used.

The data storage and management module provide access to various types of data that can be used in the configuration, training of forecasting methods and their subsequent use:

- primary input–output data and initial parameters;
- data models used by forecasting methods;
- maximum permissible error criteria set by the users;

- number and type of methods used;
- errors obtained in each of the methods;
- identifier of the most effective method;
- results from forecasting tasks, etc.

3.5. Multifactorial Multi-Step Forecasts

The forecasting module (Figure 2) provides us with two different forecasting approaches:

- simple forecasting—direct forecasting with known behavior of the factors, or if we consider the values of the target variable as a time series;
- complex forecasting—the forecasting of the target value is based on additionally created forecast data for the individual factors.

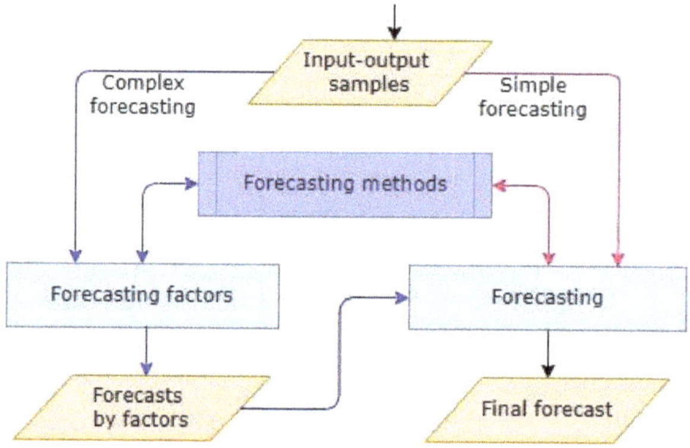

Figure 2. Forecasting module.

During the process of factor forecasting, for each of the factors the most effective method was sought, which could be used repeatedly (Figure 3). It is appropriate to forecast the individual factors in parallel and to use cloud computing to save time and resources.

The factor forecasting module receives a two-dimensional array of input data $\{x_{i,\,j}\}_{i=1\ldots k}^{j=1\ldots n_i}$, containing the sample for k in number factors, each with n_k values. Other input parameters are the number of desired forecast values $c \in N$ and the lower limit of the desired $efficiency$. For each of the factors, independent forecasts were made using the library of "prediction methods".

The set of forecasting methods $\{m_t\}_{t=1\ldots p}$, $p \in N$ was pre-set in the software system and could be extended. The methods include a variety of time series forecast models, models based on artificial neural networks in which parameters such as the number of neurons, activation functions, learning algorithms and other artificial intelligence algorithms can be changed. Models with time series and artificial neural networks are integrated in the system so far.

Applying the forecasting methods, for each of the factors there are approximating functions with the desired efficiency $\{f_t(x_{i,\,j})\}_{j=(n_i-c+1)\ldots n_i}^{t=1\ldots p}$, $i = 1, 2, \ldots k$.

In the process of evaluating the effectiveness, using the AE, MAE, and MSE errors, the efficiencies of the tested methods or variants of methods were compared. The most appropriate forecasting model was selected, including the method $m_{t_1} \in \{m_t\}_{t=1\ldots p}$ and its corresponding parameters.

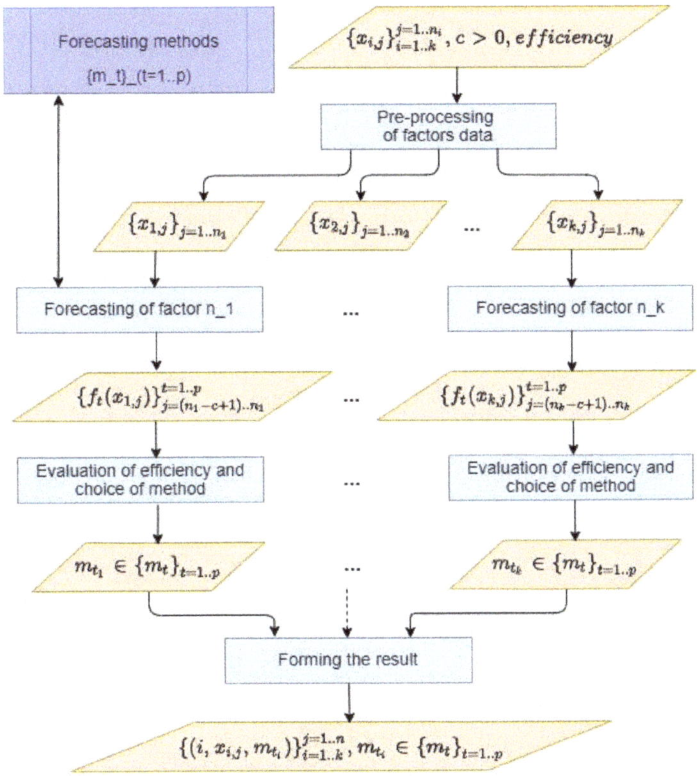

Figure 3. Choice of forecasting method for each factor.

The end result of the process is a set of ordered triples <factor, factor data, selected forecasting method and its corresponding parameters>, i.e., $\{(i,\ x_{i,\ j},\ m_{t_i})\}_{i=1\ldots k}^{j=1\ldots n}$, $m_{t_i} \in \{m_t\}_{t=1\ldots p}$. When the data change, the same selected method can be used, or the process can be restarted to search for a new method.

3.6. *Algorithm for Searching for an Optimal Artificial Neural Network*

To solve a task, many different neural networks can be built, with different numbers of neurons in the hidden layer, with greater or lesser error and different result functions, which have similar behavior in the input–output samples used. An error is not always an indicator of the complexity of the neural network. The lesser number of neurons implies faster and easier learning of the neural network and, subsequently, faster work. The optimal neural network for solving a certain task has characteristics such as a minimum number of neurons and compliance with the user-specified allowable error in training, testing, and validation with the available input-output samples. Its effectiveness in newcomer input data can be evaluated at a later stage. An algorithm providing capabilities for automated construction of multiple neural networks would help choose the optimal solution.

One of our experimented approaches for creating an optimal neural network is iteration over various parameters (Figure 4) needed to create neural networks: number of neurons in the hidden layer, activation functions (Table 1), training algorithms (Table 2), and number of training epochs. It successively changes the parameters for creating neural networks and examines the efficiency of the current neural network. The first neural network that meets the requirements set by the user is considered optimal. Therefore, an important part of the algorithm is how to change the iterative parameters. Since the creation and training of each neural network requires a certain amount of computer resources and time, the iterative approach is appropriate to use on single-processor machines only for tasks where finding the appropriate neural network is expected to have a relatively small number of iterations.

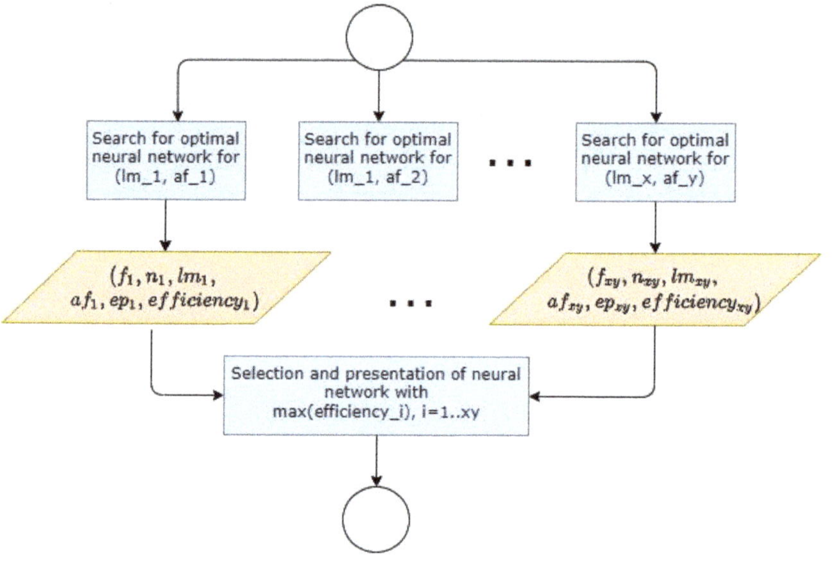

Figure 4. A variant of a parallel algorithm for constructing an optimal neural network for forecasting.

Table 1. Used activation functions.

Function
Hyperbolic tangent sigmoid transfer function
Log-sigmoid transfer function
Hard-limit transfer function
Symmetric hard-limit transfer function
Competitive transfer function
Elliot symmetric sigmoid transfer function
Elliot 2 symmetric sigmoid transfer function
Inverse transfer function
Positive linear transfer function
Radial basis transfer function
Normalized radial basis transfer function
Saturating linear transfer function
Symmetric saturating linear transfer function
Soft max transfer function
Triangular basis transfer function

Table 2. Used training methods.

Method
Levenberg–Marquardt
Bayesian Regularization
Scaled Conjugate Gradient
BFGS Quasi-Newton
Resilient Backpropagation
Conjugate Gradient with Powell/Beale Restarts
Fletcher-Powell Conjugate Gradient
Polak–Ribiére Conjugate Gradient
One Step Secant
Variable Learning Rate Backpropagation
Levenberg–Marquardt
Bayesian Regularization

Finding the optimal solution faster is associated with:

- Indication of parameters for changing the number of neurons. Reducing the limits of variation in the number of neurons will reduce the total number of iterations in the algorithm and therefore will accelerate the achievement of the final result—the creation of an optimal neural network. In some cases, the expected number of neurons can be justified mathematically [40,41].
- Arrangement of the used training methods and activation functions. Different activation functions are suitable for different tasks. To solve problems requiring the application of mathematical logic, we included the hard-limit transfer function and its variant—the Symmetric hard-limit transfer function. These functions provide an interrupted binary signal along the axon of neurons and are suitable in case there is a need to solve problems requiring binary logical thinking. On the other hand, to forecast the future development of continuous processes, we used a large number of efficient and convenient transfer functions. Priority in the order of use is given to those with a sigmoid character such as hyperbolic tangent sigmoid transfer function and log-sigmoid transfer function, useful in a very wide range of problems [42].
- Necessity to change the number of learning epochs. Sometimes, to reach the optimal neural network it is only necessary to change the number of training epochs. Increasing the iterations slows down the neural network learning process, but often even a slight change in this parameter leads to a surprising overcoming of a small plateau of neural network errors and leads to a sharp improvement in the final approximation.

The iterative algorithm for automated construction of artificial neural networks (Figure 5) has the following main steps:

(1) Preparation of input data:
  - Tensor data—input data for the factors, which are usually in the form of a one-dimensional—$\{x_i\}_{i=1\ldots k}$—or two-dimensional array—$\{x_{i,\,j}\}_{i=1\ldots k}^{j=1\ldots n_i}$.
  - Number of forecasted results—c—which is 1 for single-point forecasts or a larger integer for multi-step forecasts.
  - Desired efficiency—$efficiency$—of the trained neural network.

(2) Preparation of the parameters on which iteration is performed:
  - List of training methods $lms = \{m_i\}$, where i varies from *1* to the number of methods (Table 2). Depending on the task, to achieve the desired result faster, it is possible to arrange the methods in the list according to the expected efficiency, and some of them may even be excluded if they are considered inappropriate.
  - List of activation functions $afs = \{af_j\}$, where j varies from *1* to the number of functions (Table 1). Here, too, the functions can be arranged at the discretion of the appropriateness of their use in the specific task.
  - Minimal and maximal number of neurons—*min_n u max_n*, as well as a step by which neurons change—*step_n*. The current number of neurons we denote by n. For more elementary tasks, the number of neurons may start from *1* ($min\_n = 1$) and the step by which their number increases is also *1*. The maximum number limits the possible iterations related to the number of neurons.
  - The epochs *epochs* change from *min_ep* to *max_ep* with a step *step_ep*. Values we have experimented with are $min_{ep} = 1000$, $max_{ep} = 5000$, $step\_ep = 1000$, where usually 3–4 iterations are enough to assess whether the change of epochs affects the efficiency of the trained neural network.
  - By iterating over the number of neurons, training methods and activation functions, a neural network with their current values is created—the ordered triple $(n,\ lm,\ af)$ and the input data. The nesting of the loops for the specific task is a matter of judgment, which determines the sequence of the parameters change. In the experiments, we chose to increase the number of neurons in the outermost loop, as we wanted to find a neural network with the lowest number of neurons. Training methods and activation functions change in inner loops.

(3) After the creation of the current neural network, it is trained, tested, and validated.
(4) The desired efficiency is compared to the efficiency of the current neural network, and then:
   - If the neural network meets the condition, its data is saved and the task is completed;
   - Otherwise, attempts are made to increase the efficiency of the neural network by increasing the number of learning epochs. The information about the most efficient neural network found (with the smallest error) is saved, and it can be current or obtained in a previous iteration.
(5) The end result of the algorithm is a trained neural network having the closest possible to the specified efficiency, as well as parameters for its architecture and training—*efficiency*, number of neurons *n*, training method *lm*, activation function *af*, and epochs *ep*.

The use of the "brute force" method, by traversing all possible values of the iterative parameters and finding the neural network with the optimal ratio "number of neurons—efficiency" is not the most rational approach. Tracking changes in the specified ratio can lead to the creation of heuristic variants of the algorithm by automated changing of the order of change of the iterative parameters.

The availability of sufficient computing power predisposes to the use of different parallel algorithms for such automated search for an optimal neural network (Figure 5) in which, for example, all combinations of training methods and activation functions can be started in parallel processes—$(lm_i,\ af_j)$, where *i* and *j* are changed from *1* to the corresponding number of training methods and activation functions. After the completion of the individual processes, the efficiencies of all neural networks created during their implementation are compared and the most appropriate one is selected.

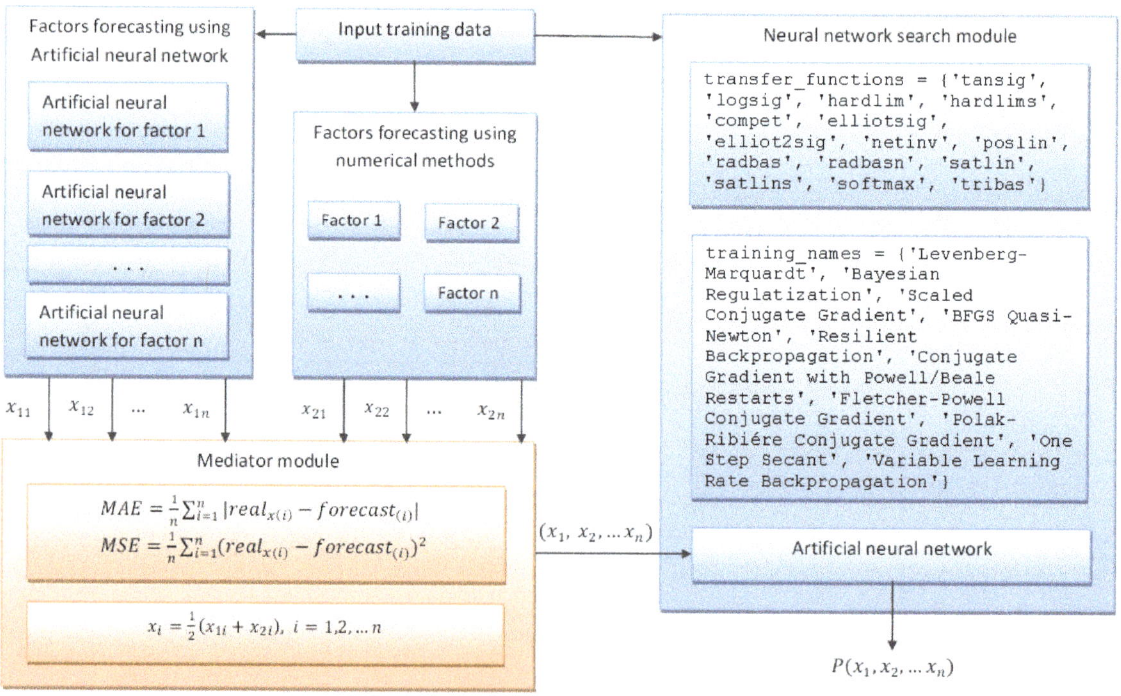

Figure 5. General scheme of forecasting processes.

4. Results and Discussion

A prototype was developed for approbation of the presented model. The program MatLab was used, with which the described basic modules and functionalities were implemented. The prototype was tested to solve prognostic tasks in the field of energy.

4.1. Setup of the Experiment

The problem of forecasting the demand and respectively—the consumption of electricity is extremely important for the planning and management of the national energy system of every country. Accurate forecasting of probable electrical loads is an important prerequisite for effective planning of production capacity, proper maintenance of the transmission and distribution network, planning of future exports or imports of electricity, the behavior and direction of energy flows both in the country and in related international networks.

The developed prototype was used to solve three tasks for forecasting electricity consumption in the National Power System of Bulgaria. The targets subject to forecasting were:

(1) Total final consumption in the national power system;
(2) Electricity consumption in the industry, the public sector, and services;
(3) Electricity consumption in households.

The forecasts were made by taking into account their dependencies on the following socio-economic factors:

(1) Gross domestic product (GDP);
(2) Energy intensity (EI);
(3) Population;
(4) Income of the population;
(5) Energy efficiency;
(6) Price of electricity.

These factors have been identified as significant in studies carried out by the Bulgarian Academy of Sciences (BAS) and Risk Management Lab [10,11]. A subsequent step would be to build a module for the automated study of correlations between the factors with the aim of minimizing their number and optimizing post-processing.

In order to forecast the target values for a specific year, we first made forecasts for the factors for the respective year.

In the process of work, data from official sources such as the National Statistical Institute [42], Information System INFOSTAT [43], Electricity System Operator [44], The World Bank Group [45], and Eurostat [46] were used. All available data on factors and target values for 17 years were used for the study.

The factor forecasting module uses different models of the time series trend. A substructure involving the use of neural networks to predict these factors provides for the possibility of further processing of the input data. Some of the data are submitted to the neural networks normalized, which is implemented by multiplying them by a specific coefficient. This reduces the radius vector of the input data and facilitates the training of this type of artificial intelligence. The general scheme showing the joint operation of the modules is shown in Figure 6:

(1) Modules for factor forecasting, respectively through time series and neural networks;
(2) Error estimation module (mediator module) for the various methods and parameters;
(3) Consumption forecasting module.

4.2. Single-Point Forecasts for the Factors

The use of the prototype for forecasting electricity consumption in single-point (annual) factor forecasting showed different results for the effectiveness of different types of trends in time series (Figure 6).

When forecasting GDP, population, and average annual income, maximum efficiency was achieved with a linear trend of the time series. The forecasting of the energy intensity, the price of the electricity for the household, and the price of the electricity for the industry achieved good results respectively in logarithmic, quadratic, and hyperbolic trends of the time series. Models for the most efficient neural networks, forecasting the same factors, are presented in Figure 7.

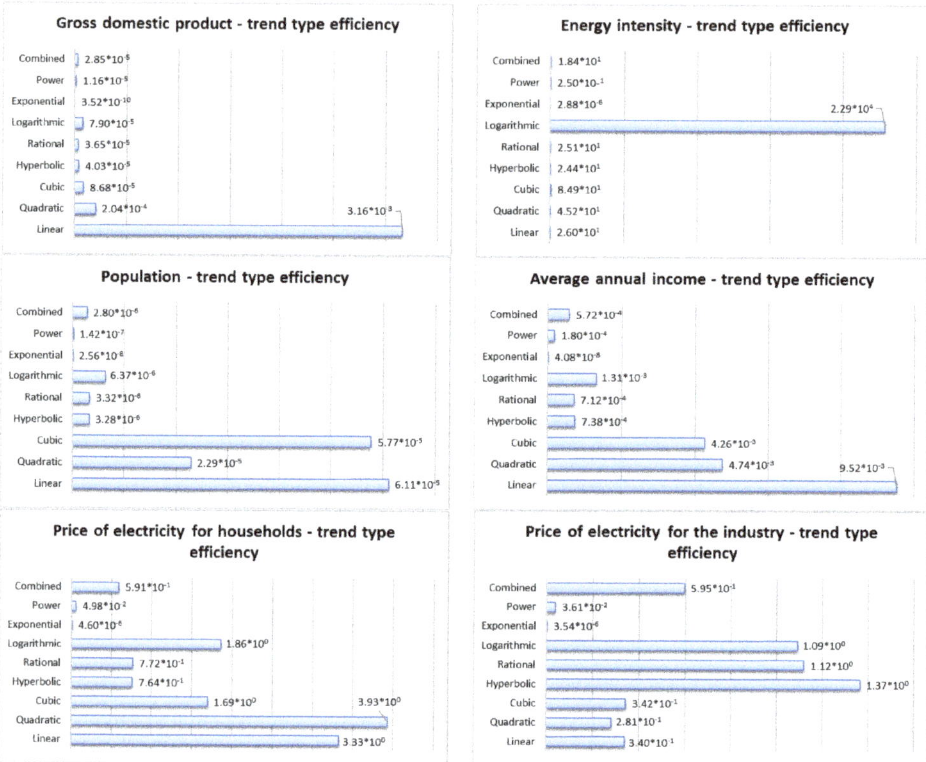

Figure 6. Comparison of efficiency in different types of time series trend.

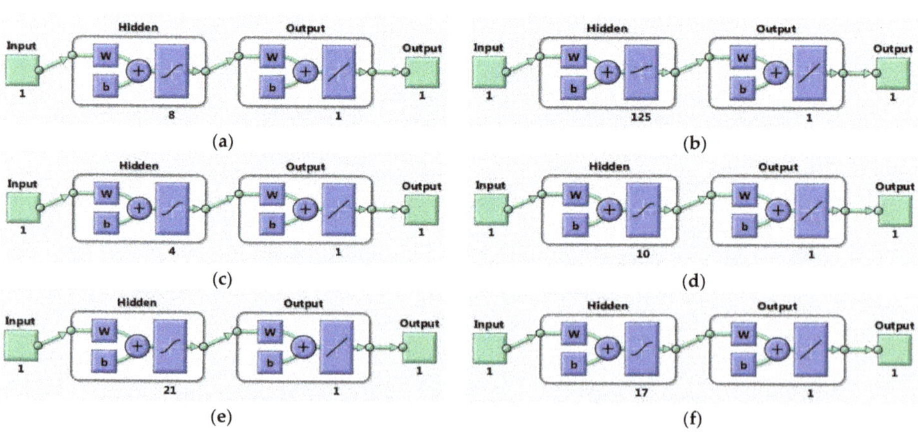

Figure 7. Forecasting the behavior of the factors influencing the electricity consumption in NEES through separate artificial neural networks. (**a**) Gross domestic product, (**b**) energy intensity of the economy (Net_Int), (**c**) population, (**d**) average annual income, (**e**) price of electricity for households, (**f**) price of electricity for the industry.

During neural network training, all factors were considered as functions of time in order to enable the comparison of the results of forecasting in time series. The type of training that proved to be most effective for current tasks is the Lavenberg–Marquardt algorithm. The most appropriate activation function of neurons in the hidden layers for Energy intensity of the economy neural network is the logarithmic sigmoid function:

$$g(x) = \sigma(x) = \frac{1}{1 - e^{-x}},$$

and for every other it is the hyperbolic tangent:

$$g(x) = \frac{e^{x} - e^{-x}}{e^{x} + e^{-x}}.$$

The activation function of the output neuron in all neural networks is linear:

$$g(x) = x.$$

The comparative analysis between the efficiency of the found neural networks in a one-year forecast and the type of trend in time series, leading to minimal error, is presented in Figure 8.

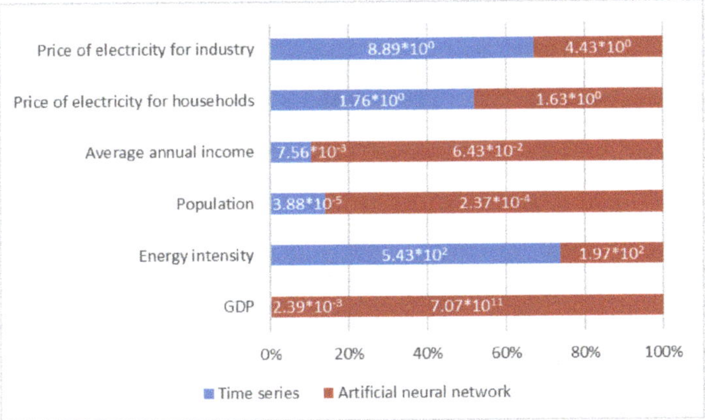

Figure 8. Comparison of efficiency between the most effective time series trend and the results of the forecasts of artificial neural networks.

The use of the presented prototype with time series and artificial neural networks showed a significant advantage in favor of neural networks in the case of forecasting GDP, population, and average annual income. In the other three cases, single-point forecast of energy intensity, price of electricity for households and its price for the economy sector, the module for comparison of errors showed some advantage in the efficiency of time series. In these cases, the type of trend was logarithmic, quadratic, and hyperbolic, respectively. Based on them, forecasts were made for the development of the values of the factors presented in Table 3.

Table 3. Single-point forecasts for the factors for the last year of the available dataset.

Factor	Predicted Value	The Most Effective Method	SMAPE (%)
GDP	102,954.73	Neural network	0.15
Energy intensity of the economy	0.42	Time series with logarithmic trend	3.65
Population	7,006,425.85	Neural network	0.65
Average annual income	6196.5	Neural network	1.57
Price of electricity for households	22.56	Time series with quadratic trend	4.98
Price of electricity for the industry	21.77	Time series with hyperbolic trend	4.8

4.3. Multifactor Single-Point Forecasts of Target Values

In the second part of the conducted experiments, the influence of the considered factors on the target values in the National Power System of Bulgaria was studied. Appropriate optimal neural structures were created for their forecast (Figure 9):

(1) Total final consumption in the National Power System—the neural network named Net_Nees, with 54 neurons in the hidden layer;
(2) Electricity consumption in industry, public sector and services—Net_Industry neural network with 35 neurons in the hidden layer;
(3) Electricity consumption on households—Net_Household neural network with 47 neurons in the hidden layer.

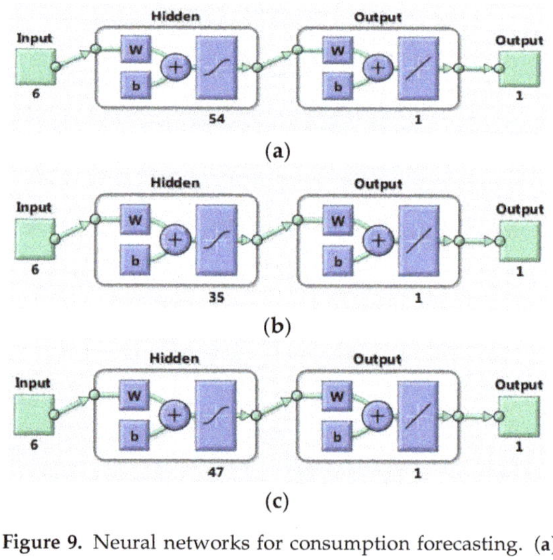

Figure 9. Neural networks for consumption forecasting. (**a**) Total final consumption in the NES (Net_Nees), (**b**) electricity consumption in industry, public sector, and services (Net_Industry), and (**c**) electricity consumption in households (Net_Households).

Activation functions of all neural networks found by the software system were hyperbolic tangent, and of the output neuron, linear.

The deviations of the forecast results from the actual data for single-point forecasts (for one-year consumption) were relatively small (Table 4) as the largest deviations in forecasting consumption in the industry are about 41 thousand tons of oil (toe) equivalent, i.e., about 1.525%. The deviation in forecasting consumption in the entire energy system was 0.0739%, and household consumption was 0.0302%.

Table 4. Forecasting of consumption with artificial neural networks, in thousand toe.

Neural Network	Actual Value	Predicted Value	Absolute Error	SMAPE (%)
Net_Nees	9737.9	9745.1	7.2	0.04
Net_Industry	2721.3	2679.8	41.48	0.77
Net_Households	2318.7	2318.0	0.7	0.02

Considering complex values as time series did not provide good results. The errors obtained with the best approximations of time series were many times worse than the corresponding neural networks (Table 5).

Table 5. The comparative table of errors in the use of these neural networks and time series.

Consumption	Method Type	Method	Absolute Error (Thousand Toe)	SMAPE (%)
Total final consumption in the National Power System	Time series	Cubic trend	199.33	1.01
	Neural network	Net_Nees	7.2	0.04
Electricity consumption in industry, public sector and services	Time series	Combined trend	111.16	2
	Neural network	Net_Industry	41.48	0.77
Electricity consumption in households	Time series	Linear trend	24.47	0.52
	Neural network	Net_House holds	0.7	0.02

The analysis of the weights of the neural networks can show us the influence of the individual factors on the predicted value of the target variable.

Let us denote by $w_{i,p}$ the weights by which the i-th factor is transmitted to the neuron p from the hidden layer of the already trained neural network, as $i,\ p \in N$. We choose the value as a criterion for the significance of factor i:

$$INPF_i = \left| \sum_{p=1}^{r} w_{i,p} \right|,$$

where r is the number of neurons in the hidden layer of the neural network.

The study showed that the largest impact on the total electricity consumption in the national network belongs to the gross domestic product of the country with $INPF_1 = 11.6$, and the least to the average annual income per capita with $INPF_4 = 1.06$. For consumption in the industry sector, things are similar. The most significant factor was GDP with $INPF_1 = 5.01$, and the one with the least importance was the price of electricity for the household—$INPF_6 = 2.3$. The most important factors in the energy consumption of households are population ($INPF_3 = 13.82$) and average annual income—$INPF_4 = 6.16$.

4.4. Multi-Step Forecasts

Using the created neural structures for factor forecasting, we created forecasts for a period of 7 years. Their accuracy can be determined over time (Table 6). Similarly, we created 7-year forecasts for the studied multifactorial values (Table 7). The forecast results showed that household electricity consumption, as well as total final consumption, will gradually increase, while electricity consumption in industry, the public sector, and services will decrease.

Table 6. Seven-year forecasting of factors.

Factor	2021	2022	2023	2024	2025	2026	2027
GDP (Million levs)	102,932	106,693	110,455	114,217	117,978	121,740	125,501
Energy intensity of the economy	0.400	0.395	0.390	0.385	0.380	0.376	0.371
Population	6,804,441	6,746,144	6,687,847	6,629,550	6,571,253	6,512,956	6,454,659
Average annual income	6671	6798	6942	7013	7105	7241	7339
Price of electricity for industry (in stotinki per kWh and without VAT)	26.79	27.62	28.46	29.29	30.12	30.96	31.79
Price of electricity for households (in stotinki per kWh and without VAT)	19.22	19.35	19.47	19.58	19.69	19.79	19.89

Table 7. Seven-year forecast for electricity consumption in the National Energy System, made through artificial neural networks.

Consumption	2021	2022	2023	2024	2025	2026	2027
Total final consumption in the National Power System	13,832	13,856	13,858	13,858	13,858	13,858	13,858
Electricity consumption in industry, public sector and services	3103	3101	3100	3099	3099	3098	3098
Electricity consumption in households	2800	2800	2800	2800	2800	2800	2800

5. Conclusions

Forecasting is a complex task. The availability of a wide variety of mathematical and statistical methods, and artificial intelligence methods, combined with the pursuit of the most accurate forecasting, usually requires a lot of time and effort. The use of software tools to automate some of the activities greatly simplifies the work.

The article proposed a model for multifactor forecasting which automatically selects the best method for forecasting significant factors and then uses the data predicted by them to create a complex multifactor forecast. The developed model was successfully experimented with to make multiple forecasts for the energy consumption of households, industry, and total consumption—for a one-year and a seven-year period. By automating the forecasting process in an indicative way, we made it easier to make predictions with fewer errors than previously.

The presented model has wide applications in various subject areas. It can be used for air quality forecasting, demographic forecasting, forecasting in industry, etc.

In the future, the system can be expanded in several directions. An important part of its development is the integration of additional forecasting methods. The development of a module to evaluate the correlation between the individual factors and the target variable would also help to optimize the forecasting process. Automated generation of graphics and documentation for each step in the overall forecasting process would be beneficial to the end user of the system.

Author Contributions: Conceptualization, K.Y., E.H. and S.H.; data curation, K.Y. and S.C.; formal analysis, K.Y., E.H. and S.H.; funding acquisition, S.H. and S.C.; investigation, K.Y., E.H. and S.C.; methodology, K.Y., E.H. and S.H.; project administration, E.H.; software, K.Y. and S.C.; supervision, E.H.; validation, K.Y., E.H. and S.H.; visualization, K.Y. and E.H.; writing—original draft, K.Y., E.H. and S.H.; writing—review and editing, S.H. and S.C. All authors have read and agreed to the published version of the manuscript.

Funding: The work is partly funded by the MU21-FMI-004 project at the Research Fund of the University of Plovdiv "Paisii Hilendarski".

Institutional Review Board Statement: Not applicable.

Informed Consent Statement: Not applicable.

Data Availability Statement: Data are available through the public databases mentioned in the text.

Conflicts of Interest: The authors declare no conflict of interest.

References

1. ReferencesHyndman, R.; Athanasopoulos, G. *Forecasting: Principles and Practice*; OTexts: Melbourne, Australia, 2021.
2. Slavova, I. *Forecasting and Planning*; New Bulgarian University: Sofia, Bulgaria, 2012.
3. Skiadas, C.H.; Bozeman, J. *Data Analysis and Applications 1: Clustering and Regression, Modeling-Estimating, Forecasting and Data Mining*; John Wiley & Sons Inc.: Hoboken, NJ, USA, 2019; Volume 2.
4. Brockwell, P.; Davis, R. *Introduction to Time Series and Forecasting*; Springer: Berlin/Heidelberg, Germany, 2016; ISBN 9783319298528.
5. Agung, I.G. *Advanced Time Series Data Analysis: Forecasting Using EViews*; John Wiley & Sons Ltd.: Hoboken, NJ, USA, 2019; ISBN 978-1-119-50471-9.
6. Ostertagová, E.; Ostertag, O. Forecasting Using Simple Exponential Smoothing Method. *Acta Electrotech. Inform.* **2012**, *12*, 64–66. [CrossRef]
7. Chen, T.; Honda, K. *Fuzzy Collaborative Forecasting and Clustering: Methodology, System Architecture, and Applications*; Springer: Berlin/Heidelberg, Germany, 2019.
8. Lazzeri, F. *Machine Learning for Time Series Forecasting with Python*; John Wiley & Sons Ltd.: Hoboken, NJ, USA, 2020; ISBN 1119682363.
9. Gridin, I. *Time Series Forecasting using Deep Learning: Combining PyTorch, RNN, TCN, and Deep Neural Network Models to Provide Production-Ready Prediction Solutions*; BPB Publications: Noida, India, 2021; ISBN 9391392571.
10. BAS Strategy. *Preparation of the National Strategy in the Field of Energy*; BAS: Sofia, Bulgaria, 2017.
11. Risk Management Lab. Forecast of the Electricity Balance of the Republic of Bulgaria 2025. 2014. Available online: https://www.bia-bg.com/uploads/files/events/Energy_balans_BG2025.pdf (accessed on 15 July 2022).
12. Mohamed, Z.; Bodger, P. Forecasting electricity consumption in New Zealand using economic and demographic variables. *Energy* **2005**, *30*, 1833–1843.
13. Lee, Y.; Gaik, T.; Yee, C.H. Forecasting Electricity Consumption Using Time Series Model. *Int. J. Eng. Technol.* **2018**, *7*, 218–223.
14. Sun, T.; Zhang, T.; Teng, Y.; Chen, Z.; Fang, J. Monthly Electricity Consumption Forecasting Method Based onX12 and STL Decomposition Model in an Integrated Energy System. *Math. Probl. Eng.* **2019**, *2019*, 9012543.
15. Chou, J.S.; Tran, D.S. Forecasting energy consumption time series using machine learning techniques based on usage patterns of residential householders. *Energy* **2018**, *165*, 709–726. [CrossRef]
16. Ding, S.; Hipel, K.; Dang, Y. Forecasting China's electricity consumption using a new grey prediction model. *Energy* **2018**, *149*, 314–328. [CrossRef]
17. Lee, Y.S.; Tong, L.T. Forecasting energy consumption using a grey model improved by incorporating genetic programming. *Energy Convers. Manag.* **2011**, *52*, 147–152. [CrossRef]
18. Huang, Y.; Wang, C.; Dang, H.; Lai, S. Evaluating performance of the DGM(2,1) model and its modified models. *Appl. Sci.* **2016**, *6*, 73. [CrossRef]
19. Chung, Y. Electricity consumption prediction using a neural-network-based grey forecasting approach. *J. Oper. Res. Soc.* **2017**, *68*, 1259–1264.
20. Chernykh, I.; Chechushkov, D.; Panikovskaya, T. Prediction of electrical loads in operational control of power systems based on neural network structures. *WIT Trans. Ecol. Environ.* **2014**, *190*, 109–117.
21. Khosravani, H.; Castilla, M.; Berenguel, M.; Ruano, A.; Ferreira, P. A Comparison of Energy Consumption Prediction Models Based on Neural Networks of a Bioclimatic Building. *Energies* **2016**, *9*, 57.
22. Yoo, S.; Myriam, H.A. Predicting residential electricity consumption using neural networks: A case study. *J. Phys. Conf. Ser.* **2018**, *1072*, 012005. [CrossRef]
23. Jahn, M. *Artificial Neural Network Regression Models: Predicting GDP Growth*; HWWI Research Paper, No. 185; Hamburg Institute of International Economics (HWWI): Hamburg, Germany, 2018.
24. Bouktif, S.; Fiaz, A.; Ouni, A.; Serhani, M.A. Optimal Deep Learning LSTM Model for Electric Load Forecasting using Feature Selection and Genetic Algorithm: Comparison with Machine Learning Approaches. *Energies* **2018**, *11*, 1636. [CrossRef]
25. Alamaniotis, M.; Bargiotas, D.; Tsoukalas, L.H. Towards smart energy systems: Application of kernel machine regression for medium term electricity load forecasting. *Springer Plus* **2016**, *5*, 58. [CrossRef]
26. Alamaniotis, M. Synergism of Deep Neural Network and ELM for Smart Very-Short-Term Load Forecasting. In Proceedings of the 2019 IEEE PES Innovative Smart Grid Technologies Europe (ISGT-Europe), Novi Sad, Serbia, 10–12 October 2019; pp. 1–5. [CrossRef]
27. Chou, J.S.; Truong, D.N. Multistep energy consumption forecasting by metaheuristic optimization of time-series analysis and machine learning. *Energy Res.* **2021**, *45*, 4581–4612. [CrossRef]
28. Kong, W.; Dong, Z.Y.; Jia, Y.; Hill, D.J.; Xu, Y.; Zhang, Y. Short-Term Residential Load Forecasting Based on LSTM Recurrent Neural Network. *IEEE Trans. Smart Grid* **2019**, *10*, 841–851. [CrossRef]

29. Moradzadeh, A.; Moayyed, H.; Zakeri, S.; Mohammadi-Ivatloo, B.; Aguiar, A.P. Deep Learning-Assisted Short-Term Load Forecasting for Sustainable Management of Energy in Microgrid. *Inventions* **2021**, *6*, 15. [CrossRef]
30. Khwaja, A.S.; Anpalagan, A.; Naeem, M.; Venkatesh, B. Joint bagged-boosted artificial neural networks: Using ensemble machine learning to improve short-term electricity load forecasting. *Electr. Power Syst. Res.* **2020**, *179*, 106080. [CrossRef]
31. Khwaja, A.S.; Zhang, X.; Anpalagan, A.; Venkatesh, B. Boosted neural networks for improved short-term electric load forecasting. *Electr. Power Syst. Res.* **2017**, *143*, 431–437. [CrossRef]
32. Cao, Z.; Wan, C.; Zhang, Z.; Li, F.; Song, Y. Hybrid Ensemble Deep Learning for Deterministic and Probabilistic Low-Voltage Load Forecasting. *IEEE Trans. Power Syst.* **2020**, *35*, 1881–1897. [CrossRef]
33. Karatasou, S.; Santamouris, M.; Geros, V. Modeling and predicting building's energy use with artificial neural networks: Methods and results. *Energy Build.* **2008**, *38*, 949–958. [CrossRef]
34. Cervená, M.; Schneider, M. Short-term forecasting of GDP with a DSGE model augmented by monthly indicators. *Int. J. Forecast.* **2014**, *30*, 498–516. [CrossRef]
35. Farsi, B.; Amayri, M.; Bouguila, N.; Eicker, U. On Short-Term Load Forecasting Using Machine Learning Techniques and a Novel Parallel Deep LSTM-CNN Approach. *IEEE Access* **2021**, *9*, 31191–31212. [CrossRef]
36. Chung, W.; Gu, Y.H.; Yoo, S.J. District heater load forecasting based on machine learning and parallel CNN-LSTM attention. *Energy* **2022**, *246*, 123350. [CrossRef]
37. Sajjad, M.; Khan, Z.A.; Ullah, A.; Hussain, T.; Ullah, W.; Lee, M.Y.; Baik, S.W. A Novel CNN-GRU-Based Hybrid Approach for Short-Term Residential Load Forecasting. *IEEE Access* **2020**, *8*, 143759–143768. [CrossRef]
38. Alamo, D.H.; Medina, R.N.; Ruano, S.D.; García, S.S.; Moustris, K.P.; Kavadias, K.K.; Zafirakis, D.; Tzanes, G.; Zafeiraki, E.; Spyropoulos, G.; et al. An Advanced Forecasting System for the Optimum Energy Management of Island Microgrids. *Energy Procedia* **2019**, *159*, 111–116. [CrossRef]
39. Botchkarev, A. A New Typology Design of Performance Metrics to Measure Errors in Machine Learning Regression Algorithms. *Interdiscip. J. Inf. Knowl. Manag.* **2019**, *14*, 45–76. [CrossRef]
40. Sheela, K.; Deepa, S.N. Review on Methods to Fix Number of Hidden Neurons in Neural Networks. *Math. Probl. Eng.* **2013**, *2013*, 425740. [CrossRef]
41. Lörke, A.; Schneider, F.; Heck, J.; Nitter, P. Cybenko's Theorem and the Capability of a Neural Network as Function Approximator. 2019. Available online: https://www.mathematik.uni-wuerzburg.de/fileadmin/10040900/2019/Seminar__Artificial_Neural_Network__24_9__.pdf (accessed on 15 July 2022).
42. National Statistical Institute. Available online: https://www.nsi.bg/ (accessed on 15 July 2022).
43. INFOSTAT Information System. Available online: https://infostat.nsi.bg/infostat/pages/external/login.jsf (accessed on 15 July 2022).
44. Electricity System Operator. Available online: http://www.eso.bg/ (accessed on 15 July 2022).
45. The World Bank Group. Available online: https://www.worldbank.org (accessed on 15 July 2022).
46. Eurostat. Your Key to European Statistic. Available online: https://ec.europa.eu/eurostat (accessed on 15 July 2022).

sustainability

MDPI

Systematic Review

Solar Panels Dirt Monitoring and Cleaning for Performance Improvement: A Systematic Review on Smart Systems

Benjamin Oluwamuyiwa Olorunfemi *, Omolola A. Ogbolumani and Nnamdi Nwulu

Center for Cyber-Physical Food, Energy and Water Systems (CCP-FEWS), University of Johannesburg, P.O. Box 524, Auckland Park, Johannesburg 2006, South Africa
* Correspondence: ben93olorunfemi@gmail.com

Abstract: The advancement in technology to manage energy generation using solar panels has proved vital for increased reliability and reduced cost. Solar panels emit no pollution while producing electricity as a renewable energy source. However, the solar panel is adversely affected by dirt, a major environmental factor affecting energy production. The intensity of light falling on the solar panel is reduced when dirt accumulates on the surface. This, in turn, lowers the output of electrical energy generated by the solar panel. Since cleansing the solar panel is essential, constant monitoring and evaluation of these processes are necessary to optimize them. This emphasizes the importance of using smart systems to monitor dirt and clean solar panels to improve their performance. The paper tries to verify the existence and the degree of research interest in this topic and seeks to evaluate the impact of smart systems to detect dirt conditions and clean solar panels compared to autonomous and manual technology. Research on smart systems for addressing dirt accumulation on solar panels was conducted taking into account efficiency, accuracy, complexity, and reliability, initial and running cost. Overall, real-time monitoring and cleaning of the solar panel improved its output power with integrated smart systems. It helps users get real-time updates of the solar panel's condition and control actions from distant locations. A critical limitation of this research is the insufficient empirical analysis of existing smart systems, which should be thoroughly examined to allow further generalization of theoretical findings.

Keywords: photovoltaic panel; remote solar plant; automated cleaning; condition monitoring; internet of things; solar panels dirt; dirt detection; dirt accumulation and removal; device management; real-time monitoring and cleaning

Citation: Olorunfemi, B.O.; Ogbolumani, O.A.; Nwulu, N. Solar Panels Dirt Monitoring and Cleaning for Performance Improvement: A Systematic Review on Smart Systems. *Sustainability* **2022**, *14*, 10920. https://doi.org/10.3390/su141710920

Academic Editors: Luis Hernández-Callejo, Sergio Nesmachnow and Sara Gallardo Saavedra

Received: 15 June 2022
Accepted: 22 August 2022
Published: 1 September 2022

1. Introduction

In many industrialized nations, electricity generation is still dependent on fossil fuels. Although these fuels are very effective in energy quality, they are not suited for long-term use because the fossil fuel source will eventually run out someday. Furthermore, fossil fuels are a considerable threat to environmental balance and create numerous ecological problems such as global warming [1,2]. Therefore, the utilization of renewable sources must be accepted as soon as possible. A significant feature of renewable electricity generation is the infinite supply [3]. Compared to conventional fossil fuel technologies, renewable electrical energy sources have a more negligible effect on the environment, considering cleanliness.

Solar panel technology is becoming more popular as a renewable electricity generation due to the growing renewable energy request [4,5]. By the end of this decade, China's solar capacity is foreseen to reach 400 GW [6]. The cumulative installed solar capacity in megawatts between 2012 and 2021 is shown in Figure 1, based on the information provided by IRENA, International Renewable Energy Agency [7].

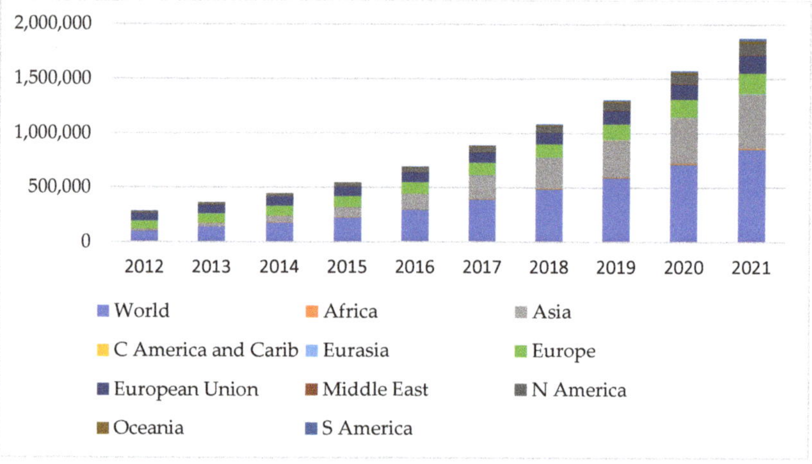

Figure 1. Cumulative installed solar capacity from 2012 to 2021 [7].

It is critical to appropriately manage solar power plants, aiming to optimize their performance and reliability for their continued use. The efficiency and stability of solar panels can be increased, while costs can be reduced [8]. Irradiation and temperature are the key environmental factors that determine the power output of a solar panel module. A decrease in the amount of irradiance and an increase in temperature decrease solar panel module efficiency [9]. Solar panels convert solar radiation into direct current electrical energy; it must constantly be exposed to the maximum amount of sunlight to maximize electricity productivity [10]. Nonetheless, irradiation decrease on the solar panel surface caused by shading due to dirt accumulation can be well controlled. This happens repeatedly and decreases the amount of sunlight reaching the panels [11]. Dirt accumulated on solar panels can include dust, snow, ice, and other organic waste [12]. Fine dust particles settle more deeply on the surface of solar panel modules, affecting their output performance more than coarse dust particles [13]. A controlled experiment conducted in [14], using spotlights to simulate solar radiation, found that the external irradiance resistance can reduce the photovoltaic performance by up to 85%. Rain can naturally wash away dust and sand, but moss requires proper cleaning [15,16]. Solar panel cleaning is one of the major challenges for solar power developers because cleaning the solar panel surface requires careful planning and resources (time, materials, and labor) and results in higher production costs. However, cleaning solar panels is an important task to ensure the long-term operational and financial success of a solar power plant [17]. Cleaning solar panels is necessary because it ensures that the solar panel surfaces are properly maintained to ensure efficient energy generation. It also prevents damage from accelerated aging or corrosion caused by weather conditions such as heavy rains, snow, hail, or high humidity [18,19].

The performance of a solar panel is mainly measured by its efficiency, which indicates how much electricity the panel produces compared to its maximum theoretical efficiency. For example, a solar panel with an efficiency of 20% means that it generates 20% more electricity than it could if left uncoated. An experiment on the cleanliness and tracking mechanism for the various conditions of a solar panel was carried out by [20]. The conditions examined are the fixed and clean panel, the dirty and fixed panel, the dirty and tracking panel, and the clean and tracking panel. Dust buildup on the solar panels' surfaces causes the efficiencies to decline even with installed sun-tracking. The high transmission rate of light on the cleaned solar panel causes an increase in efficiency [19]. Tracking a solar panel without cleaning is less efficient than keeping the solar panel fixed and cleaned, having an efficiency decrease of up to 50%. Dust deposition on solar panels reflects more loss in large-scale power plants in megawatts [21]. A 1% decrease in proficiency may meaningfully influence the Internal Rate of Return (IRR). In comparison, low-level dust accumulation might not significantly affect the production output of small-scale solar plants [22].

Solar panels are monitored with data acquisition systems, where the performance of the system is evaluated under real-world conditions [23]. Real-time monitoring and evaluation of dirt accumulated on solar panels are required to optimize the cleaning operation. Generally, monitoring dirt accumulation on solar panels can either be done online or offline [24]. Smart systems enhanced by internet connection are integrated into solar panel cleaning to improve the performance of autonomous cleaning methods. This will make the system intelligent to monitor the remote solar panel. It can detect dirty conditions and activate its removal from the solar panel surface without human control. Solar panel surface maintenance can be done at a fraction of the installation cost, and electricity generation improvements are possible [25]. There are several reviews on cleaning methods for the solar system, both manual and autonomous systems, but to the author's knowledge, none has considered a review of the different smart system approaches applied to solar panel monitoring or cleaning system approaches used in this study. Published research papers have been reviewed and analyzed. The study aims to conduct a literature review concerning the theoretical framework for smart systems as it relates to solar panel cleaning and remote monitoring to promote the concept of smart solar systems. Literature searches were conducted using specific keywords related to the paper's subject. Although there are many review papers in the literature concerning the concept of smart solar systems, there is a limited number of papers concentrating on the technology adoption aspect of smart solar systems.

1.1. Review of Solar Panels Automated Cleaning Techniques

The continuous cleaning and monitoring of solar panels after installation on a roof or at a remote solar farm is difficult [26]. Solar panels can currently be cleaned using a variety of techniques, including the traditional method of brushing away dust, coating processes, and robotic cleaning devices. This process has been automatized since cleaning with manual brushes and water is incredibly time- and labor-consuming and costly for industrial solar installations [27]. An automated cleaning system for solar panels is composed of an autonomous unit using sensors and controllers and a cleaning mechanism unit that can be watered or waterless. Solar panels can be cleaned using several methods of removing dirt [28,29]; they are robotic, heliotex, electrostatic, coating cleaning, vibrating cleaning, and forced-air cleaning. The review of the cleaning methods, listed in Table 1, compares each method's pros and cons.

1.1.1. Brush Cleaning

The brush cleaning method combines mechanical and electronic components to control the brush's movement as shown in Figure 2, for cleaning the solar panel either with or without water [38]. The turn-on and turn-off process is automated by sensing the current dust accumulation on the solar panels and comparing it with the set reference by the program. The electronic component supplies a signal to the motor for the cleaning system movement [39]. The system has to be robust with many types of complex procedures to be performed with greater precision, flexibility, and control than with conventional techniques [40]. Furthermore, the developed system improves the efficiency and output power of the solar panels as a result of improved performance [41].

Table 1. Comparison of various cleaning techniques.

Reference	Cleaning Technique	Merits	Demerits	Power Output Efficiency Compared to Clean Panels
[30]	Natural cleaning	No investment cost is involved.	Depends on the location's weather condition.	4%
[31]	Manual cleaning	Involves simple design.	Requires expensive materials and the use of human labor.	90.67%
[32]	Robotic cleaning	Effective and sustainable in all climates.	Requires complex construction.	99.5%
[33]	Heliotex cleaning	Effective for non-sticky dirt.	Requires a lot of water.	12.5%
[34]	Electrostatic cleaning	Effective for dry dust and requires no moving parts.	High voltage is required and design is costly.	3.35–11.5%
[35]	Hydrophobic and hydrophilic coating	Does not require water and labor.	Coating presence reduces screen efficiency.	6.62%
[36]	Vibrating cleaning system	Applicable for dry dirt in dry weather.	An external source is required to power the vibrating motor.	95%
[37]	Forced-air cleaning system	Applicable for dry dirt in dry weather.	An external source is required to power the air pump. Only removes small dust larger than 20 µm.	86.4%

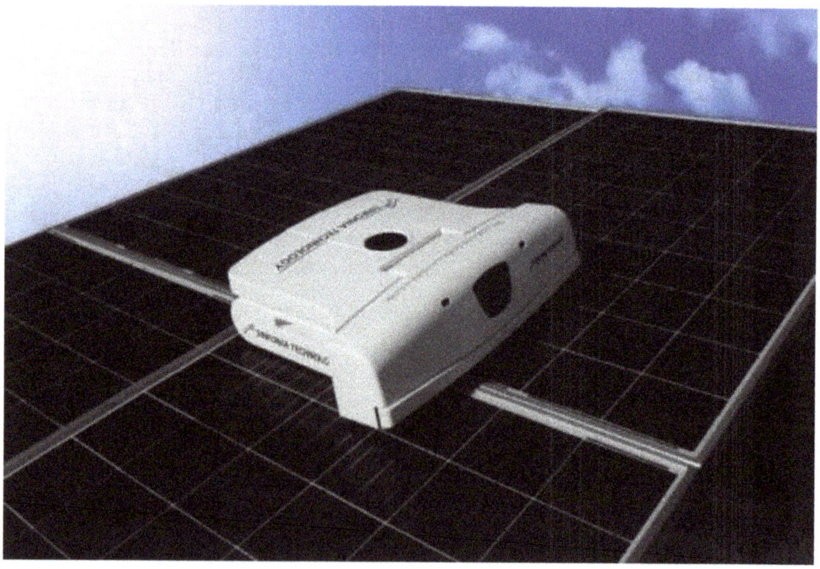

Figure 2. Robotic brush cleaning of solar panels [42].

1.1.2. Heliotex Cleaning

Heliotex cleaning involves spraying water onto the solar surfaces [43]. It is possible to program the cleaning system based on the environment whenever necessary. Further maintenance is not required, other than a periodic replacement of the water filter if it is blocked by sand and the top-ups of the cleanser. Pumps are connected via piping to a water reservoir, fixed to nozzles on the solar surface. The system is very effective and recommended for locations with no water deficiency due to the high amount of water consumed for cleaning [44]. Figure 3 demonstrates the heliotex method of cleaning. This system is not suitable for all situations [19].

Figure 3. Demonstration of the heliotex method of cleaning solar panels [45].

1.1.3. Electrostatic Cleaning

Another dust removal method is electrostatics cleaning, used on dry and dusty solar panels, as shown in Figure 4. In electrostatic precipitation (ESP), fine dust particles on the surface of the solar panel can be removed by induced electrostatic charges [46]. The solar panels are covered with transparent plastic or glass sheets of electrostatic charge material; when a high AC voltage is applied to the electrostatic material, the force acts on the dust close to it and causes the repository motion of the dust particles to shake off from the solar panel surface. The system can clean 90 percent of accumulated dust in less than two minutes [47]. A significant concern that limits the application of this method is safety. It would be unsafe since the solar panel would always remain charged even in showery weather. However, the dynamic motion of all the particles cannot be conveyed by fixed wire electrodes as experimented in [48].

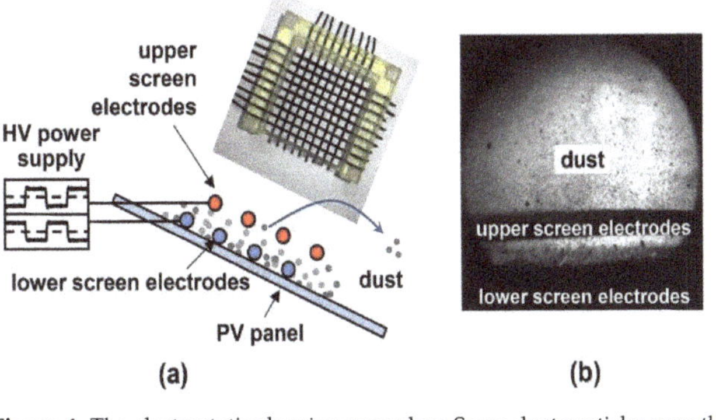

Figure 4. The electrostatic cleaning procedure Some dust particles pass through the hole in the upper screen electrode due to their inertia force and the alternating electrostatic field near the electrodes stirs up the dust particles, and a high-speed microscope camera was used to capture the results as shown in (**a**) and (**b**) respectively [49].

1.1.4. Coating Cleaning

The coating method is also a technique for cleaning solar panels using anti-soiling coating [50]. This method can be used with either a solid, liquid, or gas-based substrate. This method relies on the self-repellent action of the coating material to prevent dust particles from adhering to solar modules. Hydrophilic film and hydrophobic film are the two methods of coating cleaning [51]. The superhydrophobic coating surface method allows for self-cleaning PV panels. This has benefits, like preventing water damage and graffiti [44]. Water gets absorbed into the film in the hydrophilic and rinses the dirt away. On the other hand, the hydrophobic film repels water as it falls; due to its hydrophobic properties, water drops that reach the surface are pushed off quickly, picking up particles alongside. A specific cure time and evaporation of the solvent and drying of the nanoparticle base were required for the coatings to dry once applied as liquids with low viscosity [19]. Each of the coating samples had high transmission, low reflection, and low absorption properties in the ultraviolet (UV), visible (Vis), and near-infrared (NIR) regions.

The fundamental raw material for the coating cleaning is nano metal oxide particles and resin. The product is made by mixing chemicals [19]. Figure 5 compares a layer of hydrophilic coating causing a sliding motion to a rolling motion made by a hydrophobic coating. Despite their differences, both methods of self-cleaning serve the same end [52].

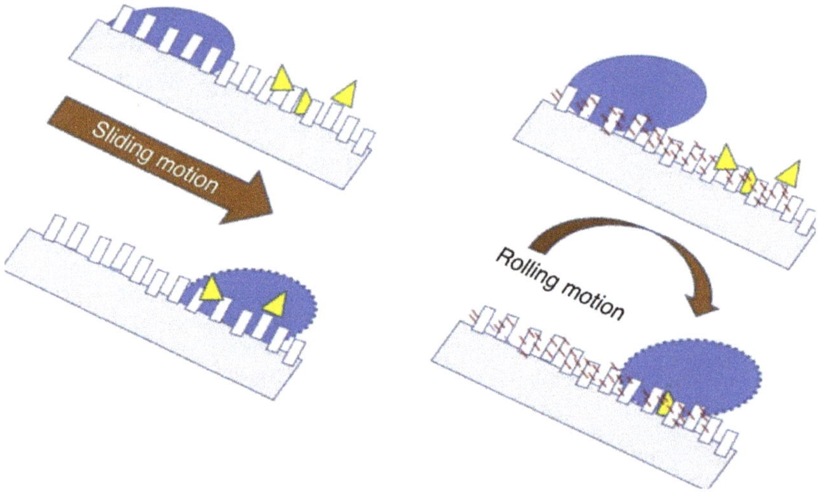

Figure 5. Droplets slide and roll during the self-cleaning process [53].

1.1.5. Vibrating Cleaning System

The vibrating cleaning method prevents solar panels from getting dirty and does not require water or manual labor [46]. To remove the adhesive force between dust particles and the solar panel surface, a mechanical vibrator attached to a panel produced harmonic excitation force. In [54], the wind energy is converted into mechanical vibration for dust removal from solar panel surfaces without consuming any energy from the solar system, thereby improving its efficiency. As vibration intensifies, the inertial force of vibration increases, which turns the dust particles' adhesion force into kinetic energy. External sources of power are usually needed to operate the vibrating motor in vibrating cleaning systems [36]. The panel's self-cleaning system is driven by a DC motor that is fastened to the rear sheet. Based on [55], a solar module was supported on four edges to simulate a system being excited by an unbalanced mass to induce vibrations (see Figure 6). As the DC motor's rotor reached the first natural frequency, a large amount of vibration was induced on the panel.

Figure 6. DC motor attached to solar panel rear for vibration cleaning [55].

1.1.6. Forced-Air Cleaning System

A forced-air cleaning system for solar panels can help to keep them clean and free of debris. This type of system uses a blower to force air through the panels, which can help to remove dirt, dust, and other debris, in addition to improving the efficiency and performance of residential and commercial solar panels. However, this method is only effective for removing dust blown by the air from solar panels [56]. Water is neither consumed nor directly contacted by the turbulent airflow generated by compressed air [37]. These results were used to construct a pilot cleaning and cooling system [37] that utilized a compressed-air unit composed of a compressor, air tank, airflow management valve, and nozzles with a thickness of 5 mm, see Figure 7. The compressor is powered by PV panels, and a valve controls the flow of compressed air from the tank to meet the needs of cleaning and cooling. A pipe assembly that can be moved around an installation as needed can transport air between the panels [19].

1.2. Evaluation of the Performance and Cost of Solar Panel Cleaning Techniques

Several studies may be done regarding cost–performance considerations after the cleaning technique is developed, implemented into use, and a cleaning frequency is defined. It is important to point out that various cleaning methods are dependent on market demands, with some common procedures employing natural, manual, mechanical, electrostatic, vibration, and coating processes. Different environmental variables and setups may be evaluated using the performance and cost. To assess the necessity for a self-cleaning system, a relationship analysis between soiling and its impacts on performance efficiency is required [57]. The amount of sunshine, dust concentration, and rain effects are a few examples of environmental variables. The effectiveness of each cleaning method and its costs may very well be forecasted for each circumstance using optimization models and machine learning. A configuration strategy and an investment plan will result from comparing the different circumstances.

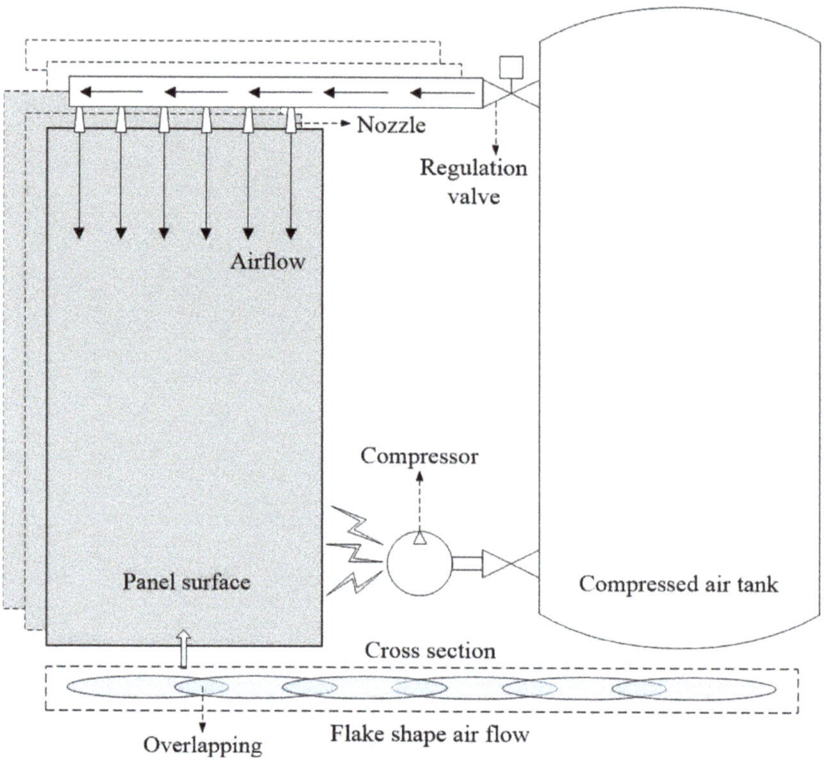

Figure 7. Regulatory mechanism for solar PV panel arrays using compressed air [37].

The configuration plan identifies the hardware and software modifications required to be made for the current platforms as well as the system architectural modifications that should be implemented. Some experts believe that the cost of restoring solar panels' capacity to capture energy should be determined exclusively using a satisfactory rate of return (ROI) [37]. The effectiveness of an integrated smart system for solar panel cleaning may be determined by this analysis [22]. The limitation of the ROI analysis is that they only evaluate the economic side of the problem. Reliability should be taken into account when determining how much to spend on solar panel cleaning methods because it is a major obstacle to effective monitoring and cleaning.

Based on the cleaning method analysis of various cleaning systems by [58], the electrostatic cleaning method is the most effective. Dust particles are removed from the surface without using water; however, spraying water on the photovoltaic cells during cleaning increases their efficiency. The economic viability of automatic self-cleaning mechanisms of solar panels is evaluated in [59] to determine their contribution to the total system cost. When comparing the power generation of PV modules with and without automated self-cleaning mechanisms, the findings reveal a difference of 35%. A domestic installation has a payback period of about five years while making an installation in a commercial setting will typically pay off after 2.25 years. Similarly, ref. [60] reported an efficiency increase of 30–33% in a solar panel array when a robotic cleaning system was used. A robot can also be programmed to fix panels of different sizes. Cleaning a complete array is extremely beneficial since the accumulation of dust on one panel can hinder the performance of the entire array. The fact that solar panel cells are usually connected in series makes it extremely important that they operate at maximum efficiency.

1.3. Review of Solar Panel Remote Monitoring

1.3.1. Condition Monitoring

It has been an important research topic to continuously check the condition of solar panels in remote areas and detect faults to provide stable power [61]. The status application captures and reports the operation, performance, and usage of the solar panel being monitored. With diagnostics applications, monitoring, troubleshooting, repairing, and maintaining networked devices are possible. IoT-enabled smart solar monitoring systems provide remote monitoring and recording. This platform monitors the solar system in real-time via the internet. Monitoring of parameters such as voltage, current, temperature, and humidity is performed by a smart solar panel cleaning system built with IoT. Solar panel performance is typically characterized by measuring the I–V curve under standard conditions (1000 W/m^2 solar irradiance and 25 °C temperature) [62].

In general, remote monitoring systems consist of three components: a sensing unit, a processing unit, and a display unit [63]. The sensing unit is located near the solar system to gather all relevant data to monitor system performance. The data from the sensing unit is carried to a processing unit using a wired or wireless (wireless sensor network—WSN) network, then to the display unit. These services are made possible by wireless sensor networks, which are cost-efficient to install, consume low power, and require little maintenance. Long-range features enable their deployment at remote sites [64]. Smart sensors are often used in a sensing unit so that the signals generated by a solar monitoring system can be handled efficiently before they are sent to a central processing unit. A plant health monitoring system utilizing IoT was proposed by [65], in which the sensors were embedded in the solar system and connected to the internet via wireless networks.

According to [66], the sensor values are crucial for determining a panel's output. A solar power plant's dirt condition can play a crucial role in monitoring the need for maintenance [22]. Despite the unpredictable nature of solar energy and the initial installation costs, research has been undertaken to discover the execution of solar energy optimization. To improve electrical systems reliability, the optimization method aims to minimize investment, operating, and maintenance costs and emissions [67,68]. Ref. [69] examined how the internet of things can be used to monitor solar panels and found its usage is crucial to the proper management of the solar system. Sensing hardware, data acquisition software, and block management modules measure data. This allows all the real-time data collection on the solar plant's electrical output variables to be viewed and stored within the block management. When the panel is not operating correctly, the smart system will offer suggestions, display errors, and send alerts when maintenance is needed.

1.3.2. Dirt Detection

Monitoring and cleaning solar systems have been studied extensively. Before the performance of the cleaning system, it can be challenging to predict the deposition rate of the organic and inorganic particles on the solar module surfaces. Therefore, to ensure that the cleaning system is as effective and efficient as possible and to make the best use of the energy yield, it is required to inspect the solar panels for dust. Besides analyzing dust effects and deposition rates, identifying the crucial information from the previous research is the purpose of this section.

A dirt detection mechanism on a solar panel was made by [39]. A weight sensor in the system continuously measures the dust. Upon receiving defined feedback from the sensors, the Arduino controller commands the dust cleaning. Solar panels are fitted with weight sensors that measure dust thickness according to changes in weight. The feedback for cleaning is that the panel weight goes more than the predefined value due to dust. A dusty solar panel will weigh more than one that is clean. The Arduino controller can reference the actual weight of the solar panel. In their findings, the continuous monitoring of the weight by the microcontroller through the load cell aids the dirt detection and cleaning of the solar panels. A mounting plate holds the load cell below the panel.

In 2019, ref. [70] developed an innovative system for monitoring solar panels' condition. Radiometric sensors are used in a condition monitoring system that links to an Arduino platform, and it works by analyzing the emissivity of a surface and recognizing a low value when the dust is present. A thermographic camera is employed for the radiometric sensor to provide reliable results. Unmanned aerial vehicles are designed to carry the system. With the Internet of Things, radiometric data can be sent to the cloud for analysis, and thermograms can be stored to be further processed. An actual solar panel measures sensor output and surface conditions from various angles and distances. Results from the radiometric sensor analysis show a high degree of accuracy, and dust is recognized in all set-ups. As part of a thermography analysis, it is found that there is consistency and regularity in the quantities. The average variation of each experiment determines accuracy. When the luminous emittance of the solar panel increases, the radiometer measures a higher temperature. The thermal images verify the results of this measurement using a radiometer sensor, the surface being measured is characterized by its temperature. Obtaining reliable measurements for thermal image processing requires knowing the surface characteristics and the ambient conditions, such as humidity, the temperature of the air, distance to an object, reflected temperature, and incident radiation [9].

A design and fabrication demonstration shows a prototype that cleans the panel's surface in [22]. The sensing unit was programmed with the regression model developed using a month's worth of data from spotless and dirty panels. By the regression model and the integrated sensing unit, the autonomous unit determines the optimal time for cleaning. The prototype autonomous unit monitors input and influencing parameters with direct or indirect impact on the solar farm's output power. They also investigated automatic cleaning in their study. Therefore, this system can determine if dust particles impact solar panels' power generation. In the prototype, the light intensity is measured with the TSL 2561 illuminance sensor, and sensors measure voltage and current to determine output power. The other manipulating parameters for solar panel output, such as temperature and humidity, are measured by DHT11 temperature and humidity sensors, whereas GP2Y1014AU0F dust sensors determine the dust density. The measured illuminance value and output power are stored in the cloud interface. Regression analysis of the processed data is carried out to determine the connection between input variables and power output.

Another study by [71] investigated robotic technology for removing solar dust from solar boards. The strategy proposes screening power generation on a mobile app and cleaning the solar surfaces in response. The input mechanism includes an Android switch unit, IP camera, voltage sensor, and current sensor in the experiment. IP cameras monitor solar panel cleaning and conditioning; it is internet-connected and displayed on a PC Windows system or an Android device. Images might be rather costly to transfer over the internet. A related study in [72] used smart cameras with "RGB" and "infrared" for night vision to continuously take pictures of solar panels. A real-time algorithm determines whether or not a panel needs to be cleaned based on the picture. An Advanced Reduced Instruction Set Computer Machines (ARM) processor, which will also be on the board, will process the incoming data and execute the algorithm. The intelligent system detects a dirty panel automatically and triggers a mechanism to clean the panel. As the panels become dusty or exposed to bird feces, but do not accumulate enough deposits to exceed the threshold, the energy drop is within the average fluctuation energy of clean panels. The dustier the panels become, the more the energy drop occurs, and the energy output can be increased by cleaning the panels. A typical cleaning interval can range from several months to less than a day, depending on weather conditions.

In the work described in [73], they used machine vision for solar power plants. A self-inspection cleaning device, fault detection systems, and combined power units using drone platforms for multi-image fusion contaminant recognition were researched. The autonomous detection and recognition function is achieved through an image recognition analysis. To collect images, an "infrared thermal imaging camera, a color visible light camera, and a black and white visible-light camera" are used; once the acquired image has been fused, it is processed. An infrared thermal imaging camera is combined with an image camera to create a visual image for inspection purposes. A recognition algorithm analyzes the image to identify hot spots and surface impurities generating fault inside the panel and pollution notification in the control system. According to [74], four generations of outdoor soiling loss monitoring systems were developed by the authors. In the fourth-generation soiling monitoring stations, a glass shutter is opened for 2 min (or less as necessary) approximately at solar noon to take Isc measurements on the clean cell. As the glass was open, the shunt voltage of this panel was also recorded. As a result of these measurements, it is possible to determine any soil accumulation on the glass surface. Positive gains occur due to thick soil layers at around 1%/mm, whereas positive gains are typically much lower than 0.2%/mm at thin soil layers. The design allows data to be sent via mobile networks from anywhere around the globe, enabling the monitoring of multiple sites.

An approach based on computer vision was presented for detecting soil and dust on solar surfaces in [75]. To sense dirt on the solar panel, physical features are extracted through the use of the Gray Level Co-occurrence Matrix (GLCM) method. Solar panel detection is the first step of the proposed solar panel classification method. A solar surface is discovered in an image. Its background is removed at this step, and the input image is stripped of extra information. By pre-processing the image, the effects of lighting can be minimized, while at the same time, fine details that represent dust and soil can be emphasized. The red, green, blue (RGB) image is converted to hue saturation value (HSV) during this process. This is followed by feature extraction, which converts the input image into a limited number of parameters. The final step in the proposed method is classification. A flexible camera orientation was used in this study to collect two hundred images under controlled conditions with variable lighting types. One-half of the collected images show clean and dry panels, while the other half shows dirty panels. Histogram equalization and a high pass filter are used to improve image contrast. The histogram equalization technique improves a picture's disparity. Based on the results of the tested images, the proposed method has a high recognition rate. It would be helpful to consider incorporating the shadow areas, broken panels, and wet panels into the pattern recognition stage in the future.

1.4. Device Management and Performance Analysis

As a device management system [76], a smart system for cleaning and monitoring can improve or enhance the solar panel's performance. In a system failure, the analyzed data are transferred to the cloud for predictive maintenance and cause assessment via the internet. In addition to using real-time monitoring data, historical data and trends can also be used to make comparisons [77]. Analytics applications such as predictive analytics, pattern recognition, and machine learning analyze data and trigger sequenced patterns of behavior based on data filtering, normalization, and transformation [78]. The site engineers can make future decisions based on historical data stored in this way. This prevents equipment failure. It also eliminates the need to keep track of upgrades and saves time and money.

The main aim of the monitoring system for the PV power plant is to transmit the data in a reliable, secure, and efficient manner. However, several issues significantly affect the performance of various monitoring technologies in terms of efficiency, range, data processing capability, sampling rate, and signal interference. There remains a clear link between dirt monitoring and dirt cleaning, especially under varying environmental test conditions [79,80]. The performance ratio of a photovoltaic system is the proportional rate between the instantaneous power generation and its rated power generation [9]. In existence are dirt accumulation monitoring and diagnostic systems, which do not take into account the instantaneous rate of dirt accumulation. It was not possible to quantify the cost of dirt contamination as there was no specific data available. The qualitative analysis revealed in this study is vital to evaluate smart system performance in monitoring and cleaning dirt from solar panels.

In an attempt to offer a systematization of the literature about optimizing the performance of smart systems monitoring solar panel cleaning, we investigated the variables and factors effective at producing the best outcomes in such systems. Parameters for the performance evaluation of solar panel monitoring and cleaning systems include the complexity of the system (hardware and software structure), the number of sensors required to provide reliable sensory feedback, the detection time for the system after an abnormal solar panel condition, and steps involved in the monitoring and cleaning action. The most important challenge in the dirt detection process is to counter the unique operating characteristics that result due to changing environmental conditions. Using environmental sensors such as weight and irradiance sensors to measure dirt accumulation rates is more susceptible to environmental effects. The complexity of these methods is lower, and they require high computational accuracy as well. In contrast, image detection is more reliable, requires less detection time, and is reported to be more robust in detecting dirt and shading on solar panels [81].

The remaining sections of the study give the methodology associated with the systematic literature review, and the results and conclusions. In conclusion, the study summarizes the results, implications, and limitations.

2. Methods

Scopus and Google Scholar were used as the fundamental database to retrieve the data for this study. Although additional sources were cited for this article, they were not included in the review, since they were only used to clarify the background of the topic. The Scopus database provides free access to STM (Scientific, technical and medical) journal articles and the references cited in those articles, research, and collection development can both be performed with the database [82]. Search topics included "solar panel", "monitoring", and "cleaning". The terms appeared in the titles, abstracts, and keywords of the publications. The period covered spanned from 2008 to 2022. Various types of publications were indexed, and the number of publications associated with smart solar panel monitoring, and cleaning was 45. According to the document types, the majority are journal articles ($n = 25$, 55.6%) and conference papers ($n = 17$, 37.8%). Similarly, conference reviews are few ($n = 3$, 6.7%). The Scopus database contains details about every publication, such as the publication year, the authors, their addresses, the title, abstract, the journal, subject categories, and references. This set of data from Scopus was exported for the analysis of publication output and growing trends, as well as geographical and institutional distribution and collaboration. The Scopus analysis feature was used to visualize the geographical and institutional distribution and collaboration, while VOSviewer was used to analyze and visualize author relationships, co-citations, and terms. VOS represents the similarity or relatedness between items according to their distance as accurately as possible [83]. This method of clustering topics into groups was used to classify them into different groups, where each group is denoted by a different color. The results section describes in detail how the visualizations were interpreted.

2.1. Search

The research scope was formulated at the intersection of the broad terms "solar panel performance improvement" and "remote monitoring and cleaning". The search string is composed of two main parts: (i) the smart solar system; (ii) the exclusions and limitations of the search scope. The structure of the search string used in the Scopus search comprises solar AND panel AND monitoring AND cleaning.

2.2. Eligibility Criteria

An overview of the hypotheses of the current study can be found in the preregistration on Open Science Framework (osf.io/rk8yj). The screening step in the evaluation process involves a deeper analysis of the publication's full-text analysis for potential research items, which is discussed accordingly. Scientific publications in the English language, including reviews, research articles, and open access documents, were considered. Moreover, the publication duplicate was checked by the EPPI-Reviewer.

After full-text analysis, two articles were excluded because they failed to meet the inclusion criteria, or did not meet the quality requirements since this review is focused more on the smart systems for solar cleaning and its implications. A total of 55 scientific publications entered the data collection phase. This is illustrated in Figure 8.

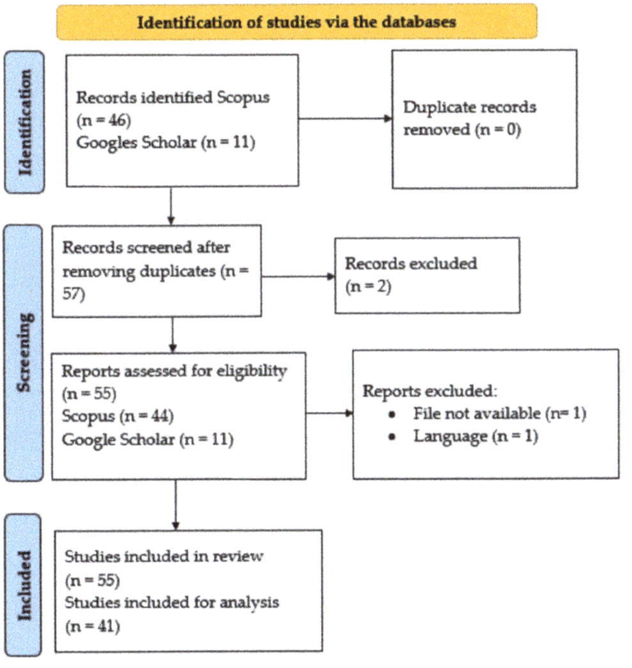

Figure 8. Process flow chart of the search.

3. Results

3.1. Overview of Selected Articles

Compared to other reviews on the monitoring and cleaning of the solar panel (e.g., [46,84]), the current review provides a relatively short bibliometric analysis. The bibliometric analysis was conducted following a systematic literature review (with PRISMA) that allowed the elimination of articles outside of the pre-defined scope (see Section 3) and work only with those within. Due to the lack of studies regarding the implementation of the Internet of Things in improving solar panel performance, the number of articles was significantly reduced during the filtering process (see Figure 8), resulting in only 41 papers for the bibliometric analysis.

3.1.1. Publication Output and Growth Trend

It is important to measure the number of publications of a scientific research discipline or subject to gauge its development trend. The number of smart systems for solar panel monitoring and cleaning publications has grown since 2008, as seen in Figure 9. In the year 2008, there was just one publication on smart systems for solar panel monitoring and cleaning. Until 2016, there were few publications on the subject (less than four publications each year). Every year since 2016 has had more than four in the number of publications, except for 2017, and 2019, when there has been a decrease. In 2020, there was a peak of publications (n = 11), followed by a downward trend (n = 4) in the first quarter of 2022.

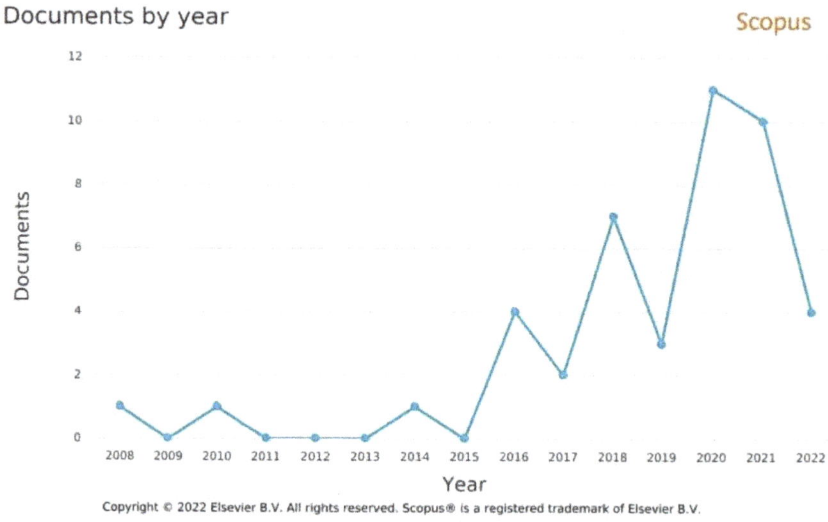

Figure 9. Publication trends related to smart systems for solar panels (between 2008 and 2022).

3.1.2. Authors and Their Collaboration

A total of 159 writers contributed to the 41 articles. Only a small set of prolific authors contributes to a considerable percentage of publications on a certain issue, which is consistent with observations in other subject areas. As shown in Figure 10, the subject areas of engineering and energy received the highest percentage of credit (22.0 percent each; approximately $n = 9/41$), while the computer science topic area accounts for 15.0 percent ($n = 6/41$). As a result of the multi-subject area publications, it can be stated that there is a lot of collaborative research in smart system technology.

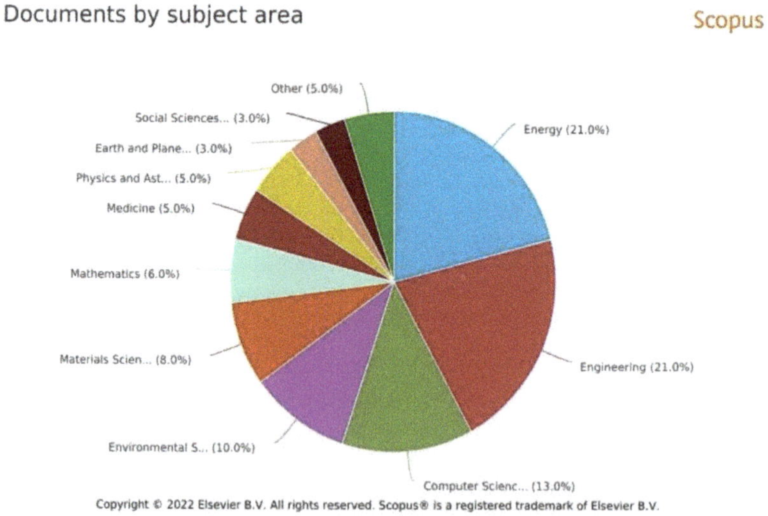

Figure 10. Analysis of subject area on smart systems for the solar panel.

One significant cluster of writers may be identified in the collaboration network in Figure 11. The average published for each of the principal researchers is in the year 2020. In terms of authorship, it is worth noting a potential bias: writers with the same name could not be separated from one another.

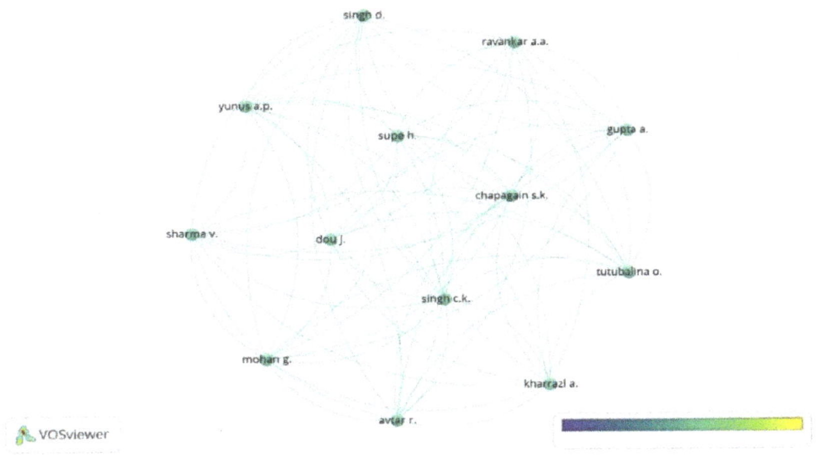

Figure 11. Author and co-citation visualization in the VOS viewer.

3.1.3. Geographical Distribution

One hundred and fifty-nine writers from various nations or territories contributed to the smart systems for solar panel publications. Elven are in India, six are in China, two are in Egypt, and one publication each came from Senegal, Morocco, and Algeria. Figure 12 depicts the global distribution of contributing countries and territories for the most productive solar panel research on smart systems technology. It is an economic investment to clean the module surfaces, but the investment must be offset by a sufficient increase in energy production [85]. Economic growth appears to encourage scientific and academic investment since the most prolific papers on smart systems for solar panel research are found in all of the world's major industrialized countries. A publication might be written by various writers from different nations or territories, or a single author can be affiliated with multiple countries or territories. When looking at the continents in Figure 12, a geographical discrepancy can be detected in the extension of the information on countries and territories. The depth of the color on the map represents the number of authors from each country.

3.2. *Integration of Smart System for Solar Panel Monitoring and Cleaning*

To keep solar panels clean, automatic connections and continuous monitoring are necessary. Smart solar monitoring and cleaning applications can overcome all of these challenges with robust and efficient cloud-based tracking systems that provide consistent and real-time monitoring from remote locations. As part of smart systems applications for solar panel cleaning, a key characteristic will be the combination of their essential functions in providing timely monitoring and device management as a solution for improving the efficiency of solar plants. Sensors and actuators would be integrated with different configurations to provide autonomous applications with a smart system that supports solar panel cleaning. Table 2 summarizes the focus of various journal papers on smart solar systems.

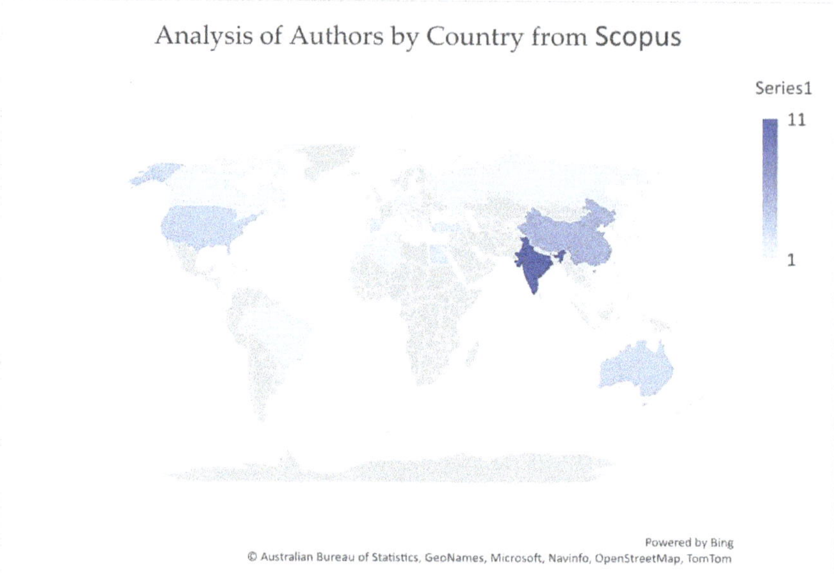

Figure 12. Geographical distribution of authors.

Table 2. Study characteristics.

S/No.	Author(s)	Year	Aim	Methodology
1.	Nalamwar H.S., et al. [64]	2017	The goal is to create and implement an IoT-based solar panel monitoring and control system.	The sensors and block management data acquisition system are placed in solar plants to collect as much data regarding the parameters of the plant as possible.
2.	Ravi K.K., and Jeswin J. [69]	2020	To explore how IoT would be implemented to monitor the various metrics related to solar panel efficiency.	A solar monitoring system that utilizes information stored in AWS to deliver actionable insights directly to clients in real-time is discussed.
3.	Abhishek P., et al. [61]	2015	To define the hardware and software required to effectively monitor solar panels continuously.	Wireless sensor nodes are used to gather and continuously store data while sending it to a central station. This collected data is then sent to the server via Ethernet.
4.	Papageorgas P., et al. [62]	2013	The purpose of this research is to provide an overview of the design process for a solar panel monitoring system.	Wired networking technologies are used to build the platform along with low-power wireless sensor nodes that have a short range. Solar panels are monitored remotely, and their performance parameters are sent to a coordinator.

Table 2. *Cont.*

S/No.	Author(s)	Year	Aim	Methodology
5.	Nur A., et al. [62]	2020	To build a solar panel dust monitoring system that accurately detects the presence and density of dust particles in real-time.	An IoT sensor was developed that could monitor dust accumulation, and the data was accessible online through smartphones and computers.
6.	Prutha M.B., et al. [65]	2018	The aim is to propose a remote wireless monitoring system to ensure the stability and efficiency of solar plants.	Sensors are used in the application for the Internet of Things (IoT), which is controlled by a CC3200 microcontroller with an ARMCortex-M4 processor and a Wi-Fi card.
7.	V. Kavitha and V. Malathi [66]	2019	To present an IoT-based solar energy monitoring system that collects and analyzes performance data to predict generation and handle unstable power generation due to environmental factors.	Blynk, as a software platform, was used together with a Wi-Fi-enabled CC3200 microcontroller featuring an embedded ARM processor. Based on this setup, the Blynk app enables real-time communication and upload of data to the cloud.
8.	Omar A., et al. [84]	2021	To analyze how solar energy systems are affected by regular operational factors and to optimize each factor's effect on the system.	A review of the main generic objectives of renewable energy system optimization was presented concerning technical, economic, social, and environmental sustainability factors.
9.	Mallikarjun G., et al. [46]	2017	To conduct a comparative study of several solar panel cleaning technologies, specifically to engineer a revolutionary new technology for dust separation using an electrostatic precipitator (ESP).	Weight-sensitive thresholds under the panel build up dust over time, and the algorithm detects the accumulation to determine whether the panel needs cleaning.
10.	Paredes-Parra, et al. [86]	2019	The goal is to provide solar power plants with a low-cost wireless system for communicating with remote areas using long-range (LoRa) technology.	The wireless communication solution is made up of the sensor layer and low power wireless area network (LPWAN) to bring a comprehensive monitoring system to the data exchange in an IoT environment.
11.	Nurhasliza H., et al. [87]	2019	To determine the most effective technology to clean solar panels, while remaining affordable and environmentally friendly.	The proposed technique uses a smartphone app to monitor power generation and clean the solar surface as needed using a robot.

Table 2. *Cont.*

S/No.	Author(s)	Year	Aim	Methodology
12.	Prasad A.A., et al. [88]	2022	To investigate a decrease in energy production by solar panels for the arid region of outer space in Australia.	Analysis of dust characteristics was performed using two reanalysis products, the Modern-Era Retrospective Analysis for Research and Applications, and the Copernicus Atmosphere Monitoring Service (CAMS), with satellite data were acquired by Himawari-8. The analysis was conducted over seasonal periods following natural sedimentation. It revealed significant reductions in the energy of up to 3%.
13.	Barker A.J., et al. [19]	2022	To apply a chemical coating to solar panels to protect the devices from sustaining damage, when the environment requires regular cleaning and disinfection.	Coatings technology was tested for glass cleaning and the incoming solar radiation was continuously monitored and logged to estimate power production capabilities and surface accumulation for each panel.
14.	Chen Y., et al. [89]	2021	To create a dust monitoring system for the increase in efficiency of solar panel generation.	The monitored real-time data include the weather, solar panel power generation, and surface images for automated aggregation by a microcontroller with transmission capabilities. Algorithms were used to process the imagery.
15.	Gupta V., et al. [90]	2022	To examine the self-cleaning of solar panels through a wireless system.	A wireless data collecting and monitoring system was used to create and test a PV system's performance. A fixed PV system with daily manual cleaning was compared to a suggested cleaning PV system for a month, and the efficiency of the proposed cleaning PV system was just 1.13 percent lower.
16.	Nattharith P., et al. [33]	2022	To create a mobile robot system for inspecting and maintaining solar panels.	An Arduino is used to control the robot while a Raspberry Pi provides an internet connection for remote users to control their cleaning system through the developed website. A webcam also gives a live stream during robot operation.
17.	Şevik S., et al. [91]	2022	To research ways on effective cleaning and maintenance of solar panels.	Thermal monitoring and snow load removal were experimented with on a connected solar panel to monitor power reduction due to dirt.

Table 2. *Cont.*

S/No.	Author(s)	Year	Aim	Methodology
18.	Sánchez-Barroso G., et al. [85]	2021	To determine the optimal period in which to clean photovoltaic panels installed at Dehesa subject to its specific environment.	Three cleaning schedules for monthly, quarterly, and semi-annual intervals were evaluated in comparison to comparable uncleaned controls.
19.	Narvios W., et al. [92]	2021	To create an Internet of Things (IoT)-based system to track, spot dust buildup, and remove dust from PV solar panel surfaces.	The dust sensor measures and detects dust on the panel. The cleaning system automatically activates when the amount of dust builds up to a certain point. The temperature and humidity sensor was used to keep track of the outside temperature.
20.	Shah M.d., et al. [93]	2021	To create a smart, Internet of Things (IoT)-based system that can spin the panel to track attributes and enable cleaning and output monitoring.	The IoT system used an Arduino Uno, a Wi-Fi module, and a smartphone to gather the data it needed for the application.
21.	Zeedan A., et al. [94]	2021	To compare output power and ambient dust density for solar panels.	Experimental data from long-term observations of various meteorological conditions and the output power of PV panels placed in Qatar University's Solar facility in Doha are used to quantify losses on solar panels.
22.	Priyadharshini N., et al. [95]	2021	To use a cooling mechanism to deal with the solar panel's temperature rising above the set point and the accumulation of dust on the panel.	Three sensors; temperature, LDR, and current are used to monitor the temperature rise and dirt on a solar panel.
23.	Pagani V.H., et al. [96]	2021	To suggest a soiling index modeling based on solar radiation and generated current for solar panel systems to establish the cleaning parameters.	The study was based on the modification of an existing mathematical model.
24.	Anilkumar G., et al. [21]	2020	To discuss potential strategies to reduce the impact of dust on the surface of solar panels.	An automated robot cleaning system was implemented using a low power wide area network (LPWAN) built on a network of ESP 8266 Node MCUs linked to a set of sensors.
25.	Jin L.C., et al. [97]	2020	To create a self-cleaning solar panel system that will increase power generation by eliminating accumulated dust from the glass surface of the panels.	The voltage, current, LDR, and IR sensors were used in the construction of the dust detecting system. The Arduino microcontroller then collected and analyzed the sensor data to launch an autonomous cleaning. For monitoring purposes, the data were also uploaded to the ThingSpeak and Blynk servers utilizing Internet of Things (IoT) technology.

Table 2. *Cont.*

S/No.	Author(s)	Year	Aim	Methodology
26.	Mohamed M., et al. [98]	2020	To achieve the optimized cleaning rate of solar panels with the least amount of energy losses and cost.	Investigated were the acquired soiling rate and cleaning PV scenarios to determine the impact of soiling density on the angle of incidence (AOI).
27.	Jaszczur M., et al. [99]	2020	To research the important factors that affect dust formation and how they interact.	Investigated were wind and rainfall as the main natural factors that affect dust buildup on solar panel surfaces and how they relate to one another.
28.	Azouzoute A., et al. [100]	2019	To research a novel method for measuring the reduction in glass transmittance on a solar panel.	The Brewster angle was utilized in the technique to assess the intensity of the reflected ray from the glass surface in the presence of various levels of dust deposition density.
29.	Arabatzis I., et al. [101]	2018	To determine how a self-cleaning solar panel with a photocatalytic and antireflective glass layer affects its effectiveness.	Utilizing UV spectroscopy and Methylene Blue's degradation as organic pollutants, respectively, the coating's optical and photocatalytic characteristics were assessed.
30.	Kama A., et al. [102]	2017	The aim is to propose a method for monitoring the performance of solar streetlights using a connected system.	A transmission mechanism and sensors are included in the proposed solar streetlight to enable real-time data collection on a distant server. The data are distributed to a Web server, where it can be viewed for monitoring reasons.
31.	Joglekar A.V., et al. [103]	2018	To propose online I-V traces for series-connected solar panels on-demand.	The I-V trace's shape analytics are utilized to identify the fault's nature and its location on a solar panel array.
32.	Nasib K., et al. [22]	2018	To demonstrate the design and construction of a solar panel cleaning prototype.	The system's prototype includes a cleaning robot and a cloud interface.
33.	Archana R., et al. [104]	2018	To develop a Smart Solar Panel Cleaning system with a primary focus on utilizing Internet of Things (IoT) technology.	The Internet of Things analyzes the solar panel's environmental factors and gives the user the ability to take appropriate action.
34.	Yousif A.A., et al. [105]	2020	To improve the efficiency of the system for cleaning and cooling solar panels.	An IoT device and a mobile application are used to remotely monitor and control the system through the internet.
35.	Mohammad A.J., et al. [106]	2015	To study and discuss the approaches and issues on solar panel dust removal.	The robot's control system is implemented using an Arduino microcontroller.
36.	Alireza G., et al. [28]	2015	To create an inexpensive automation system that can perform on-demand cleaning to maintain the effectiveness of solar panels connected in an array.	Data from individual panels were collected using wireless sensor networks. The monitored data and information are utilized to instruct a robotic device to clean the surface of dirty panels.

Table 2. *Cont.*

S/No.	Author(s)	Year	Aim	Methodology
37.	Dhanalakshmi K.S., et al. [25]	2021	To eliminate the tedious process of cleaning a solar panel using the manual method.	The autonomous robot, which is built on an Arduino platform, communicates with mobile devices through Bluetooth to remotely control the cleaning of solar panels.
38.	Neha S., et al. [107]	2018	To fully automate the maintenance of solar energy production.	Solar power generation is tracked, cleaned, and managed via an Internet of Things platform. The Raspberry Pi module serves as the processor, while the built-in Wi-Fi module facilitates data transmission to the cloud.
39.	Cova P., et al. [108]	2018	To create a model that takes into consideration the energy production losses caused by dust and other types of dirt.	Solar panels with proper measurements of current and voltage and a clean reference panel were set up to identify shading effects caused by different kinds of soiling.
40.	Mudang N., et al. [109]	2020	The goal of self-cleaning is to increase the effectiveness of solar electricity generation.	An LDR sensor detects obstructions on the solar panel, and a microcontroller determines whether to clean with Wiper and Spray water or continue charging the battery.
41.	Jaswanth Y., et al. [110]	2021	To create a technique for consistently and effectively cleaning solar panels.	The design of the cleaning robot comprises driving gear motors, a motor driver, and a second motor that powers the robot and is fitted with a cleaning membrane so that it may be washed with water. For damage and cleaning references, the camera records footage of the solar panels and sends it to the cloud.

4. Discussion and Future Prospects

The current status of smart system integration in solar panel monitoring and cleaning is summarized in this review. Through the proposed harmonized data structures, future assessments can be more efficiently planned and integrated by showing what data have already been used and what data can be used in the future. It can be especially beneficial when certain optimizations are implemented based on real-time data from a solar panel site. From this review, we have identified the following gaps and recommendations:

- Though the purpose of communication technologies and cloud platform implementation was justified in the past studies for monitoring real-time data for decision making, most do not relate to assessments of the analytical soundness, measurability, and platform deployment, as well as their linkages to one another. Furthermore, more theoretically based research is required to create reliable evidence for selecting communication technologies and implementing cloud platforms.

- The influencing score of the theoretical framework used and the impacts on the methodology are underrepresented in the smart systems for the solar panel. Research on theoretically based smart systems remains a bottleneck to future progress on smart systems for solar panel monitoring and cleaning.
- The majority of smart solar panel case studies are located in Asia, and large regions of the world possess no published assessments. Indian and Chinese institutions are the most prolific in this area.

The main gaps in the smart system for solar panel monitoring and cleaning are the optimal cleaning frequency and costs, which are yet to be proven with the monitored data. The modeling of energy output degradation is an important tool for increasing the bankability of solar plants since dirt does not accumulate uniformly over time, but rather is affected by variations in weather conditions from day to day.

5. Conclusions

In the systematic review, well-conducted studies have been shown to improve solar panel cleaning and monitoring through the inclusion of smart system integration. The findings of other reviews of smart systems for solar panels are consistent with the observation that smart systems for solar panel monitoring and maintenance are effective. The ability to visualize the solar panel dirt conditions can be instrumental in optimizing the cleaning time and operation. There were four areas of interventions our research identified: dirt detection, cleaning methods, wireless communication technologies for data gathering, and cloud platforms for IoT implementation. Currently, there is enough evidence, but more studies are needed to fill the identified knowledge gaps. Smart systems for solar panels have the potential to improve lifetime performance, reduce maintenance costs, reduce human intervention, as well as increase energy output. Ultimately, the optimal frequency and cost of cleaning must be determined with monitored data, but the evidence reviewed here can be helpful to practice, policy, and future research.

Author Contributions: Conceptualization, B.O.O. and N.N.; methodology, B.O.O. and N.N.; writing—original draft preparation, B.O.O.; writing—review and editing, B.O.O., O.A.O. and N.N.; supervision, O.A.O. and N.N. All authors have read and agreed to the published version of the manuscript.

Funding: This research received no external funding.

Institutional Review Board Statement: Not applicable.

Informed Consent Statement: Not applicable.

Data Availability Statement: Not applicable.

Conflicts of Interest: The authors declare no conflict of interest.

References

1. Shahzad, U. The Need For Renewable Energy Sources. *ITEE J. Need Renew. Energy Sources* **2012**, *2*, 1–12.
2. Ogbolumani, O.A.; Nwulu, N.I. A food-energy-water nexus meta-model for food and energy security. *Sustain. Prod. Consum.* **2021**, *30*, 438–453. [CrossRef]
3. Qazi, A.; Hussain, F.; Rahim, N.A.; Hardaker, G.; Alghazzawi, D.; Shaban, K.; Haruna, K. Towards Sustainable Energy: A Systematic Review of Renewable Energy Sources, Technologies, and Public Opinions. *IEEE Access* **2019**, *7*, 63837–63851. [CrossRef]
4. Semelane, S.; Nwulu, N.; Kambule, N.; Tazvinga, H. Economic feasibility assessment of manufacturing solar panels in South Africa—A case study of Steve Tshwete Local Municipality. *Sustain. Energy Technol. Assess.* **2021**, *43*, 100945. [CrossRef]
5. Olorunfemi, T.R.; Nwulu, N.I.; Ismail, S.O. Composites as candidate materials for photovoltaic cells. *Adv. Mater. Process. Technol.* **2022**, 1–20. [CrossRef]
6. Yang, J.; Li, X.; Peng, W.; Wagner, F.; Mauzerall, D.L. Climate, air quality and human health benefits of various solar photovoltaic deployment scenarios in China in 2030. *Environ. Res. Lett.* **2018**, *13*, 064002. [CrossRef]
7. Country Rankings. Available online: https://www.irena.org/Statistics/View-Data-by-Topic/Capacity-and-Generation/Country-Rankings (accessed on 12 July 2022).
8. Hopke, J.E.; Hestres, L.E. *Communicating about Fossil Fuel Divestment*; Oxford University Press: Oxford, UK, 2017; Volume 1. [CrossRef]

9. Fault Detection of Solar PV System Using SVM and Thermal Image Processing | Karuppiah Natarajan—Academia.edu. Available online: https://www.academia.edu/73290415/Fault_Detection_of_Solar_PV_System_Using_SVM_and_Thermal_Image_Processing (accessed on 5 July 2022).
10. Sri Venkateshwara College of Engineering. Department of Electronics and Communication Engineering, Institute of Electrical and Electronics Engineers. Bangalore Section, IEEE Computer Society, and Institute of Electrical and Electronics Engineers. In Proceedings of the 3rd IEEE International Conference on Recent Trends in Electronics, Information & Communication Technology, Karnataka, India, 18–19 May 2018.
11. Sivagami, P.; Jothiswaroopan, N.M. IOT based statistical performance improvement technique on the power output of the photovoltaic system. *J. Ambient. Intell. Humaniz. Comput.* **2020**, *12*, 5029–5043. [CrossRef]
12. Tripathi, A.K.; Aruna, M.; Ray, S.; Parida, S. Laboratory investigation of photovoltaic panel performance under the shaded condition. In Proceedings of the 1st International Conference on Power, Control and Computing Technologies, ICPC2T 2020, Chhatisgarh, India, 3–5 January 2020; pp. 273–276. [CrossRef]
13. Khan, R.A.; Farooqui, S.A.; Khan, M.H.; Sarfraz, M.; Luqman, M.; Khan, M.F. Dust Deposition on PV Module and Its Characteristics. In *The Effects of Dust and Heat on Photovoltaic Modules: Impacts and Solutions*; Green Energy and Technology; Springer: Cham, Switzerland, 2022; pp. 59–95. [CrossRef]
14. Sulaiman, S.A.; Singh, A.K.; Mokhtar, M.M.M.; Bou-Rabee, M.A. Influence of Dirt Accumulation on Performance of PV Panels. *Energy Procedia* **2014**, *50*, 50–56. [CrossRef]
15. Patil, P.A.; Bagi, J.S.; Wagh, M.M. A review on cleaning mechanism of a solar photovoltaic panel. In Proceedings of the 2017 International Conference on Energy, Communication, Data Analytics, and Soft Computing, ICECDS 2017, Tamil Nadu, India, 1–2 August 2018; pp. 250–256. [CrossRef]
16. Caron, J.R.; Littmann, B. Direct Monitoring of Energy Lost Due to Soiling on First Solar Modules in California. *IEEE J. Photovoltaics* **2012**, *3*, 336–340. [CrossRef]
17. Jamil, W.J.; Rahman, H.A.; Shaari, S.; Salam, Z. Performance degradation of photovoltaic power system: Review on mitigation methods. *Renew. Sustain. Energy Rev.* **2017**, *67*, 876–891. [CrossRef]
18. Adinoyi, M.J.; Said, S.A.M. Effect of dust accumulation on the power outputs of solar photovoltaic modules. *Renew. Energy* **2013**, *60*, 633–636. [CrossRef]
19. Barker, A.J.; Douglas, T.A.; Alberts, E.M.; IreshFernando, P.A.; George, G.W.; Maakestad, J.B.; Moores, L.C.; Saari, S.P. Influence of chemical coatings on solar panel performance and snow accumulation. *Cold Reg. Sci. Technol.* **2022**, *201*, 103598. [CrossRef]
20. Abhilash, B.; Panchal, A.K. Self-cleaning and tracking solar photovoltaic panel for improving efficiency. In Proceedings of the IEEE—2nd International Conference on Advances in Electrical, Electronics, Information, Communication and Bio-Informatics, IEEE—AEEICB 2016, Chennai, India, 27–28 February 2016; pp. 1–4. [CrossRef]
21. Anilkumar, G.; Naveen, K.; Kumar, M.T.V.S.H.K.P.; Kumar, G.V.B.; Palanisamy, K.; Patil, A.; Sachan, D. Design and development of wireless networking for solar PV panel cleaning robots. *IOP Conf. Series Mater. Sci. Eng.* **2020**, *937*, 012024. [CrossRef]
22. Khadka, N.; Bista, A.; Adhikari, B.; Shrestha, A.; Bista, D. Smart solar photovoltaic panel cleaning system. *IOP Conf. Series Earth Environ. Sci.* **2020**, *463*, 012121. [CrossRef]
23. Jiménez-Castillo, G.; Muñoz-Rodríguez, F.J.; Rus-Casas, C.; Gómez-Vidal, P. Improvements in Performance Analysis of Photovoltaic Systems: Array Power Monitoring in Pulse Width Modulation Charge Controllers. *Sensors* **2019**, *19*, 2150. [CrossRef]
24. Prasanna, J.L.; Lavanya, D.; Kumar, T.A. Condition monitoring of a virtual solar system using IoT. In Proceedings of the 2nd International Conference on Communication and Electronics Systems, ICCES 2017, Coimbatore, India, 19–20 October 2017; pp. 286–290. [CrossRef]
25. Dhanalakshmi, K.S.; Raj, S.M.; Kumar, K.S.; Keerthivash, R. Solar Panel Cleaning Robot Using Wireless Communication. 2021. Available online: http://annalsofrscb.ro (accessed on 14 June 2022).
26. Ejgar, M.; Momin, B.; Ganu, T. Intelligent monitoring and maintenance of solar plants using real-time data analysis. In Proceedings of the 2017 IEEE International Conference on Consumer Electronics-Asia, ICCE-Asia 2017, Bengaluru, India, 5–7 October 2017; pp. 133–138. [CrossRef]
27. Chakaborty, S.; Borole, Y.D.; Nanoty, A.S.; Shrivastava, A.; Jain, S.K.; Rinawa, M.L. Smart Remote Solar Panel Cleaning Robot with Wireless Communication. In Proceedings of the 2021 9th International Conference on Cyber and IT Service Management (CITSM), Bengkulu, Indonesia, 22–23 September 2021; pp. 1–5. [CrossRef]
28. Gheitasi, A.; Almaliky, A.; Albaqawi, N. Development of an automatic cleaning system for photovoltaic plants. In Proceedings of the Asia-Pacific Power and Energy Engineering Conference, APPEEC, Brisbane, Australia, 15–18 November 2015. [CrossRef]
29. Derakhshandeh, J.F.; AlLuqman, R.; Mohammad, S.; AlHussain, H.; Al Hendi, G.; Al Eid, D.; Ahmad, Z. A comprehensive review of automatic cleaning systems of solar panels. *Sustain. Energy Technol. Assess.* **2021**, *47*, 101518. [CrossRef]
30. Smith, M.K.; Wamser, C.C.; James, K.E.; Moody, S.; Sailor, D.J.; Rosenstiel, T.N. Effects of Natural and Manual Cleaning on Photovoltaic Output. *J. Sol. Energy Eng.* **2013**, *135*, 1–4. [CrossRef]
31. Optimal Cleaning Strategy of Large-Scale Solar PV Arrays Considering Non-Uniform Dust Deposition. Available online: https://repository.up.ac.za/handle/2263/79656 (accessed on 11 July 2022).
32. Parrott, B.; Aramco, S. Automated, Robotic Dry-Cleaning of Solar Panels in Thuwal, Saudi Arabia using a Silicone Rubber Brush Photocatalytic Water Splitting Device For Solar Power Generation View project Robotic Dust Mitigation in Saudi Arabia View project. 2018. Available online: https://www.researchgate.net/publication/326266398 (accessed on 11 July 2022).

33. Mondal, A.K.; Bansal, K. A brief history and future aspects in automatic cleaning systems for solar photovoltaic panels. *Adv. Robot.* **2015**, *29*, 515–524. [CrossRef]
34. Dahlioui, D.; Alaoui, S.M.; Laarabi, B. Waterless cleaning technique for photovoltaic panels on dual-axis tracker. *Res. Sq.* **2022**, in press. [CrossRef]
35. Thongsuwan, W.; Sroila, W.; Kumpika, T.; Kantarak, E.; Singjai, P. Antireflective, photocatalytic, and superhydrophilic coating prepared by facile sparking process for photovoltaic panels. *Sci. Rep.* **2022**, *12*, 1–15. [CrossRef]
36. Williams, R.B.; Tanimoto, R.; Simonyan, A.; Fuerstenau, S. Vibration Characterization of Self-Cleaning Solar Panels with Piezoceramic Actuation. In *Collection of Technical Papers—AIAA/ASME/ASCE/AHS/ASC Structures, Structural Dynamics and Materials Conference*; American Institute of Aeronautics and Astronautics Inc. (AIAA): Reston, VA, USA, 2007; Volume 1, pp. 512–520. [CrossRef]
37. Li, D.; King, M.; Dooner, M.; Guo, S.; Wang, J. Study on the cleaning and cooling of solar photovoltaic panels using compressed airflow. *Sol. Energy* **2021**, *221*, 433–444. [CrossRef]
38. Manju, B.; Bari, A.; Pavan, C.M. Automatic Solar Panel Cleaning System. *Int. J. Adv. Sci. Res. Eng.* **2018**, *4*, 26–31. [CrossRef]
39. Ronnaronglit, N.; Maneerat, N. A Cleaning Robot for Solar Panels. In Proceedings of the 5th International Conference on Engineering, Applied Sciences and Technology, ICEAST 2019, Luang Prabang, Laos, 2–5 July 2019. [CrossRef]
40. Myeong, W.; Myung, H. Development of a Wall-Climbing Drone Capable of Vertical Soft Landing Using a Tilt-Rotor Mechanism. *IEEE Access* **2019**, *7*, 4868–4879. [CrossRef]
41. Nattharith, P.; Kosum, T. Development of Mobile Robot System for Monitoring and Cleaning of Solar Panels. *GMSARN Int. J.* **2022**, *16*, 302–306.
42. Robot with Brush, Water, Wiper Tackles Solar Panel Cleaning. Available online: https://phys.org/news/2013-12-robot-wiper-tackles-solar-panel.html (accessed on 6 August 2022).
43. SKR Engineering College, Institute of Electrical and Electronics Engineers. Madras Section, and Institute of Electrical and Electronics Engineers. In Proceedings of the International Conference on Energy, Communication, Data Analytics & Soft Computing (ICECDS)-2017, Chennai, India, 1–2 August 2017.
44. Ekinci, F.; Yavuzdeğer, A.; Nazlıgül, H.; Esenboğa, B.; Mert, B.D.; Demirdelen, T. Experimental investigation on solar PV panel dust cleaning with solution method. *Sol. Energy* **2022**, *237*, 1–10. [CrossRef]
45. Heliotex's Automatic Solar Panel Cleaning System. Available online: https://newatlas.com/heliotex-solar-panel-cleaning/11009/ (accessed on 6 August 2022).
46. Hudedmani, M.G.; Joshi, G.; Umayal, R.M.; Revankar, A. A Comparative Study of Dust Cleaning Methods for the Solar PV Panels. *Adv. J. Grad. Res.* **2017**, *1*, 24–29. [CrossRef]
47. Al-Housani, M.; Bicer, Y.; Koç, M. Assessment of Various Dry Photovoltaic Cleaning Techniques and Frequencies on the Power Output of CdTe-Type Modules in Dusty Environments. *Sustainability* **2019**, *11*, 2850. [CrossRef]
48. Kawamoto, H.; Shibata, T. Electrostatic cleaning system for removal of sand from solar panels. *J. Electrost.* **2015**, *73*, 65–70. [CrossRef]
49. Kawamoto, H. Electrostatic cleaning equipment for dust removal from soiled solar panels. *J. Electrost.* **2019**, *98*, 11–16. [CrossRef]
50. Valerino, M.; Bergin, M.; Ghoroi, C.; Ratnaparkhi, A.; Smestad, G.P. Low-cost solar PV soiling sensor validation and size resolved soiling impacts: A comprehensive field study in Western India. *Sol. Energy* **2020**, *204*, 307–315. [CrossRef]
51. Sarkın, A.S.; Ekren, N.; Sağlam, S. A review of anti-reflection and self-cleaning coatings on photovoltaic panels. *Sol. Energy* **2020**, *199*, 63–73. [CrossRef]
52. Virtanen, S. Self-Cleaning Technologies on Solar Modules and Their Effect on Performance. Bachelor's Thesis, JAMK University of Applied Sciences, Jyväskylä, Finland, December 2017.
53. Syafiq, A.; Balakrishnan, V.; Ali, M.S.; Dhoble, S.J.; Rahim, N.A.; Omar, A.; Abu Bakar, A.H. Application of transparent self-cleaning coating for photovoltaic panel: A review. *Curr. Opin. Chem. Eng.* **2022**, *36*, 100801. [CrossRef]
54. Attia, O.H.; Adam, N.M.; As'Arry, A.; Rezali, K.A. Removal of Dust from the Solar Panel Surface using Mechanical Vibrator. *J. Physics Conf. Ser.* **2019**, *1262*, 012021. [CrossRef]
55. Vibration Tech for Self-Cleaning Solar Panels—PV Magazine International. Available online: https://www.pv-magazine.com/2020/04/21/vibration-tech-for-self-cleaning-solar-panels/ (accessed on 6 August 2022).
56. Minu, K.S. Self-Cleaning of Solar Photovoltaic A case Study for Improving the Performance of Solar Photovoltaic. Available online: http://troindia.in/journal/ijcesr/vol2iss7/188-191.pdf (accessed on 14 June 2022).
57. Shah, A.; Hassan, A.; Laghari, M.; Alraeesi, A. The Influence of Cleaning Frequency of Photovoltaic Modules on Power Losses in the Desert Climate. *Sustainability* **2020**, *12*, 9750. [CrossRef]
58. Saravanan, V.S.; Darvekar, S.K. Solar Photovoltaic Panels Cleaning Methods A Review. 2018. Available online: http://www.acadpubl.eu/hub/ (accessed on 14 June 2022).
59. Khan, M.U.; Abbas, M.; Khan, M.M.; Kousar, A.; Alam, M.; Massoud, Y.; Jafri, S.H.M. Modeling and design of low-cost automatic self cleaning mechanism for standalone micro PV systems. *Sustain. Energy Technol. Assess.* **2021**, *43*, 100922. [CrossRef]
60. International Conference on Devices, Circuits and Systems. In *CDCS'16: Proceedings of the 3rd International Conference on Circuits and Systems (ICDCS)—2016: 3rd to 5th—March 2016, Department of Electronics and Communication Engineering, School of Electrical Sciences*; IEEE: Piscataway, NJ, USA, 2016.

61. Parikh, A.; Pathan, F.; Rathod, B.; Shah, S. Solar Panel Condition Monitoring System based on Wireless Sensor Network Covid 19 visualization View project Hardware for Hand Electromyographic signal analysis and processing View project Solar Panel Condition Monitoring System based on Wireless Sensor Network. *Int. J. Sci. Eng. Technol. Res.* **2015**, *4*, 4320–4324. [CrossRef]
62. Papageorgas, P.; Piromalis, D.; Antonakoglou, K.; Vokas, G.; Tseles, D.; Arvanitis, K. Smart Solar Panels: In-situ Monitoring of Photovoltaic Panels based on Wired and Wireless Sensor Networks. *Energy Procedia* **2013**, *36*, 535–545. [CrossRef]
63. Onibonoje, M.O.; Nwulu, N.I.; Bokoro, P.N. A wireless sensor network system for monitoring environmental factors affecting bulk grains storability. *J. Food Process Eng.* **2019**, *42*, e13256. [CrossRef]
64. Nalamwar, H.S.; Ivanov, M.A.; Baidali, S.A. Automated Intelligent Monitoring and the Controlling Software System for Solar Panels. *J. Physics Conf. Ser.* **2017**, *803*, 012107. [CrossRef]
65. Badave, P.M.; Karthikeyan, B.; Badave, S.M.; Mahajan, S.B.; Sanjeevikumar, P.; Gill, G.S. Health Monitoring System of Solar Photovoltaic Panel: An Internet of Things Application. In *Advances in Smart Grid and Renewable Energy;* LNEE Volume 435; Springer: Singapore, 2017; pp. 347–355. [CrossRef]
66. Kavitha, V.; Malathi, V. A Smart Solar PV Monitoring System Using IoT. *Comput. Sci. Inf. Technol.* **2019**, *9*, 15. [CrossRef]
67. Nwulu, N.I.; Xia, X. A combined dynamic economic emission dispatch and time of use demand response mathematical modelling framework. *J. Renew. Sustain. Energy* **2015**, *7*, 043134. [CrossRef]
68. Nwulu, N.I.; Fahrioglu, M. A soft computing approach to projecting locational marginal price. *Neural Comput. Appl.* **2013**, *22*, 1115–1124. [CrossRef]
69. Kodali, R.K.; John, J. Smart Monitoring of Solar Panels Using AWS. In Proceedings of the 2020 International Conference on Power Electronics and IoT Applications in Renewable Energy and its Control, PARC 2020, Mathura, India, 28–29 February 2020; pp. 422–427. [CrossRef]
70. Márquez, F.P.G.; Ramírez, I.S. Condition monitoring system for solar power plants with radiometric and thermographic sensors embedded in unmanned aerial vehicles. *Meas. J. Int. Meas. Confed.* **2019**, *139*, 152–162. [CrossRef]
71. IEEE Control Systems Society. Chapter Malaysia and Institute of Electrical and Electronics Engineers. In Proceedings of the 2019 IEEE International Conference on Automatic Control and Intelligent Systems (I2CACIS 2019), Shah Alam, Malaysia, 29 June 2019.
72. Yfantis, E.; Fayed, A. A Camera System For Detecting Dust And Other Deposits On Solar Panels. *Adv. Image Video Process.* **2014**, *2*, 1–10. [CrossRef]
73. He, Z.; Zhang, Y.; Li, H. Self-inspection cleaning device for photovoltaic power plant based on machine vision. *IOP Conf. Series Earth Environ. Sci.* **2019**, *242*, 032020. [CrossRef]
74. IEEE Electron Devices Society; IEEE Photonics Society; Institute of Electrical and Electronics Engineers. *2018 IEEE 7th World Conference on Photovoltaic Energy Conversion (WCPEC 2018) (A Joint Conference of 45th IEEE PVSC, 28th PVSEC & 34th EU PVSEC), Waikoloa Village, HI, USA, 10–15 June 2018;* IEEE: Piscataway, NJ, USA, 2018.
75. Abuqaaud, K.A.; Ferrah, A. A Novel Technique for Detecting and Monitoring Dust and Soil on Solar Photovoltaic Panel. In Proceedings of the 2020 Advances in Science and Engineering Technology International Conferences (ASET), Dubai, United Arab Emirates, 4 February–9 April 2020.
76. Nwulu, N.I.; Fahrioglu, M. Power system demand management contract design: A comparison between game theory and artificial neural networks. *Int. Rev. Model. Simul.* **2011**, *4*, 104–112.
77. Kambule, N.; Yessoufou, K.; Nwulu, N. A review and identification of persistent and emerging prepaid electricity meter trends. *Energy Sustain. Dev.* **2018**, *43*, 173–185. [CrossRef]
78. Dogo, E.M.; Nwulu, N.I.; Twala, B.; Aigbavboa, C. A survey of machine learning methods applied to anomaly detection on drinking-water quality data. *Urban Water J.* **2019**, *16*, 235–248. [CrossRef]
79. Ikejemba, E.C.X.; Mpuan, P.B.; Schuur, P.C.; Van Hillegersberg, J. The empirical reality & sustainable management failures of renewable energy projects in Sub-Saharan Africa (part 1 of 2). *Renew. Energy* **2016**, *102*, 234–240. [CrossRef]
80. Gupta, V.; Sharma, M.; Pachauri, R.; Babu, K.N.D. Design and development of self-cleaning PV sliding system. *Clean Energy* **2022**, *6*, 392–403. [CrossRef]
81. Winston, D.P.; Ganesan, K.; Kumar, B.P.; Samithas, D.; Baladhanautham, C.B. Experimental investigation on output power enhancement of partial shaded solar photovoltaic system. *Energy Sources Part A Recovery Util. Environ. Eff.* **2020**, 1–17. [CrossRef]
82. Burnham, J.F. Scopus database: A review. *Biomed. Digit. Libr.* **2006**, *3*, 1–8. [CrossRef]
83. McAllister, J.T.; Lennertz, L.; Mojica, Z.A. Mapping A Discipline: A Guide to Using VOSviewer for Bibliometric and Visual Analysis. *Sci. Technol. Libr.* **2021**, 1–30. [CrossRef]
84. Al-Shahri, O.A.; Ismail, F.B.; Hannan, M.; Lipu, M.H.; Al-Shetwi, A.Q.; Begum, R.; Al-Muhsen, N.F.; Soujeri, E. Solar photovoltaic energy optimization methods, challenges and issues: A comprehensive review. *J. Clean. Prod.* **2020**, *284*, 125465. [CrossRef]
85. Sánchez-Barroso, G.; González-Domínguez, J.; García-Sanz-Calcedo, J.; Sanz, J.G. Markov chains estimation of the optimal periodicity for cleaning photovoltaic panels installed in the dehesa. *Renew. Energy* **2021**, *179*, 537–549. [CrossRef]
86. Paredes-Parra, J.M.; Jiménez-Segura, R.; Campos-Peñalver, D.; Mateo-Aroca, A.; Ramallo-González, A.P.; Molina-García, A. Democratization of PV Micro-Generation System Monitoring Based on Narrowband-IoT. *Sensors* **2022**, *22*, 4966. [CrossRef]
87. Hashim, N.; Mohammed, M.N.; AL Selvarajan, R.; Al-Zubaidi, S.; Mohammed, S. Study on Solar Panel Cleaning Robot. In Proceedings of the 2019 IEEE International Conference on Automatic Control and Intelligent Systems, I2CACIS 2019, Selangor, Malaysia, 29 June 2019; pp. 56–61. [CrossRef]

88. Prasad, A.A.; Nishant, N.; Kay, M. Dust cycle and soiling issues affecting solar energy reductions in Australia using multiple datasets. *Appl. Energy* **2022**, *310*, 118626. [CrossRef]
89. Chen, Y.; Ma, J.; Ma, Y.; Zhang, Y.; Zhou, X.; Zhou, R. The Design and Implementation of Dust Monitoring System for Photovoltaic Power Generation. In Proceedings of the IEEE 23rd International Conference on High Performance Computing & Communications, Hainan, China, 20–22 December 2021. [CrossRef]
90. Gupta, V.; Sharma, M.; Pachauri, R.; Babu, K.N.D. Performance Analysis of Solar PV system using Customize wireless data acquisition system and novel cleaning technique. *Energy Sources Part A Recover. Util. Environ. Eff.* **2022**, *44*, 2748–2769. [CrossRef]
91. Şevik, S.; Aktaş, A. Performance enhancing and improvement studies in a 600 kW solar photovoltaic (PV) power plant; manual and natural cleaning, rainwater harvesting and the snow load removal on the PV arrays. *Renew. Energy* **2021**, *181*, 490–503. [CrossRef]
92. Narvios, W.M.O.; Nguyen, Y.Q. IoT based detection, monitoring and automatic cleaning system for soiling of PV solar panel. *AIP Conf. Proc.* **2021**, *2406*, 060005. [CrossRef]
93. Shah, M.A.; Parvez, S.; Ahmed, A.; Hazari, R. IoT Based Power Monitoring of Solar Panel Incorporating Tracking System. In Proceedings of the 2021 International Conference on Automation, Control and Mechatronics for Industry 4.0, ACMI 2021, Rajshahi, Bangladesh, 8–9 July 2021. [CrossRef]
94. Zeedan, A.; Barakeh, A.; Al-Fakhroo, K.; Touati, F.; Gonzales, A. Quantification of PV Power and Economic Losses Due to Soiling in Qatar. *Sustainability* **2021**, *13*, 3364. [CrossRef]
95. Priyadharshini, N.; Kumar, S.H.; Gomathy, S.; Krithik, S.; Elakiya, V.; Shubhashree, S. A Review on Solar Panel Monitoring and Enhancing the Performance by Temperature and Dust Maintenance. In Proceedings of the 3rd IEEE International Virtual Conference on Innovations in Power and Advanced Computing Technologies, i-PACT 2021, Kuala Lumpur, Malaysia, 27–29 November 2021. [CrossRef]
96. Pagani, V.H.; Los, N.A.; Maidana, W.; Leitão, P.; Casaro, M.M.; Nascimento, C.B. Soiling Monitoring Modelling for Photovoltaic System. In *CONTROLO 2020*; Lecture Notes in Electrical Engineering; Springer: Cham, Switzerland, 2021; Volume 695, pp. 592–601. [CrossRef]
97. Jin, L.C.; Yusop, R.; Ratnadurai, D. Autonomous solar panels dry cleaning system for dust removal using microcontroller and sensors. *Int. J. Adv. Sci. Technol.* **2020**, *29*, 141–152.
98. Mohamed, M.; Attia, M.A.; Asim, A.M.; Abdelaziz, A.Y.; Kanwar, N. Optimization of cleaning frequency and dust accumulation effect on photovoltaic panels. *J. Interdiscip. Math.* **2020**, *23*, 53–68. [CrossRef]
99. Jaszczur, M.; Koshti, A.; Nawrot, W.; Sędor, P. An investigation of the dust accumulation on photovoltaic panels. *Environ. Sci. Pollut. Res.* **2019**, *27*, 2001–2014. [CrossRef]
100. Azouzoute, A.; Chouitar, M.; Garoum, M.; Bennouna, E.G.; Ghennioui, A. A new PV soiling monitoring device for optimized cleaning strategy. *AIP Conf. Proc.* **2019**, *2190*, 020068. [CrossRef]
101. Arabatzis, I.; Todorova, N.; Fasaki, I.; Tsesmeli, C.; Peppas, A.; Li, W.X.; Zhao, Z. Photocatalytic, self-cleaning, antireflective coating for photovoltaic panels: Characterization and monitoring in real conditions. *Sol. Energy* **2018**, *159*, 251–259. [CrossRef]
102. Kama, A.; Diallo, M.; Drame, M.S.; Ndiaye, M.L.; Ndiaye, A.; Ndiaye, P.A. Monitoring the Performance of Solar Street Lights in Sahelian Environment: Case Study of Senegal. In Proceedings of the International Conference on Developments in eSystems Engineering, DeSE, Paris, France, 14–16 June 2017; pp. 56–61. [CrossRef]
103. Joglekar, A.V.; Hegde, B. Online I-V Tracer for per string monitoring and maintenance of PV panels. In Proceedings of the IECON 2018—44th Annual Conference of the IEEE Industrial Electronics Society, Washington, DC, USA, 21–23 October 2018; pp. 1890–1894. [CrossRef]
104. Rao, A.; Neha; Kumar, S.S. Smart IoT based Solar Panel Cleaning System. In Proceedings of the 2018 3rd IEEE International Conference on Recent Trends in Electronics, Information and Communication Technology, RTEICT 2018, Bangalore, India, 18–19 May 2018; pp. 2357–2361. [CrossRef]
105. Al-Wajih, Y.A.; Mysorewala, M.F.; Mohammed, A.A.; Mostafa, S. Improving the Performance of Solar Panels by Automated Water Cooling System and IoT. In Proceedings of the 2020 1st International Conference of Smart Systems and Emerging Technologies, SMART-TECH 2020, Riyadh, Saudi Arabia, 3–5 November 2020; pp. 134–137. [CrossRef]
106. Jaradat, M.A.; Tauseef, M.; Altaf, Y.; Saab, R.; Adel, H.; Yousuf, N.; Zurigat, Y.H. A fully portable robot system for cleaning solar panels. In Proceedings of the ISMA 2015—10th International Symposium on Mechatronics and its Applications, Sharjah, United Arab Emirates, 8–10 December 2015. [CrossRef]
107. Deshmukh, N.S.; Bhuyar, D.L. A Smart Solar Photovoltaic Remote Monitoring and Controlling. In Proceedings of the 2018 Second International Conference on Intelligent Computing and Control Systems (ICICCS), Madurai, India, 14–15 June 2018; pp. 67–71. [CrossRef]
108. Cova, P.; Delmonte, N.; Lazzaroni, M. Photovoltaic plant maintainability optimization and degradation detection: Modelling and characterization. *Microelectron. Reliab.* **2018**, *88–90*, 1077–1082. [CrossRef]

109. Nado, M.; Taggu, A. Intelligent Automated Smart Solar Panel Using Internet of Things. In Proceedings of the 5th International Conference on Computers & Management Skills (ICCM 2019) | North Eastern Regional Institute of Science & Technology (NERIST), Nirjuli, Arunachal Pradesh, India, 6 January 2020. [CrossRef]
110. Yerramsetti, J.; Paritala, D.S.; Jayaraman, R. Design and Implementation of Automatic Robot for Floating Solar Panel Cleaning System using AI Technique. In Proceedings of the 2021 International Conference on Computer Communication and Informatics, ICCCI 2021, Coimbatore, India, 27–29 January 2021. [CrossRef]

applied sciences

MDPI

Article

Intelligent Deep-Q-Network-Based Energy Management for an Isolated Microgrid

Bao Chau Phan [1], Meng-Tse Lee [2] and Ying-Chih Lai [1,3,*]

1 Department of Aeronautics and Astronautics, College of Engineering, National Cheng Kung University, Tainan 701, Taiwan
2 Department of Automation Engineering, College of Engineering, National Formosa University, Yunlin 632, Taiwan
3 Institute of Civil Aviation, National Cheng Kung University, Tainan 701, Taiwan
* Correspondence: yingclai@mail.ncku.edu.tw; Tel.: +886-6-275-7575 (ext. 63648)

Abstract: The development of hybrid renewable energy systems (HRESs) can be the most feasible solution for a stable, environment-friendly, and cost-effective power generation, especially in rural and island territories. In this studied HRES, solar and wind energy are used as the major resources. Moreover, the electrolyzed hydrogen is utilized to store energy for the operation of a fuel cell. In case of insufficiency, battery and fuel cell are storage systems that supply energy, while a diesel generator adds a backup system to meet the load demand under bad weather conditions. An isolated HRES energy management system (EMS) based on a Deep Q Network (DQN) is introduced to ensure the reliable and efficient operation of the system. A DQN can deal with the problem of continuous state spaces and manage the dynamic behavior of hybrid systems without exact mathematical models. Following the power consumption data from Basco island of the Philippines, HOMER software is used to calculate the capacity of each component in the proposed power plant. In MATLAB/Simulink, the plant and its DQN-based EMS are simulated. Under different load profile scenarios, the proposed method is compared to the convectional dispatch (CD) control for a validation. Based on the outstanding performances with fewer fuel consumption, DQN is a very powerful and potential method for energy management.

Keywords: hybrid renewable energy system (HRES); isolated microgrid; energy management system (EMS); Deep Q Network (DQN); HOMER software

Citation: Phan, B.C.; Lee, M.-T.; Lai, Y.-C. Intelligent Deep-Q-Network-Based Energy Management for an Isolated Microgrid. *Appl. Sci.* **2022**, *12*, 8721. https://doi.org/10.3390/app12178721

Academic Editors: Luis Hernández-Callejo, Sara Gallardo Saavedra and Sergio Nesmachnow

Received: 30 July 2022
Accepted: 27 August 2022
Published: 31 August 2022

1. Introduction

The worldwide increase in energy demand leads to the consideration of using renewable energy types such as solar, wind, tidal, and geothermal. Currently, fossil fuels are still the major reliable power sources especially for rural and island electrification. On the other hand, fossil fuel price is constantly increasing, and fossil fuels are responsible for global environmental pollution. Consequently, many countries have recently opted for the long-term sustainable development of renewable energy. By 2025, the Ministry of Economic Affairs (Taiwan) aims at increasing the share of renewable energy to 20% within the total power generation, as well as phasing out nuclear energy. Several developing countries such as Philippines, Thailand, and Vietnam have changed their power development plan based on green energy. We consider them some of the most typical countries for the deployment of renewable energy power plants [1].

The recent development of solar and wind energy has recently been considered because of the available amount of solar radiation and wind distribution. These energy types are environment-friendly and cost effective, but unpredictable and uncontrollable as well due to the significant dependence on weather conditions. In order to improve the operational ability and efficiency of these power systems, the concept of a hybrid renewable energy system (HRES) was created [2]. In terms of power generation for rural and island areas, HRES is more cost effective than a grid extension. Depending on the distance from a power station, a grid extension can range from 10,000 to 50,000 USD per kilometer [3].

In a HRES, the combination for sustainable and reliable power supply of renewable energy resources, energy storage systems (ESSs), and diesel generators (DGs) can create economic, technical, environmental, and social benefits to investors. The role of ESSs is to store the excess energy from renewable energy sources. DGs can be operated when both renewable energy resources and ESSs are out of power. The configuration and topology of a hybrid HRES system can vary in several ways. The most generic classification includes on-grid and off-grid systems. According to the bus interconnection or the physical link between all components, the system can be classified as DC, AC, or hybrid DC/AC [4]. To ensure a high level of system reliability and operational efficiency, energy management algorithms are needed to manage the power flow inside the system. In particular, this algorithm has to allow for the variation of load demand and the system complexity.

Energy management system (EMS) is one of the most important components of the HRES. The main function of EMS is to balance power between the system components reducing the amount of fossil fuel used for power generation. The EMS control can be classical and intelligent [4,5]. Classical EMS is based on linear, nonlinear, or dynamic programming [6]. We can also find rule-based and flowchart methods [7]. More latest classical EMS controllers are based on proportional-integral controller [8], sliding mode controller [9], and H-infinity controller [10]. Classical EMS, which may require complicated mathematical models with various system variables, has low computational complexity. Compared to classic EMS, the intelligent one seems to be more robust and more efficient. Examples include the fuzzy logic (FL) [11], the artificial neural network controller (ANN), the Neural-Fuzzy controller (ANFIS) [12], and a model predictive controller (MPC). In addition, evolutionary algorithms-based EMS methods have been also developed, such as the Particle Swarm Optimization (PSO), the Genetic Algorithm [13], and the Modified Bat Algorithm (MBA) [14]. Recently, machine learning has been applied for EMS such as support vector machine (SVM) [15]. Among these intelligent EMS methods fuzzy logic, neural network, and ANFIS are definitely popular.

Different from classical EMS, based on the intelligent EMS, simple mathematical models are required to manage hybrid system dynamic behaviors. However, the current forms of these methods are still not able to guarantee better performance of optimal control [15]. Over time, a lot of hybrid studies have been conducted to enhance the global optimal solutions and the convergence speed. The major purpose is to find the action that optimizes the value of an objective function. In [16], a method, named as PF3SACO, was developed to improve the optimization ability and convergence speed, in which PSO and fuzzy are used to adjust system parameters. In [17], to adapt to complex scenes, the author proposed a robust tracking method based on a feature weight pool that has multiple weights for different features. In [18], a variable neighborhood search and non-dominated sorting genetic algorithm II (VNS-NSGA-II) were applied to optimally solve the routing problem with multiple time windows. In [19], a principal component analysis (PCA), a local binary pattern (LBP), and a gray wolf algorithm were combined to optimize the parameters of kernel extreme learning machine (KELM) for image classification. It can be confirmed that these hybrid methods are powerful in solving a complex optimal problem, especially since they can be used to optimize the parameters of machine-learning-based approaches. However, they would heavily depend on complex mathematical models and computational complexity.

More studies on agent-based machine learning methods for hybrid EMS have been conducted recently, such as deep learning (DL) and deep reinforcement learning (DRL) [20,21]. Instead of using a complex mathematical control model, these agent-based approaches can manage the system by learning the control policy from the environmental-interacting historical data, leading to a potential solution to energy management problems. Following the concepts of RL and DRL, the control purpose is to obtain the maximum rewards by continuously interacting with the system environment. Based on exploration-exploitation strategies, such as -greedy or softmax, the action with highest reward is taken [22]. Q-learning is a popular model-free RL algorithm. However, RL-based methods can only handle discrete control problems which may be hard to implement in practical applications. DRL-based methods combine RL with deep learning to handle continuous control problems with large state-action pair. DRL has successfully been implemented to play Go games and Atari [23]. It should be a powerful method to handle the problems of complex optimal control with large state spaces by using a deep neural network. It can also be applied in robotics [23], control of building HVAC [24], and hybrid electric cars [25].

Up to now, studies about the application of RL and DRL for energy management of a stand-alone microgrid are not common. A self-learning single neural network was proposed by Huang and Liu (2013) for EMS residential applications [26]. A two-step ahead Q-learning method was defined by Kuznetsova (2013) for scheduling the operation of the battery in a wind system. In [27], a three-step ahead Q-learning method was used to schedule battery operation a solar energy system. A Q-learning-based multi-agent for a solar system was developed in [28] to reduce the amount of energy consumption. Based on an autonomous multi-agent system in [29], it can manage RE buying and selling optically. In [30], authors proposed a multi-agent system to monitor energy generation and consumption. A Q-learning single agent system was applied to manage a solar energy system by Kofinas (2016) [31]. In [32], a Q-learning algorithm and a fuzzy reward function were introduced to improve system performance. It intends to learn about the power flow between the components of the solar system more efficiently, which includes a photovoltaic (PV), a battery, load demand, and a desalination unit (for water supply). Later, Kofinas (2018) [33] proposed a cooperative fuzzy Q-learning-based multi-agent system for the energy management of a stand-alone microgrid. The latter system included a PV, a fuel cell, a diesel generator, an electrolyzer, a hydrogen tank, battery, and a desalination plant. Each component was represented by an agent. Each agent acted as an individual learner and interacted with other agents. The simulation results from MATLAB/Simulink indicated that the controller could continuously maintain state and action space. The learning of each agent took place through exploration/exploitation with fast convergence towards a policy and with good performance. In [33], the author used fuzzy logic as the function approximation for determining the Q-values. Similar to the above approach, deep Q-learning (DQN) applies a neural network to calculate the Q-values in order to increase the learning capacity of agents. In [25], deep Q-learning was applied for the energy management of a hybrid electric vehicle. The DRL-based controller acted autonomously to learn an optimal policy without using any prediction or predefined rule.

The main goal of this study is to propose a DQN algorithm for the energy management of an isolated HRES and to present a case study about an HRES conducted at Basco island of the Philippines. It is the extended study of our previous work, which developed a DRL-based controller to track the maximum power point for PV systems under various weather and partial shading conditions [34]. DL and DRL are widely used in robotics and autonomous; however, only a few studies are about DRL application in an HRES for energy management. Thus, the advantage and novelty of this study is the application of DQN-based EMS for rural and island areas, in which the system includes battery, DG, and hydrogen system, as well as a case study with practical load demand data. The adopted power system in this study consists of a PV system, a wind turbine (WT), a battery, a DG, a fuel cell, an electrolyzer, and a hydrogen tank. Based on weather data and load demand at the applied site, we used HOMER software for determining the structure of the HRES.

The major contributions of this paper are described below:

- The implementation and simulation of a DQN-based EMS conducted based on Reinforcement Learning Toolbox of MATLAB/Simulink R2021a developed by MathWorks®.
- Defining a suitable design of the reward functions and neural networks to ensure the convergence during training process, and the trained EMS is able to respond precisely under all different weather conditions and load demand.

- Verifying the efficiency and stability of the proposed EMS system on an isolated HRES, which is designed based on HOMER software with practical data from Basco island.
- Conducting a performance comparison between the proposed method and the pre-determined-rule conventional dispatch (CD) control for validation.

The rest of the paper is organized as follows. The mathematical models of the system components are introduced in Section 2. The DQN algorithm and the CD control are introduced in Section 3. The performance of EMS controller based on DQN is simulated in Section 4. The final section describes the conclusion and future work directions.

2. Mathematical Models of the System Components

This section describes the mathematical models of the system components, which are used to calculate their power generation and consumption. In this HRES, solar and wind energy are the primary energy resources. Short-term energy storage technologies have the ability to store and discharge energy for minutes or hours after being charged. In contrast, long-term energy storage can extend the storage time between charging and discharging to weeks or seasons [35]. In HRES, FCs can be used as a long-term energy storage option [4]. However, the slow dynamics of fuel cells and their degradation due to frequent start up and shut down cycles are a major disadvantage. Hence, batteries are also needed to create a hybrid system in which they take care of the power deficit and act as a short-term energy storage medium [36]. Batteries can provide or absorb large power gradients in short time. However, due to their short lifetime, high self-discharge rate, sensitivity to environment conditions, and limited storage capacity, batteries are not suitable for long-term solution.

2.1. PV System

A *PV* system is composed of one or more solar panels integrated with inverter or other electrical and mechanical hardware, using energy from the Sun to generate electricity. The output power of the *PV* system is strongly affected by the amount of solar radiation and the ambient temperature. The expression for the *PV*-generated power is as follows [22]:

$$P_{PV} = V_{pv} I_{pv} = I_{pv} \left\{ \frac{q}{AkT} ln \left(\frac{I_{ph} - I_{pv} + I_{pvo}}{I_{pvo}} \right) - I_{pv} R_s \right\} \tag{1}$$

where k is the Boltzmann constant, A is the non-ideality factor, q is the electron charge, T is temperature, q is the light-generated current, I_{pvo} is the dark saturation current, and R_s is the series resistance.

2.2. Wind Turbine System

During wind power generation, the blow of the wind generates kinetic energy, which drives the blades allowing the turbine to rotate. The mechanical energy then gets converted into electricity by the generator. The wind turbine system is significantly influenced by the wind speed. The generated power of the *WT* system is obtained from the manufacturers as follows [3]:

$$P_{WT} = \begin{cases} 0 & if\ V < V_{in}\ or\ V > V_{out} \\ P_r \left(\frac{V - V_{in}}{V_r - V_{in}} \right)^3 & if\ V_{in} \le V < V_r \\ P_r & if\ V_r \le V \le V_{out} \end{cases} \tag{2}$$

where P_{WT} denotes the output power at a particular value of wind speed. P_r represents the rated capacity. V_{in}, V_r, V_{out} stand for the cut-in, rated, and cut-out speeds, respectively.

2.3. Battery Storage System

Among various kinds of battery storage systems such as lithium-ion battery or nickel-zinc battery, we chose lead-acid batteries for their low cost (300–600 USD per kWh). Lead-acid batteries have a good cycle efficiency of up to 90% and a low self-discharge rate of less than 0.3% [37]. They are designed to withstand more deep discharge cycles, which make them suitable for an HRES.

One of the most important parameters of the battery system is the *SOC*, which expresses the level of charge relative to its capacity. The excess power is used to charge the battery, while a power deficiency towards the load demand discharges the battery. The battery *SOC* can be defined as follows [38]:

$$SOC_{t+1} = SOC_t \pm \frac{P_{Bat}\eta_{Bat}}{P_{n,Bat}} \times 100 \tag{3}$$

where SOC_{t+1} and SOC_t contain the battery *SOC* at the next time step and the current step, respectively. P_{Bat} stands for the battery power charging or discharging (kWh), while $P_{n,Bat}$ denotes the battery rated capacity, and η_{Bat} denotes the round-trip efficiency.

When the battery is turned on during the operation, the charging and discharging rates of the battery are defined based on the amount of power required at the current time step, always satisfying:

$$P_{Bat,discharge} \leq P_{Bat} \leq P_{Bat,\ charge} \tag{4}$$

where $P_{Bat,discharge}$ with negative sign indicates the discharge rate of battery, and $P_{Bat,\ charge}$ with positive sign shows the charge rate of battery.

At any time-step, the value of *SOC* must satisfy:

$$SOC_{min} \leq SOC \leq SOC_{max} \tag{5}$$

2.4. Diesel Generator

In the HRES system, a diesel generator is used as the back-up system when the load demand cannot be met by other components. The diesel generator ensures the availability, reliability, and quality of the power system all the time. We chose the model of the *DG* system according to its fuel consumption. In [39], an approximate linear model is presented where the hourly fuel consumption is calculated from the rated capacity of the *DG* and its operating power.

$$Fuel_t = \alpha_{DG}P_{DG,t} + \beta_{DG}P_{r,t} \tag{6}$$

where $Fuel_t$ expresses the fuel consumption (l). $P_{DG,t}$ denotes the operating power, while $P_{r,t}$ denotes the rated power of the *DG* system (kW). The coefficients of the fuel consumption are $\alpha_{DG} = 0.246$ and $\beta_{DG} = 0.08145$. They were used similarly in several studies [40,41].

2.5. Fuel Cell

A fuel cell (*FC*) uses the chemical energy of hydrogen or another fuel to produce electricity. There are various types of *FCs* available in the market. The so-called proton exchange membrane fuel cell (PEMFC) is the most frequently used. The advantages of PEMFC include high-power density, low operating temperature, small size, and good performance at start up and shut down. For this reason, PEMFC was chosen for this project. The hourly hydrogen consumption can be expressed as follows [9]:

$$q_{H_2,con} = \frac{P_{FC}}{E_{low,H_2}\eta_{therm}U_f\eta_{FC}} \tag{7}$$

where P_{FC} denotes the output power supplied by the *FC*, $E_{low,H_2} = 33.35$ kWh/kg assumes the lower heating value of the hydrogen, $\eta_{therm} = 0.98$ is the thermodynamic efficiency at 289 *K*, while U_f is the fuel utilization coefficient, namely, the ratio between the mass of fuel entering the *FC* and the mass of fuel reacting in the *FC*. Finally, η_{FC} denotes the *FC* efficiency.

2.6. Electrolyzer

To supply the hydrogen fuel for the operation of the *FC*, an electrolyzer is used. It generates hydrogen from water via electrolysis. The chemical reaction in an electrolyzer is the reverse of that in an *FC*. The power absorbed by the electrolyzer and the generated hydrogen mass are related by the expression below [9]:

$$P_{EL} = Bq_{H_2,\,nom} + Aq_{H_2,\,gen} \tag{8}$$

where P_{EL} denotes the power consumed by the electrolyzer system, $q_{H_2,\,nom}$ denotes the nominal hydrogen mass flow generated by the electrolyzer, while $q_{H_2,\,gen}$ symbolizes the actual generated hydrogen mass flow (kg/h). A and B are the consumption coefficients of the electrolyzer power curve where A = 10 kW/kg and B = 40 kW/kg were used in this paper.

2.7. Hydrogen Tank

In the HRES, a hydrogen tank is used as the container of hydrogen that is generated by the electrolyzer and is consumed by the *FC* system. Hydrogen can be stored as either liquid or pressurized gas. There are three methods to store the hydrogen: compressed high-pressure gas, hydrogen-absorbing materials, and liquid storage, among which, the first one is the most common. The hydrogen level in a hydrogen tank can be determined by the following expression [9]:

$$L_{H_2}(t+1) = L_{H_2}(t) + \frac{q_{H_2,\,gen} - q_{H_2,con}}{CAP_{H_2}} \tag{9}$$

where $L_{H_2}(t+1)$ and $L_{H_2}(t)$ stand for the level of the hydrogen at the next and the current time-steps, respectively, and CAP_{H_2} denotes the capacity of the hydrogen tank (kg).

2.8. Power Balance

Power balance is the state of equality between the produced energy and the load demand. More exactly, at each time step, the total possible power generation should never fall short of the power consumption. The weather data collection for feasibility extended over one year to facilitate system analysis and to allow for scheduling the operation of the whole system. The power balance equation is expressed as follows:

$$P_{PV} + P_{WT} + P_{Bat} + P_{DG} + P_{FC} + P_{EL} = P_{Load} \tag{10}$$

3. Energy Management of an HRES Based on Deep Q-Network

3.1. Introduction of the Proposed HRES

EMS is one of the most important parts to ensure the system is in reliable and efficient operation. The main function of the EMS is to balance the power flow between the system components, and simultaneously reduce the amount of fossil fuel and cost of energy production. A proposed DC/AC-bus system for power generation is presented in Figure 1. Excess energy from PV and WT will be stored in the battery and hydrogen system by controlling the K_Battery and K_Electrolyzer switches. In case PV and WT cannot fulfill the load demand, based on the available energy levels of system components, EMS will discharge battery or turn on FC and DG by K_Fuel-Cell and K_Diesel switches, respectively.

The proposed EMS control schema is presented in Figure 2. It is a learning-based approach, so no explicit mathematical model of the system is needed. A Markov Decision Process (MDP) of the EMS is needed for the implementation of the DQN algorithm. Based on the MDP model, the objective is to find the optimal policy for dispatch control of the system components to ensure a stable operation of the power system with the lowest cost of energy. An MDP model of the EMS is firstly defined in Section 3.2, including states (S), actions (A), transition probabilities (P), and rewards (R). It is considered as a tuple S, A, P, R. In which, "S" is a finite set of states which describes the all the operating point of the system. "A" is the control action. "P" is the probability of moving from one state to another one. "R" is an immediate return given to an agent when he or she performs specific action or task. Good action will receive positive reward while bad action will get punished.

A description of the DQN algorithm for EMS control is shown in the following part. In the DQN approach, a deep neural network is designed to approximate the action-value function and the DQN algorithm is adopted to train the neural network. It takes the state of the HRES as inputs, and outputs are the signals for dispatch control of the system components. The combination of the states of K_Battery, K_Electrolyzer, K_Fuel-Cell, and K_Diesel basically determines the system modes of operations. Finally, in Section 3.4, a conventional-based EMS is also applied for the validation of the proposed method.

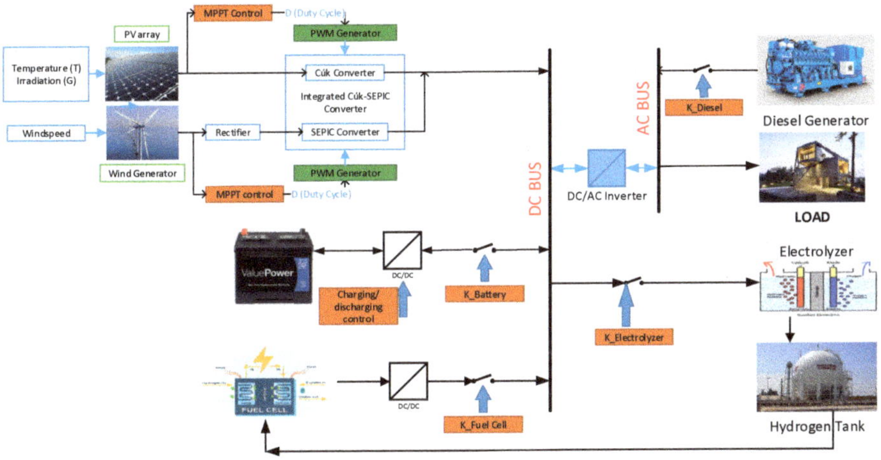

Figure 1. The diagram of the proposed HRES.

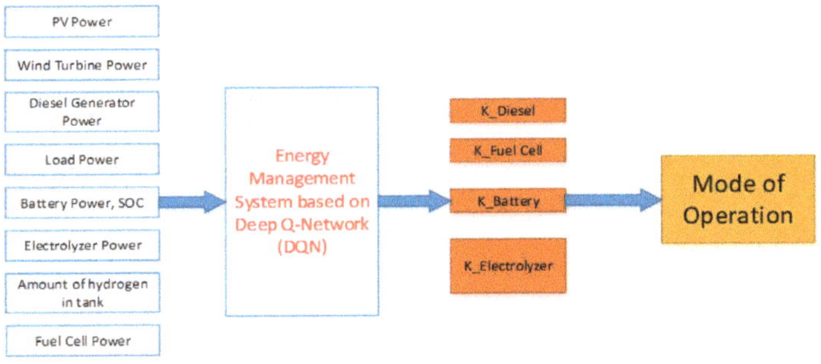

Figure 2. The proposed Deep-Q-Network-based EMS.

3.2. *Markov Decision Process Model for the EMS*

3.2.1. States and State Variables

During the operation of the HRES, the EMS controller receives a current state, it takes action, and moves to the next state based on its knowledge. The state information provides the basis for power flow control among all system components. The elements of our proposed HRES include *PV*, *WT*, *DG*, battery, and hydrogen system. The state variables are defined as combinations of the powers of load, *PV*, *WT*, *DG*, battery, fuel cell, and electrolyzer, as well as the state of charge, and the percentage of hydrogen in the tank (L_{H_2}):

$$S = \{P_{Load}, P_{PV}, P_{WT}, P_{DG}, P_{Bat}, P_{FC}, P_{EL}, SOC, L_{H_2}\} \quad (11)$$

3.2.2. Actions and Action Variables

Given the state at the current time step s_t, the EMS controller chooses an action and moves to next state by opening or dispatching the operation of following elements: *DG*, fuel cell, electrolyzer, and battery system. The action set A is formed by 4-component control signals:

$$A = \{\sigma_{Battery} \times \sigma_{DG} \times \sigma_{FC} \times \sigma_{EL}\} \quad (12)$$

The control actions of the battery system are discharging (−1), stopping (0), and charging (1), that is:

$$\sigma_{Battery} = \{-1, 0, 1\} \quad (13)$$

The control action variable of the diesel generator is in Equation (14), including stop, operating 25%, 50%, 75%, and full capacity, that is:

$$\sigma_{DG} = \{0, 0.25, 0.5, 0.75, 1\} \quad (14)$$

The control action variables of *FC* and electrolyzer are defined as σ_{FC} and σ_{EL}, respectively, including ON (0) and OFF (1), that is:

$$\sigma_{FC} = \{0, 1\} \quad (15)$$

$$\sigma_{EL} = \{0, 1\} \quad (16)$$

3.2.3. Transition Probability

Transition probability defines the probability that the agent moves from one state to another state. Given an action a_t, where t denotes the current time step, the transition probability from a current state s_t to the next state $s_{t+1} = s'$ is denoted by $P^a_{ss'}$, that is [42]:

$$P^a_{ss'} = P[S_{t+1} = s' | S_t = s, A_t = a] \quad (17)$$

In model-based energy management approaches, the transition probabilities $P^a_{ss'}$ are estimated by Monte Carlo simulation based on the prior probability distribution, or they are predicted by a short-term prediction model. However, in a model-free approach such as the DQN algorithm, they are estimated through learning from data.

3.2.4. Rewards

Reward function is used to calculate the reward from environment in response to a given state and action. It describes how the agent ought to behave. A good reward function can accelerate convergence during the training process. It can also affect the controller performance. For a simple approach, our designed reward function is the consumption of the reward from each system component as follows:

$$r_t(s_t, a_t) = r_{t,Bat} + r_{t,FC} + r_{t,EL} + r_{t,DG} \quad (18)$$

where $r_{t,Bat}$, $r_{t,FC}$, $r_{t,EL}$, and $r_{t,DG}$ are the rewards from the subsystems: battery, fuel cell, electrolyzer, and diesel generator.

The component rewards are essentially defined as follows:

$$r_{t,Bat} = \begin{cases} \frac{P_{Bat}}{P_{discharge,max}} & if \left(P_{PV} + P_{WT} + P_{DG} - \frac{P_{Load}}{\eta_{inverter}}\right) \geq 0 \\ -\frac{P_{Bat}}{P_{discharge,max}} & otherwise \end{cases} \quad (19)$$

$$r_{t,FC} = \begin{cases} \frac{2*P_{FC}}{P_{FC,max}} & if \left(P_{PV} + P_{WT} - P_{Bat} - \frac{P_{Load}}{\eta_{inverter}}\right) \leq 0 \; and \; SOC \leq 0.5 \\ \frac{P_{FC}}{P_{FC,max}} & if \left(P_{PV} + P_{WT} - P_{Bat} - \frac{P_{Load}}{\eta_{inverter}}\right) \leq 0 \\ -\frac{P_{FC}}{P_{FC,max}} & otherwise \end{cases} \quad (20)$$

$$r_{t,EL} = \begin{cases} \frac{2*P_{EL}}{P_{EL,max}} & if \left(P_{PV} + P_{WT} + P_{DG} - P_{Bat} - \frac{P_{Load}}{\eta_{inverter}}\right) \geq 0 \; and \; SOC \geq 0.9 \\ \frac{P_{EL}}{P_{EL,max}} & if \left(P_{PV} + P_{WT} + P_{DG} - P_{Bat} - \frac{P_{Load}}{\eta_{inverter}}\right) \geq 0 \\ -\frac{P_{EL}}{P_{EL,max}} & otherwise \end{cases} \quad (21)$$

$$r_{t,DG} = -\frac{Fuel_t}{Fuel_{max}} \quad (22)$$

where $\eta_{inverter}$ is the inverter efficiency, $Fuel_t$ is the fuel consumption of the diesel generator based on the actual operating power at time step t, and $Fuel_{max}$ is the fuel consumption at the maximum capacity.

As shown in Equations (19)–(22), the component reward functions are defined based on the result of the power balance function. For example, the battery will get negative reward when the sum of *PV*, *WT*, and *DG* powers is smaller than 0. Thus, the agent will learn to avoid choosing the negative-reward actions. The reward functions of *FC* and *EL* are similar to that of battery. For *DG* reward function, more fuel consumption means more negative rewards. Thus, it helps the agent to stop the operation time of the *DG* as much as possible.

In addition, the agent receives a big penalty if these parameters are out of their boundaries as shown below:

$$SOC_{min} \leq SOC \leq SOC_{max} \quad (23)$$

$$P_{Bat,discharge} \leq P_{Bat} \leq P_{Bat,charge} \quad (24)$$

$$L_{H2,min} \leq L_{H2} \leq L_{H2,max} \quad (25)$$

$$0 \leq P_{FC} \leq P_{FC,max} \quad (26)$$

$$0 \leq P_{EL} \leq P_{EL,max} \quad (27)$$

3.3. Methodology of the DQN-Based EMS

In this part, the DQN algorithm is described. Its objective is to find an optimal policy that maximizes the expected total rewards from a starting state. Figure 3 shows a graph of DQN-based EMS. The optimal policy is formulated as [42]:

$$V^{\pi^*}(s) = maxE_{\pi}\left[\sum_{t=0}^{T} \gamma^t r_{t+1} | s_0 = s\right] \quad (28)$$

where $\pi^* \in \Pi$ is the optimal policy in response to a given state and action. It is a strategy which is applied by the agent to decide the next action based on the current state. $0 < \gamma < 1$ is the discount factor used to define the importance of future reward. E_{π} denotes the expected value of reward according to the policy the agent follows.

In the DQN formulation, the optimal policy is represented by the optimal action-value function:

$$V^{\pi^*}(s) = maxQ^{\pi^*}(s,a) \quad (29)$$

where $V^{\pi^*}(s)$ is the optimal state-value function of an MDP. It is the expected return starting from state "s" following optimal π^*; $Q^{\pi^*}(s,a)$ is the optimal action-value function. It is the expected return starting from state "s", following optimal policy π^*, taking action "a". It focuses on the particular action at the particular state.

It is expressed as follows [42]:

$$Q^{\pi^*}(s,a) = E_{\pi^*}\left[\sum_{k=1}^{\infty} \gamma^{k-1} r_{t+k} | s_t = s, a_t = a\right] = E_{\pi^*}\left[r_t + \gamma maxQ^{\pi^*}(s_{t+1}, a_{t+1}) | s_t = s, a_t = a\right] \quad (30)$$

Following the optimal action-value function, the optimal policy can be determined by [42]:

$$\pi^*(s) = argmaxQ^{\pi^*}(s,a) \tag{31}$$

In the DQN algorithm as shown in Figure 4, a deep neural network is used to calculate $Q^{\pi^*}(s,a)$. It is expressed as $Q(s,a|\theta)$ network, where θ is the weight vector of the neural networks. As shown in the pseudo code in Figure 4, two separate Q-networks are used. $Q(s,a|\theta)$ represents the prediction network, while $Q(s,a|\theta')$ represents the target network [42]. To train the Q-network, a gradient descent is applied to minimize the loss function of the target and prediction networks. In every time step of the training process, the prediction Q network is updated by back-propagation method. In contrast, the target network is frozen. After a period of C time steps (C steps in the algorithm), its weights are updated by simply copying the weights from the current prediction Q network. Freezing the target Q network for a period of time helps stabilize the training process. In general, the Deep Q Network must be trained through the process in Figure 4 to ensure that EMS controller always chooses the best action. Then, EMS uses its trained Deep Q Network to calculate the Q value based on the current state information, and the next action is chosen following that Q value.

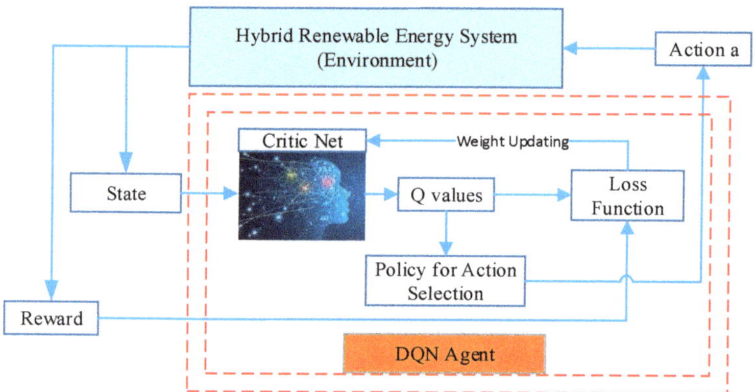

Figure 3. A graph of DQN-based EMS.

Algorithm 1: Deep Q Network (DQN)

1. Initialize replay buffer D to a fixed capacity N
2. Randomly initialize critic network $Q(s,a|\theta)$ with weight θ
3. Initialize target networks $Q'(s,a|\theta')$ with weight $\theta' \leftarrow \theta$

for episode = 1:M do
  Initialize state s
  for t = 1: T do
    With the probability smaller of equal to epsilon, select a random action a_t, otherwise select
    $a_t = argmax(Q(s_t, a_t|\theta))$
    Execute action a_t to get reward r_t and observe the next state s_{t+1}
    Store transition (s_t, a_t, r_t, s_{t+1}) in D
    Sample a random mini-batch from D
    Set

$$y_j = \begin{cases} r_j, \text{if episode terminates at the step } k+1 \\ r_j + \gamma \max_a Q(s_{t+1}, a_{t+1}|\theta'), \text{otherwise} \end{cases}$$

    Perform gradient descent to minimize the loss function for updating critic network

$$L = \frac{1}{N}\sum_j (y_j - Q(s_j, a_j|\theta^Q))^2$$

    Update the weighs of target network every C steps
  end
end

Figure 4. DQN algorithm.

3.4. Methodology of the Conventional Dispatch-Based EMS

The EMS controller chooses the operational mode of an HRES according to the power difference between generation and consumption and the available power in the energy storage system. It aims to satisfy the power demand all the time with the lowest fuel consumption. Following the work in [14], an convectional dispatch EMS method is applied in this study. It is used to compare with DQN-based method in term of system performance efficiency. The control actions of CD method are the same as DQN, including switching on/off the diesel generator, the fuel cell, and the electrolyzer, as well as charging/stopping/discharging the battery. The flow chart of the considered method is shown in Figures 5 and 6. This controller chooses the operational mode according to the power difference between generation and consumption and the available power in the energy storage system. It aims to satisfy the power demand all the time with the lowest fuel consumption.

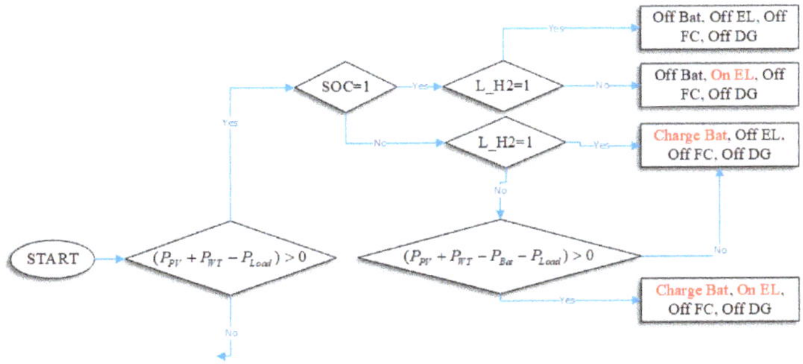

Figure 5. Flow chart of the EMS controller of our HRES based on the CD method (branch 1).

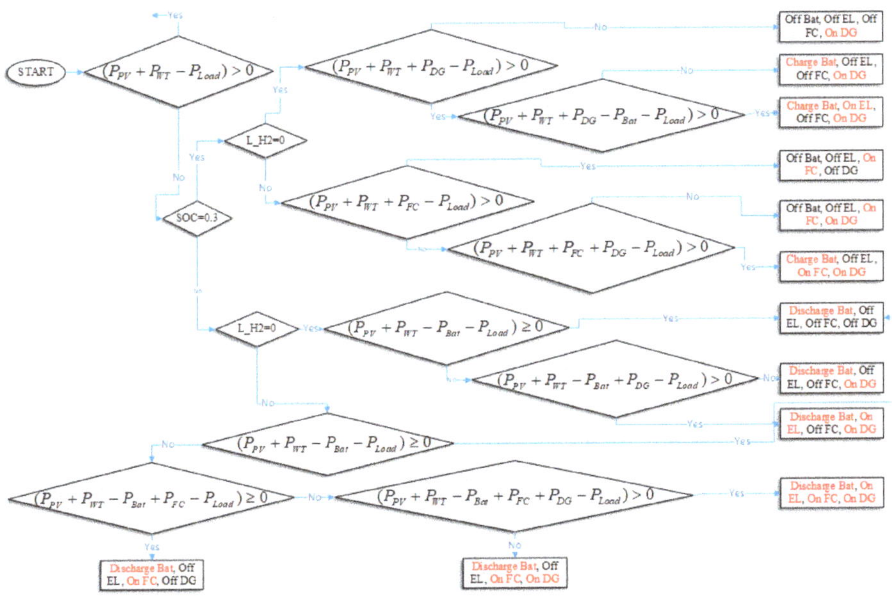

Figure 6. Flow chart of the EMS controller of our HRES based on the CD method (branch 2).

4. Results and Discussion

4.1. Site Description

Based on the weather and load data collected from this area, an optimal configuration of HRES is calculated by HOMER software [43]. Then, its simulation model is defined in MATLAB/Simulink for the implementation of our DQN-based EMS. In this part, the introduction of Basco Island is presented. This island is about 190 km away from Taiwan and is located in the northern region of the Philippines, where the major economic sectors are farming and fishing. On the island, the current source for power generation are diesel generators and fossil fuels, which require high operational costs due to the constantly increasing fuel prices and logistic costs. The location of Basco Island is excellent for marine resource management and tourism. As the government supports developing a sustainable economy, the local governors took the opportunity to invest in a more environment-friendly power system for the local community. Thus, research plays an important role in the economic development plan in this area. It ensures the continuous power supply with low cost of energy and environmental friendliness.

Figure 7 shows the diagram of our presented HRES in HOMER software (left), as well as the load profile through the year at Basco station (right) presented in HOMER software. A daily power consumption with an average demand of 700 kW every hour is shown in Figure 8. The weather data used for system simulation were taken from the database of the National Renewable Energy Lab (NREL), which can be generated by HOMER software. The average year around solar radiation is 4.44 kWh/m^2/day, while that of the wind speed is 7.22 m/s. Following the data, the energy system should supply 18 MWh a day with a peak power of 1.4 MW.

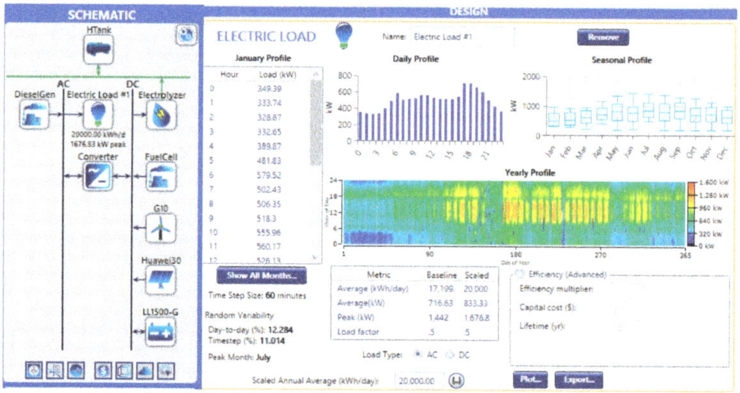

Figure 7. The proposed HRES (**left**) and the load demand at Basco station (**right**) presented in HOMER software.

Following the analysis in HOMER software, the optimal configuration of HRES in this case study is obtained [43]. It is reliable, environmentally friendly, and cost-effective. The proposed design includes a 5483 kW PV system, 236 pieces of 10 kW wind turbines, a 20,948 kWh battery system (48 V DC, 4 modules, 5237 strings), a 750 kW diesel generator, a 500 kW Fuel Cell system, a 3000 kW electrolyzer, a 500 kg hydrogen tank, and a 1575 kW converter. The Net Present Cost (NPC) of the system means the present value of the costs of investment and operation of a system over its lifetime. In this study, it was about 72.5 million USD. The Cost of Energy (COE), as the average cost per kWh of useful electrical energy produced by the system, was about 0.696 USD/kW. Furthermore, it can be concluded that the combination of the FC and the battery as the storage system is the best option for the design of HRES with lowest cost of energy. In this kind of system, FC is for a long term, while the battery is for short-term usage. Following the load demand at the applied area, the system is practical and cost-effective.

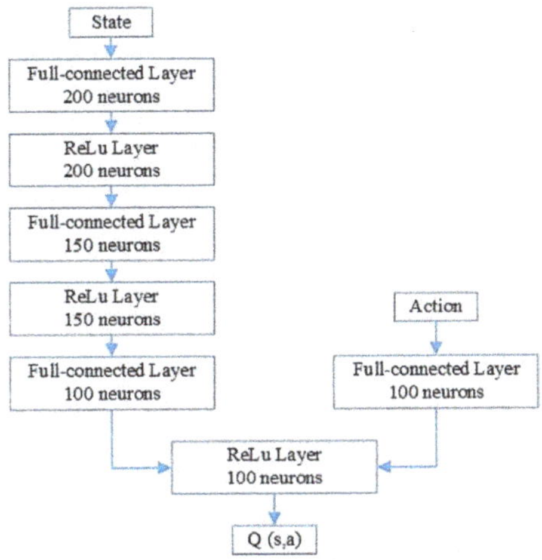

Figure 8. The structure of the critic network for the DQN-based EMS.

4.2. Implementation of DQN-Based EMS in MATLAB/Simulink

We carried out the simulation of the designed HRES in the Reinforcement Learning Toolbox of the MATLAB/Simulink environment. The time interval between two time-steps was one hour. There was a total of 5000 episodes during the training process where each episode ran for a randomly selected 48-h period. At the beginning of each episode, random initial conditions were generated including the initial state of charge and the initial amount of hydrogen in the tank.

Based on the experiences from previous publication [34] as well as trial-and-error during the training process, the structure of the network and its training parameters were determined. This is a usable reference in this area because there are not many publications that discuss details of the implementation of DRL for an HRES. In this study, the structure of the critic network applied for the DQN method is depicted in Figure 8, while the initial setting parameters for the simulation are displayed in Table 1. The amount that the network weights are updated during training is referred to as the step size or the learning rate (α). A large learning rate helps the agent to learn faster, and it could obtain the local optimal solution. On the other hand, a smaller learning rate may allow the agent to learn a global solution but may take significantly longer to train. In this study, the learning rate of the critic network is set to 0.001. It would mean that weights in the Q network are updated 0.1% of the estimated weight error each updating time. The action space of DQN comprises the combination of the actions of the four system components: battery, fuel cell, electrolyzer, and diesel generator.

Table 1. Parameters for the simulation of the DQN-based EMS.

Specifications	Value
Memory capacity	
Batch size	64
Discount factor (γ)	0.9
Exploration rate (ε)	1
Decay of exploration rate	0.001
Minimum exploration rate (ε_{min})	0.01

The discount factor (γ) affects how much weight it gives to future rewards in the value function. $\gamma = 0$ means that the agent will be completely myopic and only studies actions that produce an immediate reward. $\gamma = 1$ means that the agent will assess each of its actions based on the sum total of all of its future rewards. Exploration rate (ε) is the probability that our agent will explore the environment rather than exploit it. It is set to 1 at the beginning and reduced gradually over the training time. This ensures that the agent has enough time to explore and learn all about the environment.

4.3. Training Result

The training progress of the EMS controller based on the DQN algorithm is shown in Figure 9. The blue line represents the total reward in each episode, while average reward of total episodes at every time step is indicated by the red line. The estimation of the discounted long-term reward of critics when each episode starts, episode Q0, is marked as the yellow line in the graph. The average reward of total episodes at every time step flattens after 500 episodes. During the training process, we save the trained agents for online use when the average reward passes the design average value.

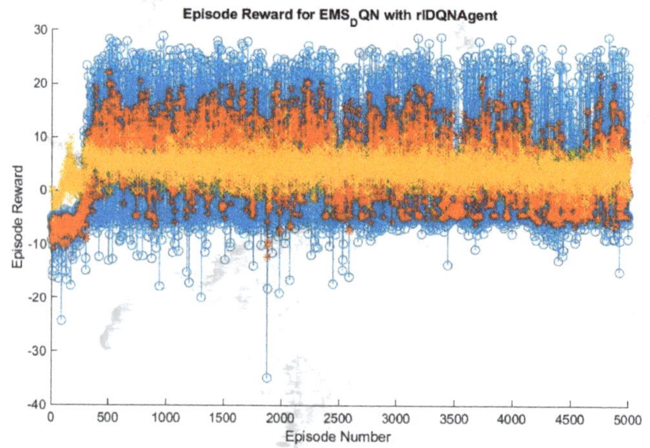

Figure 9. The training process of the DQN-based EMS.

4.4. Performance under Various Conditions

We used two scenarios for validating the performance of the proposed method. Each test also included a comparison with a conventional dispatch-based control. In the first scenarios, the operation of the diesel generator is totally turned off by the controller, because the battery and hydrogen system can fulfill the load the demand in case of not enough from solar PV and wind turbine. The second scenario is used to test for the operation of a diesel generator when all other energy resources run out of energy. It starts with less energy from PV and WT, so the operation of battery and hydrogen are required. Finally, diesel generation must be turned on to ensure the operation of the power system.

The simulation period was two days long using one-hour intervals between consecutive steps. WT, PV, and load demand were randomly generated from the year-round data. SOC and hydrogen levels were initialized with random values. Training based on random inputs shows the proposed DQN method can make effective schedules for the EMS in a deterministic environment from any initial conditions. The SOC minimum level was set to 30% in order to avoid running into deep discharging, thereby increasing battery lifetime. The minimum hydrogen level was set to 0. The simulation was implemented in the Reinforcement Learning Toolbox of MATLAB/Simulink software.

4.4.1. Scenario 1

The first scenario aimed at demonstrating the performance of the proposed DQN approach without the operation of the diesel generator. Figure 10 indicates the available power from the PV (green) and WT (blue) systems. The load demand is depicted by the red line. The simulation result is displayed in Figure 11. The three subfigures on the left apply for the DQN-based (red) EMS method, while on the right, apply for the CD-based (blue) EMS method. The first row displays the SOC of the battery. The second row displays the level of hydrogen in the tank. The third row displays the fuel consumption of the diesel generator.

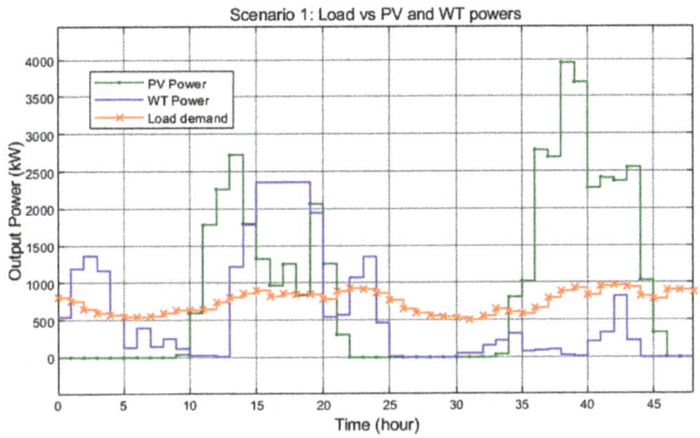

Figure 10. Load demand, PV, and WT power in Scenario 1.

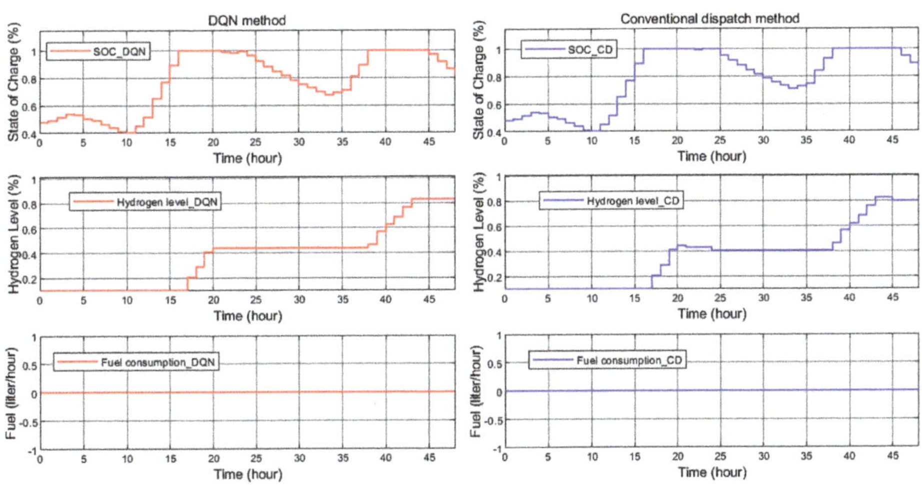

Figure 11. Comparison between the DQN and CD methods in Scenario 1.

Figure 11 shows that the diesel generators remained shut down under both methods. The SOCs on the left and right are almost identical. Between steps 0 and 5, the battery was charged by the power production of the WT system. Between steps 5 and 11, the battery switched to discharging due to no power from PV and WT. Between steps 11 and 24, more renewable power was available, so the battery was charged, and the excess power was used to run the electrolyzer. The amount of hydrogen increased between 17 and 20 h and between 38 and 43 h. Under the DQN method, the battery itself handled the problem of insufficient renewable input. Since the fuel cell remained shut down, there was no reduction of the hydrogen level in the tank over the simulation time. Under the CD approach, the fuel cell operated during steps 21–25 and 45–46, reducing the hydrogen level.

4.4.2. Scenario 2

The second scenario aimed at demonstrating the performance of the proposed DQN approach with the operation of the diesel generator. Similar to the previous case, Figure 12 shows the PV and WT productions and the load demand, while Figure 13 demonstrates the performance of the DQN- and CD-based methods. No renewable energy was available at the beginning. The level of SOC was 45%, and the amount of hydrogen in the tank was 10%. Thus, the diesel generator was forced to operate when power deficit occurred.

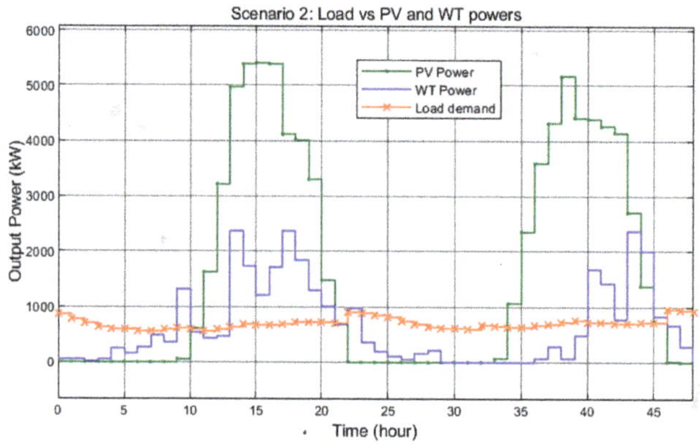

Figure 12. Load demand, PV, and WT power in Scenario 2.

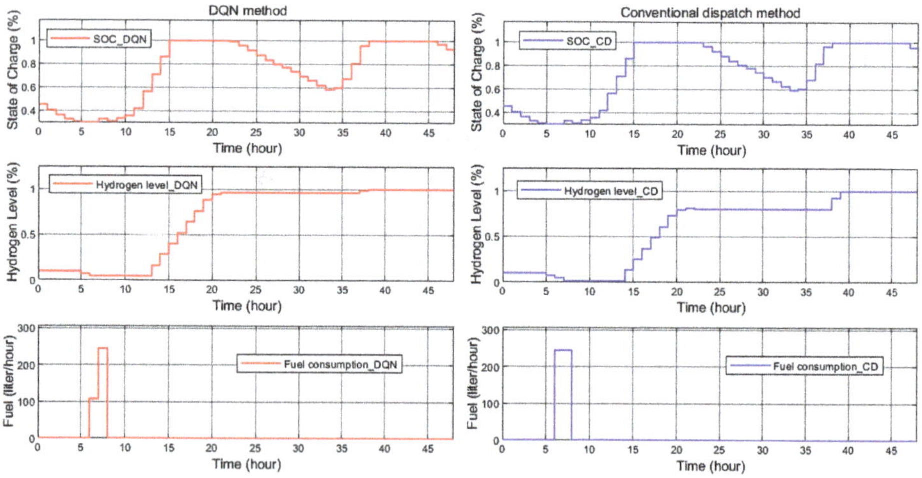

Figure 13. Comparison between the DQN and CD methods in Scenario 2.

From 0 to 4 time-steps, since no power was available from PV and WT, the battery discharged to its lower limit of 30%. The fuel cell supplied the power demand from 5 to 7 time-steps, resulting in the reduction of the hydrogen level. Since the power deficit persisted, the diesel generator turned on from 6 to 8 time-steps. The battery was charged fully from 9 to 15 time-steps when more power was produced by PV and WT. Similarly, extra power was used by the electrolyzer to generate hydrogen. After that, the battery discharged from 23 to 34 time-steps, and charged to its upper limit by step 38. The diesel generator and the fuel cell remained off due to no consumption. The operating time of the diesel generator under both methods was 2 h. However, the proposed DQN method only consumed 353 L, while the CD-based method consumed 492 L.

5. Conclusions

This study presents a DQN-based control to solve the complex problem of energy management in an HRES, where the energy flow between the HRES units is managed. The power system for case study on Basco Island, Philippines, includes a PV system, a WT system, a battery system, a diesel generator, and a hydrogen system. Due to its advantages of non-polluting power generation, the hydrogen system is considered for use in the proposed HRES. In the hydrogen system, an electrolyzer uses the excess energy from PV and WT to generate hydrogen for the operation of the fuel cell when an occasion of power shortage occurs. In the field of HRES, most of the current studies applied Q-leaning method, which has the limitation of a finite state and action space. In order to confront with continuous state space and large discrete action space, we introduced a deep neural network, allowing the agent to use function approximation to generalize across states, instead of using a Q look-up table. For any given state, the agent will choose the action with the highest value of reward and move to the next state. An MDP model of the HRES and the reward functions are formulated for the implementation of the proposed method in MATLAB/Simulink environment.

A basic rule-based EMS method named CD is considered to compare with the proposed DQN following the power efficiency. Based on this comparison, we know that the proposed method is always equal to or at least is better than the CD method. Despite only two scenarios considered for the result analysis, it can be concluded that the proposed method has good performance and outperforms the CD method under any uncertain environment. This is because the agent is trained based on the random initial conditions with random weather data and load demand, generated from the whole-year data.

The future work is to perform comparative real-time experiments with different advanced EMS methods such as Fuzzy, ANFIS, and PSO. Furthermore, to overcome the disadvantage of our proposed method, which is using a simple network structure and a basic reward function, a better study on the design of deep neural networks and gradient reward functions should be considered for fast convergence and less fluctuation of the average reward during the training process. These two factors ensure that the optimal policy for optimal EMS control of HRES is always obtained. Moreover, computational complexity should be an important metric for testing and validation. In addition, lithium-ion batteries are just as cheap as lead-acid batteries. They have lower self-discharge rates and higher lifetimes and efficiencies. Moreover, in the size category of multi-MW-storages, high temperature batteries such as sodium-sulfur batteries may be worth looking into for the future development. Instead of using two-day data, multiple-year data will be applied for the simulation.

In conclusion, we believe that deep reinforcement learning is the new potential trend in the field of energy conversion and management due to the following features: (1) the ability to learn from experience, (2) the ability to solve complex optimal control problems without prior environment knowledge, (3) the requirement of a simple mathematical model, and (4) the ability to handle problems for continuous state and action spaces.

Author Contributions: Conceptualization, B.C.P., M.-T.L. and Y.-C.L.; methodology, M.-T.L. and Y.-C.L.; software, B.C.P.; validation, B.C.P. and Y.-C.L.; formal analysis, B.C.P.; investigation, B.C.P.; resources, Y.-C.L.; data curation, B.C.P.; writing—original draft preparation, B.C.P.; writing—review and editing, B.C.P., M.-T.L. and Y.-C.L.; visualization, B.C.P. and Y.-C.L.; supervision, Y.-C.L.; project administration, Y.-C.L.; funding acquisition, Y.-C.L. All authors have read and agreed to the published version of the manuscript.

Funding: This research was funded by the Ministry of Science and Technology (MOST) under grant number MOST 111-2221-E-006-110- and MOST 111-2622-E-006-012-.

Institutional Review Board Statement: Not applicable.

Informed Consent Statement: Not applicable.

Data Availability Statement: Not applicable.

Conflicts of Interest: The authors declare no conflict of interest.

References

1. The Renewable Energy Support Programme for ASEAN (RESP) Team. *ASEAN Renewable Energy Policies*; ASEAN Centre for Energy (ACE): Jakarta, Indonesia, 2016.
2. Krishan, O.; Suhag, S. Techno-economic analysis of a hybrid renewable energy system for an energy poor rural community. *J. Energy Storage* **2019**, *23*, 305–319. [CrossRef]
3. Lin, C.E.; Phan, B.C. Optimal Hybrid Energy Solution for Island Micro-Grid. In Proceedings of the 2016 IEEE International Conferences on Big Data and Cloud Computing (BDCloud), Social Computing and Networking (SocialCom), Sustainable Computing and Communications (SustainCom) (BDCloud-SocialCom-SustainCom), Atlanta, GA, USA, 8–10 October 2016.
4. Vivas, F.J.; De Las Heras, A.; Segura, F.; Andújar Márquez, J.M. A review of energy management strategies for renewable hybrid energy systems with hydrogen backup. *Renew. Sustain. Energy Rev.* **2018**, *82*, 126–155. [CrossRef]
5. Indragandhi, V.; Subramaniyaswamy, V.; Logesh, R. Resources, configurations, and soft computing techniques for power management and control of PV/wind hybrid system. *Renew. Sustain. Energy Rev.* **2017**, *69*, 129–143.
6. Heymann, B.; Bonnans, J.F.; Martinon, P.; Silva, F.J.; Lanas, F.; Jiménez-Estévez, G. Continuous optimal control approaches to microgrid energy management. *Energy Syst.* **2018**, *9*, 59–77. [CrossRef]
7. Merabet, A.; Ahmed, K.T.; Ibrahim, H.; Beguenane, R.; Ghias, A.M.Y.M. Energy Management and Control System for Laboratory Scale Microgrid Based Wind-PV-Battery. *IEEE Trans. Sustain. Energy* **2017**, *8*, 145–154. [CrossRef]
8. Chen, Z.; Luo, A.; Wang, H.; Chen, Y.; Li, M.; Huang, Y. Adaptive sliding-mode voltage control for inverter operating in islanded mode in microgrid. *Int. J. Electr. Power Energy Syst.* **2015**, *66*, 133–143. [CrossRef]
9. Wang, F.-C.; Kuo, P.-C.; Chen, H.-J. Control design and power management of a stationary PEMFC hybrid power system. *Int. J. Hydrogen Energy* **2013**, *38*, 5845–5856. [CrossRef]
10. Jayalakshmi, N.S.; Gaonkar, D.; Nempu, P.B. Power Control of PV/Fuel Cell/Supercapacitor Hybrid System for Stand-Alone Applications. *Int. J. Renew. Energy Res.* **2016**, *6*, 672–679.
11. Roumila, Z.; Rekioua, D.; Rekioua, T. Energy management based fuzzy logic controller of hybrid system wind/photovoltaic/diesel with storage battery. *Int. J. Hydrogen Energy* **2017**, *42*, 19525–19535. [CrossRef]
12. Varghese, N.; Reji, P. Battery charge controller for hybrid stand alone system using adaptive neuro fuzzy inference system. In Proceedings of the 2016 International Conference on Energy Efficient Technologies for Sustainability (ICEETS), Nagercoil, India, 7–8 April 2016.
13. Battery charge controller for hybrid stand alone system using adaptive neuro fuzzy inference system Microgrids energy management systems: A critical review on methods, solutions, and prospects. *Appl. Energy* **2018**, *222*, 1033–1055. [CrossRef]
14. Luo, L.; Abdulkareem, S.S.; Rezvani, A.; Miveh, M.R.; Samad, S.; Aljojo, N.; Pazhoohesh, M. Optimal scheduling of a renewable based microgrid considering photovoltaic system and battery energy storage under uncertainty. *J. Energy Storage* **2020**, *28*, 101306. [CrossRef]
15. Chong, L.W.; Wong, Y.W.; Rajkumar, R.K.; Rajkumar, R.K.; Isa, D. Hybrid energy storage systems and control strategies for stand-alone renewable energy power systems. *Renew. Sustain. Energy Rev.* **2016**, *66*, 174–189. [CrossRef]
16. Zhou, X.; Ma, H.; Gu, J.; Chen, H.; Deng, W. Parameter adaptation-based ant colony optimization with dynamic hybrid mechanism. *Eng. Appl. Artif. Intell.* **2022**, *114*, 105139. [CrossRef]
17. An, Z.; Wang, X.; Li, B.; Xiang, Z.; Zhang, B. Robust visual tracking for UAVs with dynamic feature weight selection. *Appl. Intell.* **2022**. [CrossRef]
18. Wu, D.; Wu, C. Research on the Time-Dependent Split Delivery Green Vehicle Routing Problem for Fresh Agricultural Products with Multiple Time Windows. *Agriculture* **2022**, *12*, 793. [CrossRef]
19. Chen, H.; Miao, F.; Chen, Y.; Xiong, Y.; Chen, T. A Hyperspectral Image Classification Method Using Multifeature Vectors and Optimized KELM. *IEEE J. Sel. Top. Appl. Earth Obs. Remote Sens.* **2021**, *14*, 2781–2795. [CrossRef]
20. Hua, H.; Qin, Y.; Hao, C.; Cao, J. Optimal energy management strategies for energy Internet via deep reinforcement learning approach. *Appl. Energy* **2019**, *239*, 598–609. [CrossRef]
21. Cao, D.; Hu, W.; Zhao, J.; Zhang, G.; Zhang, B.; Liu, Z.; Chen, Z.; Blaabjerg, F. Reinforcement learning and its applications in modern power and energy systems: A review. *J. Mod. Power Syst. Clean Energy* **2020**, *8*, 1029–1042. [CrossRef]
22. Hsu, R.C.; Liu, C.-T.; Chen, W.-Y.; Hsieh, H.-I.; Wang, H.-L. A Reinforcement Learning-Based Maximum Power Point Tracking Method for Photovoltaic Array. *Int. J. Photoenergy* **2015**, *2015*, 496401. [CrossRef]
23. Mnih, V.; Kavukcuoglu, K.; Silver, D.; Graves, A.; Antonoglou, I.; Wierstra, D.; Riedmiller, M. Playing atari with deep reinforcement learning. *arXiv* **2013**, arXiv:1312.5602 . [CrossRef]
24. Mocanu, E.; Mocanu, D.C.; Nguyen, P.H.; Liotta, A.; Webber, M.E.; Gibescu, M.; Slootweg, J.G. On-line building energy optimization using deep reinforcement learning. *IEEE Trans. Smart Grid* **2018**, *10*, 3698–3708. [CrossRef]

25. Hu, Y.; Li, W.; Xu, K.; Zahid, T.; Qin, F.; Li, C. Energy Management Strategy for a Hybrid Electric Vehicle Based on Deep Reinforcement Learning. *Appl. Sci.* **2018**, *8*, 187. [CrossRef]
26. Huang, T.; Liu, D. A self-learning scheme for residential energy system control and management. *Neural Comput. Appl.* **2013**, *22*, 259–269. [CrossRef]
27. Leo, R.; Milton, R.S.; Sibi, S. Reinforcement learning for optimal energy management of a solar microgrid. In Proceedings of the 2014 IEEE Global Humanitarian Technology Conference-South Asia Satellite (GHTC-SAS), Trivandrum, India, 26–27 September 2014.
28. Raju, L.; Sankar, S.; Milton, R. Distributed Optimization of Solar Micro-grid Using Multi Agent Reinforcement Learning. *Procedia Comput. Sci.* **2015**, *46*, 231–239. [CrossRef]
29. Kim, H.-M.; Lim, Y.; Kinoshita, T. An Intelligent Multiagent System for Autonomous Microgrid Operation. *Energies* **2012**, *5*, 3347–3362. [CrossRef]
30. Eddy, Y.S.F.; Gooi, H.B.; Chen, S.X. Multi-Agent System for Distributed Management of Microgrids. *IEEE Trans. Power Syst.* **2015**, *30*, 24–34. [CrossRef]
31. Kofinas, P.; Vouros, G.; Dounis, A.I. Energy Management in Solar Microgrid via Reinforcement Learning. In Proceedings of the 9th Hellenic Conference on Artificial Intelligence, Thessaloniki, Greece, 18–20 May 2016; ACM: Thessaloniki, Greece, 2016; pp. 1–7.
32. Kofinas, P.; Vouros, G.; Dounis, A.I. Energy management in solar microgrid via reinforcement learning using fuzzy reward. *Adv. Build. Energy Res.* **2017**, *30*, 97–115. [CrossRef]
33. Kofinas, P.; Dounis, A.; Vouros, G. Fuzzy Q-Learning for multi-agent decentralized energy management in microgrids. *Appl. Energy* **2018**, *219*, 53–67. [CrossRef]
34. Phan, B.C.; Lai, Y.-C.; Lin, C.E. A Deep Reinforcement Learning-Based MPPT Control for PV Systems under Partial Shading Condition. *Sensors* **2020**, *20*, 3039. [CrossRef]
35. Koohi-Fayegh, S.; Rosen, M.A. A review of energy storage types, applications and recent developments. *J. Energy Storage* **2020**, *27*, 101047. [CrossRef]
36. Ahangari Hassas, M.; Pourhossein, K. Control and Management of Hybrid Renewable Energy Systems: Review and Comparison of Methods. *J. Oper. Autom. Power Eng.* **2017**, *5*, 131–138.
37. Luo, X.; Wang, J.; Dooner, M.; Clarke, J. Overview of current development in electrical energy storage technologies and the application potential in power system operation. *Appl. Energy* **2015**, *137*, 511–536. [CrossRef]
38. García, P.; Torreglosa, J.P.; Fernández, L.M.; Jurado, F. Optimal energy management system for stand-alone wind turbine/photovoltaic/hydrogen/battery hybrid system with supervisory control based on fuzzy logic. *Int. J. Hydrogen Energy* **2013**, *38*, 14146–14158. [CrossRef]
39. Skarstein, Ø.; Uhlen, K. Design consideration with respect to long term diesel saving in wind/diesel plants. *Wind. Eng.* **1989**, *13*, 72–87.
40. Ismail, M.; Moghavvemi, M.; Mahlia, T.M.I. Techno-economic analysis of an optimized photovoltaic and diesel generator hybrid power system for remote houses in a tropical climate. *Energy Convers. Manag.* **2013**, *69*, 163–173. [CrossRef]
41. Kaabeche, A.; Ibtiouen, R. Techno-economic optimization of hybrid photovoltaic/wind/diesel/battery generation in a stand-alone power system. *Sol. Energy* **2014**, *103*, 171–182. [CrossRef]
42. Fan, J.; Wang, Z.; Xie, Y.; Yang, Z. A Theoretical Analysis of Deep Q-Learning. *arXiv* **2019**, arXiv:1901.00137. [CrossRef]
43. Phan, B.C.; Lai, Y.-C. Control Strategy of a Hybrid Renewable Energy System Based on Reinforcement Learning Approach for an Isolated Microgrid. *Appl. Sci.* **2019**, *9*, 4001. [CrossRef]

applied sciences

MDPI

Article

Development of an Intelligent Solution for the Optimization of Hybrid Energy Systems

Djamel Saba [1,*], Fahima Hajjej [2], Omar Cheikhrouhou [3], Youcef Sahli [1], Abdelkader Hadidi [1] and Habib Hamam [4,5,6,7]

1 Unité de Recherche en Energies Renouvelables en Milieu Saharien, URERMS, Centre de Développement des Energies Renouvelables, CDER, Adrar 01000, Algeria
2 Department of Information Systems, College of Computer and Information Sciences, Princess Nourah bint Abdulrahman University, P.O. Box 84428, Riyadh 11671, Saudi Arabia
3 CES Laboratory, National School of Engineers of Sfax, University of Sfax, Sfax 3038, Tunisia
4 Faculty of Engineering, University de Moncton, Moncton, NB E1A3E9, Canada
5 International Institute of Technology and Management, Commune d'Akanda, Libreville P.O. Box 1989, Gabon
6 Department of Electrical and Electronic Engineering Science, School of Electrical Engineering, University of Johannesburg, Johannesburg 2006, South Africa
7 Spectrum of Knowledge Production and Skills Development, Sfax 3027, Tunisia
* Correspondence: saba_djamel@yahoo.fr

Abstract: This paper presents a proposal for the development of a new intelligent solution for the optimization of hybrid energy systems. This solution is of great importance for installers of hybrid energy systems, as it helps them obtain the best configuration of the hybrid energy system (efficient and less expensive). In this solution, it is sufficient to enter the name of the location of the hybrid energy system that we want to install; after that, the solution will show the name of the best technology from which the optimal configuration of this system can be obtained. To accomplish this goal, the study relied on the ontology approach for two reasons, one of which is related to the nature of hybrid systems, because it is characterized by a large amount of information that requires good structuring, and the second reason is the interaction of hybrid energy systems with the external environment (climate, site characteristics). Afterward, to develop the knowledge base of the ontology, many steps were followed, the first of which is related to a detailed study of the existing one and the extraction of the basic elements, such as the concepts and the relations between them, followed by the development of the rules of intelligent reasoning, which is an interaction between the elements of the ontology through which all possible cases are treated. The "Protégé" software was used to edit these elements and perform the simulation process to show the results of the developed solution. Finally, the paper includes a case study, and the results show the importance of the developed solution, and it is open to future developments.

Keywords: decision-making tool; intelligent reasoning rules; energy saving; energy domain ontology; hybrid energy system

Citation: Saba, D.; Hajjej, F.; Cheikhrouhou, O.; Sahli, Y.; Hadidi, A.; Hamam, H. Development of an Intelligent Solution for the Optimization of Hybrid Energy Systems. *Appl. Sci.* **2022**, *12*, 8397. https://doi.org/10.3390/app12178397

Academic Editors: Luis Hernández-Callejo, Sergio Nesmachnow and Sara Gallardo Saavedra

Received: 5 July 2022
Accepted: 8 August 2022
Published: 23 August 2022

1. Introduction

Air pollution, climate change, and limited fossil resources have raised awareness that sustainable development that takes care of the environment in which we live is necessary [1]. With the difficulty of connecting electricity grids to remote areas, RE presents a good alternative to fossil fuels, it does not emit greenhouse gases, and it allows the decentralized production of resources [2]. However, the random specificity of energy sources imposes special rules for the optimization and operation of energy systems. In addition, the hybridization between some of the RE sources forms a complementarity of energy production and an alternative to conventional generators generally used to produce electricity. An HES design is an important step because of its relationship to completion costs and reliability [3]. Therefore, it is necessary to provide new solutions and focus in particular on meeting consumer demands for energy while ensuring the minimum cost [4].

To meet the rapidly increasing demand for energy, all energy sources must be exploited. Renewable energies are unlimited and clean, but the biggest problem with them lies in their intermittent aspect. To overcome this problem, a combination of several energy sources is made to obtain the so-called hybrid renewable energy system [5,6]. Many works have been carried out on HES optimization methodologies. A study by Ammari et al. aimed at presenting and analyzing a literature review of recently published works in the field of hybrid renewable energy [7]. The study focuses on four basic categories of hybrid renewable energy systems, which are scaling (using software or using conventional methods), optimization (classical, synthetic, and hybrid methods), control (centralized, distributed, and hybrid control), and energy management (technical and economic objective). Furthermore, the review compares the different methods used in each category. A research work by Nourollahi et al. presents a hybrid approach to process improvement and addresses uncertainties in a residential energy system consisting of photovoltaics, fuel cells, boilers, and storage units [8]. In this regard, uncertain parameters are divided into two categories, including poorly behaved and well-behaved parameters, and then robust optimization and stochastic programming are used for modeling. Additionally, the conditional value at risk is implemented to assess the risks of well-behaved parameters. According to simulations, it has been shown that uncertainties with bad behavior have greater effects on the operation of the system. Moreover, changing the control parameters for robust optimization and conditional value at risk from 0 to 24 and from 0 to 1 increases the total cost by 5.2% and 0.47%, respectively. Comparative results also show that the proposed hybrid method makes less conservative decisions for cost optimization. The application of hybrid energy storage to distributed energy systems can greatly improve energy efficiency and reduce operating costs. However, insufficient efforts focused on investigating the integration of the two systems, configuration optimization, and systems operation strategy were found. Therefore, a new distributed energy system that combines hybrid energy storage was proposed, and the configuration of the system optimization and the operation strategy of the new system were considered simultaneously [9]. The studied hybrid system contained thermal storage and two forms of energy storage, that is, supercapacitors and a lithium battery. The impact of hybrid energy storage on distributed energy systems was fully considered. Then, a two-layer cooperative optimization method for the new system was introduced, taking into account energy efficiency, economy, and environmental protection. The new system was applied to an almost energy-free society. The results indicate that the energy-saving and equivalent pollutant-equivalent primary emission reduction rates for the new system were evaluated to be 54.8% and 63.6%, respectively, under the low-pass filter operating strategy combined with secondary feedback modulation of the supercapacitor charge state. System configuration and operating strategy can be optimized simultaneously by the two-layer collaborative optimization method. Renewable energy resources often suffer from challenges, such as erratic energy generation, as a consequence of weather and seasonal changes. Hybrid renewable energy systems are a solution for efficient energy densification of renewable resources as several of them are combined to overcome challenges arising from operating them in a stand-alone mode. A study by Sharma et al. proposed multitarget dynamic optimization of candidate hybrid renewable energy systems using a genetic algorithm in which the operating cost of hybrid renewable energy systems, nonrenewable energy use, and fuel emissions were simultaneously reduced over a limited period [10]. In optimization, three strategies are evaluated by considering the profiles of wind, solar, and 24 h load (Strategy 1), past and 1 h ahead (Strategy 2), and 1 h ahead (Strategy 3). A comparison of the results shows that the energy to be purchased from the network for Strategy 1 is 8.7% and 10.7% lower compared with Strategies 2 and 3, and also the energy sold to the network is 19% and 22% higher than for Strategies 2 and 3, respectively, while meeting the specified load profile of 100 households.

One of the most important applications of the renewable energy system is installing a well-designed HES in remote areas where grid extension is complicated and expensive. However, the proper design of such a system is a difficult task, such as coordination between different energy sources. Energy storage and load are very complex. Hybrid renewable energy system optimization is selecting appropriate components, scaling, and control strategies to provide efficient, reliable, and cost-effective alternative energy to society. A work by Fulzele and Daigavane presents a design of a hybrid renewable energy system consisting of photovoltaic cells and a wind generator with a battery and an inverter [11]. The system was optimally simulated using the HOGA (Hybrid Optimization by Genetic Algorithms) tool developed by the Department of Electrical Engineering of the University of Zaragoza, Spain. In the same context, a work by Mahmoud et al. is concerned with the application of modern optimization methods to devote the optimal configuration of hybrid renewable energy sources, which include photovoltaic panels, wind turbines with battery storage systems, and diesel generators [12]. The cost function of the optimization problem was chosen to be the energy cost and energy supply potential loss for the hybrid renewable energy system, where the main function of the optimization algorithms is to reduce this cost function. However, the optimal configuration cannot be achieved without meeting system reliability and operational limitations. A comparison of their results was made in order to determine the most effective method. Furthermore, a statistical study was conducted to determine the stability of the performance of each optimization strategy. The results revealed optimal variables, including the number of photovoltaic panels, wind turbines, batteries, and diesel generator capacity. Over the course of a year, the optimal configuration was tested against the study of capital and fuel expenditures. The statistical results also proved the robustness of the developed algorithm. Cano et al. presented a comparative study between four optimization methods for autonomous HES, which include two energy sources (PV, wind), storage batteries, and fuel cells [13]. The first method is based on mathematical equations, the second uses the SDO program, the third uses the Homer optimization software, and the last uses genetic algorithms based on the HOGA software. The results show that the HES designed by each method guarantees reliability in the energy supply, and SDO is the best HES optimization method. Other contributions use AI approaches. Maleki and Askarzadeh propose four heuristic algorithms, PSO, TS, SA, and HS, on an autonomous HES (PV, wind turbine, fuel cell) [14]. This study showed that HES with energy storage batteries is the best choice economically, and PSO is the most reliable optimization technique. In the same context, Aydin et al. were interested in the development of a methodology based on the information of a GIS to identify the preferred sites for an HES based on wind and PV [15]. This study used the principle of fuzzy logic and the MCDM approach to determine the economic feasibility of wind and solar energy. Finally, the associated maps were superimposed to obtain the most feasible places for HES. The proposed methodology can help policymakers and investors easily adapt to other types of energy resources. Luna-Rubio et al. provided an overview of optimization methods. The study presented the different architectures of stand-alone or network-connected HES [16]. This study concluded with the following submissions:

- HES that includes solar and wind energy sources requires the addition of storage batteries or integration into the grid.
- Among the methods proposed (probabilistic, analytical, iterative, and hybrid), the hybrid methods are the most powerful for HES optimizing.

Djamel Saba et al. provided an optimization solution for HES based on the intelligent reasoning rules [17,18]. Then, this solution was improved to be generic. After implementing it, the results showed its effectiveness in choosing the optimization technique. In addition, it is characterized by various advantages, such as flexibility in its operation and updating [19]. Coelho et al. proposed a MAS solution to optimize the HES [20]. Amicarelli et al. proposed to minimize the costs, maximize the use of RE, and simplify the integration into the grid [21]. Elamine et al. present an intelligent method of HES energy management [22]. The solution is based on the MAS. This concept allows the different units of the MAS to work together to achieve the objectives. Morstyn et al. studied the performance of the control strategy by verification on a microgrid that includes storage batteries and a PV generator [23]. This study allows us to know the advantages that characterize distributed management compared with centralized management. L. Raju et al. developed a MAS for managing the energy of a microgrid containing two PV systems, two wind turbines, a battery unit, and a diesel plant [24]. This solution offers the consumer the possibility of choosing the actions to increase energy efficiency. A study adapted to the west African climate was conducted by Mbodji et al. It is based on the MAS for the control, where it was applied on an HES (PV, wind, batteries) [25]. In this study, three load profiles were considered in the simulations. Then, the control made it possible to reduce battery use by 3%, 5%, and 6%, for profiles 1, 2, and 3, respectively. It was made in 1 day to obtain 35 kWh for profile 1 and 20 kWh for profile 3. From an economic aspect, the strategy applied to profile 1 was allowed a gain of EUR 6 per day and EUR 2322 per year or more than EUR 46,442 for 20 years (lifetime of the project).

The present study is intended to face the problem of the absence of a generic optimization solution for hybrid energy systems. The main objective of this work is to set up a reliable and easy solution to optimize hybrid energy systems suitable for most sites (generic). For these reasons, this study proposes a generic optimization solution. With this solution, the user is not required to know much data about optimization techniques, energy sources, and climatic data. The user only chooses the installation site, and it is up to the adopted solution that provides the appropriate sources as well as the most reliable technique to calculate the optimal configuration of the system to be installed. This contribution is based on the ontology of the field of energies. The choice of ontology is mainly due to the nature of the studied system and its environment (energy system), which is dynamic over time. In addition, this technique makes it possible to precisely present a set of knowledge in a form that can be permitted for use by a machine, which allows users to introduce the necessary updates without occurrence of any damage to the rest of the data.

2. Materials and Methods

2.1. Proposed Approach for Ontology Construction

In this work, the problem revolves around the absence of a generic optimization solution for hybrid energy systems. The main objective is to set up a reliable and easy solution to optimize hybrid energy systems suitable for most sites (generic). In the same context, and to solve any research problem, two subproblems must be initially separated and finally integrated, which are: the conceptual subproblem and the operational subproblem. For this research, the following hypothesis is employed:

- "The development of an optimization generic solution for hybrid energy systems improves the choice of better technique and simplifies the task for setup of this type of systems". According to this hypothesis, two variables can be noted, which are "the developed solution" and "the hybrid energy systems".
- "The more energy sources there are at a site, the more efficient the hybrid energy system". It is characterized by the variables: "more energy sources" and "more efficient the hybrid energy system".
- "The importance of the solution increases with the increase in the number of cases treated". Two variables can be extracted from this hypothesis: "importance of the solution increases" and "increase in the number of cases treated".

In addition, the following main questions can be asked according to this hypothesis: what is the impact of the developed solution on the optimization of hybrid energy systems? It is also possible to extract from this question some subquestions, such as:

- During bibliographic research, to what extent (year) can it be conducted?
- What are the advantages and disadvantages of these previous solutions?
- What is the best way forward to develop a solution?
- What are the techniques that can be used to test the solution?
- How to ensure the degree of relevance of the solution under development?

- Is there a relationship between the characteristics of the site and the hybrid energy system that will be installed?
- What are the reasons for using a hybrid energy system instead of a nonhybrid energy system?

2.2. Proposed Approach

The goal of this solution is to find the optimal configuration of the HES. It is based on a knowledge base (KB) and includes some steps, as shown in Figure 1:

i. Site choice: the user chooses the site of HES installation and inserts the electrical requirement energy value.
ii. Display of parameters of the chosen site: in this step, all the parameters that will be used in the next steps are displayed, mainly those related to the characteristics of the energy sources.
iii. The development of intelligent reasoning rules: in this step, we develop intelligent reasoning rules based on ontology. These rules permit to select the appropriate energy sources and the best energy optimization technique.
iv. List of energy sources: the results of the previous step are displayed (the best and available energy sources).
v. Chosen optimization technique: the most appropriate technique is selected.

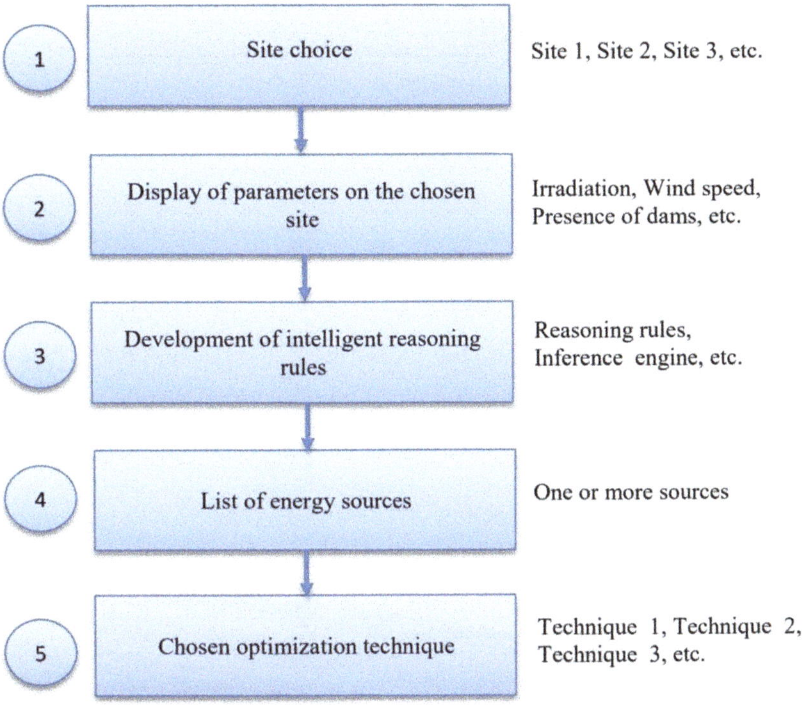

Figure 1. Proposed solution flowchart.

In addition, each optimization technique in the literature has been developed for a specific situation (types of energy sources, type of load, duration of application, etc.). The proposed solution at the start chooses the most appropriate optimization technique according to the data previously presented (concept, attribute, relation, etc.) and recorded in the knowledge base of the solution ontology. The second step concerns the import and execution of its algorithm to finally obtain the best optimal configuration.

2.3. Construction of Ontology

The modeling process consists of several steps, as itemized below [19,26]:

2.3.1. Define the Ontology Domain

This step is reserved for defining the ontology subject or domain. It includes a full understanding of the ontology subject. The ontology in this work would focus on the management of the energy required to model and optimize the HES.

2.3.2. Reuse of Existing Ontologies

After consulting previous works, many contributions concern the optimization of HES [27]. These works have achieved very encouraging results, but they still require some development through the introduction (or removal) of concepts and the development of intelligent reasoning [28–30]. For this work, the creation of the ontology was performed from the beginning, and we did not use previous ontologies.

2.3.3. Interesting Concepts for the Ontology

The main concepts of this ontology are sources (photovoltaic, wind energy, etc.), load, and climatic data (temperature, wind speed, etc.).

2.3.4. Explain Classes and Their Hierarchy

The ontology concepts form a hierarchy [30]. There are some methods to developing a class hierarchy [31]:

- Development of the ontology hierarchy from top to bottom
- A bottom–up development process
- A combination of the two approaches, top–down and bottom–up

In this work, the first method (top–down) was chosen (Table 1).

Table 1. An extract from the classes of the system.

Class	Description
Source	Energy source
OptimizationTechnique	Optimization technique
Load	Load
ClimateData	Climate data
Site	HES installation site
PhotovoltaicSource	Photovoltaic energy source

2.3.5. The Properties of Classes and the Facets of Attributes

After the presentation of the concepts, it is essential to describe their interior structure [28,29]. Then, the attribute facet describes the types of values that can be assigned to the attribute [32] (Table 2).

Table 2. An extract of the attributes.

Attribute	Description	Type	Class
SiteName	Installation site name	Alphabetical	Site
HasRadiation	Irradiation that characterizes the site	Digital	Radiation
SocBat	Battery charge level	Digital	Battery
SocBatMin	Minimum battery charge level	Digital	Battery
EfficiencyCharg	Battery charging efficiency	Digital	Battery

2.3.6. Design Instances and Relationships

The last step concerns the creation of instances and relationships for the proposed solution (Tables 3 and 4). These two elements are necessary for the development of the ontology [30].

Table 3. An extract from the ontology relations.

Relationship	Classes	Description
FeedSourceLoad	Source, Load	Energy supply to the load by energy source
FeedStorageLoad	Storage, Load	Supply of the load by stored energy
LoadSourceStorage	Source, Storage	Energy source charges the batteries

Table 4. Examples of instances.

Classes	Instances
Site	Alger, Adrar, Annaba, Oran, Ouargla, Tamanrasset
Month	January, February, March, April,..., December
Radiation	3378 Watt/m^2, 3350 Watt/m^2, 2650 Watt/m^2
WindSpeed	4, 4.5, 6 km/h
SoilTemperature	15, 38, 45 °C

2.3.7. Intelligent Reasoning of the Solution

All intelligent reasoning rules are constructed through a detailed study of the internal and external environments of the hybrid energy systems (the main elements of a hybrid energy system, environmental data, geographic data, etc.). We mainly focused our attention on related aspects that influence the HES operation. In addition, through this study, we extracted a series of information, namely, optimization techniques, installation site, climate data, load, and so on. Finally, depending on the predicate logic, the programming language Python, and the essential elements of ontology (concepts, attributes, relationships, and instances), we can formulate the intelligent reasoning rules associated with choosing the optimization technique (Figure 2, Table 5).

R1: Site (x) ∧ Radiation (x, y) ∧ ≥ (y, r) ∧ WindSpeed (x, z) ∧ ≥ (z, w) ∧Load (LoadDaily)
⇨ Technique (t)
R2: Site (x) ∧ Radiation (x, y) ∧ ≥ (y, r) ∧ WindSpeed (x, z) ∧ ≥ (z, w) ∧ Load (LoadMonthly)
⇨ Technique (t)
R3: Site (x) ∧ Radiation (x, y) ∧ ≥ (y, r) ∧ WindSpeed (x, z) ∧ WindSpeed (x, z) ∧ Load (LoadAnnual)
⇨ Technique (t)

Figure 2. Examples of the solution rules.

Table 5. Description of R1, R2, and R3.

Rule	Description
R1	If a site (s) is characterized by radiation greater than or equal to r Wh/m^2, the wind speed is greater than or equal to w m/s and the load is of the daily type, then the reasoning proposes the optimization technique (t) [33].
R2	If a site (x), characterized by radiation greater than or equal to r W/m^2 and a wind speed greater than or equal to w m/s, and the load is of the monthly type, then the rule proposes the optimization technique (t) [34].
R3	If a site (s) is characterized by radiation greater than or equal to r W/m^2, a wind speed greater than or equal to w m/s, and the load is of annual type, then the reasoning proposes the optimization technique (t) [34].

2.4. *Ontology Editing and Presentation of a Scenario*

2.4.1. Choice of Editor

For the editing of ontology, the "Protégé OWL 3.4.4" software was used because of the following advantages [19,35]:

- It is compatible with standard languages.
- It has a modular interface, which allows it to be enriched with additional modules (plugins).
- It provides a comfortable expression editor.
- It provides an API (or GUI) that allows the manipulation of the ontology created by the "Protégé" editor in Java code. It also provides a Java API, allowing developers to integrate with their Protégé OWL applications, import or export the ontology in different languages, and implement the ontology.

2.4.2. Choice of Reasoning Tools

Most inference engines can process rules added to the ontology. Rules-specific engines can be used, such as the Jess engine [36]. Most inference engines can process rules added to the ontology. Rules-specific engines can be used, such as the Jess engine [34]. The latter has a language for the expression of knowledge in the form of rules. It can be used from "Protégé-OWL API," thanks to the existence of a bridge that allows for the translation and execution of an ontology model in the language of Jess to retrieve the result in the "Protégé" software.

2.4.3. SWRL

This is a rules language for the semantic web combining OWL-DL and RuleML [37]. It makes it possible to enrich the semantics of an ontology presented in OWL and to manipulate instances by variables (? X, ? Y, ? Z). It simply allows you to add relations according to the values of the variables and the satisfaction of the rule. However, SWRL does not allow the generation of concepts or relationships.

2.4.4. Editing the Ontology

The ontology elements edited in Protégé are the concepts, attributes, relationships, and instances (Figure 3).

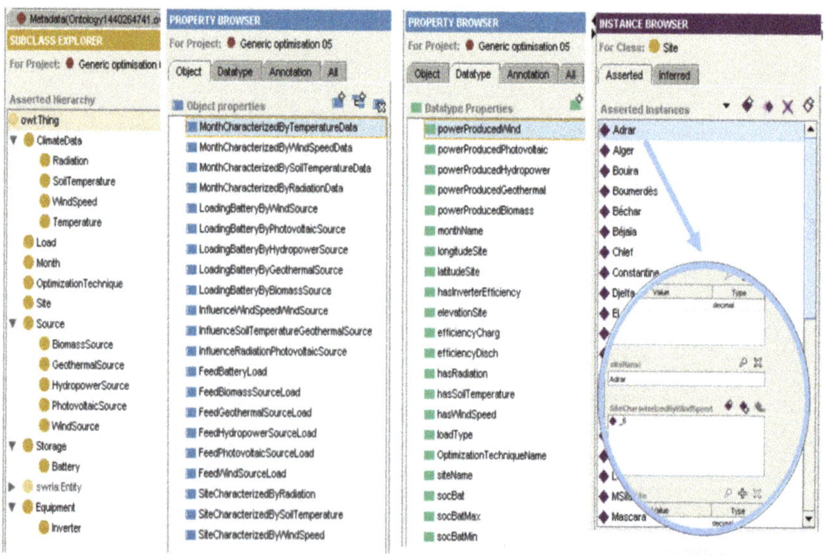

Figure 3. Elements of ontology in Protégé-OWL 3.4.4.

2.4.5. Implementation of Intelligent Reasoning Rules

The SWRL rules editor operates in Protégé OWL and provides a very interactive interface for editing rules (supporting all the features of the SWRL language). Rule engines, such as Jess, can be integrated with this editor, which helps provide richer rule-based reasoning (Figure 4).

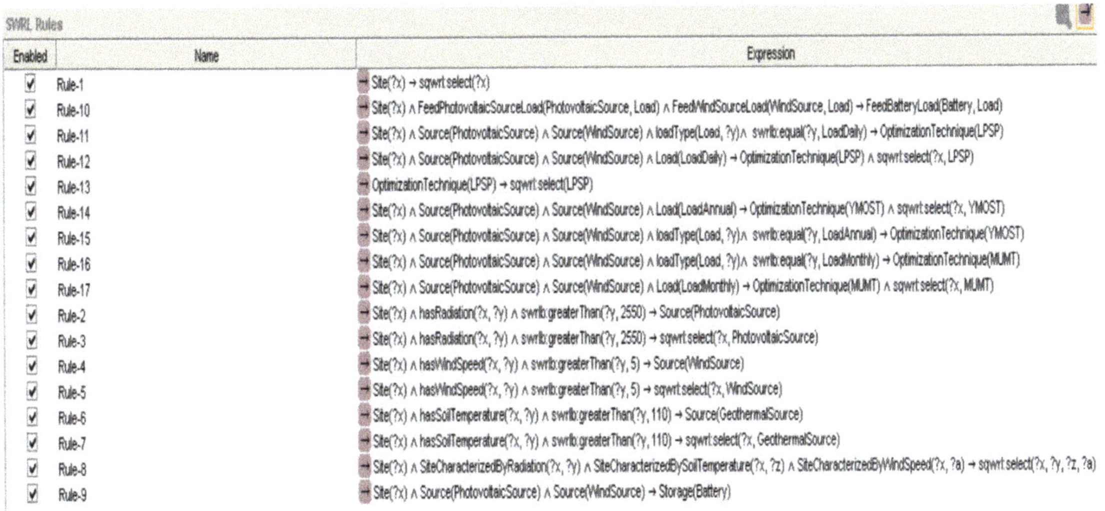

Figure 4. Rules of inference in Protégé-OWL.

2.4.6. Presentation of a Scenario

In this scenario, three Algerian cities were taken—Adrar, Annaba, and Illizi—which have different climatic characteristics. The three cities are presented with the average and annual radiation, the average and annual wind speed, the average and annual soil temperature, and the type of energy load (Tables 6 and 7).

Table 6. Example of KB data.

Site	Average Annual Radiation (Wh/m^2)	Average Annual Wind Speed (m/s)	Average Annual Soil Temperature (°C)	Load Type
Adrar	3378	6	65	LoadDaily
Annaba	2550	4	65	LoadAnnual
Illizi	3350	4.5	45	LoadMonthly

Table 7. Criteria for selecting energy sources.

Criteria	Energy Source
Average and annual radiation => 2550 Wh/m^2	Photovoltaic
Average and annual wind speed => 5 m/s	Wind
Average and annual soil temperature => 110 °C	Geothermal

The criteria for selecting renewable energy sources are defined (Table 7):

Regarding the values of 2550 Wh/m^2 for solar radiation, 5 m/s for wind speed, and 110 °C for soil temperature, they are considered the minimum acceptable values for electrical energy production for solar energy source, wind energy source, and geothermal energy source, respectively. For this reason, they were selected for the case study.

In this case study, we tested three techniques: LPSP, MUMT, and YMOST. In addition, the first technique is applied to loads of the "Load Daily" type. The second technique is applied to loads of the "Load Monthly" type. The third technique is applied to loads of the "Load Annual" type.

The execution of the solution includes a set of steps:

Step 1:

This step is reserved for the choice of the site based on the rule form presented in Figure 5.

```
R4: Site(?x) ⇒ sqwrl:select(?x)
```

Figure 5. Rule form.

In this example, the choice is Adrar City.

Step 2:

The objective in this step is to choose the energy sources. This operation is realized based on the intelligent reasoning rules (Figure 6).

```
R5: Site(?x) ∧ hasRadiation(?x, ?y) ∧ swrlb:greaterThan(?y, 2550)
⇒ sqwrl: select(?x, PhotovoltaicSource)
R6: Site(?x) ∧ hasWindSpeed(?x, ?y) ∧ swrlb:greaterThan(?y, 5)
⇒ sqwrl:select(?x, WindSource)
R7: Site(?x) ∧ hasSoilTemperature(?x, ?y) ∧ swrlb:greaterThan(?y, 110)
⇒ sqwrl:select(?x, GeothermalSource)
R8: Site (x) ∧ Source (PhotovoltaicSource) ∧ Source (WindSource)
⇒ Storage (Battery)
```

Figure 6. Examples of intelligent reasoning rules.

- In rule (R5), by replacing the variables "x" and "y" with "3378" and "Adrar" sequentially, it can be seen that "y" is greater than 2550, which means that the Adrar site is characterized by a very high solar potential. The reasoning of the solution proposes the first source of energy, which is the solar photovoltaic.
- By replacing the variables (x = Adrar, y = 6) in rule (R6), we notice that "y" is greater than 5, which means that Adrar is characterized by an interesting wind potential. The reasoning of the solution proposes the second source of energy, which is the wind turbine.
- By replacing the variables x = Adrar and y = 65 in the rule (R7), we notice that "y" is less than 110, which means that the geothermal source is not interesting for the production of electricity on the site of Adrar.
- Based on the previous results and rule (R8), the reasoning of the solution proposes energy storage batteries as an essential element to ensure the good reliability of HES.

Step 3:

In this step, the objective is to choose the optimization technique by using the R1, R2, and R3 reasoning rules, then by replacing the variable "y = DailyLoad" in the three rules, (R1), (R2), and (R3). The "LPSP" technique is proposed by the reasoning of the solution.

Due to the complexity of using and mastering the software and optimization tools for HES, this solution was proposed. It allows the selection of the appropriate tool to calculate the optimal configuration. This selection is made in a simple, easy way. It also makes it possible to relieve the user of any knowledge of energy sources or optimization techniques.

The use of other methods requires knowledge of several aspects (sites, techniques, energy sources). On the other hand, with this solution, the user only chooses the site concerned with the installation, and the rest of the steps are ensured by the developed solution.

3. Results and Discussion

Evaluation of the Proposed Solution

To have an alignment of a proposed solution of a research question, which presents the key to evaluating solutions from a research point of view, requires the good formulation and use of hypotheses and conceptual subproblems. Research is only feasible if we can establish the validity of the hypotheses and conceptual subproblems proposed through a good review of the literature. Additionally, the correct answers to the conceptual subproblems allow a good alignment.

Through the results obtained in this work, we can see that all responses to the elements mentioned in subsection "2.1. *Proposed Approach for Ontology Construction*", respond to the sub-questions stated in the same sub-section (Table 8).

Table 8. Samples of evaluation of the proposed solution.

Hypothesis/Question	Response
The development of an optimization generic solution for HES facilitates the choice of an appropriate technique and simplifies the task for the setup of this type of systems. (Hypothesis)	The proposed solution allows choosing the best technology to improve HES and is easy to use.
The more energy sources there are at a site, the more effective the hybrid energy system is. (Hypothesis)	Through this work, it was confirmed that there is a direct relationship between the number of energy sources available at the site and the efficiency of the hybrid energy system.
The importance of the solution increases with the increase in the number of cases treated. (Hypothesis)	The proposed solution can be applied to the majority of cases (despite changing locations) and is, therefore, a generic solution.
What is the impact of the solution developed on the optimization of hybrid energy systems? (Main question)	According to the obtained results, the optimization of hybrid energy systems using the developed solution makes it possible to gain a system that is more efficient, reliable, and economical.
To what extent can research be performed in previous works? (Subquestion)	Bibliographic research focused on recent works concerning the optimization of hybrid energy systems. Consequently, this research was very positive and allowed us to elaborate on the problematic with precision.
What is the appropriate approach (technique) that can be used to develop a solution in this work? (Subquestion)	Through the use of the ontological approach, we discovered that it is a very suitable tool for representing all knowledge as well as in the processes of developing rules of intelligent reasoning. Then, ontology-based solutions make it possible to perform all the necessary updates on the knowledge base without damaging the overall structure, unlike classic databases.
Is there a relationship between the characteristics of the site and the hybrid energy system that will be installed? (Subquestion)	Through the work carried out, it is confirmed that there is a strong relationship between the characteristics of the site (available energy sources, climatic data, geographical location, etc.) and the characteristics of the hybrid energy system that will be installed.
What are the reasons for using a hybrid energy system instead of a nonhybrid energy system? (Subquestion)	The use of a hybrid energy system has several benefits, including ensuring the continuity of supply and energy needs.

To assess the hypotheses proposed in this work, the following steps are realized:

(1) Proposing the null hypothesis (the set of hypotheses and questions that have been proposed);
(2) Suggesting methodologies and choosing the necessary means to develop the solution (programs, algorithms, etc.);
(3) Testing the solution using data for real sites located in Algerian cities, where the results for this test were very positive, and all the questions that were set were answered.

4. Conclusions

This work focused on the problem of optimizing HES. In this context, it was proposed to develop an optimal solution based on ontology.

The HES are systems that combine multiple energy sources to meet electricity needs. They allow improvements in terms of energy efficiency and increasing integration of RE sources. However, the process of optimizing these systems is complex, especially with its distributed and interactive nature with the external environment.

In an HES, energy sources and storage tools are combined to meet consumer demand for electricity while ensuring the lowest cost. These tasks are the main goals for HES optimization. However, because of the intermittent nature of RE, the optimization of the HES is difficult, which depends principally on the data of energy sources and climate. Responding to the optimization problem is precisely the objective of this work.

There is a set of tools for HES optimization, such as algorithms and software, where each is based on one or hybridization between several approaches (probability, linear programming, fuzzy logic, neural networks, etc.). There is also some software for HES optimization, the most popular being Hybrid, TRNSYS [38], Hybrid2, Homer [39], and HOGA [40]. However, the choice between these tools is considered a real problem. First, a generic optimization solution for HES was developed. This contribution shows its reliability following an explanatory scenario and based on real data from Algerian sites. The results indicate the importance of the proposed solution for real uses, where the user can benefit from many advantages, such as saving time, accuracy, and ease of use.

For future works, this solution needs collaborative work with experts in many fields, which will allow the development of the existing solution in many points, as well as the introduction of some other technologies, such as multiagent systems, especially the mobile agent, which can navigate to the source of information to take it instead of staying in a fixed address and communicate with the rest of the agents based on network protocols, which makes the solution more effective. Finally, extensive testing must be performed for other case studies to find the limits of the use of this solution.

Author Contributions: Formal analysis, D.S.; writing—original draft, D.S.; software, D.S. and H.H.; methodology, D.S. and O.C.; validation, D.S. and F.H.; funding acquisition, O.C., F.H. and H.H.; supervision, D.S., O.C. and Y.S.; visualization, F.H. and H.H.; writing—review and editing, Y.S., A.H. and H.H.; project administration, D.S. All authors have read and agreed to the published version of the manuscript.

Funding: Princess Nourah bint Abdulrahman University, Riyadh, Saudi Arabia, Project Number PNURSP2022R236.

Institutional Review Board Statement: Not applicable.

Informed Consent Statement: Not applicable.

Data Availability Statement: Not applicable.

Acknowledgments: This project was supported by Princess Nourah bint Abdulrahman University researchers supporting Project Number (PNURSP2022R236), Princess Nourah bint Abdulrahman University, Riyadh, Saudi Arabia.

Conflicts of Interest: The authors declare no conflict of interest.

Abbreviation

AI	artificial intelligence
API	application programming interface
GIS	geographic information system
GUI	graphical user interface
HES	hybrid energy system
HOGA	Hybrid Optimization by Genetic Algorithms
HS	harmony search
KB	knowledge base
kWh	kilowatt hour
LPSP	loss of power supply probability
MCDM	multicriteria decision making
MUMT	most unfavorable month technique
OWL-DL	Web Ontology Language–Language Description
PSO	particle swarm optimization
PV	photovoltaic
RE	renewable energy
RuleML	Rule Markup Language
SA	simulated annealing
SDO	Simulink Design Optimization
SWRL	Semantic Web Rule Language
TS	tabu search
YMOST	yearly monthly overage sizing

References

1. Bo, M.; Mercalli, L.; Pognant, F.; Cat Berro, D.; Clerico, M. Urban air pollution, climate change and wildfires: The case study of an extended forest fire episode in northern Italy favoured by drought and warm weather conditions. *Energy Rep.* **2020**, *6*, 781–786. [CrossRef]
2. Nazir, M.S.; Mahdi, A.J.; Bilal, M.; Sohail, H.M.; Ali, N.; Iqbal, H.M.N. Environmental impact and pollution-related challenges of renewable wind energy paradigm—A review. *Sci. Total Environ.* **2019**, *683*, 436–444. [CrossRef] [PubMed]
3. De Doile, G.N.D.; Rotella Junior, P.; Rocha, L.C.S.; Bolis, I.; Janda, K.; Coelho Junior, L.M. Hybrid Wind and Solar Photovoltaic Generation with Energy Storage Systems: A Systematic Literature Review and Contributions to Technical and Economic Regulations. *Energies* **2021**, *14*, 6521. [CrossRef]
4. Olówósejéjé, S.; Leahy, P.; Morrison, A.P. Optimising photovoltaic-centric hybrid power systems for energy autonomy. *Energy Rep.* **2021**, *7*, 1943–1953. [CrossRef]
5. Saba, D.; Laallam, F.Z.; Hadidi, A.E.; Berbaoui, B. Optimization of a Multi-source System with Renewable Energy Based on Ontology. *Proc. Energy Procedia* **2015**, *74*, 608–615. [CrossRef]
6. Maouedj, R.; Benmedjahed, M.; Saba, D.; Mamemri, A.; Barbaoui, B.; Bezari, S. Experimental analysis of a stand-alone wind-photovoltaic hybrid system in the Sahara Desert. *Eur. J. Electr. Eng.* **2019**, *21*, 93–97. [CrossRef]
7. Ammari, C.; Belatrache, D.; Touhami, B.; Makhloufi, S. Sizing, optimization, control and energy management of hybrid renewable energy system—A review. *Energy Built Environ.* **2021**, *3*, 399–411. [CrossRef]
8. Nourollahi, R.; Tabar, V.S.; Zadeh, S.G.; Akbari-Dibavar, A. A hybrid optimization approach to analyze the risk-constrained operation of a residential hybrid energy system incorporating responsive loads. *Comput. Chem. Eng.* **2022**, *157*, 107603. [CrossRef]
9. Guo, J.; Zhang, P.; Wu, D.; Liu, Z.; Ge, H.; Zhang, S.; Yang, X. A new collaborative optimization method for a distributed energy system combining hybrid energy storage. *Sustain. Cities Soc.* **2021**, *75*, 103330. [CrossRef]
10. Sharma, R.; Kodamana, H.; Ramteke, M. Multi-objective dynamic optimization of hybrid renewable energy systems. *Chem. Eng. Process. Process Intensif.* **2022**, *170*, 108663. [CrossRef]
11. Fulzele, J.B.; Daigavane, M.B. Design and Optimization of Hybrid PV-Wind Renewable Energy System. *Mater. Today Proc.* **2018**, *5*, 810–818. [CrossRef]
12. Mahmoud, F.S.; Diab, A.A.Z.; Ali, Z.M.; El-Sayed, A.-H.M.; Alquthami, T.; Ahmed, M.; Ramadan, H.A. Optimal sizing of smart hybrid renewable energy system using different optimization algorithms. *Energy Rep.* **2022**, *8*, 4935–4956. [CrossRef]
13. Cano, A.; Jurado, F.; Sánchez, H.; Fernández, L.M.; Castañeda, M. Optimal sizing of stand-alone hybrid systems based on PV/WT/FC by using several methodologies. *J. Energy Inst.* **2014**, *87*, 330–340. [CrossRef]
14. Maleki, A.; Askarzadeh, A. Comparative study of artificial intelligence techniques for sizing of a hydrogen-based stand-alone photovoltaic/wind hybrid system. *Int. J. Hydrogen Energy* **2014**, *39*, 9973–9984. [CrossRef]
15. Aydin, N.Y.; Kentel, E.; Duzgun, H.S. GIS-based site selection methodology for hybrid renewable energy systems: A case study from western Turkey. *Energy Convers. Manag.* **2013**, *70*, 90–106. [CrossRef]

16. Luna-Rubio, R.; Trejo-Perea, M.; Vargas-Vázquez, D.; Ríos-Moreno, G.J. Optimal sizing of renewable hybrids energy systems: A review of methodologies. *Sol. Energy* **2012**, *86*, 1077–1088. [CrossRef]
17. Saba, D.; Laallam, F.Z.; Hadidi, A.E.; Berbaoui, B. Contribution to the Management of Energy in the Systems Multi Renewable Sources with Energy by the Application of the Multi Agents Systems "MAS". *Energy Procedia* **2015**, *74*, 616–623. [CrossRef]
18. Saba, D.; Laallam, F.Z.; Berbaoui, B.; Abanda, F.H. *An Energy Management Approach in Hybrid Energy System Based on Agent's Coordination*; Hassanien, A., Shaalan, K., Gaber, T., Azar, A., Tolba, M., Eds.; Springer: Cham, Switzerland, 2017; Volume 533, ISBN 9783319483078.
19. Saba, D.; Degha, H.E.; Berbaoui, B.; Maouedj, R. *Development of an Ontology Based Solution for Energy Saving through a Smart Home in the City of Adrar in Algeria*; Hassanien, A., Tolba, M., Elhoseny, M., Mostafa, M., Eds.; Springer: Cham, Switzerland, 2018; Volume 723, ISBN 9783319746890.
20. Coelho, V.N.; Weiss Cohen, M.; Coelho, I.M.; Liu, N.; Guimarães, F.G. Multi-agent systems applied for energy systems integration: State-of-the-art applications and trends in microgrids. *Appl. Energy* **2017**, *187*, 820–832. [CrossRef]
21. Amicarelli, E.; Tuan, T.Q.; Bacha, S. Multi-agent system for day-ahead energy management of microgrid. In Proceedings of the Power Electronics and Applications EPE'16 ECCE Europe, 2016 18th European Conference, Karlsruhe, Germany, 5–9 September 2016; pp. 1–10.
22. Elamine, D.O.; Nfaoui, E.H.; Jaouad, B. Multi-agent system based on fuzzy control and prediction using NN for smart microgrid energy management. In Proceedings of the 2015 Intelligent Systems and Computer Vision (ISCV), Fez, Morocco, 25–26 March 2015; pp. 1–6.
23. Morstyn, T.; Hredzak, B.; Agelidis, V. Network Topology Independent Multi-Agent Dynamic Optimal Power Flow for Microgrids with Distributed Energy Storage Systems. *IEEE Trans. Smart Grid* **2016**, *9*, 3419–3429. [CrossRef]
24. Raju, L.; Rajkumar, I.; Appaswamy, K. Integrated energy management of micro-grid using multi agent system. In Proceedings of the 2016 International Conference on Emerging Trends in Engineering, Technology and Science (ICETETS), Pudukkottai, India, 24–26 February 2016; pp. 1–8.
25. Mbodji, A.K.; Ndiaye, M.L.; Ndiaye, P.A. Decentralized control of the hybrid electrical system consumption: A multi-agent approach. *Renew. Sustain. Energy Rev.* **2016**, *59*, 972–978. [CrossRef]
26. Saba, D.; Maouedj, R.; Berbaoui, B. Contribution to the development of an energy management solution in a green smart home (EMSGSH). In Proceedings of the 7th International Conference on Software Engineering and New Technologies, Hammamet, Tunisia, 26–28 December 2018; ACM Press: New York, NY, USA, 2018; pp. 1–7.
27. Saba, D.; Cheikhrouhou, O.; Alhakami, W.; Sahli, Y.; Hadidi, A.; Hamam, H. Intelligent Reasoning Rules for Home Energy Management (IRRHEM): Algeria Case Study. *Appl. Sci.* **2022**, *12*, 1861. [CrossRef]
28. Saba, D.; Sahli, Y.; Abanda, F.H.; Maouedj, R.; Tidjar, B. Development of new ontological solution for an energy intelligent management in Adrar city. *Sustain. Comput. Inform. Syst.* **2019**, *21*, 189–203. [CrossRef]
29. Saba, D.; Sahli, Y.; Hadidi, A. An ontology based energy management for smart home. *Sustain. Comput. Inform. Syst.* **2021**, *31*, 100591. [CrossRef]
30. Saba, D.; Laallam, F.Z.; Degha, H.E.; Berbaoui, B.; Maouedj, R. Design and Development of an Intelligent Ontology-Based Solution for Energy Management in the Home. In *Machine Learning Paradigms: Theory and Application*; Hassanien, A., Ed.; Springer: Berlin/Heidelberg, Germany; Cham, Switzerland, 2019; Volume 801, ISBN 978-3-030-02357-7.
31. Uschold, M.; Gruninger, M. Ontologies: Principles, methods and applications. *Knowl. Eng. Rev.* **1996**, *11*, 93–136. [CrossRef]
32. Saba, D.; Sahli, Y.; Maouedj, R.; Hadidi, A.; Medjahed, M.B. *Contribution to the Realization of a Smart and Sustainable Home*; Hassanien, A., Bhatnagar, R., Darwish, A., Eds.; Springer: Cham, Switzerland, 2021; Volume 912, ISBN 978-3-030-51920-9.
33. Abdelkader, A.; Rabeh, A.; Mohamed Ali, D.; Mohamed, J. Multi-objective genetic algorithm based sizing optimization of a stand-alone wind/PV power supply system with enhanced battery/supercapacitor hybrid energy storage. *Energy* **2018**, *163*, 351–363. [CrossRef]
34. Saba, D.; Berbaoui, B.; Degha, H.E.; Laallam, F.Z. *A Generic Optimization Solution for Hybrid Energy Systems Based on Agent Coordination*; Hassanien, A., Shaalan, K., Gaber, T., Tolba, M., Eds.; Springer: Cham, Switzerland, 2018; Volume 639, ISBN 9783319648606.
35. Saba, D.; Zohra Laallam, F.; Belmili, H.; Henry Abanda, F.; Bouraiou, A. Development of an ontology-based generic optimisation tool for the design of hybrid energy systemsDevelopment of an ontology-based generic optimisation tool for the design of hybrid energy systems. *Int. J. Comput. Appl. Technol.* **2017**, *55*, 232–243. [CrossRef]
36. Degha, H.E.; Laallam, F.Z.; Said, B.; Saba, D. Onto-SB: Human Profile Ontology for Energy Efficiency in Smart Building. In Proceedings of the 2018 3rd International Conference on Pattern Analysis and Intelligent Systems (PAIS), Tebessa, Algeria, 24–25 October 2018.
37. Abanda, F.H.; Tah, J.H.M.; Duce, D. PV-TONS: A photovoltaic technology ontology system for the design of PV-systems. *Eng. Appl. Artif. Intell.* **2013**, *26*, 1399–1412. [CrossRef]
38. Figaj, R.; Żołądek, M.; Goryl, W. Dynamic Simulation and Energy Economic Analysis of a Household Hybrid Ground-Solar-Wind System Using TRNSYS Software. *Energies* **2020**, *13*, 3523. [CrossRef]

39. Antonio Barrozo Budes, F.; Valencia Ochoa, G.; Obregon, L.G.; Arango-Manrique, A.; Ricardo Núñez Álvarez, J. Energy, Economic, and Environmental Evaluation of a Proposed Solar-Wind Power On-grid System Using HOMER Pro®: A Case Study in Colombia. *Energies* **2020**, *13*, 1662. [CrossRef]
40. Guelleh, H.O.; Mujtaba, I.M.; Patel, R. Computational Tools used in Hybrid Renewable Energy Systems Optimization-An overview. *Comput. Aided Chem. Eng.* **2020**, *48*, 1477–1482.

sustainability

MDPI

Article

A Hybrid Generative Adversarial Network Model for Ultra Short-Term Wind Speed Prediction

Qingyuan Wang [1,†], Longnv Huang [1,†], Jiehui Huang [1,†], Qiaoan Liu [2], Limin Chen [1,*], Yin Liang [1], Peter X. Liu [1,3] and Chunquan Li [1]

1 The School of Information Engineering, Nanchang University, Nanchang 330031, China; 7904119023@email.ncu.edu.cn (Q.W.); 6105119006@email.ncu.edu.cn (L.H.); 7803018161@email.ncu.edu.cn (J.H.); liangyin@ncu.edu.cn (Y.L.); xpliu@sce.carleton.ca (P.X.L.); lichunquan@ncu.edu.cn (C.L.)

2 The School of Future Technology, Nanchang University, Nanchang 330031, China; 5811121050@email.ncu.edu.cn

3 The Department of Systems and Computer Engineering, Carleton University, Ottawa, ON K1S 5B6, Canada

* Correspondence: chenlimin@ncu.edu.cn

† These authors contributed equally to this work.

Abstract: To improve the accuracy of ultra-short-term wind speed prediction, a hybrid generative adversarial network model (HGANN) is proposed in this paper. Firstly, to reduce the noise of the wind sequence, the raw wind data are decomposed using complete ensemble empirical mode decomposition with adaptive noise (CEEMDAN). Then the decomposed modalities are entered into the HGANN network for prediction. HGANN is a continuous game between the generator and the discriminator, which in turn allows the generator to learn the distribution of the wind data and make predictions about it. Notably, we developed the optimized broad learning system (OBLS) as a generator for the HGANN network, which can improve the generalization ability and error convergence of HGANN. In addition, improved particle swarm optimization (IPSO) was used to optimize the hyperparameters of OBLS. To validate the performance of the HGANN model, experiments were conducted using wind sequences from different regions and at different times. The experimental results show that our model outperforms other cutting-edge benchmark models in single-step and multi-step forecasts. This demonstrates not only the accuracy and robustness of the proposed model but also the applicability of our model to more general environments for wind speed prediction.

Keywords: wind speed forecast; OBLS; data preprocessing; optimized hyper-parameters

Citation: Wang, Q.; Huang, L.; Huang, J.; Liu, Q.; Chen, L.; Liang, Y.; Liu, P.X.; Li, C. A Hybrid Generative Adversarial Network Model for Ultra Short-Term Wind Speed Prediction. *Sustainability* **2022**, *14*, 9021. https://doi.org/10.3390/su14159021

Academic Editors: Luis Hernández-Callejo, Sergio Nesmachnow and Sara Gallardo Saavedra

Received: 5 June 2022
Accepted: 20 July 2022
Published: 22 July 2022

1. Introduction

Energy demand has always been one of the main problems of human development since the increasing consumption of energy with the improvement of living standards. In recent years, renewable energy has gradually become a research hotspot. Wind energy is valued for its clean, pollution-free, renewable, and abundant availability. However, wind is highly random and volatile, which may affect the stability of the power system and hinder the efficient use of wind energy [1]. Accurate ultra-short-term wind speed prediction models are therefore crucial in power dispatch planning and power market operations [2]. Thus, reliable wind speed prediction has drawn a lot of interest.

The three common wind speed prediction models are physical models, statistical models, and hybrid models. Physical models take into account the physical conditions and locations of wind farms, which require abundant meteorological data. Numerical weather prediction is a typical physical model, as it takes into account temperature pressure and obstacles for wind speed prediction, so it has a long calculation period [3]. Physical models are efficient and accurate for long-term forecasting, but they are computationally intensive and expensive for small-scale forecasting.

Statistical models make better use of historical wind speed data to predict future wind speeds than physical models. Statistical models include both traditional statistical models and neural network-based models. Traditional statistical models include the autoregressive moving average model [4], the autoregressive integrated moving average model [5], the Bayesian model [6], etc. However, the non-linear nature of wind makes it difficult for traditional statistical models to extract the deeper features of wind speed data. Neural networks are introduced into the field of wind speed prediction for their ability to fit the non-linear part of the data well. Neural network-based models can extract deeper features from wind speed data than traditional statistical models—for example, BP [7], RBF [8], artificial neural network [9], SVR [10], etc. To improve the learning ability and predictive ability of predictive models, deep neural networks are introduced into wind speed prediction, such as the deep belief network [11], RNN [12], GNN [13], and LSTM [14].

In recent years, hybrid models have gradually become the mainstream wind speed prediction models. Hybrid models typically use one or more auxiliary strategies to assist the main forecasting network in wind speed prediction. Therefore, hybrid models can achieve better prediction performance than physical models and statistical models. The auxiliary strategies involved in hybrid models include data preprocessing techniques, optimization algorithms, error correction, and weighting strategies.

(1) Data preprocessing techniques. Zhang et al. [15] used EMD for data pre-processing of wind speed, which effectively reduced the volatility of the wind speed series. However, EMD suffers from the problem of modal confusion, which leads to unsatisfactory decomposition results. Santhosh et al. [16] used EEMD to process the raw wind speed series, which effectively mitigated the EMD problem. However, EEMD has a noise residual problem affected by noise residuals. Wang et al. [17] used CEEMD for wind speed prediction. CEEMD cancels out the residual noise with a pair of white noises, effectively improving the efficiency of the calculation. Ren et al. [18] experimentally demonstrated that the CEEMDAN-based model always performs best compared to the EMD-based model.

(2) Optimization algorithms. Optimization algorithms can be used to optimize the hyper-parameters, weights, network structure, and thresholds of predictive models. Li et al. [19] used PSO to optimize two hyper-parameters of LSTM, which solved the problem of wide intervals caused by interval superposition and thus improved the wind speed prediction accuracy. Tian [20] used PSO to optimize the weight coefficients of each prediction model, and the experimental results demonstrate the necessity of introducing the weight coefficient optimization strategy. Liu et al. [21] used GA to optimize the internal parameters of LSTM, which improved the efficiency and accuracy of the prediction model. Cui et al. [22] used the Bat algorithm to optimize the thresholds of BP networks, effectively improving the generalization ability and nonlinear mapping ability of BP networks.

(3) Error correction. Error correction is a post-processing technique for wind forecasting. It predicts the residuals and superimposes the predictions on the original predictions to obtain the final predictions. Duan et al. [23] used improved CEEMDAN to decompose the errors, and the experimental results showed that the error decomposition correction method can significantly enhance the prediction accuracy. Liu et al. [24] proposed an adaptive multiple error correction method, which makes full use of the deeper predictable components and effectively improves the reliability and accuracy of the model. Zhang et al. [25] demonstrated experimentally that the final predicted values after Markov chain correction are closer to the original wind field data, which proves that the use of the Markov chain is effective.

(4) Weighting strategies. To scientifically determine the weights of different prediction networks in a hybrid model, many scholars have proposed different weighting strategies. To alleviate the adverse effects of multi-collinearity in combinatorial prediction models, Jiang et al. [26] used a GMDH neural network to automatically identify the weights of three nonlinear models. The application of GMDH can significantly improve the predictive

capability compared to the widely used equal-weighting scheme. Wang et al. [27] used MTO to minimize the error sum of squares of the IOWA operator, which obtains the optimal weight vector for the combined prediction model and ensures the stability of the prediction results. Altan et al. [28] optimized the weighting coefficients for each IMF using the gray wolf optimizer algorithm.

Although the above models achieve good predictive performance, they still have some problems. Methods involving deep neural networks [27] cause huge computational costs. Hybrid methods based on weighting strategy [28] may have the problem of multicollinearity, which reduces the prediction accuracy. The performance of hybrid methods based on parameter optimization [26] is largely influenced by the understanding of the researcher of the optimization algorithm.

Considering the above issues, we propose a hybrid model combining data preprocessing techniques and optimization algorithms for ultra-short-term wind speed prediction. We design the hybrid generative adversarial network (HGANN) as the prediction master network for the proposed hybrid model. The contributions and innovations of this research are concluded as follows:

(1) A hybrid generative adversarial network model (HGANN) is proposed for ultra-short-term wind speed prediction, which learns the distribution of wind data and predicts it through a continuous game between generators and discriminators.
(2) To improve the error convergence of the model, the OBLS was developed as a generator for HGANN. The IPSO was used to optimize the hyperparameters of the OBLS. To maintain the stability of the generated samples, we used the discriminator of WGAN as the discriminator of HGANN.
(3) A wind data decomposition and denoising process was carried out using CEEMDAN to reduce the randomness and instability in the original wind series.

The rest of this article is organized as follows. Section 2 introduces the model framework and methods involved in this article in detail. In Section 3, the experimental cases and prediction results are elaborated in detail, which verifies the validity of the framework we propose. Section 4 contains a discussion of the results of the experiment. The conclusions are presented in Section 5.

2. Proposed Predictive Framework

2.1. Overall Framework of HGANN

Generative adversarial networks (GANs) [29] are deep learning networks, which are composed of a generator and discriminator that confront each other. The role of the generator is to generate false samples that are close to the real ones. The role of the discriminator is to distinguish between true and false samples as correctly as possible. However, GANs often suffer from the problem of target confusion. Our proposed HGANN alleviates this problem to a great extent.

We developed a hybrid generative adversarial network model (HGANN) for ultra-short-term wind speed prediction, which uses the two networks to compete with each other to achieve highly accurate wind speed predictions. The proposed model is shown in Figure 1. First, CEEMDAN decomposes the raw wind speed data into multiple modalities. These modalities are separately fed into the generator of HGANN, the OBLS. The generator is used to obtain virtual samples that are similar to real samples. The virtual samples and real samples are then fed into the discriminator, which consists of convolutional layers and fully connected layers. The discriminator extracts the high-dimensional features of the input samples through the convolutional layer, and then further extracts the effective features by the fully connected layer. The outputs scalars "1" or "0" of the discriminator are passed to the generator and the discriminator to perform the iterative update of HGANN. Via the continuous iterative update, OBLS obtains the best parameters and performs wind speed prediction. Finally, the final wind speed forecast can be obtained by stacking all forecast values.

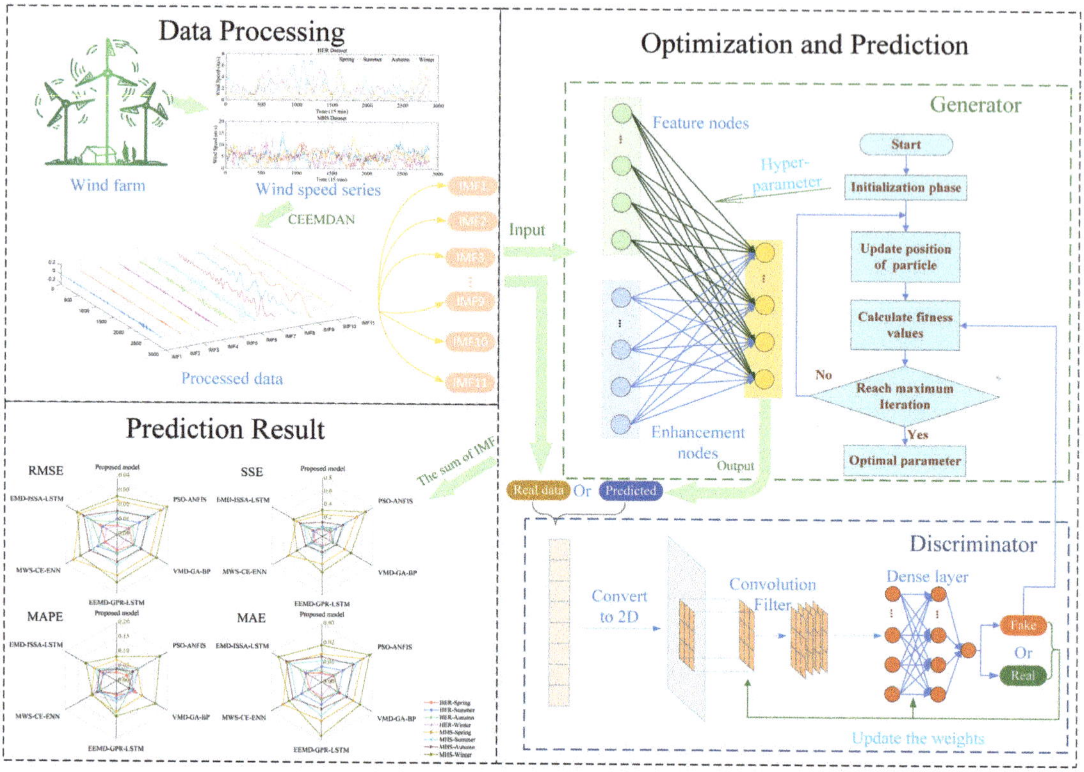

Figure 1. The proposed short-term wind speed forecasting framework. In the data processing step, CEEMDAN turns the wind data into multiple modalities. The HGANN network consisting of a generator and discriminator predicts these modalities. The final wind speed prediction result can then be obtained by stacking the prediction results of all modalities.

2.2. CEEMDAN Model

Due to the high volatility of the wind speed series, CEEMDAN [30] is introduced to smooth the wind speed data. CEEMDAN decomposes a signal into some modalities.

The original wind speed series is defined as $X(n)$. CEEMDAN decomposes $X(n)$ into $\overline{IMF_j}(n), j = 1, 2, 3, \ldots J$ and residue $r_j(n)$. Figure 2 shows the flow chart of the CEEMDAN algorithm. The specific steps of the algorithm are as follows.

Randomly generate white noise with (0, 1), which is defined as $w^i(n)$, $i = 1, 2, \ldots I$. Define an operator $E\{*\}$, which generates the IMFs by EMD. We set the noise standard deviation to $\varepsilon = 0.2$ and the ensemble size to $I = 500$.

Add $w^i(n)$ to the $X(n)$ and generate a new series with noisy signals $X(n) + \varepsilon_0 w^i(n)$. For $j = 1$, the first-order $\overline{IMF_1}(n)$ that is decomposed by EMD is expressed as:

$$\overline{IMF_1}(n) = \frac{1}{I}\sum_{i=1}^{I} E\left\{X(n) + \varepsilon_0 w^i(n)\right\} \quad (1)$$

The first-order residue is computed as follows:

$$r_1(n) = X(n) - \overline{IMF_1}(n) \quad (2)$$

For $j = 2, 3, \ldots J$, calculate the $\overline{IMF_j}(n)$ and the jth residue as follows:

$$\overline{IMF_j}(n) = \frac{1}{I}\sum_{i=1}^{I} E\left\{ r_{j-1}(n) + \varepsilon_{j-1}w^i(n)\right\} \tag{3}$$

$$r_j(n) = r_{j-1}(n) - \overline{IMF_j}(n) \tag{4}$$

Decompose $E\{r_{j-1}(n) + \varepsilon_{j-1}w^i(n)\}$ until the residue $r_j(n)$ cannot be decomposed and has only one extreme value. Then we can get $\overline{IMF_j}(n) = \frac{1}{I}\sum_{i=1}^{I} E\{ r_{j-1}(n) + \varepsilon_{j-1}w^i(n)\}$ and the final residue $r_j(n) = X(n) - \sum_{j=1}^{J} \overline{IMF_j}(n)$.

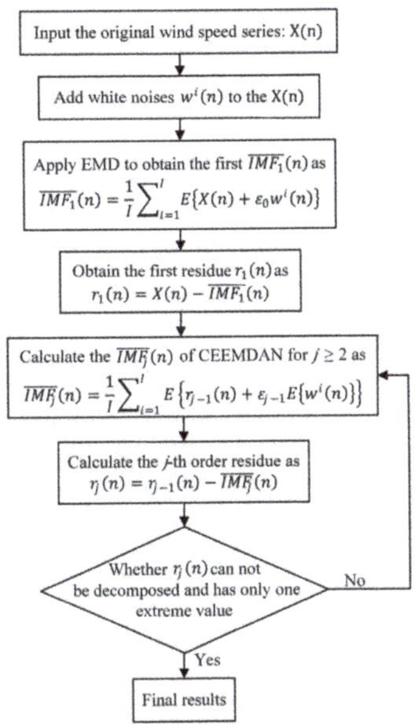

Figure 2. Flowchart of the CEEMDAN algorithm procedure.

The original wind speed time series $X(n)$ can be decomposed as $X(n) = \sum_{j=1}^{J} \overline{IMF_j}(n)$ $+r_j(n)$, where $\overline{IMF_j}(n)$ or $r_j(n)$ can represent different features of the wind speed.

2.3. Generator OBLS for HGANN

BLS [31] can provide incremental structural learning. It achieves better forecasting results in time-series forecasting. Furthermore, because of its shallow network structure, BLS has higher error convergence performance than CNN. Compared with BLS, OBLS can provide both higher convergence performance and predictive accuracy. This is because OBLS uses IPSO to improve the network hyper-parameter of optimization. Therefore, OBLS has faster convergence and higher error convergence than CNN. Therefore, instead of using CNN as the generator of GAN, we use OBLS as the generator of HGANN to solve the problem of target confusion during HGANN training, which can improve the generalization ability and error convergence of HGANN, and thus make HGANN more suitable for wind speed prediction. The following is the detailed process of the OBLS algorithm.

Randomly generate n particles so that the dimensions of the particles are a three-dimensional vector $\{NF, NW, NE\}$ corresponding to the three parameters of BLS, respectively. Initialize the particle position $x_{id} \in (1, 100)$, and speed $v_{id} \in (-1, 1)$. Determine the learning factors $c_1 = 1.5$ and $c_2 = 1.5$, inertia weights $w_{max} = 1.0$ and $w_{min} = 0.4$, and the maximum number of iterations $iter_{max} = 100$.

Assume the input wind speed series data $X(n)$ and project the data using $\varnothing_i(X(n)W_{ei}+\beta_{ei})$ to represent ith mapped feature Z_i, where W_{ei} represents random weight with the proper dimensions. The jth group of enhancement nodes $\varnothing_j\left(Z_iW_{hj}+\beta_{hj}\right)$ is denoted as H_i. $\varnothing_i$ and $\varnothing_j$ can be different functions. The ith mappings can be denoted as:

$$Z_i = \varnothing_i(X(n)W_{ei}+\beta_{ei}), i = 1, 2, \ldots n \tag{5}$$

The feature nodes are denoted as $Z^n \triangleq [Z_1, Z_2 \ldots Z_n]$, where W_{hj} and β_{hj} are random weights. The enhanced nodes are denoted as:

$$H_j = \varnothing_j\left(Z_iW_{hj}+\beta_{hj}\right), j = 1, 2, \ldots m \tag{6}$$

Let $H^m \triangleq [H_1, H_2 \ldots H_m]$ where the symbol $\triangleq$ means "noted as"; then the output of the BLS can be denoted as:

$$Y = \{Z^n|H^m\}W^n \tag{7}$$

where the W^n is the final target weight needed by OBLS and is obtained through the ridge regression algorithm, that is, $W^n \triangleq \{Z^n|H^m\}^+Y$.

Let $\{M\} = \{Z^n|H^m\}$; then $\{Z^n|H^m\}^+$ can be expressed as follows:

$$\{Z^n|H^m\}^+ = \lim_{\lambda \to 0}\left\{\lambda I + \{M\}\{M\}^T\right\}^{-1}\{M\}^T \tag{8}$$

where λ is l_2 regularization.

The IPSO [32] is introduced to iterate to optimize the parameters of BLS: $\{NF, NW, NE\}$. When the iteration of IPSO is consistently performed, the position and speed of the particles are continually updated through the following equation:

$$v_{id} = wv_{id} + c_1r_1(p_{id}-x_{id}) + c_2r_2\left(p_{gd}-x_{gd}\right) \tag{9}$$

$$x_{id} = x_{id} + \gamma v_{id} \tag{10}$$

Here, γ is the velocity coefficient; the value of the inertia weight w is $w = w_{max} - (w_{max} - w_{min}) * 1/iter$. When reaching the maximum iterative number $iter_{max}$, the iteration is stopped and the optimal value of $\{NF, NW, NE\}$ can be obtained.

The generator takes the wind speed subsequence $\{x_1, x_2, \ldots x_n\}$ as input, which is generated by CEEMDAN. Then the generator generates a new wind speed sequence $\{y'_1, y'_2, \ldots y'_n\}$, which is statistically similar to the wind speed sequence $\{y_1, y_2, \ldots y_n\}$.

From Equations (8)–(10), OBLS does not require layer-to-layer coupling. Since there are no multi-layer connections, OBLS does not need to use gradient descent to update the weights, so the computational cost of OBLS is significantly lower than that of deep learning. When the accuracy of OBLS does not meet the requirements, its accuracy can be improved by increasing the "width" of the network nodes. Compared with the increase in the amount of calculation by increasing the number of layers in the deep network, the increase in that by increasing the "width" of the network nodes in OBLS is negligible.

2.4. Discriminators for HGANN

To maintain the stability of the generated samples, we used the discriminator of WGAN [33] as the discriminator of HGANN. In HGANN, the discriminator takes $\{x_i, y'_i\}$ or $\{x_i, y_i\}$ as input. The training goal of discriminator is to discriminate $\{x_i, y'_i\}$ as false and $\{x_i, y_i\}$ as true. The discriminator is trained by minimizing the distance function ($\mathcal{L}_D$) (loss function), which is defined as follows:

$$\mathcal{L}_D = \mathcal{L}(D(\{x_i, y_i\}), 1) + \mathcal{L}(D(\{x_i, y'_i\}), 0) + GP \tag{11}$$

$$P = \frac{\breve{}}{m}\sum_{i}^{m}\left[\nabla_{x_i,y_i'} D\left(x_i,y_i'\right)^{i}\right]^2 \quad (12)$$

here, $D(*)$ represents the output of the D; $\mathcal{L}$ is the binary cross entropy, defined as:

$$\mathcal{L} = -[klog(s) + (1-k)log(1-s)] \quad (13)$$

Based on this loss function, the discriminator can achieve an output of 1 when the input is $\{x_i, y_i\}$ and an output of 0 when the input is $\{x_i, y_i'\}$, and then discriminates the wind speed sequence $\{y_1, y_2, \ldots y_n\}$.

The discriminator outputs a scalar of "0" or "1." The scalar of "0" or "1" has two purposes: (1) It can influence and then adjust the weights of the neural network in the discriminator and maximize Equation (14) through a backpropagation algorithm. (2) It can be passed to the generator to assist the PSO algorithm to find the optimal hyper-parameters of the OBLS and then calculate the value of the fitness function F_c, which is defined as follows:

$$F_c = \frac{1}{n}\sum_{i=1}^{n} D(G(x(n))) \quad (14)$$

where $G(*)$ represents the output of the generator.

2.5. Prediction Steps of the Proposed HGANN Model

We propose the HGANN model for ultra-short-term wind speed prediction. The flow chart of the prediction process of the proposed model is shown in Figure 3. CEEMDAN is used to decompose the raw wind speed data $\{x_1, x_2, \ldots x_n\}$ into multiple modes $\overline{IMF_j}(n)$. These $\overline{IMF_j}(n)$ are separately sent into the generator (OBLS) of HGANN to obtain virtual samples $\{y_1', y_2', \ldots y_n'\}$. The discriminator (WGAN) takes $\{x_i, y_i'\}$ or $\{x_i, y_i\}$ as input and then outputs scalars "1" or "0." The scalars "1" or "0" are passed to the generator (OBLS) and the discriminator (WGAN) to participate in iterative model updates. Through the continuous iterative update, OBLS obtains the optimal value of $\{NF, NW, NE\}$. The final wind speed forecasting values $\{y_1, y_2, \ldots y_n\}$ can be obtained by stacking all forecast values.

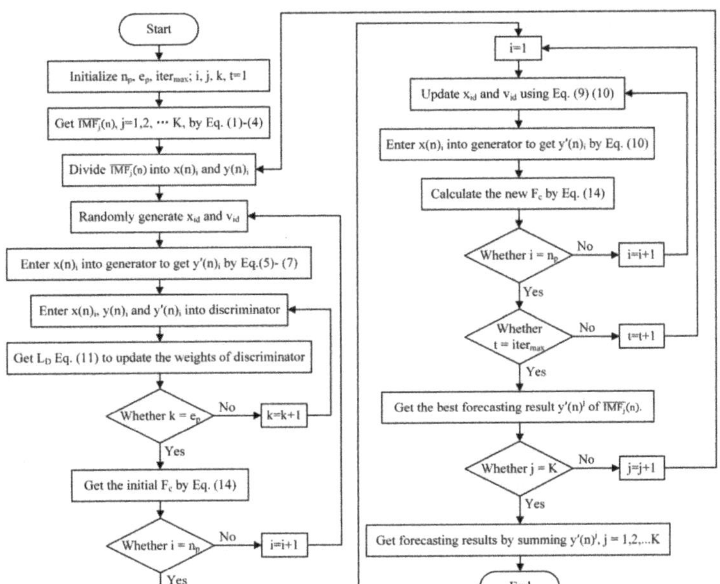

Figure 3. Flowchart of the proposed HGANN model prediction procedure.

3. Case Analysis

3.1. Data Description

To demonstrate the applicability of the proposed model in different locations, we used datasets from the 50Hertz wind farm in Germany and the Mahuangshan wind farm in China [26]: HER and MHS, respectively. The HER datasets are freely available at http://www.netztransparenz.de/ (accessed on 5 October 2021). Both data sets are recorded for one year and wind speeds are measured in 15 min intervals. We selected wind speed series from both HER and MHS datasets for March, June, September, and December, representing spring, summer, autumn, and winter, respectively. Experiments using the four wind speed series of spring, summer, autumn, and winter can verify the applicability of our model at different periods. Each series contains 2880 samples. Table 1 shows information on the selected wind speed data for spring, summer, autumn, and winter.

Table 1. Seasonal statistics of the wind speed data.

Season	Mean (m/s)	Median (m/s)	Max (m/s)	Min (m/s)	Standard Deviation (m/s)
HER data					
Spring	1.89	1.53	7.25	0.03	1.47
Summer	0.82	0.63	4.27	0	0.75
Autumn	1.23	0.83	6.65	0	1.19
Winter	1.90	1.60	5.68	0.07	1.35
MHS data					
Spring	5.78	5.70	16.50	0.40	2.25
Summer	5.46	5.45	12.50	0.40	2.21
Autumn	5.18	5.20	16.50	0.40	2.33
Winter	4.50	4.27	14.07	0	2.28

In our experiments, the first 80% of the wind speed sequence was used as the training set, and the rest was used as the test set for ultra-short-term wind prediction. Table 1 displays the information of the four datasets. The experiments were implemented in MATLAB R2021b on a 64-bit personal computer with Intel(R) core i5-9300 CPU/16.00 GB RAM.

3.2. Evaluation Index

To comprehensively evaluate the prediction performance of HGANN, four evaluate indicators were given. MAE can accurately reflect the average value of the absolute error. MAPE divides the absolute error by the corresponding actual value. RMSE represents the sample standard deviation between the predicted value and the actual observation value, which has a very sensitive reflection and can reflect the accuracy of the prediction well. SSE represents the total error of the model. Their definitions are as follows:

$$MAE = \frac{1}{N}\sum_{i=1}^{N}|y_i - \hat{y}_i| \tag{15}$$

$$MAPE = \frac{1}{N}\sum_{i=1}^{N}\left|\frac{y_i - \hat{y}_i}{y_i}\right| \tag{16}$$

$$RMSE = \sqrt{\frac{1}{N}\sum_{i=1}^{N}(y_i - \hat{y}_i)^2} \tag{17}$$

$$SSE = \sum_{j=1}^{n}(y_i - \hat{y}_i)^2 \tag{18}$$

where $\hat{y}_i$ is the predicted value and y_i is the actual value.

3.3. Comparable Methods

To verify the prediction performance of the proposed HGANN, it was compared with 10 advanced predictive models, involving PSO-ANFIS [34], VMD-GA-BP [35], EEMD-GPR-LSTM [36], EMD-ISSA-LSTM [37], MWS-CE-ENN [20], CNN [38], WGAN [39], BLS [31], OBLS, and WGAN-OBLS. Table 2 lists the parameter settings of six comparison methods. BLS, WGAN, PSO-BLS, and PSO-WGAN-OBLS are the same as those of HAGNN to perform the ablation experiment for HAGNN.

Table 2. Parameter settings of the models.

Model	Parameter Setting
PSO-ANFIS	$iter_{max} = 300,\ n_p = 40,\ c_1 = 1.0,\ c_2 = 2.0,\ n_r = 4,\ n_v = 52$
VMD-GA-BP	$k = 11,\ iter_{max} = 150,\ e_p = 100,\ n_p = 40,\ l_r = 0.1,\ n_{b1} = 9$
EEMD-GPR-LSTM	$k = 11,\ e_p = 200,\ n_{b1} = 100,\ n_{b2} = 100,\ s_1 = 50,\ \sigma = 20$
MWS-CE-ENN	$e_p = 1000,\ l_r = 0.1,\ p_r = 0.000001,\ n_p = 40,\ n_i = 5,\ n_{b1} = 6,\ n_o = 1,\ iter_{max} = 100,\ n_{std} = 0.2$
EMD-ISSA-LSTM	$e_p = 100,\ l_r = 0.005,\ n_p = 10,\ iter_{max} = 20,\ v_e = 0.6,\ p_d = 0.7,\ p_e = 0.2$
Proposed Model	$n_{std} = 0.01, n_p = 40,\ iter_{max} = 100,\ c_1 = 1.5,\ c_2 = 1.5,\ e_p = 50,\ l_r = 0.002,\ \lambda = 10^{-30}$, $n_{b1} = 48,\ n_{b2} = 96,\ n_{b3} = 384$

In Table 2, $iter_{max}$ is the iterative number; e_p is the number of network iterations; n_p is population size; c_1 and c_2 are personal and global learning coefficients, respectively; n_r, n_v, n_i, and n_o are the number of rules, variables, input nodes, and output nodes, respectively; k is the decomposition number of VMD/EEMD; n_{bi} is the number of the ith hidden nodes; l_r is the learning rate of the network; p_r is the training requirement accuracy; n_{std} is the noise standard deviation in ICEEMDAN/CEEMDAN; v_e is the early warning value; p_d and p_s are the proportion of discoverers and sparrows aware of danger, respectively; and ˇ is the regularization parameter for ridge regression.

3.4. Experimental Results

(1) Experiment I: Comparison Between Different Forecasting Methods

We experimentally verified the effectiveness and advancement of the proposed HGANN by comparing it with PSO-ANFIS, VMD-GA-BP, EEMD-GPR-LSTM, MWS-CE-ENN, and EMD-ISSA-LSTM. Considering that wind data characteristics show strong seasonality, experiments were conducted using wind series from multiple seasons to further validate the predictive performance of the model. We chose the HER and MHS datasets for March, June, September, and December for this experiment. The training and testing processes of each of the compared models were repeated 10 times. The experimental results of the different datasets are presented in Tables 3 and 4, where the first-best predictions are highlighted. Figure 4 depicts the wind speed prediction results of the proposed model for the HER dataset.

Table 3. Forecast results of different models for the HER data.

Season	Metrics	Proposed Model	PSO-ANFIS	VMD-GA-BP	EEMD-GPR-LSTM	MWS-CE-ENN	EMD-ISSA-LSTM
Spring	RMSE	**0.0065**	0.0092	0.0128	0.0105	0.0093	0.0131
	SSE	**0.0180**	0.0365	0.0817	0.0471	0.0370	0.0737
	MAPE	**0.0300**	0.0402	0.0732	0.0493	0.0632	0.0536
	MAE	**0.0048**	0.0068	0.0108	0.0083	0.0071	0.0096
Summer	RMSE	**0.0112**	0.0172	0.0134	0.0126	0.0152	0.0117
	SSE	**0.0537**	0.1279	0.0768	0.0679	0.0989	0.0586
	MAPE	**0.0420**	0.0641	0.0632	0.0596	0.0730	0.0500
	MAE	**0.0081**	0.0123	0.0099	0.0114	0.0119	0.0093

Table 3. *Cont.*

Season	Metrics	Proposed Model	PSO-ANFIS	VMD-GA-BP	EEMD-GPR-LSTM	MWS-CE-ENN	EMD-ISSA-LSTM
Autumn	RMSE	**0.0088**	0.0120	0.0137	0.0126	0.0132	0.0204
	SSE	**0.0331**	0.0615	0.0806	0.0676	0.0746	0.1779
	MAPE	**0.0422**	0.0514	0.0524	0.1177	0.0619	0.0851
	MAE	**0.0059**	0.0077	0.0092	0.0089	0.0082	0.0129
Winter	RMSE	**0.0063**	0.0090	0.0067	0.0093	0.0104	0.0086
	SSE	**0.0172**	0.0354	0.0190	0.0370	0.0463	0.0317
	MAPE	**0.0575**	0.0880	0.0683	0.0722	0.0923	0.0620
	MAE	**0.0043**	0.0064	0.0047	0.0069	0.0073	0.0056

Table 4. Forecast results of different models for the MHS data.

Season	Metrics	Proposed Model	PSO-ANFIS	VMD-GA-BP	EEMD-GPR-LSTM	MWS-CE-ENN	EMD-ISSA-LSTM
Spring	RMSE	**0.0237**	0.0344	0.0264	0.0276	0.0339	0.0291
	SSE	**0.2404**	0.5054	0.2977	0.3250	0.4918	0.3624
	MAPE	**0.0724**	0.0830	0.1007	0.1055	0.0990	0.0866
	MAE	**0.0146**	0.0244	0.0187	0.0198	0.0229	0.0203
Summer	RMSE	**0.0156**	0.0267	0.0179	0.0205	0.0189	0.0236
	SSE	**0.0104**	0.3045	0.1375	0.1799	0.1529	0.2384
	MAPE	**0.0433**	0.0857	0.0588	0.0697	0.0680	0.0765
	MAE	**0.0110**	0.0200	0.0136	0.0171	0.0149	0.0192
Autumn	RMSE	**0.0166**	0.0245	0.0190	0.0183	0.0214	0.0262
	SSE	**0.1179**	0.2570	0.1543	0.1433	0.1960	0.2937
	MAPE	**0.0410**	0.0641	0.0505	0.0439	0.0762	0.0658
	MAE	**0.0131**	0.0185	0.0149	0.0137	0.0170	0.0211
Winter	RMSE	**0.0266**	0.0394	0.0309	0.0325	0.0286	0.0312
	SSE	**0.3034**	0.6645	0.4075	0.4521	0.3501	0.4166
	MAPE	**0.0828**	0.1698	0.1540	0.1215	0.0937	0.1200
	MAE	**0.0190**	0.0284	0.0228	0.0276	0.0193	0.0257

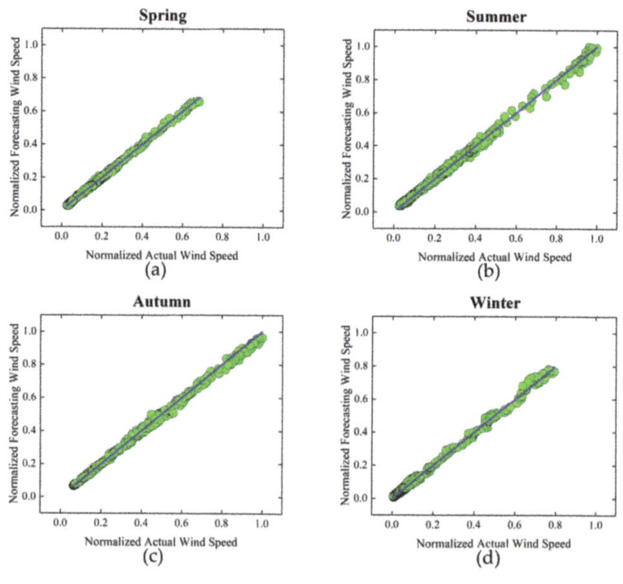

Figure 4. Forecasting results of HER wind speed data sets: (**a**) experiment results of spring wind speed sequences; (**b**) experiment results of summer wind speed sequences; (**c**) experiment results of autumn wind speed sequences; (**d**) experiment results of winter wind speed sequences.

Interestingly, it can be seen from Tables 3 and 4 that the proposed HGANN had the best prediction performance for each of the RMSE, SSE, MAPE, and MAE indicators on four datasets among all models. Wind speed forecasts at different times or at different locations may yield different results. Notably, our model achieved promising predictive results on both the geographically distinct German dataset HER and the Chinese dataset MHS. Furthermore, our model showed competitive prediction performance for wind series in different seasons. This indicates that our model can be extended to more general environments for wind speed prediction.

The abscissa and ordinate in Figure 4 represent the actual wind speed and the predicted wind speed, respectively; the blue line indicates that the predicted value is equal to the actual value. The ordinate of the green point is the predicted value, so the fit of the green point to the straight line reflects the accuracy of the prediction. As can be seen from Figure 4, the green points are very close to the blue line, which indicates that our model can predict wind speed effectively.

(2) Experiment II: Multi-Step Prediction Experiment

Multi-step forecasting can be built based on single-step forecasting. Compared to single-step forecasting, multi-step forecasting is more practical for power systems. Therefore, in wind speed prediction, multi-step prediction is of high practical value. The experiment aimed to demonstrate the predictive performance of the HGANN model in multi-step forecasting. We selected 2880 samples from 23 August to 22 September from the HER dataset for the one-step, two-step, and three-step experiments. Performance metrics involved RMSE, SSE, MAPE, and MAE. Benchmark models covered PSO-ANFIS, VMD-GA-BP, EEMD-GPR-LSTM, MWS-CE-ENN, and EMD-ISSA-LSTM. The training and testing processes of each model were repeated 10 times. The experimental results of the proposed model and the benchmark models are shown in Table 5.

Table 5. Multi-step prediction results for 15 min wind speed.

Model	RMSE			SSE			MAPE			MAE		
	1-Step	**2-Step**	**3-Step**	**1-Step**	**2-Step**	**3-Step**	**1-Step**	**2-Step**	**3-Step**	**1-Step**	**2-Step**	**3-Step**
Proposed	**0.0091**	**0.0136**	**0.0178**	**0.0356**	**0.0679**	**0.1361**	**0.0463**	**0.0712**	**0.1087**	**0.0061**	**0.0098**	**0.0134**
PSO-ANFIS	0.0120	0.0174	0.0238	0.0615	0.1299	0.2428	0.0514	0.0885	0.1270	0.0077	0.0119	0.0161
VMD-GA-BP	0.0137	0.0151	0.0203	0.0806	0.0976	0.1764	0.0524	0.0721	0.1123	0.0092	0.0101	0.0213
EEMD-GPR-LSTM	0.0126	0.0164	0.0192	0.0676	0.1151	0.1578	0.1177	0.1353	0.1536	0.0089	0.0129	0.0157
MWS-CE-ENN	0.0132	0.0176	0.0213	0.0746	0.1326	0.1942	0.0619	0.0826	0.1121	0.0082	0.0172	0.0224
EMD-ISSA-LSTM	0.0109	0.0139	0.0183	0.0508	0.0827	0.1433	0.0503	0.0919	0.1325	0.0086	0.0134	0.0192

From Table 5, it can be seen that the one-step, two-step, and three-step prediction results of the proposed HGANN model provided lower RMSE, SSE, MAPE, and MAE values than those of the benchmark models. For instance, the proposed HGANN provided 0.0091 (one-step), 0.0136 (two-step), and 0.0178 (three-step) on RMSE, compared with EMD-ISSA-LSTM, which provided the predictive results of 0.0109 (one-step), 0.0139 (two-step), and 0.0183 (three-step). Furthermore, we also provide clear visual results of multi-step predictions for the six models in Figure 5. The results from Table 5 and Figure 5 indicate that the proposed HGANN model had the best robustness and the highest wind speed prediction accuracy among all compared models.

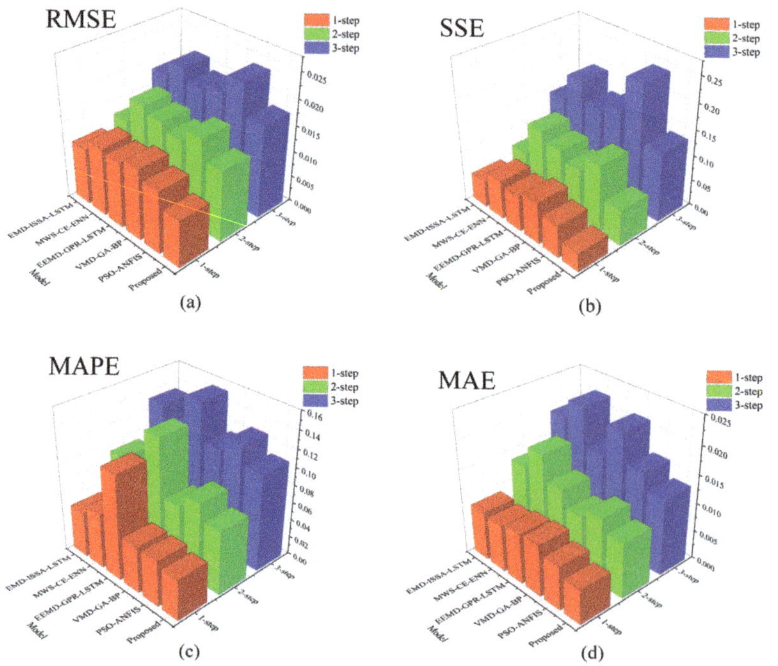

Figure 5. Multi-step forecasting experiments under RMSE, SSE, MAPE, and MAE indicators: (**a**) experiment results on the RMSE indicator; (**b**) experiment results on the SSE indicator; (**c**) experiment results on the MAPE indicator; (**d**) experiment results on the MAE indicator.

(3) Experiment III: Ablation Experiment Between Single Models and Hybrid Models.

To verify the rationality of the proposed HGANN model, it was compared with WGAN-OBLS, OBLS, WGAN, BLS, and CNN on the HER dataset. The generator of HGANN is OBLS and its discriminator is the discriminator of WGAN. To emphasize the effectiveness of OBLS, it was compared with the generator of WGAN, namely, CNN. Similarly, all compared models were repeatedly trained and tested 10 times. In the HER dataset, 2880 data from 23 August to 22 September were selected for this experiment. The experimental results are shown in Table 6, where the first-best predictions are highlighted with dark gray backgrounds. The forecast results for 22 September 2019 are plotted in Figure 6, which also shows the forecast errors in superimposed shades.

Table 6. Forecasting performances of the proposed model and reference models.

Indicators	Proposed Model	WGAN-OBLS	OBLS	WGAN	CNN	BLS
RMSE	**0.0088**	0.0119	0.0122	0.0138	0.0221	0.0308
SSE	**0.0331**	0.0603	0.0642	0.0819	0.2090	0.4058
MAPE	**0.0422**	0.0506	0.0546	0.0571	0.0601	0.0648
MAE	**0.0059**	0.0076	0.0080	0.0086	0.0161	0.0164

Figure 6 shows that among all the compared models, our proposed model had the best curve fitting and the smallest predicted error. The suggested model consistently outperformed WGAN-OBLS, PSO-BLS, WGAN, OBLS, CNN, and BLS, as shown in Table 6. This further demonstrates the advantages of our proposed model, as it combines CEEMDAN, OBLS, and WGAN.

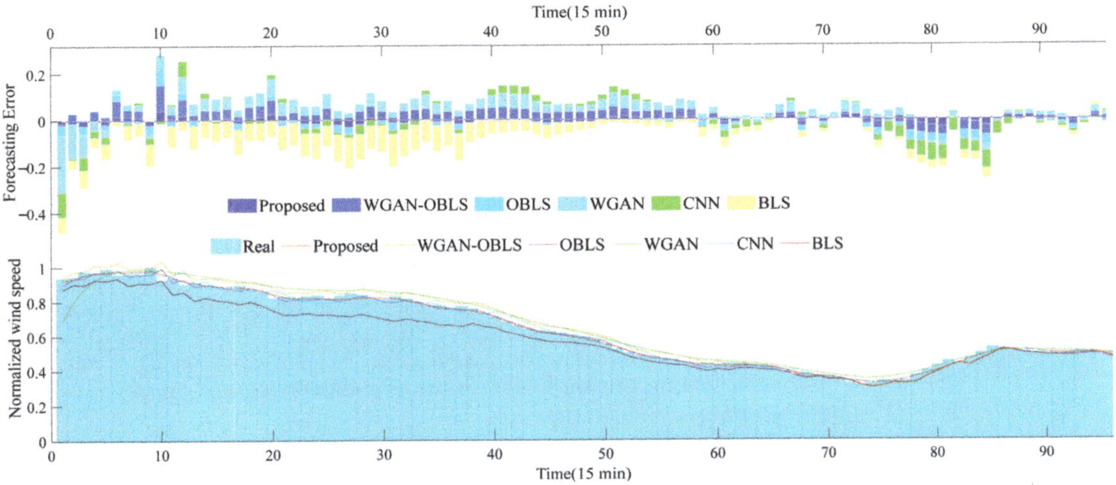

Figure 6. Forecasting results in HER dataset (22 September 2019).

Furthermore, first, compared with WGAN-OBLS without CEEMDAN, the proposed model had better predictive performance due to its covering CEEMADN and WGAN-OBLS, thus showing the effectiveness of CEEMDAN. Second, compared with WGAN or OBLS, WGAN-OBLS provided better predictive performance due to combing OBLS and WGAN, thus showing the effectiveness of OBLS and WGAN in WGAN-OBLS. Third, compared with BLS, OBLS provided better predictive performance due to using the improved PSO, thus showing the effectiveness of PSO in OBLS. Fourth, OBLS had better predictive results compared to the generator of WGAN, namely, CNN. This demonstrates the advantage of OBLS over CNN as a generator. This may be due to the flexible structure and better error convergence of OBLS.

4. Discussion

Our model was compared with five advanced models to evaluate its performance and advantages in various wind sequence experiments. Experimental results show that the proposed model had better predictive performance. The reasons behind this fact are given as follows.

First, the wind speed data were one-year data from wind farms in Germany and China, which cover complex fluctuation characteristics. Therefore, our HGANN model uses CEEMDAN to smoothen the volatility of the data and improve the predictive performance.

Second, HGANN uses OBLS as the generator to provide a special shallow broad incremental learning network structure, which can not only be beneficial for improving prediction accuracy for one-dimensional wind speed prediction compared to CNN but also greatly decrease computational cost using pseudo-inverse operations to determine the network weights instead of using convolution operations.

Third, in our HGANN model, the proposed OBLS uses an improved PSO to optimize the hyper-parameters of its network, which can search in a wider range and obtain the optimal parameters over BLS. Therefore, OBLS has better generalization ability than BLS.

Finally, HGANN can better extract the deeper features of wind speed data by playing a minimum–maximum game between the generator and discriminator for wind speed prediction.

5. Conclusions

Although existing various hybrid predictive models have provided competitive performance in ultra-short-term wind speed prediction, they still need to be further improved—

for instance, how to effectively reduce the computational cost of hybrid predictive models, and how to effectively deal with the multicollinearity problem of the hybrid forecasting model based on weighted strategy, which leads to the problem of reduced forecasting accuracy. To enhance the predictive power and decrease the computational cost, this paper proposes the HGANN model for ultra-short-term wind speed forecasting. HGANN is a generative adversarial network in which the generator and discriminator play against each other to obtain wind speed predictions with high accuracy. In HGANN, we developed OBLS and the convolutional structures as the generator and the discriminator, respectively, which enables them to obtain effective synergies to improve predictive performance. Particularly, OBLS involves a special shallow broad incremental learning network structure, which can effectively deal with one-dimensional wind speed data. Furthermore, the shallow network structure of OBLS can also significantly decrease computational cost via using pseudo-inverse operations rather than convolution operations. In addition, the proposed OBLS applies an improved PSO to obtain the optimal network hyper-parameters. CEEMDAN performs noise reduction and decomposition of the wind data. Via the above rational combination, the proposed HGANN provides high predictive accuracy and generalization ability with low computational cost in ultra-short-term wind speed prediction. The experimental results indicate the above fact. For instance, the RMSE predictive errors of the proposed model were 29.35%, 49.22%, 38.09%, and 30.10% compared to the four state-of-art predictive models PSO-ANFIS, VMD-GA-BP, EEMD-GPR-LSTM, and MWS-CE-ENN on the spring wind data of the HER dataset, respectively. In the future, we plan to use parallel computing to speed up the process of PSO optimization of BLS during training. Furthermore, the proposed HGANN will be extended to a wider range of applications, such as financial time-series forecasting, electricity-load forecasting, traffic forecasting, etc.

Author Contributions: Conceptualization, Q.W. and L.H.; methodology, Q.W.; software, J.H.; validation, Q.W., J.H. and L.H.; formal analysis, L.H.; investigation, Q.W. and Q.L.; resources, J.H.; data curation, Q.W. and Q.L.; writing—original draft preparation, L.H.; writing—review and editing, L.C. and Y.L.; visualization, Q.W.; supervision, P.X.L.; project administration, P.X.L.; funding acquisition, C.L. All authors have read and agreed to the published version of the manuscript.

Funding: This work was supported in part by the National Natural Science Foundation of China under grants 62173176 and 61863028, and in part by the Science and Technology Department of Jiangxi Province of China under grant 20204ABC03A39.

Institutional Review Board Statement: Not applicable.

Informed Consent Statement: Not applicable.

Data Availability Statement: Not applicable.

Conflicts of Interest: The authors declare no conflict of interest.

Nomenclature

ANFIS	Adaptive-network-based fuzzy inference system
BLS	Broad learning system
BP	Back propagation
CEEMD	Complementary ensemble empirical mode decomposition
CEEMDAN	Complete ensemble empirical mode decomposition with adaptive noise
CNN	Convolutional neural network
EEMD	Ensemble empirical mode decomposition
EMD	Empirical mode decomposition
GA	Genetic algorithm
GMDH	Group method of data handling neural network
GNN	Graph neural network
GAN	Generative adversarial network
GPR	Gaussian process regression
ICEEMDAN	Improved CEEMDAN

IMF	Intrinsic mode function
IOWA	Induced ordered weighted averaging
LSTM	Long short-term memory
MAE	Mean absolute error
MAPE	Mean absolute percentage error
MTO	Multi-tracker optimizer
PSO	Particle swarm optimization
OBLS	Optimized broad learning system
RBF	Radial basis function
RMSE	Root mean square error
SSE	Sum of squared error
SVR	Support vector regression
VMD	Variational mode decomposition
WGAN	Wasserstein generative adversarial network

References

1. Zhen, Z.; Qiu, G.; Mei, S.; Wang, F.; Zhang, X.; Yin, R.; Li, Y.; Osório, G.J.; Shafie-khah, M.; Catalão, J.P.S. An ultra-short-term wind speed forecasting model based on time scale recognition and dynamic adaptive modeling. *Int. J. Electr. Power Energy Syst.* **2022**, *135*, 107502. [CrossRef]
2. Wang, F.; Tong, S.; Sun, Y.; Xie, Y.; Zhen, Z.; Li, G.; Cao, C.; Duić, N.; Liu, D. Wind process pattern forecasting based ultra-short-term wind speed hybrid prediction. *Energy* **2022**, *255*, 124509. [CrossRef]
3. Wang, H.; Han, S.; Liu, Y.; Yan, J.; Li, L. Sequence transfer correction algorithm for numerical weather prediction wind speed and its application in a wind power forecasting system. *Appl. Energy* **2019**, *237*, 1–10. [CrossRef]
4. Do, D.-P.N.; Lee, Y.; Choi, J. Hourly Average Wind Speed Simulation and Forecast Based on ARMA Model in Jeju Island, Korea. *J. Electr. Eng. Technol.* **2016**, *11*, 1548–1555. [CrossRef]
5. Yunus, K.; Thiringer, T.; Chen, P. ARIMA-Based Frequency-Decomposed Modeling of Wind Speed Time Series. *IEEE Trans. Power Syst.* **2016**, *31*, 2546–2556. [CrossRef]
6. Wang, Y.; Wang, H.; Srinivasan, D.; Hu, Q. Robust functional regression for wind speed forecasting based on Sparse Bayesian learning. *Renew. Energy* **2019**, *132*, 43–60. [CrossRef]
7. Wang, S.; Zhang, N.; Wu, L.; Wang, Y. Wind speed forecasting based on the hybrid ensemble empirical mode decomposition and GA-BP neural network method. *Renew. Energy* **2016**, *94*, 629–636. [CrossRef]
8. Zhang, Y.; Zhang, C.; Zhao, Y.; Gao, S. Wind speed prediction with RBF neural network based on PCA and ICA. *J. Electr. Eng.* **2018**, *69*, 148–155. [CrossRef]
9. Son, Y.; Yoon, Y.; Cho, J.; Choi, S. Cloud Cover Forecast Based on Correlation Analysis on Satellite Images for Short-Term Photovoltaic Power Forecasting. *Sustainability* **2022**, *14*, 4427. [CrossRef]
10. Hameed, S.S.; Ramadoss, R.; Raju, K.; Shafiullah, G.M. A Framework-Based Wind Forecasting to Assess Wind Potential with Improved Grey Wolf Optimization and Support Vector Regression. *Sustainability* **2022**, *14*, 4235. [CrossRef]
11. Khodayar, M.; Wang, J.; Manthouri, M. Interval Deep Generative Neural Network for Wind Speed Forecasting. *IEEE Trans. Smart Grid* **2019**, *10*, 3974–3989. [CrossRef]
12. Beigi, M.; Beigi Harchegani, H.; Torki, M.; Kaveh, M.; Szymanek, M.; Khalife, E.; Dziwulski, J. Forecasting of Power Output of a PVPS Based on Meteorological Data Using RNN Approaches. *Sustainability* **2022**, *14*, 3104. [CrossRef]
13. Khodayar, M.; Wang, J. Spatio-Temporal Graph Deep Neural Network for Short-Term Wind Speed Forecasting. *IEEE Trans. Sustain. Energy* **2019**, *10*, 670–681. [CrossRef]
14. Zhu, Q.; Chen, J.; Shi, D.; Zhu, L.; Bai, X.; Duan, X.; Liu, Y. Learning Temporal and Spatial Correlations Jointly: A Unified Framework for Wind Speed Prediction. *IEEE Trans. Sustain. Energy* **2020**, *11*, 509–523. [CrossRef]
15. Zhang, Y.; Pan, G.; Zhang, C.; Zhao, Y. Wind speed prediction research with EMD-BP based on Lorenz disturbance. *J. Electr. Eng.* **2019**, *70*, 198–207. [CrossRef]
16. Santhosh, M.; Venkaiah, C.; Kumar, D.M.V. Short-term wind speed forecasting approach using Ensemble Empirical Mode Decomposition and Deep Boltzmann Machine. *Sustain. Energy Grids Netw.* **2019**, *19*, 100242. [CrossRef]
17. Wang, J.; Yang, Z. Ultra-short-term wind speed forecasting using an optimized artificial intelligence algorithm. *Renew. Energy* **2021**, *171*, 1418–1435. [CrossRef]
18. Ren, Y.; Suganthan, P.N.; Srikanth, N. A Comparative Study of Empirical Mode Decomposition-Based Short-Term Wind Speed Forecasting Methods. *IEEE Trans. Sustain. Energy* **2015**, *6*, 236–244. [CrossRef]
19. Li, Y.; Chen, X.; Li, C.; Tang, G.; Gan, Z.; An, X. A Hybrid Deep Interval Prediction Model for Wind Speed Forecasting. *IEEE Access* **2021**, *9*, 7323–7335. [CrossRef]
20. Tian, Z. Modes decomposition forecasting approach for ultra-short-term wind speed. *Appl. Soft Comput.* **2021**, *105*. [CrossRef]
21. Liu, Z.; Hara, R.; Kita, H. Hybrid forecasting system based on data area division and deep learning neural network for short-term wind speed forecasting. *Energy Convers. Manag.* **2021**, *238*, 114136. [CrossRef]

22. Cui, Y.; Huang, C.; Cui, Y. A novel compound wind speed forecasting model based on the back propagation neural network optimized by bat algorithm. *Environ. Sci. Pollut. Res. Int.* **2020**, *27*, 7353–7365. [CrossRef]
23. Duan, J.; Zuo, H.; Bai, Y.; Duan, J.; Chang, M.; Chen, B. Short-term wind speed forecasting using recurrent neural networks with error correction. *Energy* **2021**, *217*, 119397. [CrossRef]
24. Liu, H.; Yang, R.; Wang, T.; Zhang, L. A hybrid neural network model for short-term wind speed forecasting based on decomposition, multi-learner ensemble, and adaptive multiple error corrections. *Renew. Energy* **2021**, *165*, 573–594. [CrossRef]
25. Zhang, Y.; Han, J.; Pan, G.; Xu, Y.; Wang, F. A multi-stage predicting methodology based on data decomposition and error correction for ultra-short-term wind energy prediction. *J. Clean. Prod.* **2021**, *292*, 125981. [CrossRef]
26. Jiang, Y.; Liu, S.; Zhao, N.; Xin, J.; Wu, B. Short-term wind speed prediction using time varying filter-based empirical mode decomposition and group method of data handling-based hybrid model. *Energy Convers. Manag.* **2020**, *220*, 113076. [CrossRef]
27. Wang, J.; Wang, Y.; Li, Z.; Li, H.; Yang, H. A combined framework based on data preprocessing, neural networks and multi-tracker optimizer for wind speed prediction. *Sustain. Energy Technol. Assess.* **2020**, *40*, 100757. [CrossRef]
28. Altan, A.; Karasu, S.; Zio, E. A new hybrid model for wind speed forecasting combining long short-term memory neural network, decomposition methods and grey wolf optimizer. *Appl. Soft Comput.* **2021**, *100*, 106996. [CrossRef]
29. Goodfellow, I.; Pouget-Abadie, J.; Mirza, M. Generative adversarial nets. In Proceedings of the International Conference on Neural Information Processing Systems, Kuching, Malaysia, 8–13 December 2014; pp. 2672–2680.
30. Torres, M.; Colominas, M.; Schlotthauer, G.; Flandrin, P. A complete ensemble empirical mode decomposition with adaptive noise. In Proceedings of the IEEE International Conference on Acoustics, Speech and Signal Processing (ICASSP), Prague, Czech Republic, 22–27 May 2011; pp. 4144–4147. [CrossRef]
31. Chen, C.L.P.; Liu, Z. Broad Learning System: An Effective and Efficient Incremental Learning System without the Need for Deep Architecture. *IEEE Trans. Neural Netw. Learn. Syst.* **2018**, *29*, 10–24. [CrossRef]
32. Zhang, Y.; Sun, H.; Guo, Y. Wind Power Prediction Based on PSO-SVR and Grey Combination Model. *IEEE Access* **2019**, *7*, 136254–136267. [CrossRef]
33. Arjovsky, M.; Chintala, S.; Bottou, L. Wasserstein GAN. *arXiv* **2017**, arXiv:1701.07875v3.
34. Dokur, E.; Yuzgec, U.; Kurban, M. Performance Comparison of Hybrid Neuro-Fuzzy Models using Meta-Heuristic Algorithms for Short-Term Wind Speed Forecasting. *Electrica* **2021**, *21*, 305–321. [CrossRef]
35. Zhang, Y.; Pan, G.; Chen, B.; Han, J.; Zhao, Y.; Zhang, C. Short-term wind speed prediction model based on GA-ANN improved by VMD. *Renew. Energy* **2020**, *156*, 1373–1388. [CrossRef]
36. Huang, Y.; Liu, S.; Yang, L. Wind Speed Forecasting Method Using EEMD and the Combination Forecasting Method Based on GPR and LSTM. *Sustainability* **2018**, *10*, 3693. [CrossRef]
37. Tian, Z.; Chen, H. A novel decomposition-ensemble prediction model for ultra-short-term wind speed. *Energy Convers. Manag.* **2021**, *248*, 114775. [CrossRef]
38. Huang, L.; Li, L.; Wei, X.; Zhang, D. Short-term prediction of wind power based on BiLSTM–CNN–WGAN-GP. *Soft Comput.* **2022**, *26*, 1–15. [CrossRef]
39. Wang, H.; Yi, H.; Peng, J.; Wang, G.; Liu, Y.; Jiang, H.; Liu, W. Deterministic and probabilistic forecasting of photovoltaic power based on deep convolutional neural network. *Energy Convers. Manag.* **2017**, *153*, 409–422. [CrossRef]

MDPI
St. Alban-Anlage 66
4052 Basel
Switzerland
Tel. +41 61 683 77 34
Fax +41 61 302 89 18
www.mdpi.com

MDPI Books Editorial Office
E-mail: books@mdpi.com
www.mdpi.com/journal/books

www.ingramcontent.com/pod-product-compliance
Lightning Source LLC
Chambersburg PA
CBHW061510220926
51850CB00004B/216
* 9 7 8 3 0 3 6 5 7 6 4 4 2 *